W0055315

PROGRESS IN COLLOID & POLYMER SCIENCE

Editors: F. Kremer (Leipzig) and G. Lagaly (Kiel)

Volume 106 (1997)

Formation and Dynamics of Self-Organized Structures in Surfactants and Polymer Solutions

Guest Editors:

K. Kawasaki (Chubu), B. Lindman (Lund) and H. Okabayashi (Nagoya)

Springer

SPRINGER-VERLAG BERLIN
HEIDELBERG GMBH

Die Deutsche Bibliothek –
CIP-Einheitsaufnahme

Progress in colloid & polymer science. –
Früher Schriftenreihe
Vol. 106. Formation and dynamics of
self-organized structures in surfactants
and polymer solutions. – 1997
**Formation and dynamics of self-organized
structures in surfactants and polymer
solutions** / guest ed.: K. Kawasaki ... –
(Progress in colloid & polymer science ;
Vol. 106)

ISBN 978-3-662-15704-6   ISBN 978-3-7985-1659-5 (eBook)
DOI 10.1007/978-3-7985-1659-5

This work is subject to copyright. All
rights are reserved, whether the whole or
part of the material is concerned,
specifically those rights of translation,
reprinting, reuse of illustrations,
recitation, broadcasting, reproduction on
microfilms or in other ways, and storage
in data banks. Duplication of this
publication or parts thereof is only
permitted under the provisions of the
German Copyright Law of September 9,
1965, in its version of June 24, 1985,
and a copyright fee must always be paid.
Violations fall under the prosecution act
of the German Copyright Law.

The use of registered names, trademarks,
etc. in this publication does not imply,
even in the absence of specific statement,
that such names are exempt from the
relevant protective laws and regulations
and therefore free for general use.

© 1997 by Springer-Verlag Berlin Heidelberg
Originally published by Dr. Dietrich Steinkopff
Verlag GmbH & Co. KG, Darmstadt in 1997
Softcover reprint of the hardcover 1st edition 199'

Chemistry Editor:
Dr. Maria Magdalene Nabbe;
Production: Holger Frey, Ajit Vaidya.

Typesetting and Copy-Editing:
Macmillan Ltd., Bangalore, India

Progr Colloid Polym Sci (1997) V
© Steinkopff Verlag 1997

The International Symposium on Colloids and Polymer Science – Formation and Dynamics of Self-Organized Structures in Surfactants and Polymer Solutions – Recent Advances (ISCPS), was held at Nagoya Institute of Technology, Nagoya, Japan, from October 10–13, 1996. The aim of the symposium was to bring together scientists who are working in the various fields of Colloid and Polymer Science, with special emphasis on formation and dynamics of self-organized structures, including technical developments, applications, general theory, and results of investigations. An important feature of the symposium was the informal discussion sessions on various issues relevant to different topics of current interest, which yielded many fruitful suggestions for future research in the field of organized assemblies. The program covered both the fundamental aspects as well as technological applications of micelles, microemulsions, monolayers, biocolloids, etc. The scientific sessions were opened by Professor K. Kawasaki (Chubu, Japan), and Plenary Lectures on main themes of the symposium were given by B. Lindman (Lund, Sweden), S. Komura (Hiroshima, Japan), B. Farago (Grenoble, France), U. Olsson (Lund, Sweden), K. Shinoda (Lund, Sweden), and J. Sunamoto (Kyoto, Japan). In all, 60 papers were presented orally and 61 posters were displayed during the four days of the meeting. Lively discussions took place during the sessions and around the posters. Our thanks go to all of those who contributed to the cordial scientific atmosphere of the symposium. The symposium was attended by 157 people.

The contributions to those proceedings were reviewed by internationally leading scientists. We are very grateful to them for their cooperation. Thus, these proceedings are not simply a collection of unreviewed papers; rather peer reviewing was an integral part of the total editing process.

The conference was kindly sponsored by the Ministry of Education, Japan (MONBUSHO) and Nagoya Institute of Technology. We would like to express our appreciation to the members of the international advisory committee, Dr. Charmian J. O'Connor (The University of Auckland, New Zealand), Dr. Peter Schurtenberger (Eidgenössische Technische Hochschule, Switzerland), and Dr. Gordon J. T. Tiddy (Unilever Research, Port Sunlight Laboratory, U.K.). Special appreciation should be expressed to Charmian, who served to check the final proofs of the conference circulars. Our thanks are due to the members of the Local Organizing Committee (in particular, Keijiro Taga and Tanoj Kumar Jain) who zealously carried out the various chores demanded by this conference. The financial support of MONBUSHO and the DAIKO Foundation and generous donations made by corporate friends and organizations all helped to finance the symposium. Our warmest thanks go to these organizations, as well as to a number of people who helped with the day to day organization of the meeting before, during, and after the symposium.

Progr Colloid Polym Sci (1997) VII
© Steinkopff Verlag 1997

CONTENTS

Progr Colloid Polym Sci (1997) 106:1–5
© Steinkopff Verlag 1997

S. Komura
H. Seto
T. Takeda

Self-organization, phase transition and dynamics in amphiphilic systems

Received: 13 October 1996
Accepted: 4 February 1997

Dr. S. Komura (✉) · H. Seto · T. Takeda
Faculty of Integrated Arts and Sciences
Hiroshima University
1-7-1 Kagamiyama
Higashi-hiroshima 739 Japan

Abstract Recent studies on the structures, critical phenomena and slow dynamics in amphiphiles/water/oil ternary systems using small-angle neutron scattering (SANS) and neutron spin echo (NSE) are reviewed. From the fitting of the scattering profiles of SANS to Teubner and Strey's and Gompper and Schick's formula we could obtain several structural and interaction parameters of a system. We observed critical scattering near the phase boundary into the decomposition of a system by SANS. The observed crossover from mean field to 3d-Ising behavior is interpreted in terms of an expression by Belyakov and Kiselev with an appropriate Ginzburg reduced temperature. We observed NSE signal from a system and obtained effective diffusion coefficients for the shape fluctuations of the interfacial films in lamellar phase which has two decaying processes. The powerfulness of neutron-scattering methods to characterize the structure and dynamics of amphiphilic systems is illustrated.

Key words Self-assembly – phase transition – slow dynamics – amphiphiles – neutron scattering

Introduction

Amphiphiles are intriguing materials that self-assemble in various forms when mixed with oil and water. In disordered phases they take a structure of spherical droplets of water-in-oil (or oil-in-water) or a bicontinuous structure, in which both water and oil are separated from each other by intertwined interfacial films. In ordered phases they take a rather regular structure of hexagonal arrays of cylindrical tubules made of oil (or water) in surrounding water (or oil) or parallel arrays of alternative oil and water sheets. In these mixtures amphiphile molecules are mostly located at the interface of oil and water with their polar heads directed into water media and with apolar acyl tails into oil media, thus forming interfacial monolayers. The phase diagram of these structures depends strongly on temperature, pressure, kinds and concentrations of the three constituents and addition of fourth or fifth elements such as salts or cosurfactants.

In order to probe the microscopic structures in such systems experiments using small-angle X-ray scattering (SAXS) and small-angle neutron scattering (SANS) are powerful tools. It is possible to obtain structure functions $S(q)$ of the system, which can be compared with those calculated from various proposed models. From the comparisons we can determine not only the structural parameters that characterize the system but also interaction parameters among different molecules inherent in the system. In order to probe the dynamic aspects of the structures experiments using dynamical light scattering (DLS) and neutron spin echo (NSE) are useful. Since we expect highly diffusive nature of shape fluctuation in the systems, these methods provide us with intermediate correlation functions $I(q, t)$, whose Fourier transform with respect to the correlation time t is the dynamic structure factor $S(q, \omega)$.

2

S. Komura et al.
Self-organization, phase transition and dynamics in amphiphilic systems

The present paper describes our recent studies by means of SANS and NSE on the structures, critical phenomena and slow dynamics in amphiphiles/water/oil ternary systems and illustrates the powerfulness of such methods to illucidate the structure and dynamics of amphiphilic systems.

Self-organization

Teubner and Strey [1] have proposed the following equation for X-ray or neutron scattering intensities from surfactant/oil/water ternary system on the basis of phenomenological Ginzburg–Landau free-energy expansion:

$$I^{TS}(q) = \frac{1}{A + Bq^2 + Cq^4} \,, \tag{1}$$

where $A > 0$, $B < 0$, $C > 0$ for bicontinuous system with almost equal oil and water volume fractions resulting in a peak in the scattering profile.

On the other hand, Gompper and Schick [2] have proposed the following equation for scattering on the basis of spin-lattice model of ternary systems:

$$I^{GS}(q) = \frac{1}{a + b \cos(qd) + c \cos(2qd)} \,, \tag{2}$$

where d is the lattice size and the constants a, b, c are related to A, B, C by the following equations for small q:

$$a = A + \frac{15}{6d^2} B + \frac{6}{d^4} C \,,$$

$$b = -\frac{8}{3d^2} B - \frac{8}{d^4} C \,,$$

$$c = \frac{1}{6d^2} B + \frac{2}{d^4} C \,. \tag{3}$$

Once the fitting parameters A, B, C are determined from the scattering profiles, one can determine the structural parameters such as the repeat distance D between oil and water, and the structural correlation length ξ and the area a_H per head of the surfactant molecule in the interface by the following equations [1, 3]:

$$D = 2\pi \left[\frac{1}{2} \left(\sqrt{\frac{A}{C}} - \frac{B}{2C} \right) \right]^{-1/2} \,, \tag{4a}$$

$$\xi = \left[\frac{1}{2} \left(\sqrt{\frac{A}{C}} + \frac{B}{2C} \right) \right]^{-1/2} \,, \tag{4b}$$

$$a_H = 4\beta(1 - \beta) \frac{v_S}{\xi \phi_S} \,. \tag{4c}$$

From a, b, c one can also determine the interaction parameters such as repulsive energy J between oil and

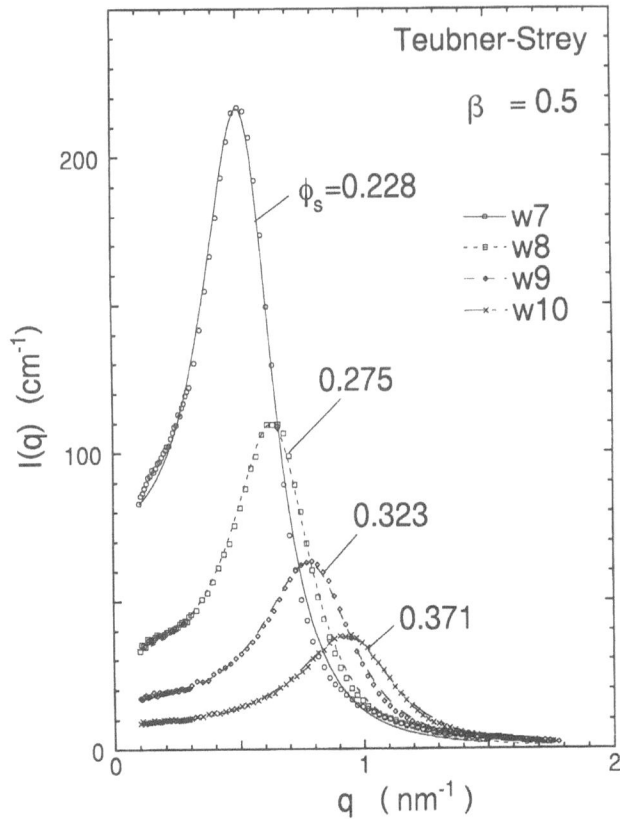

Fig. 1 Neutron-scattering cross section $I(q)$ from AOT/D$_2$O/ n-decane bicontinuous microemulsion systems with equal volume fraction of D$_2$O and decane at different AOT volume fractions ϕ_S at room temperature. Curves are fittings to Eq. (1)

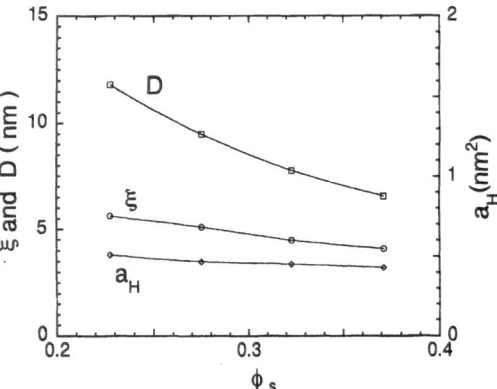

Fig. 2 Structure parameters obtained from the data in Fig. 1, by Eq. (4), such as repeat distance D, structure correlation length ξ and area per head a_H of surfactant molecules

water and the amphiphilicity L of surfactant molecule toward oil and water molecule as a ratio to the thermal energy T and the critical surfactant volume fractions ϕ_S^c below which an emulsification failure occurs by the

Progr Colloid Polym Sci (1997) 106:1–5
© Steinkopff Verlag 1997

following equations [2]:

$$\frac{J}{T} = -\frac{b}{6a(1-\phi_s)}, \tag{5a}$$

$$\frac{|L|}{T} = \frac{c}{6\phi_s a(1-\phi_s)}, \tag{5b}$$

$$\phi_s^c = -\frac{b\phi_s}{4c}, \tag{5c}$$

where ϕ_s is the volume fraction of the surfactants.

We have studied [3] the structural changes associated with a change of the concentrations of oil against water or the concentrations of amphiphiles against the rest in an ionic amphiphilic AOT/water/n-decane system in bicontinuous microemulsion phase by means of SANS at room temperature. The results of neutron-scattering intensity $I(q)$ for different surfactant volume fraction ϕ_s at equal volume fraction of oil and water $\beta = \phi_0/(\phi_0 + \phi_w) = 0.5$ are shown in Fig. 1. The fitting to the Teubner and Strey's Eq. (1) are also shown in the same figure. From the fitting we could obtain the repeat distance D between oil and water, the correlation length ξ and the interfacial area a_H per amphiphile molecule by Eq. (4). They have a general trend to become small as ϕ_s increases as shown in Fig. 2. The same data are further analyzed by Eqs. (3) and (5) in order to obtain the strength of segregation J, the amphiphilicity L of the amphiphile and the critical volume fraction ϕ_s^c. The results are shown in Fig. 3. They maintain almost constant values as ϕ_s is changed. The structural and interaction parameters are all in reasonable order of magnitude.

We could obtain similar results for different oil volume fractions against water from $\beta = 0.1$–0.4 at constant $\phi_s = 0.181$. From these results we can probe the relation between the structure and the interaction constants.

One of our interesting findings is that there is a scaling relation of the scattering profiles such that

$$I(q, \phi_s) = [q_m(\phi_s)]^{-3} \tilde{I}(q/q_m(\phi_s), k\xi), \tag{6}$$

where $-q_m(\phi_s)$ is the value of q that gives the maximum of the profile for a system with ϕ_s and $k = 2\pi/D$. Figure 4 demonstrates that $[q_m(\phi_s)]^3 I(q, \phi_s)$ plotted against $x = q/q_m$ collapses into a single curve $\tilde{I}(x, k\xi)$ for a fixed value of $k\xi$ supporting the scaling relation (6).

Phase transition

Microemulsion systems of water-in-oil (or oil-in-water) droplets often undergo a decomposition into two phases with different droplet number densities usually separated from each other by a meniscus. The associated phase transitions is of second order that accompanies fluctuation

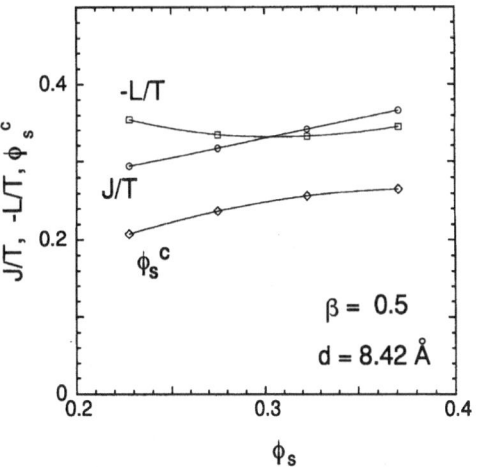

Fig. 3 Interaction parameters obtained from the data in Fig. 1 by Eq. (5), such as the strength J of segregation between water and decane, the amphiphilicity $L < 0$ of surfactant molecules and the critical concentrations ϕ_s of surfactant

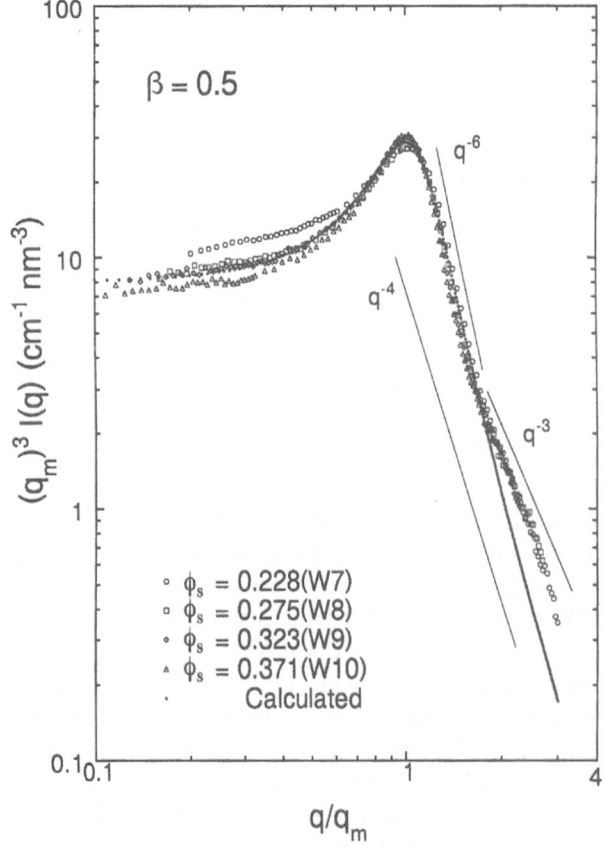

Fig. 4 Plotting $[q_m(\phi_s)]^3 I(q, \phi_s)$ as a function of $x = q/q_m(\phi_s)$, which collapses into a single function $\tilde{I}(x, k\xi)$ in accordance with a scaling relation Eq. (6)

4

S. Komura et al.
Self-organization, phase transition and dynamics in amphiphilic systems

of the susceptibilities S_0 of the number density of the droplets against the osmotic pressure and the correlation length ξ of the number density. The standard fluctuation theory of critical phenomena describes the behaviors of S_0 and ξ according to the power laws

$$S_0 \approx \tau^{-\gamma}, \qquad \xi \approx \tau^{-\nu}, \tag{7}$$

where $\tau = (T^{-1} - T_c^{-1})/T_c^{-1}$ is the reduced temperature with respect to the critical temperature T_c. The indices γ and ν are universal critical exponents specific to the universality class to which the system belongs and are related to each other by a scaling law $\gamma = 2\nu$. The decomposition of the droplet microemulsion into two phases is considered to belong to 3D Ising universality class for which $\gamma = 1.25$ and $\nu = 0.63$ like ordinary gas–liquid phase transition.

However, such description of the fluctuation applies only near the critical point where the correlation length ξ exceeds the range of interactions dominating the phase separation among the constituents, i.e. droplets in case of microemulsion systems. In the reverse case away from the critical point, where ξ is still smaller than the range of interactions, the classical mean field theory applies, for which the critical exponents are $\gamma = 1.0$ and $\nu = 0.5$. The value of τ at which the crossover from mean field to 3D Ising occurs is the Ginzburg number τ_G. Obviously, the value of τ_G is small if the range of interaction is large and the opposite holds for short-range interactions. Therefore, it is very interesting to determine the value of τ_G for microemulsion systems in order to know the range of the interactions among droplets.

Belyakov and Kiselev [4] and Anisimov et al. [5] have proposed the following system-independent universal relation between the renormalized susceptibility $\hat{S}_0 = S_0 a_0 \tau_G$ and the renormalized temperature $\hat{\tau} = \tau/\tau_G$ applicable to the crossover region, the two limits of which leads to either the critical fluctuation or the mean field extreme:

$$\hat{\tau} = (1 + \chi_0 \hat{S}_0^{\Delta/\gamma})^{(\gamma-1)/\Delta} [\hat{S}_0^{-1} + (1 + \chi_0 \hat{S}_0^{\Delta/\gamma})^{-\gamma/\Delta}], \tag{8}$$

where $\chi_0 = 2.333$ and a_0 the coefficient of the second-order term in the Landau free-energy expansion in terms of the order parameter and particularly for 3D Ising case $\gamma = 1.24$ and $\Delta = 0.51$.

We have studied critical density fluctuations of water-in-oil droplets in AOT/water/n-decane system that has a lower critical solution temperature (LCST) decomposing at higher temperature by SANS below and near the phase boundary of decomposition into two phases [6, 7]. We observed critical scattering $I(q)$ growing toward the phase boundary at small q. From the extrapolation of $I(q)$ to zero of q, we get $I(0)$ which is proportional to the susceptibility S_0. The observed behavior of S_0 revealed a crossover from mean field to 3D Ising and is well described in terms of asymptotic expression Eq. (8) as depicted

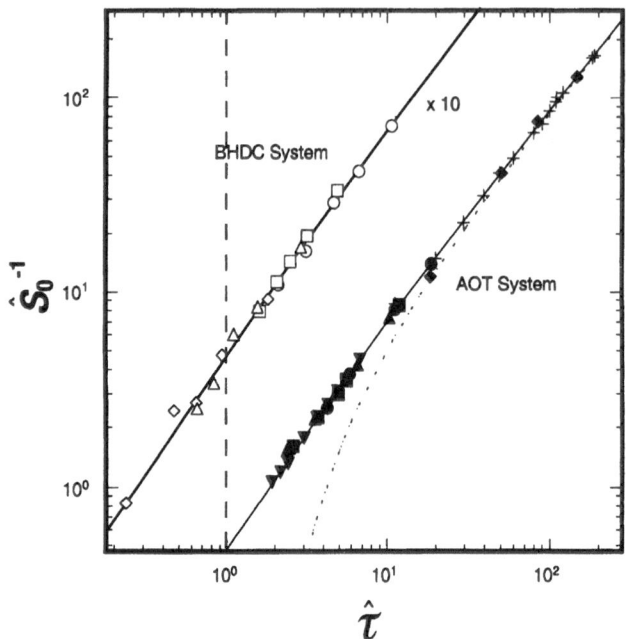

Fig. 5 Renormalized susceptibility $\hat{S}_0 = S_0 a_0 \tau_G$ as a function of renormalized temperature $\hat{\tau} = \tau/\tau_G$ in accordance with the crossover function Eq. (8) in water-in-oil droplet systems of both AOT/D_2O/n-decane (LCST) and BHDC/water/benzene (UCST) near the phase boundary. The vertical dotted line shows a boundary $\hat{\tau} = 1$ below which the critical theory applies. The curved dotted line shows the calculated mean-field behavior

in Fig. 5, where $\hat{S}_0^{-1}$ is plotted against $\hat{\tau}$. The observed Ginzburg reduced temeparture $\tau_G = 0.001$, below which the critical regime appears, is one order of magnitude smaller than that for simple liquids of low molecular weight.

On the contrary we have also studied critical fluctuations of water-in-oil droplets in BHDC/benzene/water system that has an upper critical solution temperature (UCST) decomposing at lower temeprature [8]. The behavior of the susceptibility is also well described by Eq. (8) and is depicted in Fig. 5 with a Ginzburg reduced temperature $\tau_G = 0.01$ which is ten times more than that of AOT system. Since the range of the interaction in BHDC system is expected to be less than that of the LCST system involving AOT, the obtained results for τ_G are reasonable.

Dynamics

By NSE spectroscopy one can probe the intermediate dynamical correlation function $I(q, t)$ such that

$$I(q, t) = N^{-1} \sum_{k,l} \langle \exp\{i\mathbf{q} \cdot \mathbf{r}_k(0)\} \exp\{-i\mathbf{q} \cdot \mathbf{r}_l(t)\} \rangle, \tag{9}$$

where $\langle \ \rangle$ is the ensemble average at the temperature $T \cdot \mathbf{r}_k(t)$s are the position of kth atom at time t. N is the number of the total atoms.

Progr Colloid Polym Sci (1997) 106:1–5
© Steinkopff Verlag 1997

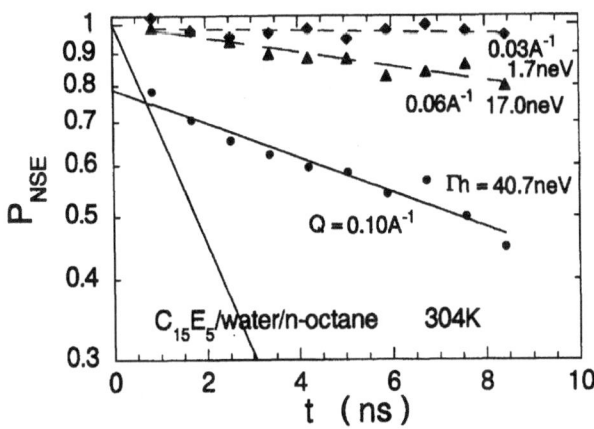

Fig. 6 Intermediate correlation function $I(q, t)$ from $C_{12}E_5$/water/n-octane lamellar systems at 31 °C with equal volume fraction of water and octane at $C_{12}E_5$ volume fraction of $\phi_S = 0.2$. It is expressed by a sum of two exponential functions Eq. (10)

For bulk contrast of water against oil and surfactant the sums in Eq. (9) are taken only for atoms belonging to water molecules giving rise to water–water partial correlation function $I^{ww}(q, t)$. Likewise for film contrast the sums are taken only for atoms belonging to surfactant molecules to get $I^{ss}(q, t)$. However, in asymmetric case where surfactant prefers oil to water, the diffusive motion of the local order parameter $\beta(r, t) = \phi_0(r, t)/(\phi_0(r, t) + \phi_w(r, t))$ describing oil volume fraction against water is not decoupled from that of the local order parameter $\phi_S(r, t)$ describing surfactant volume fraction even for equal average volume fraction of oil and water, i.e. $\beta = 0.5$. Accordingly there are two basis dynamical correlation function $I_1 = \langle \beta(-q, 0)\beta(q, t)\rangle$ and $I_2 = \langle \phi_S(-q, 0)\phi_S(q, t)\rangle$, respectively.

In this case both $I^{ww}(q, t)$ and $I^{ss}(q, t)$ are expressed by a linear combination of two exponential functions

$$I(q, t) = f_1 \exp(-\Gamma_1 t) + f_2 \exp(-\Gamma_2 t) , \qquad (10)$$

where f_1 and f_2 ($f_1 + f_2 = 1$) are fractions of the contribution of the two diffusion processes of the order parameters $\beta(r, t)$ and $\phi_S(r, t)$ each with corresponding decay time Γ_1^{-1} and Γ_2^{-1}. The diffusion process is realized by the translation, undulation and shape fluctuation of the interfacial films.

On non-ionic $C_{12}E_5$/water/n-octane system in bulk contrast of water against oil and surfactant with $\beta = 0.5$ and $\phi_S = 0.2$ we could measure the intermediate correlation function $I(q, t)$ by the use of NSE spectrometer recently built in JRR-3M at Tokai [9]. Figure 6 displays the $I(q, t)$ for a bulk contrast sample in lamellar phase at 31 °C at different q's [10]. From this data one can deduce $f_1 = 0.8$, $\Gamma_1^{-1} = 16$ ns and $f_2 = 0.2$, $\Gamma_2^{-1} = 2.5$ ns for $q = 0.10$ Å^{-1}. It is tempting to consider that the process with longer decay time corresponds to the diffusion process of the order parameter β and that with shorter decay time to the order parameter ϕ_S. However, in order to assess the conclusion, further studies on the system are required.

Conclusions

Neutron-scattering methods such as SANS and NSE have proved to be powerful tools to characterize the structure and dynamics and to extract microscopic interactions in amphiphilic systems as illustrated below.

In terms of interaction parameters J and L within the spin-lattice model, Gompper and Schick have calculated the structure function of bicontinuous microemulsion systems. Comparing the structure functions obtainable from SANS with those calculated ones by Gompper and Schick, we could determine the interaction parameters in right order of magnitude.

In water-in-oil droplet microemulsion systems undergoing a phase separation, we observed the crossover behavior of Belyakov and Kiselev and could determine the Ginzburg temperature τ_G which is smaller in LCST case than and equal in UCST case to that of gas–liquid-phase transitions of low molecular weight as expected.

In three-component microemulsion systems there are two order parameters that describe the fluctuation of the system; i.e. oil volume fraction against water $\beta = \phi_0/(\phi_0 + \phi_S)$ and surfactant volume fraction ϕ_S. The two decaying processes observed in NSE spectra might be correlated to the relaxation of these two order parameters.

References

1. Teubner M, Strey R (1987) J Chem Phys 87:3195–3200
2. Gompper G, Schick M (1990) Phys Rev B 41:9148–9162
3. Komura S, Seto H, Takeda T, Nagao M, Itoh Y, Imai M (1996) J Chem Phys 105:3264–3277
4. Belyakov MY, Kiselev SB (1992) Physica A 190:75–94
5. Anisimov MA, Kiselev SB, Sengers JV, Tang S (1992) Physica A 188:487–525
6. Seto H, Schwahn D, Mortensen K, Komura S (1993) J Chem Phys 99:5512–5519
7. Seto H, Schwahn D, Nagao M, Yokoi E, Komura S, Imai M, Mortensen K (1996) Phys Rev E 54:629–633
8. Seto H, Wignall RD, Triolo R, Chillura-Martino D, Komura S (1997) Progr Colloid Polym Sci 106:104–107
9. Takeda T, Komura S, Seto H, Nagai M, Kobayashi H, Yokoi E, Zeyen C, Ebisawa T, Tasaki S, Ito Y, Takahashi S, Yoshizawa H (1995) Nucl Instr and Meth A 364:186–192
10. Takeda T et al. (1998) to be published

Progr Colloid Polym Sci (1997) 106:6–13
© Steinkopff Verlag 1997

U. Olsson
H. Bagger-Jörgensen
M. Leaver
J. Morris
K. Mortensen
R. Strey
P. Schurtenberger
H. Wennerström

Stable, metastable and unstable oil-in-water droplets

Received: 13 October 1996
Accepted: 4 February 1997

Dr. U. Olsson (✉) · H. Bagger-Jörgensen
M. Leaver · J. Morris · H. Wennerström
Physical Chemistry 1
Center for Chemistry
and Chemical Engineering
Lund University
P.O. Box 124
22100 Lund, Sweden

K. Mortensen
Physics Department
Risø National Laboratory
4000 Roskilde, Denmark

R. Strey
Max-Planck-Institut
für Biophysikalische Chemie
Postfach 28 41
37018 Göttingen, Germany

P. Schurtenberger
Institut für Polymere
ETH Zentrum
8092 Zürich, Switzerland

Abstract In this paper we bring together some recent results concerning the stability and properties of O/W microemulsion droplets in a ternary system composed of water, decane and the nonionic surfactant pentaethylene glycol dodecylether ($C_{12}E_5$). Stable microemulsion droplets can be prepared when the spontaneous curvature has a finite but not too low value. Near the limit of maximum oil solubilisation the droplets adopt a spherical shape with low poly-dispersity. Experimental results obtained from low shear viscosity, collective and long time self-diffusion and static light scattering show that the spherical droplets interact to a very good approximation as hard spheres over a large range of volume fractions. A supersaturated micro-emulsion can be prepared by a rapid temperature quench (drop) into the two-phase area where a smaller droplet size coexists with excess oil. In the two-phase area, we can distinguish a region near the microemulsion phase boundary where the droplets are metastable, from a region further away from the boundary where the droplets are unstable and the oil-phase nucleates instantaneously. Treating the initial phase separation as a homogeneous nucleation it is possible to calculate an activation energy within the curvature energy approach.

Key words Microemulsion – hard sphere – metastability – nucleation

Introduction

Microemulsions are thermodynamically stable liquid mixtures of water, oil and surfactant. While being macroscopically homogeneous, they are locally structured into polar and apolar domains separated by a surfactant-rich dividing surface. Due to the many ways of dividing space, microemulsions may show a large variation in microstructure. Under certain conditions, it is possible to stabilize spherical droplets of, say, oil in water with a low polydis-persity and concentration invariant size. The conditions are a finite but not too low spontaneous curvature of the surfactant film and that the system is saturated with the dispersed oil [1, 2].

One such system, with the nonionic surfactant pentaethylene glycol dodecyl ether ($C_{12}E_5$), water and decane has recently been investigated in detail [3–5]. Spherical oil droplets were prepared with a surfactant-to-oil ratio $\phi_s/\phi_o = 0.815$, where ϕ_s and ϕ_o are the surfactant and oil volume fraction, respectively, over a large range of droplet volume fractions $\phi = \phi_s + \phi_o$. The surfactant consists of

Progr Colloid Polym Sci (1997) 106: 6–13
© Steinkopff Verlag 1997

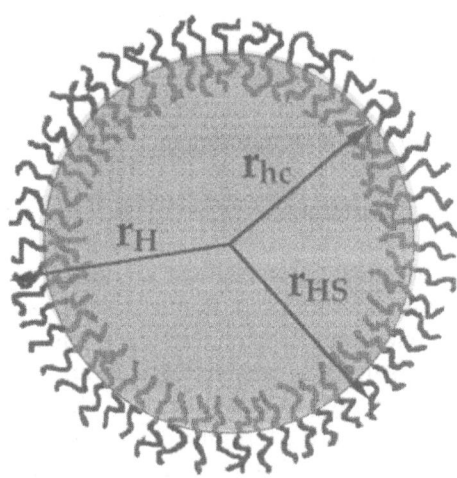

Fig. 1 Ilustration of a spherical O/W droplet stabilized by the nonionic surfactant $C_{12}E_5$. The radius r_{hc} encloses the oil and the alkyl chain of the surfactant. The droplet properties are also characterized by a hard-sphere radius (r_{HS}) and a hydrodynamic radius (r_H) with $r_{hc} < r_{HS} < r_H$

a dodecyl alkyl chain connected to an oligo ethylene oxide block of five ethylene oxide units. The microemulsion particles can be considered as spherical oil droplets of (hydrocarbon) radius r_{hc} covered by a dense brush of end-grafted penta ethylene oxide chains, as illustrated in Fig. 1, where the "grafting density" is approximately 45 Å^2 per chain. This is a situation quite analogous to sterically stabilized "solid" colloid particles.

In this paper we bring together and review the results from different experimental techniques on this microemulsion. The results, which are compared with theory and with results from "solid" colloidal particles, demonstrate the existence of spherical droplets that, to a very good approximation, interact as hard spheres over a large range of volume fractions. At the end we will also briefly discuss the properties of the system when it is supersaturated with oil, i.e. when a sample is brought by a temperature change into the two-phase area corresponding at equilibrium to a saturated microemulsion coexisting with an excess oil phase.

Solubilisation limit from curvature energy

In theoretical analyses of the various self-assembly microstructures of nonionic surfactant–water–oil systems, the flexible surface model, using the curvature energy concept [6], has been found to be very useful [1, 2, 7–10]. Here, the relative stability of a given phase and microstructure results from an interplay of the curvature energy of the

surfactant film and entropy, but where the mean-curvature energy appears to be particularly important [1]. The local-curvature energy density, g_c, is often written to second order in the curvatures [6]:

$$g_c = 2\kappa(H - H_0)^2 + \bar{\kappa}K . \tag{1}$$

Here H is the mean curvature, H_0 the spontaneous curvature, K the Gaussian curvature and κ and $\bar{\kappa}$ are the bending and saddle splay moduli, respectively.

Safran et al. [11] discussed the stability requirements for various shapes, comparing spheres, cylinders and planes. The sphere is the shape having the highest volume for a given interfacial area. Spherical oil-in-water droplets are therefore found at the phase boundary when the microemulsion is saturated with oil. Approximating the phase boundary as corresponding to the oil chemical potential $\mu_0 = 0$, the radius of the spheres is given by [2]

$$r = \frac{1}{H_0}\left(1 + \frac{\bar{\kappa}}{2\kappa}\right) + \frac{k_B T}{\kappa}f(\phi) , \tag{2}$$

where the last term comes from the entropy of mixing, ϕ being the volume fraction of micelles.

The interface separating oil and the alkyl chain of $C_{12}E_5$ from the ethylene oxide chain and water is known to have an essentially invariant area, a_s, per molecule, independent of the curvature. It is therefore useful to define the curvature at this particular interface, and the corresponding sphere radius, which we denote the hydrocarbon radius, r_{hc}, becomes

$$r_{hc} = \frac{3(\phi_0 + 0.5\phi_s)l_s}{\phi_s} . \tag{3}$$

Here $l_s = v_s/a_s$ is the surfactant volume to area ratio where $v_s \approx 700$ Å^3. The factor $0.5\phi_s$ comes from the alkyl chain volume of $C_{12}E_5$ having approximately half of the total molecular volume.

Phase diagram

The system under investigation is the $C_{12}E_5$/water/decane system. The partial phase diagram (data taken from ref. [3]) at constant surfactant to oil ratio, $\phi_s/\phi_0 = 0.815$, is presented in Fig. 2, plotted as temperature vs. the volume fraction of surfactant plus oil. The phase diagram is made with heavy water (D$_2$O). In normal water the phase boundaries are shifted upwards by 1.5–2 °C, otherwise the phase diagrams are equivalent. We note that the phase diagram is very similar to that of the binary $C_{12}E_5$–water system [12]. At higher water contents we see the sequence of a liquid micellar phase (L_1), here oil swollen micelles, and lamellar phase (L_α) and a L_3 (sponge) phase with

8

U. Olsson et al.
Stable, metastable and unstable droplets

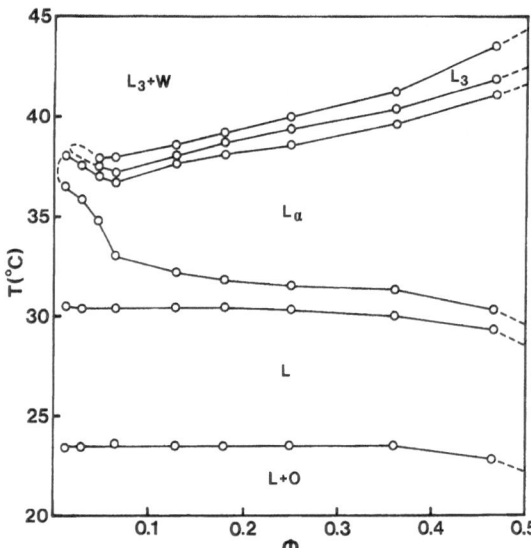

Fig. 2 Partial phase diagram of the $C_{12}E_5$–water (D_2O)–decane system at a constant surfactant-to-oil ratio ϕ_s/ϕ_o. L_1 denotes the microemulsion phase, L_α the lamellar liquid-crystalline phase. In addition there is the L_3 ("sponge") phase at higher temperatures. Of particular interest here is the lower phase boundary of the microemulsion phase ($\approx 23.5\,^\circ\text{C}$ with D_2O and $\approx 25.0\,^\circ\text{C}$ with H_2O) below which the microemulsion coexists with excess oil ($L_1 + O$)

increasing temperature. At lower temperatures there is an incomplete solubilisation of the oil and the L_1 phase exists in equilibrium with excess oil.

As in the binary $C_{12}E_5$ system, the micelles in the L_1 phase grow with increasing temperature [13, 14], although in the ternary system the growth is only minor due to the fact that the spherical micelle is already relatively large (for the given surfactant-to-oil ratio it contains approximately 1500 surfactant and 4000 oil molecules). Increasing the temperature further, lamellar and L_3 phases are formed where the aggregates have essentially an infinite aggregation number. This counter intuitive behavior, which goes against the preference of entropy, we today understand as the result of a decreasing spontaneous curvature, H_0, of the surfactant film with increasing temperature. When the film is saturated by both water and oil the temperature dependence of H_0 is found to be linear over a large temperature interval [1, 15, 16]:

$$H_0 = a(T_0 - T)\,, \tag{4}$$

where $a \approx 10^{-3}\,\text{Å}^{-1}\,\text{K}^{-1}$.

The lower phase boundary can be identified as the solubility limit as discussed in the previous section. The phase boundary, which is temperature independent up to approximately $\phi = 0.5$, corresponds to a dilution line of

concentration invariant spherical O/W microemulsion droplets.

Stable droplets: a hard sphere microemulsion

Small-angle neutron scattering

Figure 3 shows the small-angle neutron scattering (obtained at Risø National Laboratory, Denmark) from a dilute sample of $\phi = 0.02$ where the oil and water has been contrast matched resulting in coherent scattering from the surfactant film alone [5]. A fit to the data, shown as a solid line, corresponds to the form factor of a spherical shell of radius $r_{\text{hc}} = 75\,\text{Å}$ and a relative polydispersity $\sigma/r_{\text{hc}} = 0.16$. This polydispersity contains contributions from both size (volume) and shape polydispersity [5]. Their relative contributions are not accurately known, however, contrast match experiments indicate that it is the shape polydispersity which dominates.

Static light scattering

In a static light scattering experiment the effective structure factor, $S(0)$ at zero scattering vector was obtained from the extrapolated forward scattering. In Fig. 4 is shown the variation of the excess Rayleigh ratio $\Delta R(0)$ extrapolated zero scattering vector with the volume fraction of droplets, ϕ [3]. In the monodisperse case, $S(0)$ is linked to the osmotic compressibility for which, in the case of hard spheres, an accurate expression exists due to Carnahan and Starling [17]. The experimental data were

Fig. 3 SANS spectrum from a sample of $\phi = 0.02$, where the scattering length density of the oil is matched to that of water. The solid line is the best two parameter fit with a form factor of a shell yielding the radius $r_{\text{hc}} = 75\,\text{Å}$ and Gaussian relative standard deviation $\sigma/r_{\text{hc}} = 0.16$. Data taken from ref. [5]

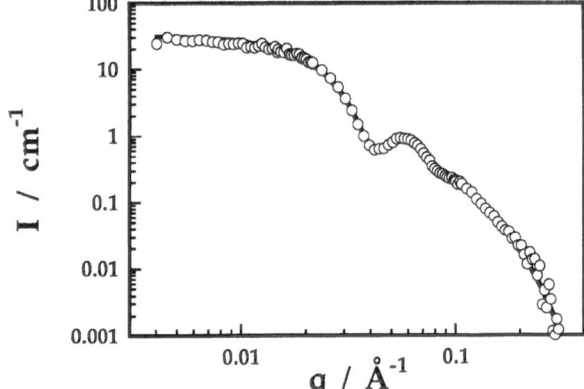

Progr Colloid Polym Sci (1997) 106:6–13
© Steinkopff Verlag 1997

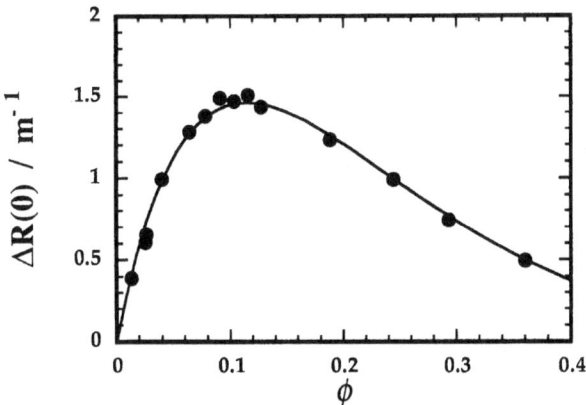

Fig. 4 Variation of the excess Rayleigh ratio, ΔR, extrapolated to zero scattering vector, as a function of the droplet volume fraction, ϕ. The solid line is the best two parameter fit using the Carnahan–Starling equation for the osmotic pressure. The two fitted parameters were $r_{HS} = 86$ Å and $r_{hc} = 76$ Å. Data taken from ref. [3]

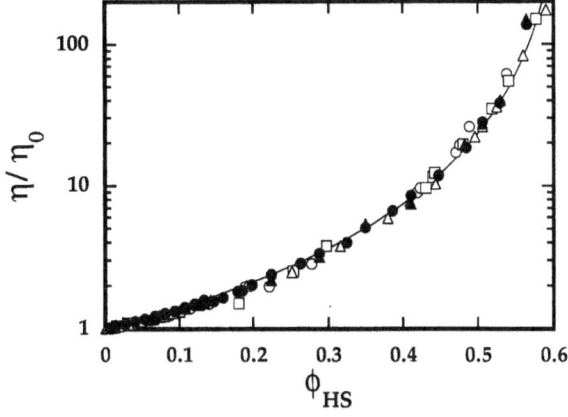

Fig. 5 Variation of the normalized low shear viscosity η/η_0 with the hard-sphere volume fraction ϕ_{HS}. Samples from the microemulsion (data from ref. [4]) was measured in a capillary (●) or in a cone and plate rheometer (▲). Shown with open symbols are data of different radii of coated silica spheres in oil, taken from ref. [18]. The solid line shows the prediction of Eq. (1)

fitted with the Carnahan–Starling equation for $S(0)$ using two adjustable parameters, namely the hard-sphere radius, r_{HS}, and the hydrocarbon radius, r_{hc}. The fit, shown in Fig. 4 as a solid line, results in an almost perfect match with the data. The parameters were found to be $r_{HS} = 86$ Å and $r_{hc} = 76$ Å, the latter being in excellent agreement with the SANS data. From r_{HS} we also obtain the corresponding hard-sphere volume fraction $\phi_{HS} = 1.14\,\phi$. The fact that we are able to describe the system so well assuming monodisperse spheres indicates that polydispersity effects are minor.

Low shear viscosity

The low shear viscosity, η, was measured using capillary and, at higher concentrations, a cone-plate rheometer [4]. The two techniques gave equivalent results in the overlapping concentration range. The variation of the normalised low shear viscosity η/η_0, where η_0 is the water solvent viscosity, with the hard-sphere volume fraction ϕ_{HS} is shown in Fig. 5. For comparison, we have also plotted data from van der Werff and de Kruif [18] for hard-sphere silica dispersions of three different sizes. As can be seen, there is a perfect agreement between the microemulsion and silica data. The solid line in Fig. 5 shows the Quemada expression [19]

$$\eta/\eta_0 = (1 - \phi_{HS}/\phi_m)^{-2} \tag{5}$$

with $\phi_m = 0.63$, which provides an accurate description of the data.

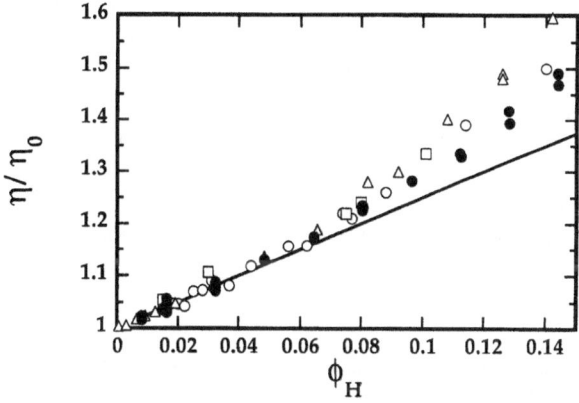

Fig. 6 Concentration dependence of the normalized low shear viscosity in the dilute regime, plotted as a function of ϕ_H. The solid line is according to Eq. (2). Symbols are the same as in Fig. 4

In Fig. 5, which mainly highlights the concentrated regime, we saw that the relevant parameter for describing the concentration was the hard-sphere volume fraction ϕ_{HS}. At high dilution, on the other hand, the viscosity is governed by the hydrodynamic volume fraction ϕ_H, which in the microemulsion corresponds to 1.36ϕ (see below). In Fig. 6 we have plotted the concentration dependence of η/η_0 in the dilute regime, where the concentration is now represented by ϕ_H. The microemulsion and silica data are in good agreement at lower concentrations. The solid line in Fig. 5 is the Einstein relation [20]

$$\eta/\eta_0 = 1 + \tfrac{5}{2}\,\phi_H \ . \tag{6}$$

As is seen, the data agree well with the Einstein relation below $\phi_H \approx 0.05$. At $\phi_H \approx 0.09$, the data from the microemulsion begins to deviate from the silica system. Above this concentration, direct interactions become important and ϕ_H is no longer the relevant concentration variable for the microemulsion particle. ϕ_H overestimates the effective concentration which is now better described by ϕ_{HS}.

Diffusion

The concentration dependence of the collective (D_c) and self-diffusion (D_s) coefficients of the droplets are presented in Fig. 7, where we have plotted D_c/D_0 and D_s/D_0 as a function of ϕ_{HS} [3]. Here, D_0 denotes the diffusion coefficient extrapolated to infinite dilution. Due to the liquid nature of the droplets their NMR spectrum is in the motional narrowing regime with long transverse relaxation times. For this reason, the long time self-diffusion coefficients are conveniently and accurately measured using the pulsed gradient spin-echo (PGSE) technique. D_c was obtained from dynamic light scattering. From $D_0 = 2.0 \times 10^{-11}\,\mathrm{m^2\,s^{-1}}$ (using heavy water, D_2O, as solvent and 23.5 °C) a hydrodynamic radius of $r_H = 95$ Å is obtained from the Stokes–Einstein relation. From this value, we obtain a hydrodynamic volume fraction $\phi_H = 1.36\phi$, which was used to describe the low shear viscosity in the dilute regime above. In Fig. 7, we also show, for comparison, D_c data on silica dispersions taken from Kops-Werkhoven and Fijnaut [21] and D_s data from

Fig. 7 Variation of the normalized collective (D_c/D_0) (▲) and long time self-diffusion (D_s/D_0) (●) coefficients with the hard-sphere volume fraction ϕ_{HS}. All filled symbols refer to microemulsion data (taken from ref. [3]) D_c/D_0 data shown as open triangles correspond to silica spheres, taken from ref. [21]. D_s/D_0 data shown as open circles correspond to the self-diffusion of traces of silica spheres in a dispersion of pmma spheres (data taken from ref. [22]). The broken line is the equation $D_c/D_0 = 1 + 1.3\,\phi_{HS}$. The solid line shows the relation $D_s/D_0 = (1 - \phi_{HS}/0.63)^2$

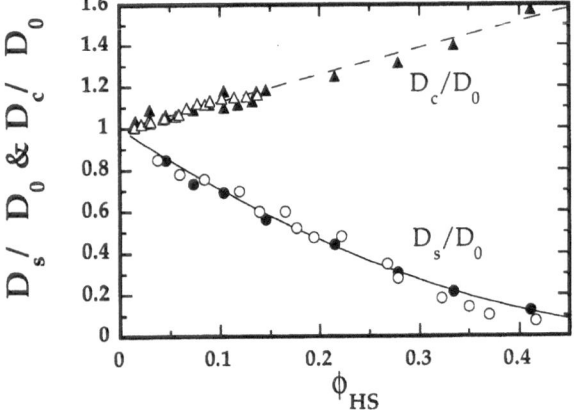

van Megen and Underwood measured on traces of silica particles in a dispersion of poly(methylmetacrylate) spheres [22]. As is seen, there is a good agreement, both in the collective and self-diffusion data. D_c increases approximately linearly with ϕ_{HS}, $D_c/D_0 \approx 1 + 1.3\,\phi_{HS}$, up to high volume fractions.

It is known from simple liquids that there is a correlation between D_s and η, where the product $D_s\eta$ is approximately constant upon variations in pressure or temperature [23]. A similar correlation was also found in a colloidal hard-sphere system [24], with $D_s/D_0 = (\eta/\eta_0)$, which implies that the variation of D_s/D_0 can be described by the inverse of the relation describing η/η_0;

$$D_s/D_0 = (1 - \phi_{HS}/\phi_m)^2 . \tag{7}$$

This equation is shown as a solid line in Fig. 7 and as is seen, it provides a good description of the self-diffusion data. The correlation between the low shear viscosity and the long time self-diffusion data becomes even clearer when diffusion and viscosity data are plotted together as in Fig. 8.

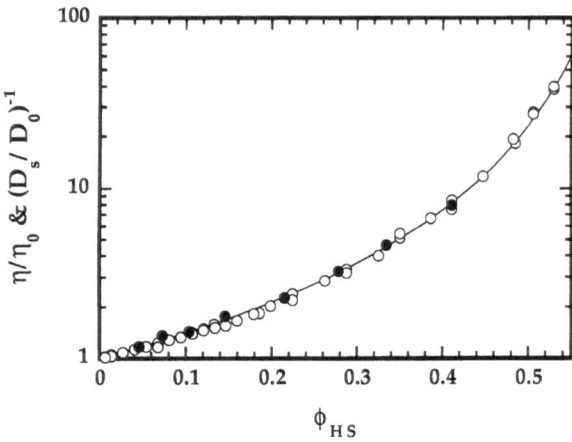

Fig. 8 Plot of the inverse normalized self-diffusion coefficient $(D_s/D_0)^{-1}$ and the normalized low shear viscosity η/η_0 as a function of hard-sphere volume fraction ϕ_{HS}. Open circles correspond to η/η_0 and filled circles to $(D_s/D_0)^{-1}$. The solid line is the Quemada function: $(1 - \phi_{HS}/0.63)^{-2}$

Metastable and unstable droplets: nucleation of the oil phase

As shown above, the droplets on the lower phase boundary are spherical and behave as hard spheres. If the temperature is lowered a smaller radius is preferred and the droplets expel excess oil to a separate phase. Thus, the equilibrium situation in the two-phase area $L_1 + O$ corresponds to smaller droplets in equilibrium with an oil film on top of the sample. The deeper one goes into this

Progr Colloid Polym Sci (1997) 106: 6–13
© Steinkopff Verlag 1997

two phase area, the more oil is expelled and the smaller are the droplets.

An interesting question concerns the mechanistic pathway to the new equilibrium state when a stable microemulsion is quenched into the two phase area by a drop in temperature. A sample containing $\phi = 0.12$ was investigated by measuring the turbidity continuously with time. After equilibration at approximately 28 °C the temperature was dropped to a new value while recording the turbidity. When lowering the temperature within the L_1 phase the turbidity decreases to a new equilibrium value which is time independent. The reason for the temperature dependence is the temperature dependence of the droplet size. Near the lower phase boundary the turbidity becomes temperature independent as the droplets become spherical.

In Fig. 9 we show the turbidity trace (data taken from ref. [25]) when the temperature is dropped to a value below the lower phase boundary in the $L_1 + O$ two phase area. In Fig. 9a the new temperature is 22.4 °C in (b) it is 21.9 °C and in (c) it is 20.3 °C. At the highest of these three temperatures the turbidity reaches a value after the temperature drop which is essentially identical to that measured at the phase boundary and which does not vary with time for at least an hour. The microemulsion is supercooled and the sample is apparently metastable. At 21.9 °C (Fig. 9b) we observe a small but significant increase in the turbidity with time showing that phase separation occurs instantaneously after the temperature drop although the kinetics are slow. At 20.3 °C (Fig. 9c) the turbidity raises strongly with time after the quench.

Thus, we have a temperature interval from the phase boundary at 25 °C down to approximately 22 °C where the droplets are metastable, and a region below 22 °C where the droplets are unstable. Measurements performed on a lower ($\phi = 0.05$) and a higher ($\phi = 0.24$) concentration showed the same behavior as the $\phi = 0.12$ sample. In particular, the temperature separating the metastable and unstable regions was essentially identical (22 °C).

Now, how can we understand the fact that there is a metastable and an unstable region and that the temperature boundary between the two is independent of the droplet concentration? If we focus on the initial stage of the phase separation, the oil phase can nucleate homogeneously in bulk or heterogeneously at the surface of the container or for example a dust particle. The results are reproducible which indicates that we can ignore any influence from impurities like dust particles. Also, the turbidity experiment probes essentially the bulk, and it is therefore reasonable to believe that what we observe in the experiment is a homogeneous nucleation.

The driving force for the phase separation is the increase in spontaneous curvature and the wish of the droplets to decrease in size. In the case of homogeneous

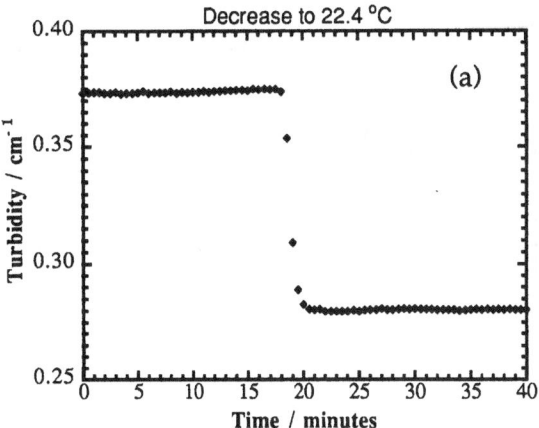

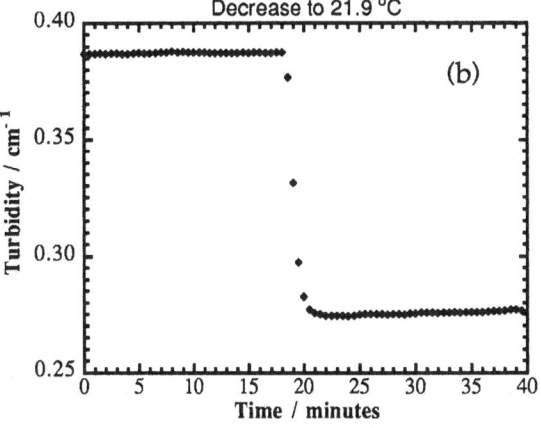

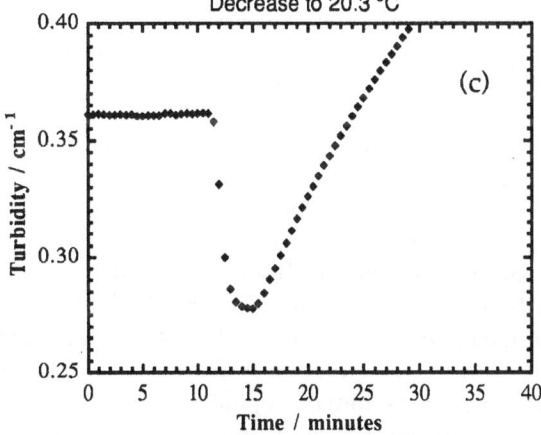

Fig. 9 Turbidity as a function of time (t) from a sample with $\phi = 0.12$. The sample is equilibrated at ≈ 28 °C and $t = 0$ and approximately $t = 20$ min the temperature is lowered to (a) 22.4 °C, (b) 21.9 °C and (c) 20.3 °C

nucleation however, one or more droplets has to be sacrificed and allowed to grow in order to let the majority of droplets decrease in size. The sacrifice of one or more droplets implies the existence of an activation energy,

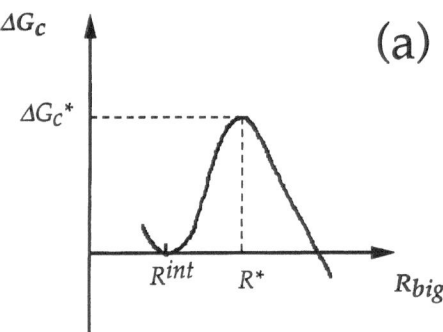

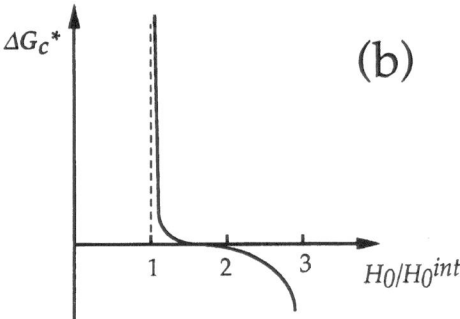

Fig. 10 Schematic variation of ΔG_c with R_{big} (a) and of ΔG_c^* with H_0/H_0^{int} (b)

which if it is sufficiently large, will explain the observed metastability.

The activation energy can be estimated from a model of N droplets of initial radius R^{int}, where we let one droplet grow in size and the remaining droplets shrink coherently (all small droplets have the same size). Conservation of volume and area implies that the number of droplets increases. The curvature energy difference, ΔG_c, between the state with one large droplet and the initial state with all droplets having the initial radius is first of all, in the limit of large N, found to be independent of N. This is due to a cancellation involving the number of droplets and the amount by which the small droplets shrink. Second, ΔG_c shows a maximum as a function of R_{big} at a certain value $R_{\mathrm{big}} = R^* > R^{\mathrm{int}}$. The situation is illustrated in Fig. 10. Thus, the process is initially uphill with an activation energy ΔG^*. Solving ΔG_c^* from $\partial \Delta G_\mathrm{c}/\partial R_{\mathrm{big}} = 0$, we find

that ΔG_c^* only depends on the elastic constants and spontaneous curvature ratio H_0/H_0^{int} [26]

$$\Delta G_\mathrm{c}^* = 8\pi k \frac{(1 - \frac{2}{3}(H_0/H_0^{\mathrm{int}}))^3}{(1 - (H_0/H_0^{\mathrm{int}}))^2} , \tag{8}$$

where $k = \kappa + \bar{\kappa}/2$.

As expected ΔG_c^* diverges at the phase boundary where $H_0 = H_0^{\mathrm{int}}$, but we also see that it decreases with increasing H_0/H_0^{int} and vanishes at $H_0/H_0^{\mathrm{int}} = \frac{3}{2}$. This is consistent with the experiments where we see a crossover from metastable to unstable droplets when we increase the quench depth. At sufficiently deep quenches (large H_0/H_0^{int}) the model predicts correctly that the nucleation process is only downhill and therefore occurs instantaneously.

H_0 is proportional to $(T_0 - T)$, and we have

$$\frac{H_0}{H_0^{\mathrm{int}}} = \frac{T_0 - T}{T_0 - T^{\mathrm{int}}} , \tag{9}$$

where $T^{\mathrm{int}} = 25\,^\circ\mathrm{C}$ is the phase-boundary temperature. Since $T_0 = 38\,^\circ\mathrm{C}$ we find $T = 18.5\,^\circ\mathrm{C}$ as the temperature where $\Delta G_\mathrm{c}^* = 0$. This is slightly lower than the temperature $22\,^\circ\mathrm{C}$ estimated experimentally for the metastable to unstable transition. This may at first seem as a discrepancy, however, we can expect to observe nucleation in our experiment already for a finite but sufficiently small ΔG_c^*. Furthermore, as is seen in Eq. (8), $H_0/H_0^{\mathrm{int}} = \frac{3}{2}$ is a triple root and ΔG_c^* varies slowly in its vicinity. In fact, if we define $\Delta G_\mathrm{c}^* = k \approx k_\mathrm{B}T$ as the metastability limit we find that this occurs at $H_0/H_0^{\mathrm{int}} \approx 1.3$ and hence a temperature of approximately $21\,^\circ\mathrm{C}$ which is close to that experimentally found.

Conclusion

As been shown above, it is possible to prepare a hard-sphere microemulsion up to a rather high volume fraction ($\phi \approx 0.5$) of droplets. The recipe for spheres we obtain from the curvature energy model which predicts spheres at the solubility limit with a size dictated primarily by the spontaneous curvature. It appears also that we can understand the intriguing kinetic properties of the droplets, when quenched into the supersaturated state, within the curvature-energy approach.

References

1. Olsson U, Wennerström H (1994) Adv Colloid Interface Sci 49:113–146
2. Safran SA (1994) Statistical Thermodynamics of Surfaces, Interfaces, and Membranes. Addison-Wesley, Reading, MA
3. Olsson U, Schurtenberger P (1993) Langmuir 9:3389–3394
4. Leaver MS, Olsson U (1994) Langmuir 10:3449–3454
5. Bagger-Jörgensen H, Olsson U, Mortensen K (1997) Langmuir 13:1413–1421
6. Helfrich W (1973) Z Naturforsch 28c:693–703
7. Pieruschka P, Safran SA (1993) Europhys Lett 22:625–630
8. Porte G, Appell J, Bassereau P, Marignan L (1989) J Phys France 50:1335–1347

Progr Colloid Polym Sci (1997) 106:6–13
© Steinkopff Verlag 1997

9. Porte G, Delsant M, Billard I, Skouri M, Appell J, Marignan J, Debeauvais F (1991) J Phys France II 1:1101–1120
10. Daicic J, Olsson U, Wennerström H, Jerke G, Schurtenberger P (1995) J Phys II France 5:199–215
11. Safran SA, Turkevich LA, Pincus PA (1984) J Phys Lett 45:L69
12. Strey R, Schomäcker R, Roux D, Nallet F, Olsson U (1990) J Chem Soc Faraday Trans 86:2253–2261
13. Leaver MS, Olsson U, Wennerström H, Strey R (1994) J Phys II 4:515–531
14. Leaver M, Furó I, Olsson U (1995) Langmuir 11:1524–1529
15. Rajagopalan V, Bagger-Jörgensen H, Fukuda K, Olsson U, Jönsson B (1996) Langmuir 12:2939–2946
16. Strey R (1994) Colloid Polym Sci 272:1005–1019
17. Carnahan NF, Starling KE (1969) J Chem Phys 51:635
18. van der Werff JC, de Kruif CG (1989) J. Rheol 33:421–454
19. Quemada D (1977) Rheol Acta 16:82–94
20. Einstein A (1956) Investigations on the Theory of the Brownian Motion. Dover, New York
21. Kops-Werkhoven MM, Fijnaut HM (1981) J Chem Phys 74:1618
22. van Megen W, Underwood SM (1989) J Chem Phys 91:552
23. Packhurst Jr. HJ, Jonas J (1975) J Chem Phys 63:2698, 2705
24. van Bladeren A, Peetermans J, Maret G, Dhont JKG (1992) J Chem Phys 96:4591
25. Morris J, Olsson U, Wennerström H (1997) Langmuir 13:606–608
26. Morris J, Olsson U, Wennerström H (submitted)

Progr Colloid Polym Sci (1997) 106:14–23
© Steinkopff Verlag 1997

S.-M. Choi
S.-H. Chen

The relationship between the interfacial curvatures and phase behavior in bicontinuous microemulsions – a SANS study

Received: 13 October 1996
Accepted: 4 February 1997

Abstract We introduce a new technique using small-angle neutron scattering (SANS) to measure the average Gaussian curvature and the average square-mean curvature of the oil–water interface in a three-component, nearly isometric (equal volume fractions of water and oil) ionic microemulsion system. The microemulsion is composed of AOT/brine/decane. SANS measurements are made as a function of both the volume fraction of surfactant and salinity at a constant temperature, 45 °C, within the one-phase channel. The temperature is chosen at the hydrophile–lipophile balance (HLB) temperature for a salinity of 0.49%. The SANS data taken with an oil–water contrast are analyzed by using a random-wave model with an appropriate spectral function. The spectral function is an inverse eighth-order polynomial in wave number k, containing three length scales $1/a$, $1/b$, and $1/c$, and has finite second and fourth moments. This three-parameter spectral function is then used in conjunction with Cahn's clipping scheme to obtain the Debye correlation function for the microphase-separated bicontinuous microemulsions. The model shows good agreement with the intensity data in an absolute scale. We then use the three parameters so obtained to calculate the average Gaussian curvature and the average square-mean curvature of the interface. We determine the variation of these curvatures as functions of the surfactant volume fraction and salinity and discuss their implication on the degree of local order of the bicontinuous structure. We also show a 3-D morphology of the microemulsion at the contact point of the three-phase and the one-phase region of the phase diagram generated by this model.

Key words Interfacial curvatures – microemulsion – SANS

Dr. S.-M. Choi (✉) · S.-H. Chen
Department of Nuclear Engineering 24-209
Massachusetts Institute of Technology
Cambridge, Massachusetts 02139, USA

Introduction

A microemulsion is a thermodynamically stable, transparent or translucent three-component molecular liquid mixture having a mesoscopic-scale self-organized internal structure. It is composed of two originally immiscible liquids, water and oil, made miscible by an addition of amphiphilic molecules. An amphiphilic molecule, or more conventionally called a surfactant, has one end of it compatible to water (or hydrophilic) and the other end compatible to oil (or hydrophobic). Upon mixing, majority of the surfactant molecules position themselves at the interfaces between mesoscopic water and oil domains thus drastically reducing the interfacial tension between the water and oil. At equilibrium, free energy of the system is then

Progr Colloid Polym Sci (1997) 106: 14–23
© Steinkopff Verlag 1997

largely controlled by the term containing the entropy. Under this condition, the system would like to organize in such a way as to maximize its entropy. As a result, the microemulsion usually adopts a disordered microstructure which is seen to be consisting of interpenetrating mesoscopic domains of water and oil [1]. At a temperature where the hydrophilicity and hydrophobicity of the surfactant molecules are balanced, this type of microstructure is in fact the most favored one for isometric compositions with equal volume fractions of water and oil and having dilute surfactant concentrations. This kind of microstructure is often called a bicontinuous structure in the literature. As the volume fraction of the surfactant increases the microstructure tends to favor a more ordered lamellar structure.

In this paper we choose to study isometric microemulsions made of an ionic surfactant called AOT (sodium-bis-ethylhexylsulpho-succinate), water and decane. This three-component system can easily form microemulsions in the vicinity of room temperature at wide range of compositions. It has been shown, however, that the pure ternary system forms a water-in-oil droplet microemulsion only at room temperature because of a preferential curvature of the surfactant film toward water [2]. Thus, the pure ternary system normally does not form a bicontinuous structures. In order to realize a bicontinuous structure in this system one needs to add small amounts of salt [3]. For ease of visualization, the phase prism of an isometric microemulsion system can be projected onto the temperature–surfactant volume fraction plane. In this regard it is useful to remark that in the case of the AOT-based microemulsion system one can also represent the phase diagram equally well in the salinity–surfactant volume fraction plane if one chooses a temperature close but above the HLB (hydrophilic–lipophilic balanced) temperature of the system. The physical reason for this is that the spontaneous curvature of the AOT monolayer at the oil–water interface can be varied continuously by either changing temperature at a fixed salinity or by changing salinity at a fixed temperature. The hydrophilicity of an AOT molecule increases with increasing temperature but diminishes upon higher salinity.

Figure 1 shows the phase diagram of AOT/D_2O (NaCl)/H-decane system in the salinity–surfactant volume fraction plane at a temperature 45 °C. For small surfactant volume fractions, less than 0.04, and low salinity, the system shows a $\underline{2}$-phase with a coexisting excess oil layer on the top and an oil-in-water microemulsion at the bottom. As salinity increases, the system goes through a three-phase region, where a middle-phase microemulsion is in coexistence with an excess oil layer on the top and an excess water layer in the bottom, and at high salinity it transforms to a $\overline{2}$-phase with an excess water layer in the

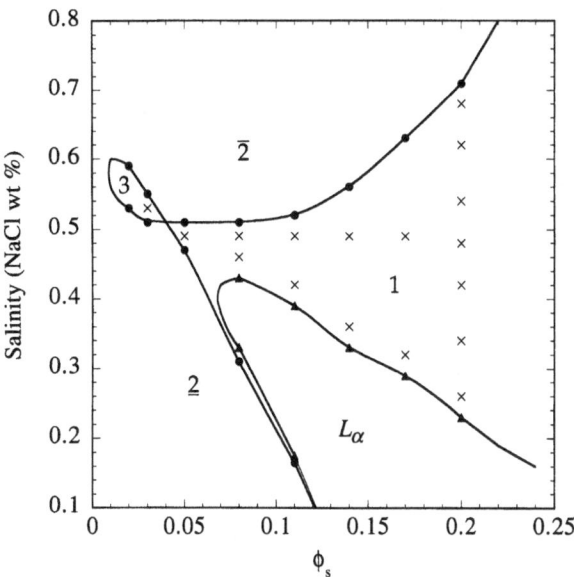

Fig. 1 Phase diagram of an $AOT/D_2O(NaCl)/H$-decane microemulsion system at $T = 45$ °C

bottom and water-in-oil microemulsion on the top. For the surfactant volume fractions larger than 0.035, as salinity increases, the system transforms from a $\underline{2}$-phase to a $\overline{2}$-phase through a single-phase region. In the lower part of the one-phase channel, one finds a region of a lamellar phase of which the salinity interval increases with increasing volume fraction of AOT. It should be noted from this phase diagram that the microstructure of the microemulsion inverts from an oil-in-water structure at low salinity to a water-in-oil structure at high salinity at a fixed composition. It is thus expected that somewhere in the middle of the one phase region there should exist a series of disordered bicontinuous microemulsions with zero mean curvature.

Experiments

The phase behavior of $AOT/D_2O(NaCl)/H$-decane system shown in Fig. 1 was mapped out by varying the salinity and the volume fraction of surfactant at a constant temperature of 45 °C. The phase diagram was prepared for performing bulk contrast SANS experiments in which we match the scattering length density of decane with that of hydrocarbon tail of AOT. For this purpose, the volume fraction of D_2O plus head group of AOT was maintained at 0.5 for all surfactant volume fractions at a certain reference salinity $S = 0$, i.e. $\varphi_1 = \varphi_{D_2O} + \beta(S = 0)\varphi_S = 0.5$ $(\varphi_2 = \varphi_{Decane} + (1 - \beta(S = 0))\varphi_S = 0.5)$. $\beta(S = 0)$ was taken to be 0.1064 [4]. In this paper we treat β as a function of salinity S. Therefore, the microemulsions are isometric at the reference salinity and become non-isometric if the salinity changes from the reference value. In our SANS measurements, a suitable mixture of H-decane (94.99 wt%) and D-decane (5.01 wt%) was used to match the scattering length density of decane with that of the hydrocarbon tail of AOT (-0.1845×10^{10} cm^{-2}).

The SANS measurements were performed at 45 °C for the points indicated as crosses in the phase digram. The point along the boundary between one-phase and lamellar phase are to measure the average Gaussian curvature and the average square-mean curvature as a function of surfactant volume fraction. Since it is well known that the average mean curvature of lamellar phase is zero, the average mean curvature at the points close to lamellar phase are expected to be very close to zero. The points at $\varphi_S = 0.20$ with different salinity are to measure the average Gaussian curvature and the average square-mean curvature as a function of salinity.

SANS experiments were carried out at the small-angle neutron diffractometer (SAND) at IPNS in Argonne National Laboratory. SAND uses a spallation neutron source generated by a 500 MeV H-accelerator. After moderation we have a pulse of white neutron beam with an effective wave length range from 1 to 14 Å. In SAND all these neutrons are utilized by encoding their individual time-of-flight and their scattering angles determined by their detected position in a 2D area detector. The 2D area detector has an active area of 40×40 cm^2 and the sample to detector distance is 2 m. This configuration allows a maximum scattering angle of about 9°. The reliable Q-range covered in the measurements were from 0.004 to 0.4 Å^{-1}. Sample liquid was contained in a flat quartz cell with 1 mm path length. Temperature of the sample was set by a thermostated circulating water bath to an accuracy of 0.1 °C. Measured intensities were corrected for background and empty-cell contributions and normalized by a reference scattering intensity of 1 mm water at room temperature.

Theory of scattering from a bicontinuous microemulsion in bulk contrast

The intensity distribution of SANS from an isotropic, disordered two-component porous material can be calculated generally from a Debye correlation function $\Gamma(r)$ by the following formula [5]:

$$I(Q) = \langle \eta^2 \rangle \int_0^\infty dr\, 4\pi r^2 j_0(Qr)\Gamma(r)\,, \tag{1}$$

where $\langle \eta^2 \rangle = \varphi_1\varphi_2(\rho_1 - \rho_2)^2$, is the mean-square fluctuation of local scattering length density or so-called "invariant". φ_1 and φ_2 refer to the volume fractions of components 1 and 2, and ρ_1 and ρ_2 to the corresponding scattering length densities. There are two physical boundary conditions that the Debye correlation function, a function of a scalar distance r, must satisfy: it is normalized to unity at the origin and it should go to zero at infinity. Due to the normalization condition, the invariant can also be written as:

$$\langle \eta^2 \rangle = \frac{1}{2\pi^2} \int_0^\infty dQ\, I(Q)Q^2 \tag{2}$$

which is a practical way for calculating this quantity. When the intensity is measured over a sufficiently large Q range, the measured intensity $I(Q)$ can be divided by the invariant calculated according to Eq. (2) to remove the uncertainty in the absolute intensity calibration.

We shall describe a method for calculating the Debye correlation functions used in this paper. The most important property of the correlation function for the bulk contrast case with a sharp boundary between two regions having different scattering length densities is that it has a linear and cubic terms in small r expansion of the form

$$\Gamma_B(r \to 0) = 1 - ar + br^3 + \cdots$$
$$= 1 - \frac{1}{4\varphi_1\varphi_2}\frac{S}{V}r\left(1 - \frac{b}{a}r^2\right) + \cdots\,, \tag{3}$$

where $a = (S/V)/4\varphi_1\varphi_2$ is a factor proportional to the total interfacial area per unit volume and the ratio of the coefficient of the cubic term to the linear term has been given by Kirste and Porod [6] in terms of curvatures as

$$\frac{b}{a} = \frac{1}{8}\langle H^2 \rangle - \frac{1}{24}\langle K \rangle\,. \tag{4}$$

The average Gaussian curvature $\langle K \rangle$, the mean curvature $\langle H \rangle$, and the average square-mean curvature $\langle H^2 \rangle$, are defined, respectively, as

$$\langle K \rangle = \left\langle \frac{1}{R_1 R_2} \right\rangle\,, \tag{5}$$

$$\langle H \rangle = \left\langle \frac{1}{2}\left(\frac{1}{R_1} + \frac{1}{R_2}\right) \right\rangle\,, \tag{6}$$

and

$$\langle H^2 \rangle = \left\langle \frac{1}{4}\left(\frac{1}{R_1} + \frac{1}{R_2}\right)^2 \right\rangle\,, \tag{7}$$

where R_1 and R_2 are the radius of curvatures along two principal axes at an arbitrary point on the interface and the averages are performed over the whole surface.

In the random wave model of Berk [7], a Gaussian random field $\psi(\mathbf{r})$ is constructed by superposition of a large number N of cosine waves with random phases:

$$\psi(\mathbf{r}) = \sqrt{\frac{2}{N}} \sum_{i=1}^{N} \cos(\mathbf{k}_i \cdot \mathbf{r} + \varphi_i) \,, \qquad (8)$$

where directions of the wave vector $\mathbf{k}_i$ are assumed to distribute isotropically over a unit sphere and the phase φ_i distributed randomly over an interval $(0, 2\pi)$. In constructing the sum, the magnitude of the $\mathbf{k}_i$ vector is sampled from a distribution $f(k)$.

The statistical properties of a Gaussian random process is completely characterized by giving its two-point correlation function. We define a two-point correlation function $g(|\mathbf{r}_1 - \mathbf{r}_2|) = \langle \psi(\mathbf{r}_1)\psi(\mathbf{r}_2) \rangle$ and the associated spectral function $f(k)$ by a Fourier transform relation

$$g(|\mathbf{r}_1 - \mathbf{r}_2|) = \int_0^\infty 4\pi k^2 j_0(k|\mathbf{r}_1 - \mathbf{r}_2|)f(k)\,\mathrm{d}k \,. \qquad (9)$$

This continuous random process $\psi(\mathbf{r})$, varying between $+1$ and -1, having a mean-square value of unity, is then clipped and transformed into a two-state discrete random process. By clipping we mean assigning a constant value $+1$ to the function whenever the Gaussian random field at that point is above a certain level called α, and a constant 0 whenever its value is below α. This transformation can be defined as follows:

$$\zeta(\mathbf{r}) = \Theta_\alpha (\psi(\mathbf{r})) = \begin{cases} 1 & \text{when } \psi(\mathbf{r}) \geq \alpha \,, \\ 0 & \text{otherwise} \,, \end{cases} \qquad (10)$$

where Θ_α is a step function. Then the Debye correlation function for this discrete random process is given exactly as

$$\Gamma_\mathrm{B}(\mathbf{r}) = \frac{\langle \zeta(0)\zeta(\mathbf{r})\rangle - \langle \zeta \rangle^2}{\langle \zeta \rangle - \langle \zeta \rangle^2} \,, \qquad (11)$$

where the quantities on the right-hand side are given by Teubner [8] as

$$\langle \zeta \rangle = \frac{1}{2} - \frac{1}{\sqrt{2\pi}} \int_\alpha^0 \exp(-x^2/2)\,\mathrm{d}x \,, \qquad (12)$$

$$\langle \zeta(0)\zeta(\mathbf{r})\rangle = \langle \zeta \rangle - \frac{1}{2\pi} \int_0^{\cos^{-1}(g(r))} \exp\left(-\frac{\alpha^2}{1+\cos\theta}\right)\mathrm{d}\theta \,. \qquad (13)$$

The average value of the clipped Gaussian random field, $\langle \zeta \rangle$, and $1 - \langle \zeta \rangle$ correspond to volume fractions of the majority and minority phases. Using Eq. (12) and $\langle \zeta \rangle = \varphi_1$, Eq. (11) can be rewritten as

$$\Gamma_\mathrm{B}(\mathbf{r}) = 1 - \frac{1}{2\pi\varphi_1(1-\varphi_1)} \int_0^{\cos^{-1}(g(r))} \exp\left(-\frac{\alpha^2}{1+\cos\theta}\right)\mathrm{d}\theta \,. \qquad (14)$$

For a small α, meaning a slight deviation from an isometric case, Eq. (13) can be approximated as

$$\Gamma_\mathrm{B}(\mathbf{r}) \cong 1 - \frac{1}{2\pi\varphi_1(1 - \varphi_1)}$$
$$\times \left[\cos^{-1}(g(r)) - \alpha^2 \tan\left(\frac{\cos^{-1}(g(r))}{2}\right) \right], \qquad (15)$$

where the volume fraction φ_1 can be approximated as

$$\varphi_1 \cong \frac{1}{2} - \frac{\alpha}{\sqrt{2\pi}} \,. \qquad (16)$$

For an isometric microemulsion, i.e. $\varphi_1 = 0.5$ and $\alpha = 0$, Eq. (15) reduces to

$$\Gamma_\mathrm{B}(\mathbf{r}) = \frac{2}{\pi} \sin^{-1}(g(r)) \qquad (17)$$

which is same as the result by Teubner [8].

From the relation in Eq. (9), one has a small r expansion of $g(r)$ of the form

$$g(r) = \int_0^\infty 4\pi k^2 \left[1 - \frac{1}{6} k^2 r^2 + \frac{1}{120} k^4 r^4 + \cdots \right] f(k)\,\mathrm{d}k$$
$$= 1 - \frac{1}{6}\langle k^2 \rangle r^2 + \frac{1}{120}\langle k^4 \rangle r^4 + \cdots , \qquad (18)$$

where we used the normalization condition $g(0) = 1$, and $\langle k^2 \rangle$ and $\langle k^4 \rangle$ denote the second and fourth moment of the spectral function. Note that this expansion has a quadratic term followed by a quartic term. Using the result of Eq. (18) in Eq. (14), we obtain a small r expansion of the form

$$\Gamma_\mathrm{B}(r \to 0) = 1 - \frac{1}{2\pi\sqrt{3\varphi_1\varphi_2}}(\langle k^2 \rangle)^{1/2}\mathrm{e}^{-\alpha^2/2}\,r$$
$$\times \left[1 - \left(\frac{1}{40}\frac{\langle k^4 \rangle}{\langle k^2 \rangle} - \frac{1}{72}\langle k^2 \rangle(\alpha^2 - 1)\right)r^2 \right]. \qquad (19)$$

Comparing this with Eq. (3), we arrive at a useful relation connecting the second moment of the spectral function to the fundamental quantity of a porous material, the interfacial area per unit volume,

$$\frac{S}{V} = \frac{2}{\pi\sqrt{3}}(\langle k^2 \rangle)^{1/2}\mathrm{e}^{-\alpha^2/2} \,. \qquad (20)$$

This relation also implies that one of the basic requirements for the physically acceptable spectral function is that the second moment be finite.

We next consider a random surface generated by the level set

$$\psi(\mathbf{r}) = \psi(x, y, z) = \alpha \,, \qquad (21)$$

where the function $\psi(x, y, z)$ is defined by Eq. (8). Teubner [8] has proved a remarkable theorem that for this random surface the average mean, Gaussian and square-mean curvatures are given by

$$\langle H \rangle = \frac{\alpha}{2} \sqrt{\frac{\pi}{6}} \langle k^2 \rangle , \tag{22}$$

$$\langle K \rangle = \frac{1}{6} \langle k^2 \rangle (\alpha^2 - 1) , \tag{23}$$

$$\langle H^2 \rangle = \frac{1}{6} \langle k^2 \rangle (\alpha^2 + v^2) , \tag{24}$$

where

$$v^2 = \frac{6}{5} \frac{\langle k^4 \rangle}{\langle k^2 \rangle^2} - 1 . \tag{25}$$

So the second requirement of the physically acceptable spectral function is that it has a finite fourth moment also. Since for a bicontinuous microemulsion the level surface defined by Eq. (21) is approximately the mid-plane passing through the surfactant monolayer in a bulk contrast experiment, the average Gaussian and average square-mean curvatures of the surfactant monolayer can be computed once a physically acceptable spectral function can be found.

Choice of the spectral function can be based on a criterion that when it is substituted into Eqs. (11), (9) and (1), it would give an intensity distribution which agrees with SANS data. A suitable form of the spectral function have been proposed by Chen and Lee [4], which is an inverse

sixth-order polynomial in k and which contains three parameters a, b, and c.

The first two parameters a and b have their approximate correspondences in the Teubner–Strey theory [9]. In the T–S theory, the Debye correlation function is given by

$$\Gamma_{TS}(r) = e^{-r/\xi} [\sin(2\pi r/d)/(2\pi r/d)] . \tag{26}$$

The correspondences are $a \approx 2\pi/d$ and $b \approx 1/\xi$, where d is the inter-domain (water–water or oil–oil) repeat distance and ξ the coherence length of the local order [10, 11]. The parameter c controls transition from the small Q peak to the large Q behavior of the scattering intensity distribution, the existence of which is essential for the good agreement of the theory and experiment at large Q [4].

In this paper, we choose a similar spectral function for which both the second and fourth moments exist. This function is an inverse eighth-order polynomial containing

three parameters, which are the minimal set for the physical situation under study.

$$f(k) = \frac{bc(a^2 + (b + c)^2)^2/(b + c)\pi^2}{(k^2 + c^2)^2(k^4 + 2(b^2 - a^2)k^2 + (a^2 + b^2)^2)} . \tag{27}$$

The second and fourth moments of this spectral function are given by

$$\langle k^2 \rangle = \frac{c(a^2 + b^2 + bc)}{(b + c)} , \tag{28}$$

$$\langle k^4 \rangle = \frac{c(a^4 + 2a^2b^2 + b^4 + 4a^2bc + 4b^3c + 4b^2c^2 + bc^3)}{(b + c)} .$$

The corresponding two-point correlation function is

$$g(r) = \frac{4bc(a^2 + (b + c)^2)^2}{(b + c)r} \left[e^{-cr} \left(\frac{(a^2 - b^2 + c^2)}{(4a^2b^2 + (a^2 - b^2 + c^2)^2)^2} \right. \right.$$
$$\left. + \frac{r}{4c(a^2 + b^2)^2 + 2(a^2 - b^2)c^2 + c^4)} \right)$$
$$+ \frac{e^{-br}}{ab} \left(\frac{-8a^2b^2 + (a^2 + b^2)^2 + 2(a^2 - b^2)c^2 + c^4}{4(4a^2b^2 + (a^2 - b^2 + c^2)^2)^2} \sin(ar) \right.$$
$$\left. \left. - \frac{ab(a^2 - b^2 + c^2)}{(4a^2b^2 + (a^2 - b^2 + c^2)^2)^2} \cos(ar) \right) \right] \tag{29}$$

which has only even powers of r in the small r expansion.

Explicit expressions for the curvatures are given in terms of the three parameters a, b, and c as

$$\langle K \rangle = -\frac{1}{6} \frac{c(a^2 + b^2 + bc)}{(b + c)} , \tag{30}$$

$$\langle H^2 \rangle = \frac{a^4(6b + c) + a^2(12b^3 + 26b^2c + 14b^2) + 6b^5 + 25b^4c + 38b^3c^2 + 25b^2c^3 + 6bc^4}{30(b + c)(a^2 + b^2 + bc)} . \tag{31}$$

This model fits all the bulk contrast data nicely so we use Eqs. (30) and (31) to calculate the respective curvatures.

Data analysis

The scattering intensity in absolute scale obtained after the standard data normalization procedure contains about 10% of uncertainty in calibration using 1 mm water. This may cause an unnecessary uncertainty in the determination of parameters, a, b, and c. This uncertainty factor, however, can be eliminated by normalizing the scattering intensity by the invariant $\langle \eta^2 \rangle$ calculated according to Eq. (2). In the calculation of the invariant, the interval of integration was divided into three parts, $0 < Q < Q_{min}$, $Q_{min} < Q < Q_{max}$, and $Q_{max} < Q < \infty$ where Q_{min} and Q_{max} are the minimum and maximum values of Q in the

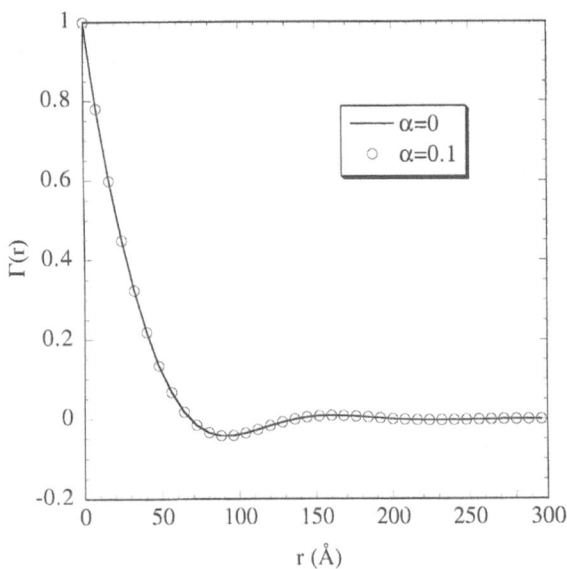

Fig. 2 Debye correlation functions with $a = 0.4664$ Å^{-1}, $b = 0.0141$ Å^{-1} and $c = 0.2966$ Å^{-1}. The cases with $\alpha = 0$ and $\alpha = 0.1$ are compared

measurements. In the range, $Q_{min} < Q < Q_{max}$, the measured intensity was used for $I(Q)$. In the range, $0 < Q < Q_{min}$, it was assumed that $I(Q)$ is constant at $I(Q_{min})$. Porod's law was assumed for the Q-range, $Q_{max} < Q < \infty$.

Figure 2 shows the sensitivity of Debye correlation function to the change of parameter α which is defined in Eq. (10). As we explained in Section on Theory of scattering, $\alpha = 0$ corresponds to an isometric two-component system and $\alpha \neq 0$ a non-isometric system. When $\alpha = 0.1$ a two-component system becomes a non-isometric system with volume fractions, $\varphi_1 = 0.46$ and $\varphi_2 = 0.54$. The Debye correlation functions for two cases, $\alpha = 0$ and $\alpha = 0.1$, were calculated with a set of representative values of a, b and c. Figure 2 shows that a small deviation from isometry does not affect shape of the Debye correlation function. All our samples were prepared so that they are isometric at a reference salinity, and the change of an effective volume fraction as a function of salinity is expected less than 10%. Therefore, we treat the parameter α as effectively zero in all data analysis.

Figure 3a shows the scattering intensities measured at a series of points close to the one-phase and lamellar phase

Fig. 3 Analyses of bulk contrast scattering intensities as a function of surfactant volume fraction. The scattering intensities were measured at points close to lamellar and single phase boundary where the average mean curvature is nearly zero. (a) Raw data after background correction, (b)–(d) three representative curves comparing the theory and the experimental data. Solid lines are the fits using modified Berk theory

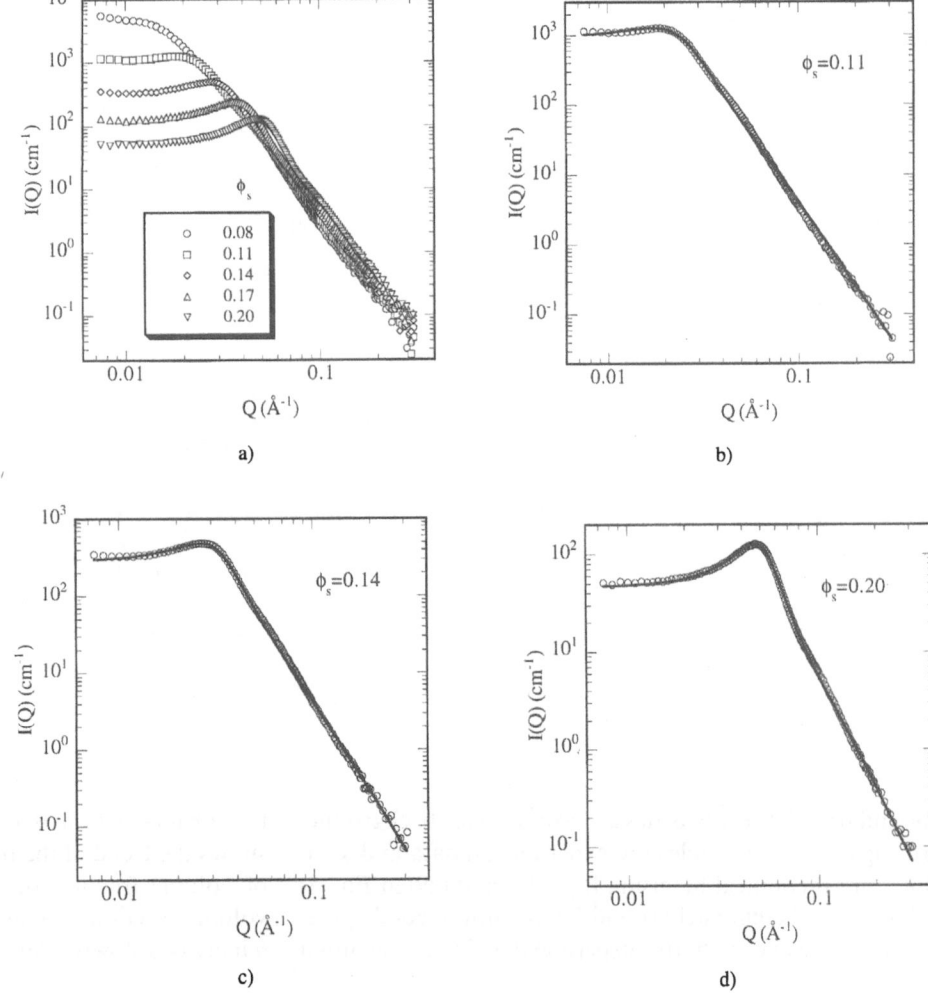

Table 1 Fitted parameters and calculated interfacial curvatures along the two horizontal lines in the phase diagram (one close to lamellar-single phase boundary, and the other at a fixed salinity 0.49 wt%)

ϕ_S	NaCl [wt%]	a [Å^{-1}]	b [Å^{-1}]	c [Å^{-1}]	$\langle\eta^2\rangle$ [10^{20} cm^{-4}]	bgd [cm^{-1}]	$\langle K\rangle$ [10^{-4} Å^{-2}]	$\langle H^2\rangle$ [10^{-4} Å^{-2}]	χ^2
0.03	0.53	0.00389	0.01211	0.02171	21.62	0.385	−0.454	2.36	130.5
0.05	0.49	0.00613	0.01104	0.01710	25.92	0.385	−0.353	1.62	113.1
0.08	0.46	0.01317	0.01198	0.04682	13.07	0.383	−1.165	6.35	113.6
0.08	0.49	0.01273	0.01181	0.04895	12.87	0.383	−1.181	6.80	49.9
0.11	0.42	0.02139	0.01120	0.09114	10.86	0.393	−2.382	16.78	68.2
0.11	0.49	0.02043	0.01136	0.09228	11.37	0.390	−2.367	17.59	85.5
0.14	0.36	0.03005	0.01236	0.1096	9.33	0.382	−3.160	21.65	38.1
0.14	0.49	0.02909	0.01219	0.1215	10.47	0.394	−3.750	26.56	44.4
0.17	0.32	0.03939	0.01346	0.1530	9.30	0.352	−5.809	37.67	19.1
0.17	0.49	0.03726	0.01309	0.1835	10.21	0.383	−6.163	55.11	43.7
0.20	0.26	0.04968	0.01459	0.2096	9.00	0.321	−8.943	64.85	18.1
0.20	0.48	0.04664	0.01410	0.2966	8.73	0.321	−10.430	137.00	36.1

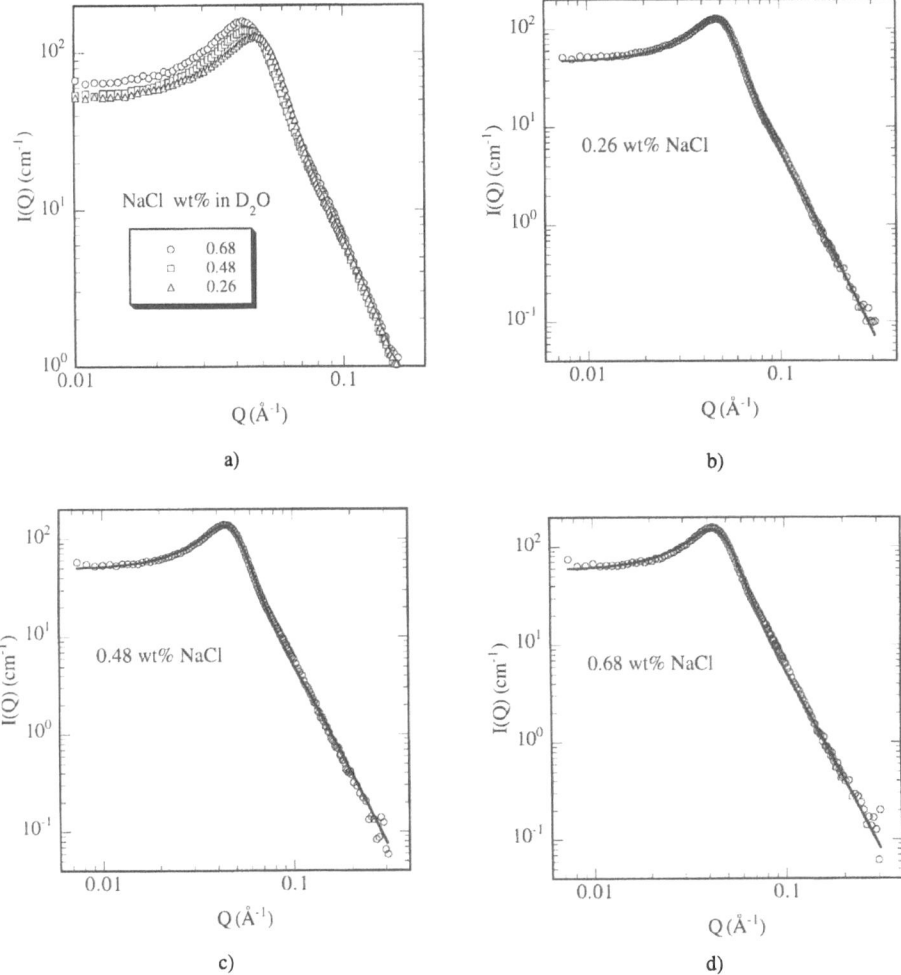

Fig. 4 Analyses of bulk contrast scattering intensities as a function of salinity at a fixed surfactant volume fraction, $\phi_S = 0.20$. (a) Raw data after background correction, (b)–(d) three representative curves comparing the theory and the experimental data. Solid lines are the fits using modified Berk theory

boundary. All the intensities are corrected for background. In Figs. 3b–d, the circles are experimental data and solid lines are theoretical fits using Eqs. (15) and (29) in Eq. (1) plus an incoherent background. These show a good agreement of theory with the experimental data. The fitted parameters a, b, and c are listed in Table 1. Figure 5A shows the trend of the parameters a, b, and c as a function of volume fraction of the surfactant. As the surfactant volume fraction increases a and c increase rapidly while b increases slowly. Considering the relations $d \approx 2\pi/a$ and

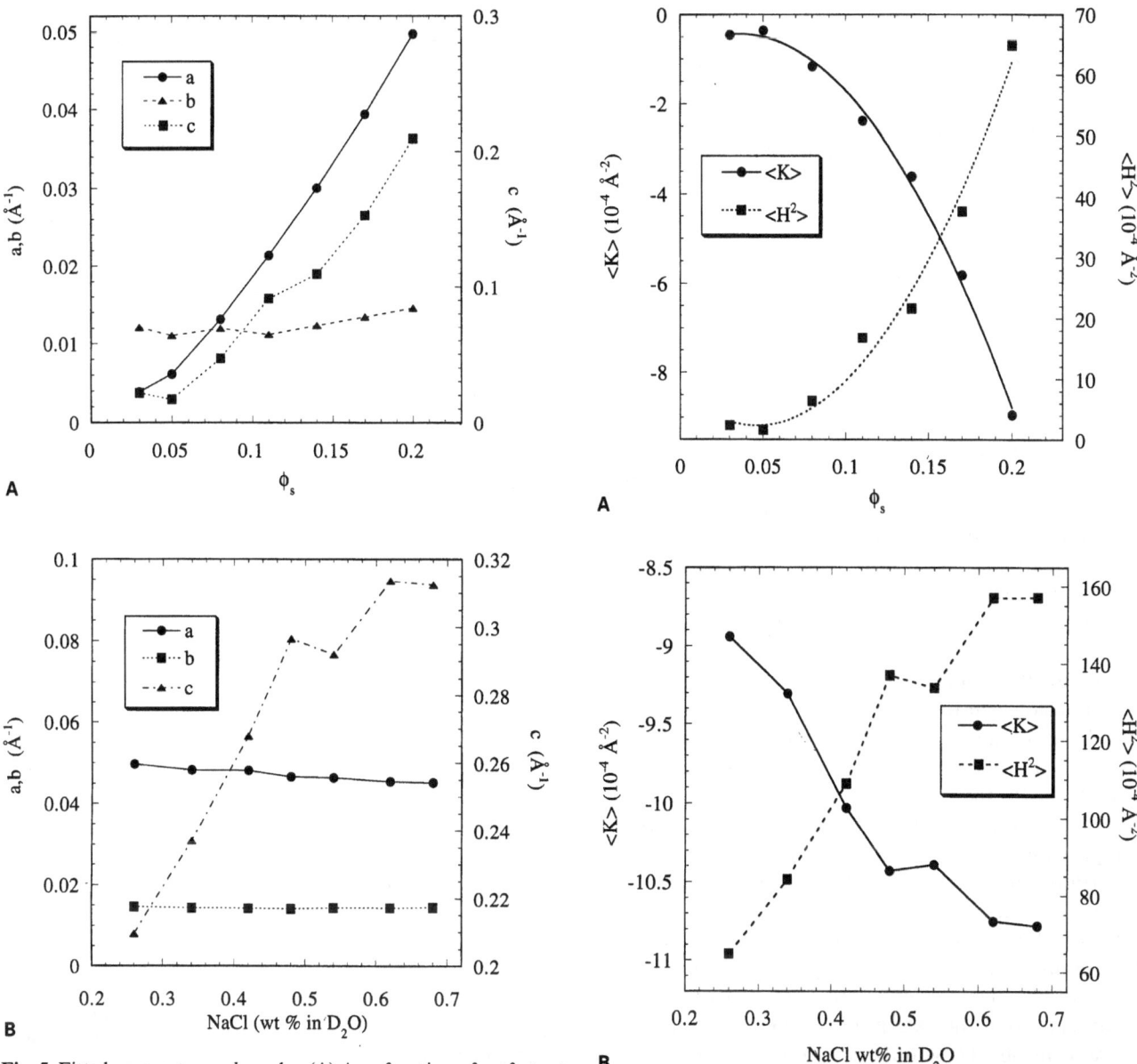

Fig. 5 Fitted parameters a, b, and c. (A) As a function of surfactant volume fraction (along the points close to lamellar and single phase boundary), (B) as a function of salinity at a fixed surfactant volume fraction, $\phi_S = 0.20$

Fig. 6 Average Gaussian curvature and average square-mean curvature. (A) As a function of surfactant volume fraction (along the points close to lamellar and single phase boundary), (B) as a function of salinity at a fixed surfactant volume fraction, $\phi_S = 0.20$. Solid circles are the calculated average Gaussian curvature using Eq. (30) and solid boxes are the calculated square-mean cuvatures using Eq. (31). In (A), the solid line is a fit with a phenomenological expression, Eq. (28) and the dash line is a fit for the square mean curvature with an equation similar to Eq. (28)

$\xi \approx 1/b$, the inter-domain size d and the coherence length ξ decreases. This makes sense because we create more surfaces per unit volumes as the number of surfactant molecules increase and therefore the inter-domain size should decrease. Considering that the ratio ξ/d increases with the surfactant volume fraction ($\xi/d = 0.088$ at $\phi_S = 0.05$ and $\xi/d = 0.54$ at $\phi_S = 0.20$), the bicontinuous microemulsion becomes more disordered at a smaller surfactant volume fraction.

The average Gaussian and square mean curvatures calculated by using Eqs. (30) and (31), respectively, are

given in Fig. 6A. The solid line is a fit using a phenomenological parabolic equation

$$\langle K \rangle = c_0 + c_1 (\phi_S - \phi_o)^2 \tag{32}$$

and the dashed line is a fit using a similar equation

$$\langle H^2 \rangle = c_0' + c_1' (\phi_S - \phi_o)^2 . \tag{33}$$

Table 2 Fitting parameters and calculated interfacial curvatures along the vertical line in the phase diagram ($\phi_S = 0.20$ with different salinity)

ϕ_S	NaCl [wt%]	a [$\mathring{A}^{-1}$]	b [$\mathring{A}^{-1}$]	c [$\mathring{A}^{-1}$]	$\langle \eta^2 \rangle$ [10^{20} cm^{-4}]	bgd [cm^{-1}]	$\langle K \rangle$ [$10^{-4}\mathring{A}^{-2}$]	$\langle H^2 \rangle$ [$10^{-4}\mathring{A}^{-2}$]	χ^2
0.20	0.26	0.04968	0.01459	0.2096	8.996	0.321	−8.94	64.88	18.1
0.20	0.34	0.04827	0.01428	0.2370	9.060	0.321	−9.30	84.18	15.5
0.20	0.42	0.04814	0.01424	0.2679	8.988	0.321	−10.03	108.9	31.5
0.20	0.48	0.04664	0.01410	0.2966	8.734	0.321	−10.43	137.0	36.1
0.20	0.54	0.04640	0.01432	0.2920	9.138	0.321	−10.39	133.8	50.0
0.20	0.62	0.04547	0.01427	0.3136	9.159	0.321	−10.75	157.0	61.5
0.20	0.68	0.04521	0.01446	0.3124	9.299	0.321	−10.78	157.1	27.6

Results of the fit are $c_0 = -0.434 \times 10^{-4}\,\mathring{A}^{-2}$; $c_1 = -310.2 \times 10^{-4}\,\mathring{A}^{-2}$; $\phi_0 = 0.036$ and $c'_0 = 2.35 \times 10^{-4}\,\mathring{A}^{-2}$; $c'_1 = 2530 \times 10^{-4}\,\mathring{A}^{-2}$; $\phi_0 = 0.46$. In both cases, ϕ_0 turns out to be very close to the volume fraction at the fish tail of the phase diagram shown in Fig. 1. Therefore, we can say that c_0 and c'_0 are the average Gaussian curvature and the average square-mean curvature at the fish tail which is the lowest volume fraction of surfactant where the one-phase microemulsion first forms. Magnitudes of average Gaussian curvatures obtained in this experiment are comparable to that obtained for the $C_{10}E_4/D_2O/Octane$ microemulsion near the fish tail before [12]. The quadratic dependence of the average Gaussian curvature on the volume fraction of surfactant is understandable. According to Eq. (20), the magnitude of the average Gaussian curvature is proportional to the square of the interfacial area per unit volume. The interfacial area is in turn proportional to the amount of surfactant added to the system. Since the average mean curvature is expected to be zero for microemulsions that we studied at these phase points, the average square mean curvature is the variant of the fluctuation of the mean curvature. One also observes that the variant of the fluctuation increases quadratically as the volume fraction increases. This is reasonable because as more surfactant is added to the system, more interfacial area is created and the surfactant film has to bend around to accommodate itself in the liquid.

The scattering intensities measured at $\phi_S = 0.20$ with various salinities are shown in Fig. 4 with three representative data fitting. The fitted parameters a, b, and c are given in Table 2 and their trends are shown in Fig. 5B. As the salinity increases, a decreases slightly and b remains almost constant while c increases by a factor of 1.5. The increase of c corresponds to the decrease in interfacial length scale $1/c$ which measures the persistence length of the interface. This is reasonable as salinity increases the surfactant film tends to curve toward water and the persistence length of the surfactant film decreases.

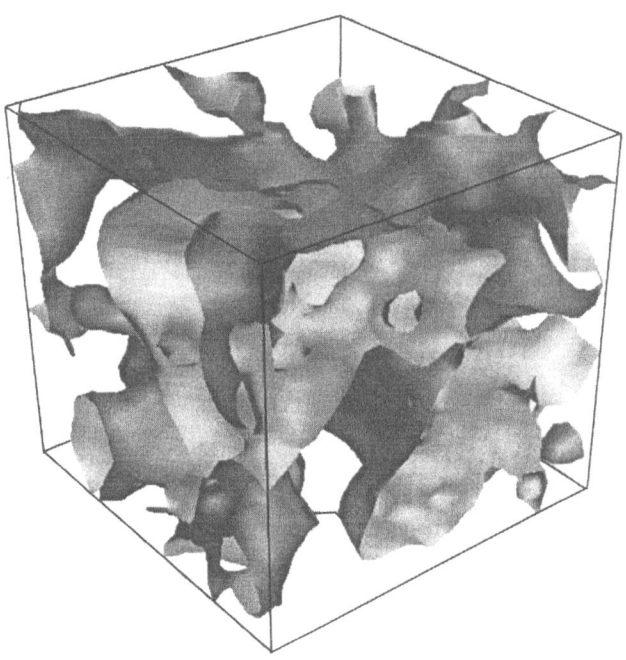

Fig. 7 3-D morphologies of two disordered bicontinuous microemulsion at $\phi_S = 0.05$ and a salinity 0.49 wt%. The size of the box is $960 \times 960 \times 960\ \mathring{A}^3$

As shown in Fig. 6B, the average Gaussian curvature decreases with salinity while the square mean curvature increases, both linearly. The lowest salinity we are starting with, corresponds to a phase-point very close to lamellar phase and the highest salinity corresponds to a point where the curving of interface toward water is maximum within the one-phase channel. In the lamellar phase the interface is flat and both curvatures are zero. As one deviates from this structure the absolute magnitude of these quantities should increase.

In Fig. 7 we show a 3-D construction of the level surface defined by Eq. (21) for a microemulsion at 5% surfactant volume fraction.

Discussion and conclusion

We have presented a method of analysis of SANS data, taken from samples having bulk contrast, by which quantitative information on the curvatures of nearly isometric bicontinuous microemulsions can be obtained. Specifically, we can extract both the average Gaussian curvature and the average square-mean curvature of the oil–water interface when the deviation from an isometric composition is small. The method relies on a choice of a physically reasonable spectral function of the underlying Gaussian random field which upon clipping generates a bicontinuous structure that produces the correct scattering pattern. By a physically reasonable spectral function we mean that the second and fourth moments of the spectral function must be finite to ensure a finite surface area per unit volume and a finite average square-mean curvature. The method has been put to test in the case of a nearly isometric $AOT/D_2O(NaCl)/H$-decane microemulsion system as a function of surfactant volume fraction and as a function of salinity. The theoretical intensities can be made to agree with the scattering data in an absolute scale by adjusting three fitting parameters a, b, and c. The commonly used two-parameter Teubner–Strey model can also be used to fit the scattering data in the small Q region. But the fit is distinctly worse compared to our three parameter theory for Q region after the peak. The parameter a can be approximately identified as k and b as $1/\xi$ of the corresponding parameters in the T–S theory. The third parameter c is necessary for ensuring a smooth transition of scattering intensity from small Q behavior near the peak to large Q behavior dominated by the Porod's law. Thus $2\pi/a$ is a measure of the average inter-domain (water–water or oil–oil) distance, $1/b$ the coherence length of the local order surrounding a given domain and $1/c$ the switching parameter which control the transition from the small Q behavior, reflecting the global feature of the local order, to the large Q behavior, which relates to the length scale controlling the structure in the interfacial region. We can thus say that three length scales are necessary for a complete description of microstructure of a bicontinuous microemulsion. The random wave model with an appropriate choice of the spectral function is shown to be an excellent tool for quantitative analyses of small-angle scattering data for nearly isometric bicontinuous microemulsions or any other micro-phase-separated system, such as a symmetric binary mixture of polymers at late stage of spinodal decomposition. It not only gives information on curvatures of the interface, which cannot be obtained by other means, but can also generate the morphology of 3-D microstructure as shown in Fig. 7.

Acknowledgment This research is supported by a grant from Materials Science Division of US Department of Energy. We are grateful to the Intense Pulse Neutron Source Division of Argonne National Laboratory for neutron beam time at the SAND Low-Angle Diffractometer and to Dr. Chwan-Yuan Ku for technical assistance.

References

1. Jahn W, Strey R (1988) J Phys Chem 92:2294–2301
2. See for example, a review article by Chen SH, Lin TL, Huang JS (1987) p 285–313 In: Safran SA, Clark NA (eds) Physics of Complex and Supermolecular Fluids. Wiley, New York
3. Chen SH, Chang SL, Strey R (1990) J Chem Phys 93:1907–1917
4. Chen SH, Lee DD, Chang SL (1993) J Mol Structure 296:259–264
5. Debye P, Anderson HR Jr, Brumberger H (1957) J Appl Phys 28:679–683
6. Kirste Von R, Porod G (1962) Kolloid-Z&Z f Polym 184:1–7
7. Berk NF (1987) Phys Rev Lett 58: 2718–2721
8. Teubner M (1991) Europhys Lett 14(5): 403–408
9. Teubner M, Strey R (1987) J Chem Phys 87:3195–3200
10. Chen SH, Chang SL, Strey R (1990) Progr Colloid Polym Sci 81:30–35
11. Chen SH, Chang SL, Strey R (1991) J Appl Cryst 24:721–731
12. Chen SH, Lee DD, Kimishima K, Jinnai H, Hashimoto T (1996) Phys Rev E, to be published
13. Lee DD, Chen S-H (1994) Phys Rev Lett 73:106–109
14. Pieruschka P, Marcelja S (1992) J de Phys II 2:235–245
15. Auvray L, Cotton JP, Ober R, Taupin C (1984) J Phys France 45:913–922

Progr Colloid Polym Sci (1997) 106:24–33
© Steinkopff Verlag 1997

P.M. Claesson
A. Dedinaite
M. Fielden
M. Kjellin
R. Audebert

Polyelectrolyte–surfactant interactions at interfaces

Received: 13 October 1996
Accepted: 4 February 1997

Dr. P. Claesson (✉) · A. Dedinaite
M. Fielden · M. Kjellin
Royal Institute of Technology
and Institute for Surface Chemistry
P.O. Box 56 07
114 86 Stockholm, Sweden

R. Audebert
Laboratoire de Physico-Chimie
Macromoléculaire
Université Pierre et Marie Curie
CNRS URA 278
ESPCI 10 rue Vauquelin
75231 Paris, Cedex 05, France

Abstract Interactions between
negatively charged surfaces coated
with cationic polyelectrolytes across
solutions containing an anionic
surfactant, sodium dodecyl sulphate
(SDS) have been studied. Polyelectrolytes with charge densities between
100% and 10%, counted per
monomer unit, were used. At low
ionic strength the polyelectrolytes
adsorb in a flat conformation to
neutralize the negative mica surface
charge. The higher the linear charge
density of the polyelectrolyte, the
thinner adsorbed layers are obtained.
In no case could any desorption be
detected when the polyelectrolyte
containing solution was replaced with
an aqueous polyelectrolyte-free
solution. The presence of SDS at
concentrations considerably below
the critical micellar concentration
(cmc) does in all cases result in
a recharging and a considerable
swelling of the adsorbed layer. This is

due to a cooperative association of
surfactants in the preadsorbed
polyelectrolyte layer. In case of the
100% charged PCMA, the force
versus distance profile displays clear
oscillations. We interpret these
oscillations as being caused by the
spatial arrangement of SDS micelles
induced by the polyelectrolyte. The
oscillations in the force curve remain
as the SDS concentration is increased
to twice the cmc. No similar oscillations in the force distance curve are
observed when the surfaces are
precoated with less charged polyelectrolytes. In these cases a strong
swelling of the polyelectrolyte layer is
observed once the surfactant concentration reaches a critical value (well
below the cmc).

Key words Polyelectrolyte–surfactant
association – surface forces – steric
forces – critical association
concentration – adsorption

Introduction

Mixtures of surfactants and polyelectrolytes are used in
many applications, e.g. for rheology control or to facilitate
particle deposition, for emulsification, and colloidal stabilization or flocculation. The association between ionic surfactants and polyelectrolytes in bulk solution has been
studied for a long time, and this process can largely be
understood by considering electrostatic and hydrophobic

forces [1–6]. However, also factors influencing how easily
a polyelectrolyte can bend around a micelle, such as the
stiffness of the polyelectrolyte chain and possible cross-
links of chains, are important parameters [7–9]. In particular, it is found that in bulk solution the initial binding
of a charged surfactant to an oppositely charged polyelectrolyte is electrostatically driven. At a higher concentration a cooperative association takes place that is driven by
the hydrophobic interaction (mainly between the surfactant tails). The surfactant concentration at which this

Progr Colloid Polym Sci (1997) 106:24–33
© Steinkopff Verlag 1997

cooperative association step occurs, the critical association concentration, is lower than the critical micellar concentration. This is largely due to a difference in the change in counterion entropy following formations of free micelles and surfactant aggregates associated with polyelectrolytes [7, 9]. At saturation, micelle-like structures are formed by the surfactant along the polyelectrolyte chain in a bead-and-necklace structure [5].

Very few studies of adsorption from, and surface interactions in, polyelectrolyte–surfactant mixtures have been conducted [10, 11] and our understanding of these systems remains rather limited. In this paper, we review the work carried out up to this point in our laboratory. The systems studied consist of a strongly negatively charged surface, muscovite mica, and preadsorbed cationic polyelectrolytes with different charge densities (100–10% of the monomers being charged). The effect of adding anionic surfactants on surface interactions and adsorbed layer structures was then assessed. Some of the data have been published previously [11], but most of the material that appear in this article is unpublished. These data will be summarized here with more detailed publications to follow. Here we focus primarily on how the charge density of the polyelectrolyte influences the association between surfactants and polyelectrolytes at the solid–liquid interface.

Experimental

Materials

The three polyelectrolytes used in this investigation have 100, 30 and 10% of the monomers charged, respectively. The fully charged polyelectrolyte (referred to in the text as PCMA) is made from CMA (2-acryloxyethyl)-trimethyl-ammonium chloride monomers. The 30% charged polyelectrolyte is a copolymer of the monomers {3-(2-methylpropionamide)propyl}-trimethylammonium chloride (MAPTAC) and acrylamide (AM) and referred to as AM–MAPTAC-30 in the following text. Finally, the 10% charged polyelectrolyte, AM–CMA-10 is a copolymer of AM (90%) and CMA (10%). The structures of these monomers are provided in Fig. 1. The mean molecular weight of PCMA and AM–CMA-10 is about 1.5×10^6 g mol^{-1}, whereas the AM–MAPTAC-30 has a mean molecular weight of about 7.8×10^5 g mol^{-1}. The sodium dodecyl sulphate (SDS) used in the experiments was obtained from BDH and it was used without further purification. The water used in the experiments was purified using a Milli-RO 10PLUS reverse osmosis pretreatment unit and a Milli-Q PLUS185 unit followed by filtration through a 0.2 μm filter.

Surface force measurements

Interactions between polyelectrolyte-coated surfaces in the absence and presence of SDS were examined using the interferometric surface force technique of either the Mark II type [12] or the Mark IV type [13]. Muscovite mica was obtained from Mica New York Corp. in New York (green mica) and from M. Watanabe & Co. in Tokyo (ruby mica). It was cleaved to 1–3 μm thin pieces and silvered on one side. The pieces were then glued (using Epon 1004 from Shell Chemicals) onto two half-cylindrical silica discs with the silvered side down. The surfaces were mounted in a crossed cylinder configuration inside the surface force apparatus with the upper surface connected to a piezo-electric tube and the lower one on a double cantilever spring. The distance between the surfaces in changed by means of motors or by means of the piezo-electric crystal.

White light directed normally through the surfaces is multiply reflected between the silver layers. The

Fig. 1 The structure of the monomers CMA, (2-acryloxyethyl)-trimethylammonium chloride, MAPTAC, {3-(2-methylpropionamide)propyl}-trimethylammonium chloride and AM, acrylamide

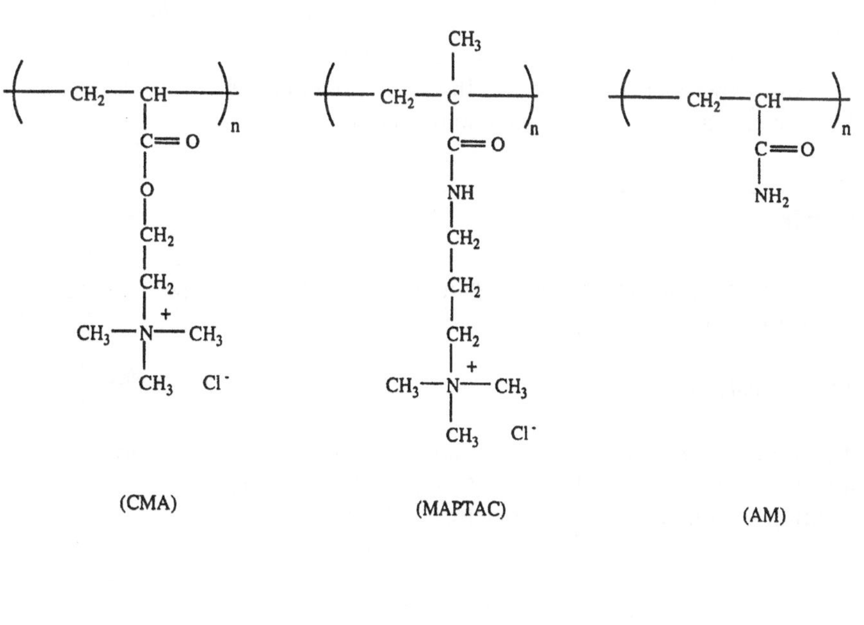

(CMA)   (MAPTAC)   (AM)

26

P.M. Claesson et al.
Polyelectrolyte–surfactant interactions

wavelengths of the standing waves that are created are determined in a spectrometer. From these data the distance between the surfaces is calculated with an accuracy of about 2 Å. The force, $F(D)$, is measured from the deflection of the spring and it is normalized by the local geometric mean radius of the surfaces, R, which is approximately 2 cm. According to the Derjaguin approximation [14] this quantity is related to the free energy of interaction per unit area between flat surfaces, $G_f(D)$:

$$\frac{F(D)}{R} = 2\pi G_f(D) \tag{1}$$

with the condition that $D \ll R$. Since D is of the order of 10^{-6} m or less this condition is satisfied. Another requirement is that the radius of the surfaces should be independent of D, but this is not always the case since the surfaces deform under strongly repulsive or attractive forces [15]. When the gradient of the force $\partial F/\partial D$ exceeds the spring constant of the double cantilever spring the system becomes unstable and the surfaces jump together to the next stable region of the force curve.

The experiments were started by bringing the surfaces into contact first in dry air and then in pure water or 10^{-4} M KBr. This establishes the zero separation, and from the adhesion force and from the shape of the interference fringes it can be judged that the surfaces are clean. In the next step, polyelectrolytes were added to the solution to a concentration of 20 ppm (PCMA and AM–MAP-TAC-30) or 50 ppm (AM–PCMA-10). These conditions were used since the adsorption of the polyelectrolytes at these concentrations results in a close to net zero charge of the polyelectrolyte coated mica surface [11, 16, 17]. After allowing at least 12 h for the adsorption to take place the force–distance curve was measured and then the polyelectrolyte containing solution was replaced with a polyelectrolyte-free aqueous solution. This does not result in any significant desorption of the polyelectrolyte [11, 16, 17]. In the final step, SDS was added to the solution. Hence, *no polyelectrolytes are present in the solution* in these experiments, except for the data presented in Fig. 8 where a different experimental procedure was used.

Results

Forces across polyelectrolyte-free solutions of SDS, preadsorbed polyelectrolyte layers

PCMA and SDS

The forces acting between mica surfaces precoated by a layer of PCMA (adsorbed from a 20 ppm 10^{-4} M KBr solution overnight) across a polyelectrolyte-free aqueous

10^{-4} M KBr solution are shown in Fig. 2. No double-layer force acts between the surfaces which demonstrates that the polyelectrolyte-coated surface carries a zero net charge. A strong attractive force acts between the surfaces once the separation is less than about 150 Å. This force is caused by polyelectrolyte chains crossing the midplane between the surfaces, giving rise to an entropic bridging attraction [16]. An enthalpic contribution from polyelectrolytes directly bound to both surfaces is also likely considering the thin adsorbed layer. The strong attraction pulls the surfaces into a separation of about 10 Å. Hence, under the influence of the bridging attraction the adsorbed layers adopt a very flat conformation. The normalized pull-off force that has to be overcome when separating the surfaces is large, 65 ± 15 mN/m. The long-range forces observed are within experimental error identical to those acting across a 10^{-4} M KBr solution also containing 20 ppm PCMA [18]. However, the observed pull-off force is significantly smaller than reported in a previous study [18].

Addition of SDS to a concentration of 0.01 or 0.02 cmc (cmc = 8.3×10^{-3} M) does not result in any change in the measured long-range interaction (Fig. 2) or pull-off force. However, as the SDS concentration is increased further to 0.1 cmc a long-range repulsive double-layer force appears, showing that SDS is incorporated in the adsorbed layer. The repulsive force is overcome by an attraction at a separation of 110 Å. This attraction pulls the surface inwards to a separation of 40 Å. A further increase in the compressional force hardly affects the surface separation, indicating a dense layer structure. Clearly, the layers on the

Fig. 2 Force normalized by radius as a function of surface separation between mica surfaces precoated with PCMA. The forces were measured in 10^{-4} M KBr (♦), and in 10^{-4} M KBr solutions also containing SDS at concentrations of 0.01 cmc (■), 0.02 cmc (▲), and 0.1 cmc (□). The arrows indicate inward jumps due to the presence of strongly attractive forces. The dashed line represents the layer thickness at SDS concentrations between 0 and 0.02 cmc, and the solid vertical line the layer thickness at an SDS concentration of 0.1 cmc

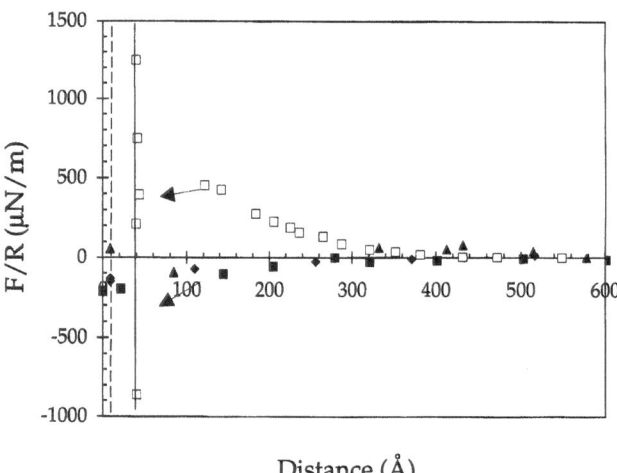

surfaces are considerably thicker when the SDS concentration in bulk is 0.1 cmc compared to at lower surfactant concentrations. The normalized pull-off force is in this case about 25–30 mN/m.

An increase in surfactant concentration results in an increased incorporation of SDS in the adsorbed layer and at an SDS concentration of 0.2 cmc pronounced oscillations appear in the force curve (Fig. 3). It is possible to identify three oscillations. The innermost one is located at a separation between 40 and 50 Å, the next one in the distance interval 70–90 Å, and the outermost one at a separation of 120–130 Å. The oscillations thus have a periodicity of about 40 Å, and it is observed that both the repulsive and the attractive branch increase in magnitude as the surfaces are moved from an outer to an inner oscillation. The reason for the presence of these oscillations are discussed below. At separations larger than 130 Å a repulsive double-layer force dominates the interaction.

A further increase in surfactant concentration to 2 cmc does not affect the qualitative behavior of the interaction (Fig. 4). At large separations a repulsive double-layer force dominates the interaction, and at separations of about 120, 80 and 40 Å are three clear oscillations observed. Hence, the periodicity of the oscillations remains unchanged when the SDS concentration is increased. One or two less steep and less-pronounced oscillations in the force profile can also be distinguished at larger separations.

AM–MAPTAC-30 and SDS

The forces acting between mica surfaces carrying a pre-adsorbed layer of AM–MAPTAC-30 (obtained by adsorp-

Fig. 3 Force normalized by radius as a function of surface separation between mica surfaces precoated with PCMA. The forces were measured in solutions containing 10^{-4} M KBr and SDS at a concentration of 0.2 cmc. The two symbols represent data obtained in two different experiments

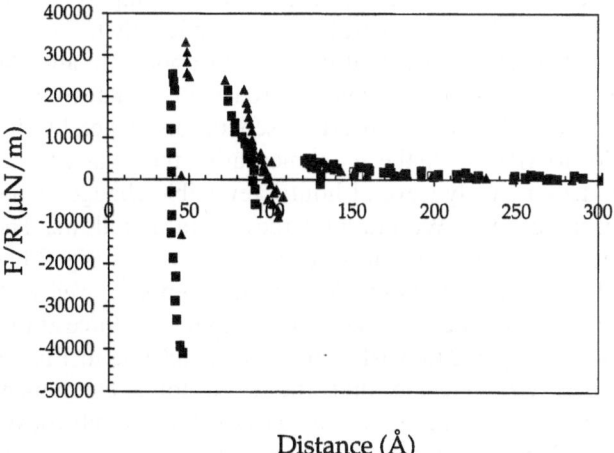

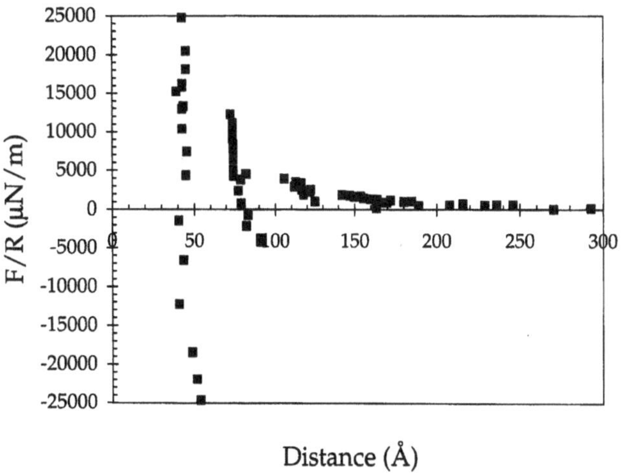

Distance (Å)

Fig. 4 Force normalized by radius as a function of surface separation between mica surfaces precoated with PCMA. The forces were measured in solutions containing 10^{-4} M KBr and SDS at a concentration of 2 cmc

tion from a 20 ppm polyelectrolyte 10^{-4} M KBr solution overnight) after addition of SDS to a concentration of 0.005 cmc are shown in Fig. 5. The forces are similar to those observed before addition of SDS (data not shown). A very weak electrostatic double-layer force dominates the interaction at large separations, demonstrating a nearly perfect match between the surface charge and the charges in the polyelectrolyte layer. From a separation of about 150 Å an attractive force causes the surfaces to jump inwards to a separation of about 40 Å. This force is most likely due to bridging. The normalized pull-off force measured upon separation is about 1 mN/m.

A tenfold increase in SDS concentration to 0.05 cmc results in a significant change in the adsorbed layer and the measured forces. A strong repulsive force is dominating the long-range interaction. The decay length of this force is not consistent with a double-layer repulsion and we conclude that both steric and electrostatic forces contribute to the interaction. The increased repulsion demonstrates that SDS has associated with the polyelectrolyte layer and that this results in a recharging and an increase in the length (and most likely number of) polymer tails. The final layer thickness reached under a high compressive force is about 120 Å (60 Å per surface). It is also noted that the forces measured on a first approach are slightly more long-ranged than the forces measured on subsequent approaches, showing that the act of measuring the forces results in a more compact layer. However, if the surfaces are left apart for a long time, the adsorbed layer expands again.

A further increase in SDS concentration to 0.2 or 2 cmc does not have any dramatic effect on the measured interaction (Fig. 5), but the long-range force remains dominated

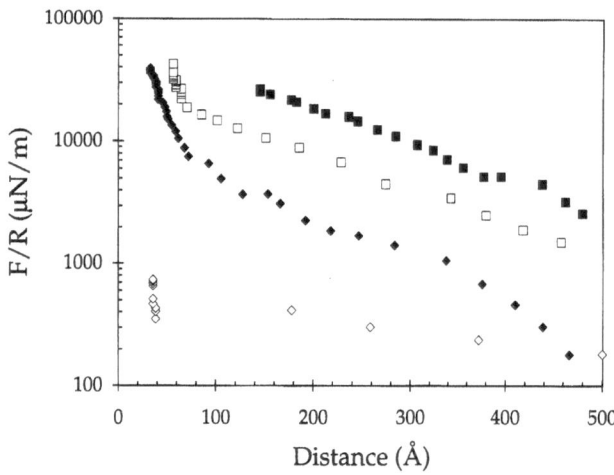

Fig. 5 Force normalized by radius as a function of surface separation between mica surfaces precoated with AM–MAPTAC-30. The forces were measured in solutions containing 10^{-4} M KBr and SDS at a concentration of 0.005 cmc ($\diamond$), 0.05 cmc ($\blacksquare$), 0.2 cmc ($\square$), and 2 cmc ($\blacklozenge$)

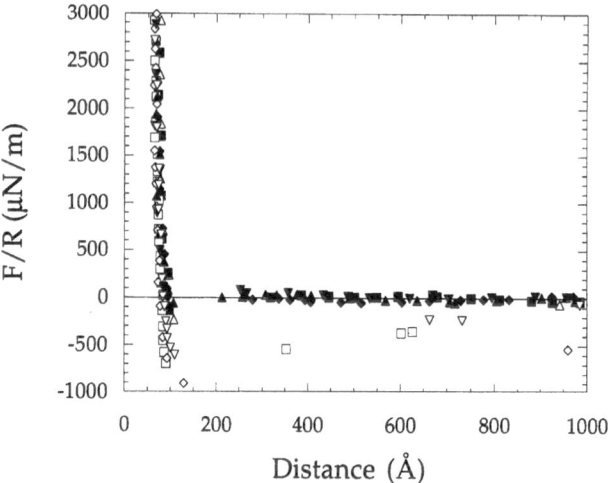

Fig. 6 Force normalized by radius as a function of surface separation between mica surfaces precoated with a 10% charged CMA/AM copolymer. The forces were measured across pure water ($\blacksquare$, $\square$, $\blacklozenge$, $\diamond$) and after addition of SDS to a concentration of 0.01 cmc ($\blacktriangle$, $\triangle$, $\blacktriangledown$, $\triangledown$). Filled and unfilled symbols represent forces measured on approach and on separation, respectively

by steric and electrostatic forces. However, we note that the compressed layer thickness becomes thinner at higher surfactant concentrations, indicating that some desorption of AM–MAPTAC-30 takes place at these SDS concentrations. More details about the interactions between AM–MAPTAC-30 coated mica surfaces in the presence of SDS will be reported in a forthcoming publication [19].

AM–PCMA-10 and SDS

When mica surfaces are immersed in a 50 ppm AM–CMA-10 solution overnight the surface charge is completely neutralized by the charges on the adsorbed polyelectrolyte [20]. No change in interaction is observed when the polyelectrolyte solution is replaced with a polyelectrolyte-free solution. The forces at this stage of the experiment are shown in Fig. 6. No force is detected until the surfaces are at a separation of about 250 Å. From this separation an attractive force pulls the surfaces into a separation of 100 Å. The layer thickness at zero force is about 50 Å on each surface. A steeply rising repulsive force, due to compression of the adsorbed polyelectrolyte layer, is present at shorter separations. When the surfaces are taken apart a slowly decaying attraction is observed. This indicates that bridging polyelectrolytes connect the surfaces until the separation has been increased to about 2000 Å. The absence of any long-range electrostatic double-layer force confirms that the mica surface charge remains neutralized by the preadsorbed polyelectrolytes. Addition of SDS to a concentration of 0.01 cmc does not affect the interaction (Fig. 6).

When the SDS concentration is increased further to 0.02 cmc (0.166 mM) the adsorbed layers expand drastically and the surface forces become much more long-ranged (see Fig. 7). This expansion is due to binding of SDS to the adsorbed polyelectrolytes. At this and higher SDS concentrations the details of the surface force profiles are no longer reproducible. However, the general features are always the same, as described below. The most long-range part of the force curve decays roughly exponentially with separation as seen in Fig. 7. At smaller separations, here around 600 Å (the logarithm of) the repulsive force increases less rapidly and sometimes a sudden inward "jump" takes place. At even smaller separations, below about 100 Å, the repulsive force increases rapidly again. The final layer thickness reached under a high compressive force is similar to that observed before adding SDS. The forces measured on separation are significantly less repulsive than those measured on the first approach, and the forces measured on a second approach are similar to those experienced during the previous separation. Hence, like for AM–MAPTAC-30 the act of measuring the surface forces induces a slowly (several hours) reversible change in the adsorbed layer. We interpret these data as follows: The outermost part of the force curve displayed in Fig. 7 has a decay length of about 250 Å, which compares well with the theoretical decay lengths of a double-layer force at this ionic strength, 240 Å (shown by the solid line in Fig. 7). Hence, we conclude that the main force component at large separations is an electrostatic double-layer force. However, we have noted that also at these large

Progr Colloid Polym Sci (1997) 106: 24–33
© Steinkopff Verlag 1997

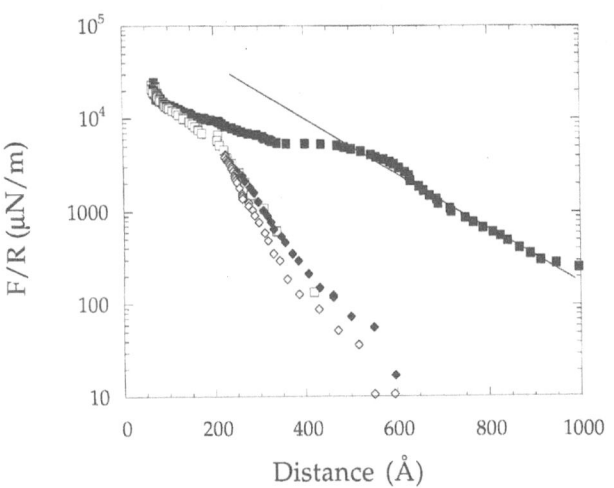

Fig. 7 A typical force–distance profile obtained for mica surfaces precoated with a 10% charged CMA/AM copolymer across a solution containing SDS at a concentration of 0.02 cmc. (■) and (□) represent the first approach and separation respectively, (◆) and (◇) the following ones. The solid line corresponds to the theoretical decay length at the present ionic strength (based on the surfactant concentration) according to the Poisson–Boltzmann theory

separations there is a small hysteresis in the force curve between the inward and outward curve, indicating that steric interactions due to rearrangement of the layer also contribute to the interaction. The decreasing slope at 600 Å in Fig. 7 is partly due to an inward shift of the plane of charge due to a decreasing length of the most extended polymer tails. However, we propose that also a desorption of SDS from the polyelectrolyte-coated surfaces takes place. This hypothesis is supported by the fact that the thickness of the layer reached under a high force is similar to that observed before SDS was added to the solution. Some surfactant readsorbs when the pressure is decreased during the separation process as seen by a rapid increase in distance with decreasing force between about 100 and 200 Å. The hysteresis effect between approach and separation shows that the measurements do not, unlike at lower surfactant concentrations, represent equilibrium forces.

Forces across polyelectrolyte–surfactant solutions

PCMA and SDS

We have recently initiated studies concerned with interactions between surfaces across solution containing both polyelectrolytes and oppositely charged surfactants. These studies are complicated by the fact that the solution phase separates into a polymer and surfactant rich phase and a phase depleted of these components. This renders the solution cloudy and the interference pattern less sharp, making accurate measurements more difficult. For this reason we did not fill the whole measuring chamber with the solution but rather placed a droplet of solution with the required total composition between the surfaces. Results for such measurements using mica and a solution containing 20 ppm PCMA, 10^{-4} M KBr and SDS at a total concentration of 0.04 cmc are shown in Fig. 8. At this concentration the polyelectrolyte–surfactant aggregates formed in bulk solution are close to uncharged. Clearly, the forces are very different when polyelectrolytes are present in solution compared to when the polyelectrolyte is preadsorbed (cf. Figs. 3 and 8). When the polyelectrolyte is present in solution a very thick layer is formed and a repulsive force is noticeable out to a separation of more than 3000 Å. This demonstrates that the phase separated polymer–surfactant phase accumulates at the surfaces. On approach the repulsive force increases monotonically with decreasing surface separation, but unlike the situation with preadsorbed PCMA layers no oscillations are observed at this PCMA/SDS composition. When the surfaces are separated an attractive force is observed. The attractive force increases moderately in strength as the separation is carried out from smaller separations (Fig. 8). This attractive force is attributed to interlayer polyelectrolyte–micelle complexes.

Further studies of the interaction between surfaces in solutions containing both polyelectrolyte and surfactant are in progress and more detailed results will be reported in a forthcoming publication. However, we can already now state that at low surfactant concentrations (≤ 0.1 cmc) the results are qualitatively similar to those reported here for preadsorbed PCMA layers. At high

Fig. 8 The forces as a function of separation between mica surfaces immersed in a solution containing 20 ppm PCMA, 10^{-4} M KBr and SDS at a concentration of 0.04 cmc. Three different force curves where the surfaces have been compressed to different separations are shown

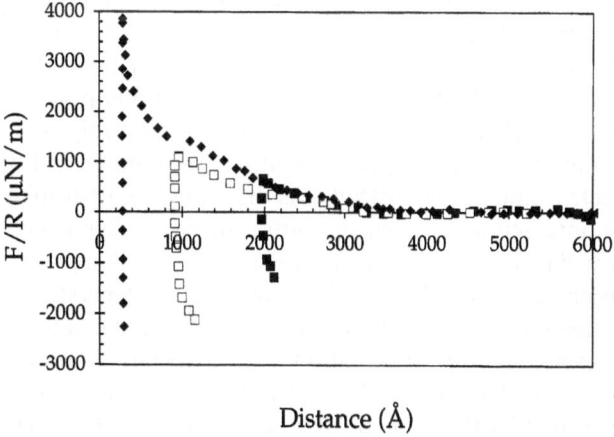

surfactant concentrations (0.4 cmc) oscillations with a periodicity similar to, but with smaller magnitude than, those shown in Fig. 4 are observed.

Discussions

The mica-polyelectrolyte-surfactant system

Before we discuss in more detail some aspects of the data reported in this article it is worthwhile to recapitulate what is known about the systems studied. The surface, muscovite mica, is a layered aluminosilicate mineral. It consists of approximately 10 Å thick sheets. Each sheet is strongly negatively charged due to isomorphous substitution of aluminium for silicon. This negative lattice charge is in the crystal neutralized by potassium ions (and some sodium ions) located between the sheets. When a mica sheet is immersed in an aqueous solution the surface ions are dissolved and partly replaced by other ions present in solution, such as H_3O^+ [21]. The number of exchangeable ions, and thus the number of negative sites on the mica surface is 2.1×10^{18} m^{-2}, corresponding to a surface charge density of -0.34 C m^{-2}. The surface charge obtained in simple electrolyte solutions is normally orders of magnitude lower due to adsorption of cations present in solution [22, 23].

The adsorption of polyelectrolytes to the mica surface is electrostatically driven. It has been found that the total charge carried by adsorbed PCMA at the charge neutralization point is slightly less than needed to neutralize all of the negative sites on the mica basal plane [18]. This demonstrates that some small ions (protons and potassium in this case) also are adsorbed to the mica surface at the charge neutralization point. The co-adsorption of proteins and small ions on mica [24] and other surfaces [25], [26] has also been demonstrated.

The polyelectrolyte coated surface is uncharged, or in the case of AM–MAPTAC-30 weakly positively charged, before SDS is added. When SDS associates with an uncharged polyelectrolyte layer it will result in a recharging of the interface which results in the development of an electrostatic double layer and a less favorable polyelectrolyte-surface interaction. Thus, the polyelectrolyte-surfactant association at the mica surface is counteracted by electrostatic forces. Instead, it is driven by the hydrophobic interaction between the surfactant tails. This is confirmed by the cooperative nature of the association process observed for all polyelectrolyte–surfactant systems studied in this report. It is well known that hydrophobic interactions are very important also for the association between polyelectrolytes and surfactants in bulk solutions as demonstrated by the cooperativity of this process [1–6]. The

role of the polyelectrolyte is simply that it acts as a very efficient counterion to the micelle. This is supported by the observation that surfactants associate more readily with flexible polyelectrolytes than with stiff ones [9].

The effect of the polyelectrolyte charge density on the association with surfactants

The critical association concentration between a polymer and a surfactant is defined as the free surfactant concentration at which the cooperative adsorption is initiated [1–6]. This concentration, that depends on ionic strength [3, 27] and polyelectrolyte concentration [28] can be determined from e.g. the adsorption isotherm. For the polyelectrolyte-surfactant mixtures studied here we have not yet determined the adsorption isotherm but instead we estimate an upper limit of the critical association concentration in bulk solution (cac$_b$) as the total (i.e. bound + free) surfactant concentration needed to give a significant increase in turbidity due to formation of large flocs in the polyelectrolyte–surfactant solution. The values obtained for the 100%, 30% and 10% charged polyelectrolyte (20 ppm polyelectrolyte solution, 0.1 mM KBr as background electrolyte) were about 0.005, 0.01 and 0.01 cmc. The free surfactant concentration at cac$_b$ is thus lower than these values.

It is also possible to determine a critical association concentration between surfactants and polyelectrolytes at the surface (cac$_s$) which will depend on the properties of the surfactant, the polyelectrolyte and the surfaces. The values for (cac$_s$) found from surface force measurements in this study are 0.1, 0.05 and 0.02 cmc for the 100%, 30% and 10% charged polyelectrolyte, respectively. Hence, for our system the trend cmc > cac$_s$ > cac$_b$ is observed. The reasons for the difference in association between SDS and cationic polyelectrolytes in bulk solution and at a negatively charged surface are: first, the polyelectrolyte at the surface is less flexible than the polyelectrolyte in bulk solution, second, less polyelectrolyte charges are available for the surfactant to bind to since most of the charged polyelectrolyte segments are in close contact with the negatively charged surface, and third, in bulk solution the counterion entropy increases when polyelectrolytes associate with surfactants. On the other hand, association at an uncharged polyelectrolyte coated surface results in a decrease in counterion entropy, due to the recharging of the surface that confines the counterions in the diffuse electrical double layer. No similar restriction of the counterion distribution results when uncharged polyelectrolyte-surfactant complexes are formed in bulk solution [7, 9].

It is expected and often found that for a given surfactant the cac in bulk solution decreases and the cooperativity in the adsorption increases with increasing polyelectrolyte charge density [29]. That the cooperativity of the surfactant adsorption increases with increasing charge density of the polyelectrolyte in our system is indicated by measurements of the mobility of the aggregates formed by polyelectrolytes and surfactants (20 ppm polyelectrolyte solution). A zero mobility was reached at a total SDS concentration of 0.04, 0.15 and 0.15 cmc for the 100%, 30% and 10% charged polyelectrolyte, respectively. This surfactant concentration also closely correspond with the maximum in turbidity. Again, the free surfactant concentration is lower than these values. At lower SDS concentrations the mobility of the surfactant–polyelectrolyte complex was positive and at higher SDS concentrations it was negative. The reason for the difference between the different polyelectrolytes is that the more highly charged polyelectrolyte can act as a more efficient counterion. At a surface the situation is different and we have two opposing effects. Just as in bulk solution the more highly charged polyelectrolyte has the potential to be the most effective counterion. However, counteracting this trend is the fact that the more highly charged polyelectrolyte forms a thinner and more compact layer than the low charge density one. This makes it harder for the negatively charged surfactants to successfully compete with the negatively charged surface sites for binding to the cationic segments. It is hard to evaluate a priori which effect is most important. However, from the data so far it is clear that the cac_s for the 30 and 10% charged polyelectrolytes are about the same and significantly lower than for the 100% charged one. It is possible that for even less charged polyelectrolytes the cac_s will increase again. This is a topic for future investigation.

How the pull-off force is affected by the polyelectrolyte charge density and the concentration of SDS is illustrated in Fig. 9. Two conclusions are immediately apparent. First, the pull-off force in the absence of SDS decreases very rapidly with decreasing polyelectrolyte charge density. This is due to a decreasing importance of bridging attraction [16]. Second, when SDS starts to associate with the polyelectrolyte coated surface a sharp drop in the pull-off force is observed. An increase in SDS concentration above that of the cac_s has a negligible effect on the pull-off force, providing us with an indication that the saturation amount of surfactant in the polyelectrolyte layer is reached at a concentration not very much above the cac_s. Above cac_s a strong pull-off force is seen for the most highly charged polyelectrolyte, whereas no, or a very weak, adhesion is observed in presence of the other polyelectrolytes. This demonstrates that the strongest interlayer polyelectrolyte–surfactant bridges are formed for the most highly charged polyelectrolyte.

The variation in compressed layer thickness and the range of the interaction as a function of SDS concentration

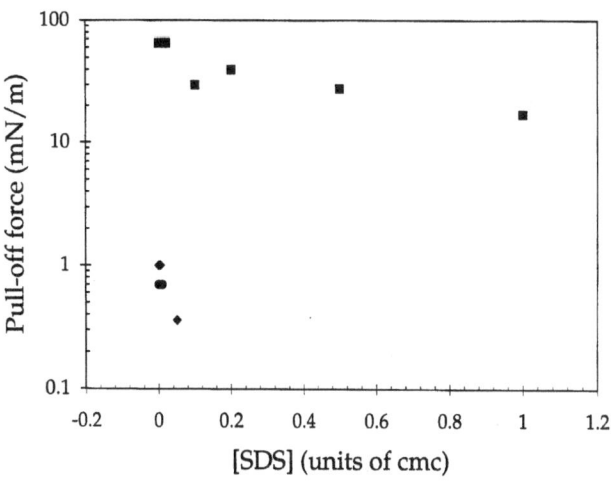

Fig. 9 The pull-off force between polyelectrolyte-coated mica surfaces as a function of SDS concentration expressed in units of cmc. Data are shown for PCMA (■), AM–MAPTAC-30 (♦), and AM–CMA-10 (●)

is shown in Fig. 10. Let us first consider the compressed layer thickness, which we define as the surface separation at which point a hard-wall-like repulsion in encountered. This definition is not very precise but it serves our purpose when we only want to illustrate changes in the adsorbed layer qualitatively. For PCMA, a sudden increase in compressed layer thickness occurs at the cac_s, demonstrating the incorporation of surfactants in the adsorbed layer. No further change is observed at higher surfactant concentrations. For AM–MAPTAC-30 an increase in compressed layer thickness is observed at cac_s and at higher SDS concentrations it decreases again, indicating a slight desorption of the polyelectrolyte. The AM–CMA-10 polyelectrolyte showed no change in compressed layer thickness up to 0.1 cmc. This polyelectrolyte was not studied at higher SDS concentrations but based on the AM–MAPTAC-30 results we expect that it may partially desorb at higher SDS concentrations. Previously we have shown that a preadsorbed layer of lysozyme desorbs once the SDS concentration reaches between 0.5 and 1 cmc [30].

The range of the interaction, arbitrarily defined as the point where the repulsive force has reached a magnitude of 1000 μN/m, also increases suddenly at cac_s. A further increase in surfactant concentration results in a decrease in the range of interaction for AM–MAPTAC-30, and less clearly also for PCMA. This is partly due to a screening of electrostatic forces and partly to some desorption of the polyelectrolytes. Again, AM–CMA-10 was not studied at surfactant concentrations close to cmc. Another definition of the range of the interaction would change the numbers in Fig. 10, but not the general trend.

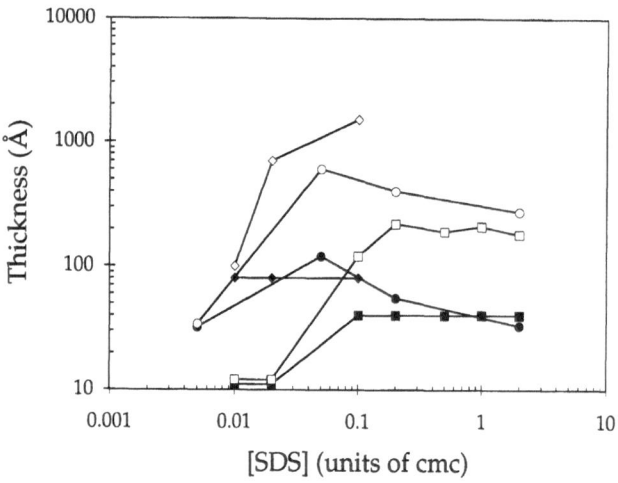

Fig. 10 The compressed layer thickness obtained when an essentially hard wall repulsion has been reached (filled symbols) and an estimate of the uncompressed layer, defined as the point where the force divided by radius equals 100 $\mu N/m$ (unfilled symbols). Data are shown for PCMA (squares), AM–MAPTAC-30 (circles) and AM–CMA-10 (diamonds)

Oscillating forces

Finally, we should mention the reason for the oscillations in the force curve observed above cac_s for the PCMA. It has been argued previously [11] that the oscillations are due to the spatial arrangement of polyelectrolyte associated micelles in the gap between the surfaces. At a separation of 120 Å, three layers can be accommodated, at 80 Å two layers and at 40 Å only one layer. The oscillations seen here is, thus, a type of structural force similar in nature to that observed in micellar solutions well above the cmc [31, 32].

Conclusions

We have shown that surfactants associate with oppositely charged polyelectrolytes preadsorbed onto surfaces having the same charge as the surfactant. The association is, just like in bulk solution, a cooperative process driven by the hydrophobic interaction between the surfactant side chains and made more favorable by the presence of polyelectrolytes acting as counterions to the surfactant aggregates. It is possible to define a critical association

concentration at the surface, cac_s, that in general will depend on ionic strength, type of polyelectrolyte, type of surfactant and properties of the interface. We found that for highly charged cationic polyelectrolytes and anionic surfactant, SDS, the cac_s at the negatively charged mica surface is higher than the cac in bulk solution. This can be rationalized by considering the counterion entropy, chain stiffness and competition between surfactants and charged surface sites for the association with oppositely charged polyelectrolyte segments. The cac_s is higher, about 0.1 cmc, for the 100% charge PCMA than for the lower charged ones, about 0.02 cmc, studied here. This is an opposite trend to what normally is found in bulk solution. We suggest that this is due to formation of less and shorter loops and trains for more highly charged polyelectrolytes adsorbed to an oppositely charged surface. This suggestion is supported by the very thin layers formed by the highly charged polyelectrolyte at the charge neutralization point. The expansion of the adsorbed layer increases with decreasing charge density. This may partly be due to an increased size of the polyelectrolyte corona surrounding the micelle. Oscillations with a periodicity of about 40 Å are observed in the force distance profile at SDS concentrations between 0.2 and 2 cmc for the 100% charged case, but not for the other polyelectrolytes. It has previously been argued that the oscillations arise from the spatial arrangement of polyelectrolyte associated micelles [11]. It is not surprising that such oscillations are seen only for PCMA considering that the most highly charged polyelectrolyte once the tails no longer are attached to the surface will be able to bind most closely to the micellar surface and thus, permit the most efficient spatial packing of the polyelectrolyte–micelle complexes. The polyelectrolyte layer is only partially desorbed even at an SDS concentration of 2 cmc. Some preliminary measurements have been carried out in a system where both PCMA and SDS are present in bulk solution. A thick adsorbed layer is formed under conditions when a phase separation occurs in bulk solution. The forces measured on approach are purely repulsive whereas attractive forces, due to formation of interlayer polyelectrolyte–micelle bridges, are observed on separation.

Acknowledgements This work was partly supported by the Human Capital and Mobility program (contract #CHRX-CT94-0655). Matthew Fielden acknowledges a post-doc grant from the Swedish Natural Science Research Council (NFR).

References

1. Satake I, Yang JT (1976) Biopolymers 15:2263–2275
2. Hayakawa K, Santerre JP, Kwak CT (1983) Macromolecules 16:1642–1645
3. Hayakawa K, Kwak JCT (1984) J Phys Chem 87:506–509
4. Santerre JP, Hayakawa K, Kwak JCT (1985) Colloids Surf 13:35–45
5. Lindman B, Thalberg K (1993) In: Goddard ED, Ananthapadmanabhan KP (eds) Interactions of Surfactants with Polymers and Proteins. CRC Press, Boca Raton, FL

Progr Colloid Polym Sci (1997) 106: 24–33
© Steinkopff Verlag 1997

6. Malovikova A, Hayakawa K, Kwak JCT (1984) J Phys Chem 88:1930–1933
7. Khokhlov AR, Kramarenko EY, Makhaeva EE, Starodubtzev SG (1992) Macromolecules 25:4779–4783
8. Gong JP, Osada Y (1995) J Phys Chem 99:10971–10975
9. Wallin T, Linse P (1996) Langmuir 12: 305–314
10. Shubin V, Petrov P, Lindman B (1994) Colloid Polym Sci 272:1590–1601
11. Claesson PM, Dedinaite A, Blomberg E, Sergeyev VG (1996) Ber Bunsenges Phys Chem 100:1008–1013
12. Israelachvili JN, Adams GE (1978) J Chem Soc Faraday Trans 1 174: 975–1001
13. Parker JL, Christenson HK, Ninham BW (1989) Review Sci Instrum 60: 3135–3138
14. Derjaguin B (1934) Kolloidn J 69: 155–164

15. Parker JL, Attard P (1992) J Phys Chem 96:10398–10405
16. Dahlgren MAG, Waltermo Å, Blomberg E, Claesson PM, Sjöström L, Åkesson T, Jönsson B (1993) J Phys Chem 97: 11769–11775
17. Dahlgren MAG (1994) Langmuir 10: 1580–1583
18. Dahlgren MAG, Claesson PM, Audebert R (1994) J Colloid Interface Sci 166:343–349
19. Fielden ML, Schillén K, Claesson PM, Brown W, Audebert R (in preparation)
20. Kjellin MUR, Claesson PM, Audebert R (1997) J Colloid Interface Sci 190: 476–484
21. Claesson PM, Herder PC, Stenius P, Eriksson JC, Pashley RM (1986) J Colloid Interface Sci 109:31–39
22. Pashley RM (1981) J Colloid Interface Sci 80:153–162
23. Pashley RM (1981) J Colloid Interface Sci 83:531–546

24. Blomberg E, Claesson PM, Fröberg JC (1997) Biomaterials, in press
25. Van Dulm P, Norde W, Lyklema J (1981) J Colloid Interface Sci 82:77–82
26. Norde W (1986) Adv Colloid Interface Sci 25:267–340
27. Hayakawa K, Kwak JCT (1982) J Phys Chem 86:3866–3870
28. Thalberg K, Lindman B, Bergfeldt K (1991) Langmuir 7:2893–2898
29. Satake I, Takahashi T, Hayakawa K, Maeda T, Aoyagi M (1990) Bull Chem Soc Jpn 63:926–928
30. Blomberg E (1993) Surface force studies of adsorbed proteins. PhD dissertation Royal Institute of Technology, Stockholm, Stockholm
31. Richetti P, Kékicheff P (1992) Phys Rev Lett 68:1951–1954
32. Bergeron V, Radke CJ (1992) Langmuir 8:3020–3026

Progr Colloid Polym Sci (1997) 106:34–41
© Steinkopff Verlag 1997

O. Söderman

Pulsed field gradient NMR studies of emulsions: droplet sizes and concentrated emulsions

Received: 13 October 1996
Accepted: 4 February 1997

Dr. O. Söderman (✉)
Department of Physical Chemistry 1
Center for Chemistry and
Chemical Engineering
Lund University
P.O. Box 124
22100 Lund, Sweden

Abstract In this contribution we suggest that nuclear magnetic resonance constitutes a promising technique for studying essential features of emulsions. The background of the method is discussed, and it is emphasized that the method determines the mean-squared displacements of molecules over distances of order 10^{-6} m. It is pointed out that such distances correspond to typical emulsion droplet sizes. As a consequence, the method, when applied to emulsions, yields information on droplet sizes and the presence of diffusional barriers. To exemplify this, two particular examples are discussed. The first pertains to the determination of droplet size distributions of emulsion, using data from two different systems, viz. a low-calorie spread and a multiple emulsion, as examples. The second example deals with concentrated emulsions, and it is shown that both dynamical (long-time diffusion and lifetimes in the droplets) and structural (droplet sizes) information can be obtained.

Key words Nuclear magnetic resonance – pulsed field gradients – diffusion – emulsions – droplet size – concentrated emulsions

Introduction

Emulsions are of great fundamental as well as technical importance. They occur in a multitude of situations ranging from biological systems such as in the digestion of fats to the extraction of crude oil. Therefore, it is not surprising that the practical knowledge about emulsions is quite extensive. People engaged in the production generally know how to produce emulsions with desired properties such as droplet size distributions and shelf life. The same level of empirical knowledge is at hand for the opposite process of breaking an emulsion. When it comes to the basic scientific understanding of emulsions, we are a little worse off. It is quite clear that several fundamental properties of emulsions, such as what factors determine the stability of emulsions, what is the importance of the properties of the continuous phase and so on, are not fully understood. To some extent, we also lack or have not yet applied suitable techniques in the study of emulsions.

The majority of emulsions are stabilized by surfactants and Bancroft, one of the pioneers in emulsion science, realized that the stability of an emulsion was related to the properties of the surfactant film. This insight has become important during recent years when attempts have been made to draw on the rather detailed and profound understanding that exists today about bulk surfactant systems in the description of emulsions [1].

This high level of understanding about bulk surfactant systems stems not the least from the application of modern physico–chemical techniques such as scattering methods and nuclear magnetic resonance (NMR).

It seems reasonable to expect that the application of these techniques to emulsion systems would lead to an increased basic understanding of such systems. This contribution gives examples from a few years work by the

Progr Colloid Polym Sci (1997) 106:34–41
© Steinkopff Verlag 1997

author and coworkers in which a particular NMR method, namely, that of pulsed field gradient NMR, has been applied to emulsion systems. This NMR application has proven very useful in the study of bulk surfactants [2] and there is no reason why the same should not be true for emulsions.

We start by summarizing the pulsed field gradient method and then describe two relevant topics: the first deals with droplet size determination of emulsions and the second deals with the characterization of concentrated emulsions. The former is a well-developed area, which has been used for a long time not only by the present author, while the latter constitutes a novel field for the application of the method. We present some (very) preliminary data from such a system.

Pulsed field gradient NMR method

Measurements of self-diffusion coefficients and flow by means of pulsed field gradient (PFG) techniques have evolved to become a very important tool in the study of a multitude of different problems. The technique as such has recently been described in a review article [2], in which other relevant references may also be found, so here we will merely state that the technique requires no isotopic labeling (avoiding possible disturbances due to the addition of probes) and that it gives component-resolved self-diffusion coefficients with great precision in a minimum of measuring time. The main nucleus studied is the proton, but other nuclei, such as Li, F, Cs and P are also of interest.

The method monitors transport over macroscopic distances (typically in the micrometer regime). Therefore, when the method is applied to the field of surface and colloid chemistry, the diffusion coefficients determined reflect the aggregate sizes and obstruction effects of the colloidal particles. This is the origin of the success the method has had in the study of microstructures of surfactant solutions and also forms the basis of its applications to emulsion systems.

The fact that the information is obtained without invoking complicated models, as is the case for the NMR relaxation approach, is particularly important. In this context it should be stressed that the PFG approach measures the self-diffusion rather than the collective diffusion coefficient, which is measured by, for instance, light scattering methods.

In its simplest version, the method consists of two equal and rectangular gradient pulses of magnitude g and length δ, sandwiched on either side of the 180° RF-pulse in a simple Hahn echo experiment. For molecules undergoing free (Gaussian) diffusion characterized by a diffusion

coefficient of magnitude D, the echo attenuation due to diffusion is given by [3, 4]

$$E(\Delta, \delta, g) = E_0 \exp\left(-\gamma^2 g^2 \delta^2 \left(\Delta - \frac{\delta}{3}\right) D\right), \quad (1)$$

where Δ represents the distance between the leading edges of the two gradient pulses, γ is the magnetogyric ratio of the monitored spin and E_0 denotes the echo intensity in the absence of any field gradient. By varying either g, δ or Δ (while at the same time keeping the distance between the two RF-pulses constant) D can be backed out by fitting Eq. (1) to the observed intensities.

As mentioned above, one of the key foundations of PFG diffusion experiments is the notion that the transport of molecules is measured over a time Δ, which we are free to choose at our own will in the range from a few ms to several seconds. This means that the length scale over which we are measuring the molecular transport in the micrometer regime for low molecular-weight liquids. When the molecules experience some sort of boundary with regard to their diffusion during the time Δ, the molecular displacement is lowered as compared to free diffusion, and the outcome of the experiment becomes drastically changed [5–7].

One situation where this is encountered, is for the case of restricted motion inside an emulsion droplet. In this case the molecular displacements do not exceed the droplet size, which, indeed, often is in the micrometer regime. In actual fact, the exact equations governing the echo amplitude as a function of the relevant parameters (corresponding to Eq. (1) above for free diffusion) are not known for any other case than for free diffusion and free diffusion superposed on flow (and for the case of a particle confined to a harmonic potential [8]). Therefore, one has to rely on various levels of approximations.

In one of these, one considers gradient pulses which are so narrow that no transport during the pulse takes place. This has been termed the short gradient pulse (SGP) (or narrow gradient pulse, NGP) limit. This case leads to a very useful formalism whereby the echo attenuation can be written as

$$E(\delta, \Delta, g) = \iint \rho(\mathbf{r_0}) P(\mathbf{r_0}|\mathbf{r}, \Delta) \exp[i\gamma g \delta \cdot (\mathbf{r} - \mathbf{r_0})] \, d\mathbf{r} \, d\mathbf{r_0} , \quad (2)$$

where $P(\mathbf{r_0}|\mathbf{r}, \Delta)$ is the propagator which gives the probability of finding a spin at position $\mathbf{r}$ after a time Δ if it was originally at position $\mathbf{r_0}$. For free diffusion $P(\mathbf{r_0}|\mathbf{r}, \Delta)$ is a Gaussian function and if this form is inserted in Eq. (2), Eq. (1) above with the term $(\Delta - \delta/3)$ replaced with Δ is obtained, which is the SGP result for free diffusion. For other cases than free diffusion, alternate expressions for $P(\mathbf{r_0}|\mathbf{r}, \Delta)$ have to be used. Tanner and Stejskal [9] solved the problem of reflecting planar boundaries, while the case of interest to us in the context of emulsion droplets, i.e. that of molecules confined to a spherical cavity of radius R, was presented by Balinov et al. [10]. The result is

$$E(\delta, \varDelta, g) = \frac{9\left[\gamma g\delta R \cos(\gamma g\delta R) - \sin(\gamma g\delta R)\right]^2}{(\gamma g\delta R)^6} + 6(\gamma g\delta R)^2$$

$$\times \sum_{n=0}^{\infty} \left[j'_n(\gamma g\delta R)\right]^2 \sum_m \frac{(2n+1)\alpha_{nm}^2}{\alpha_{nm}^2 - n^2 - n}$$

$$\times \exp\left(-\frac{\alpha_{nm}^2 D\varDelta}{R^2}\right) \frac{1}{\left[\alpha_{nm}^2 - (\gamma g\delta R)^2\right]^2}\,, \qquad (3)$$

where $j_n(x)$ is the spherical Bessel function of the first kind and α_{nm} is the mth root of the equation $j'_n(\alpha) = 0$. D is the bulk diffusion of the entrapped liquid, and the rest of the quantities are defined above. The main point to notice about Eq. (3) is that the echo decay does indeed depend on the radius, and thus the droplet radii can be obtained from the echo decay for molecules confined to the sphere, provided that the conditions underlying the SGP approximation are met.

The second approximation used is the so-called Gaussian phase distribution. Originally introduced by Douglass and McCall [11], the approach rests on the approximation that the phases accumulated by the spins on account of the action of the field gradients are Gaussian distributed. Within this approximation and for the case of a steady-gradient, Neuman [12] derived the echo attenuation for molecules confined within a sphere, a cylinder and between planes. For spherical geometry Murday and Cotts [13] derived the equation for pulsed field gradients in the Hahn echo experiment described above. The result is

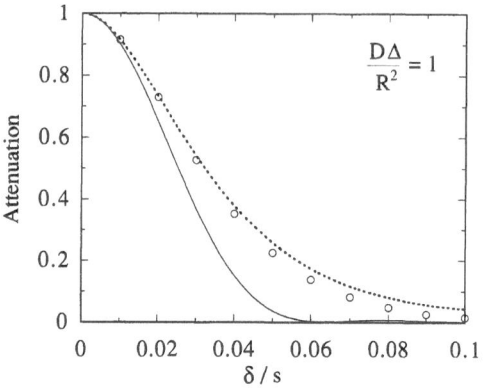

Fig. 1 Results of a simulation of the diffusion of water molecules in an emulsion droplet of radius R, given as the echo amplitude vs. the duration δ of the field gradient pulse. The ratio $D\varDelta/R^2$ is 1. The dotted line is the prediction of the Gaussian phase approximation (Eq. (4)) while the solid line is the prediction of the short-gradient pulse (Eq. (3)). Figure adapted from [10]

Recently, Balinov et al. [10] performed accurate computer simulations aimed at further testing its applicability over a wide range of parameter values. An example is shown in Fig. 1. The conclusion reached in [10] was that Eq. (4) never deviates by more than 5% in predicting the echo attenuation for typically used experimental parameters. Thus, it is a useful approximation and we shall use it in the next section of this paper.

As pointed out above, the NMR echo signal E depends on the droplets radius which can be estimated by measuring E at different durations δ of the pulse gradient. A typical $E(\delta)$ plot, generated by using Eq. (4), is presented in

$$\ln\left[E(\delta, \varDelta, g)\right] = -\frac{2\gamma^2 g^2}{D} \sum_{m=1}^{\infty} \frac{\alpha_m^{-4}}{\alpha_m^2 R^2 - 2}$$

$$\times \left\{2\delta - \frac{2 + \exp\left[-\alpha_m^2 D(\varDelta - \delta)\right] - 2\exp(-\alpha_m^2 D\delta) - 2\exp(-\alpha_m^2 D\varDelta) + \exp\left[-\alpha_m^2 D(\varDelta + \delta)\right]}{\alpha_m^2 D}\right\}, \qquad (4)$$

where α_m is the mth root of the Bessel equation $1/(\alpha R)J_{3/2}(\alpha R) = J_{5/2}(\alpha R)$. Again, D is the bulk diffusion coefficient of the entrapped liquid.

Thus, we have at our disposal two equations with which to interpret PFG data from emulsions in terms of droplet radii, neither of which are exact for all values of experimental and system parameters. As the conditions of the SGP regime are technically demanding to achieve, Eq. (4) (or limiting forms of it) have been used in most cases to determine the droplet radii. A key question is then under what conditions Eq. (4) is valid. That it reduces to the exact result in the limit of $R \to \infty$ is easy to show and also obvious from the fact that we are then approaching the case of free diffusion, in which the Gaussian phase approximation becomes exact.

Fig. 2 which demonstrates the sensitivity of the NMR self-diffusion experiment to resolve micrometer droplet sizes.

In conclusion, the PFG method puts at our disposal a method by which diffusion coefficient for molecules can be measured. The values of these diffusion coefficients carry important information with regard to solution microstructure in surfactant systems which often constitute the continuous phase in emulsions. For molecules which do not undergo Gaussian diffusion but whose diffusion processes are perturbed by the presence of barriers which impede their diffusion, the echo attenuation will depend on the geometry of these barriers, and also, as we shall see, on such properties as the lifetime of the dispersed

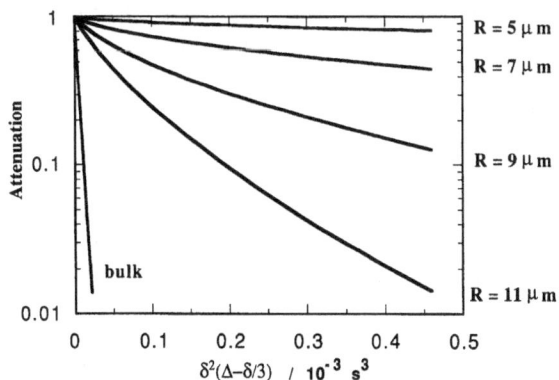

Fig. 2 The echo attenuation as a function of $\delta^2(\Delta - \delta/3)$ for different radii of emulsion droplets (according to Eq. (4)) with $\Delta = 0.100$ s, $\gamma g = 10^7$ rad m^{-1} s^{-1} and $D = 2 \times 10^{-9}$ m^2 s^{-1}

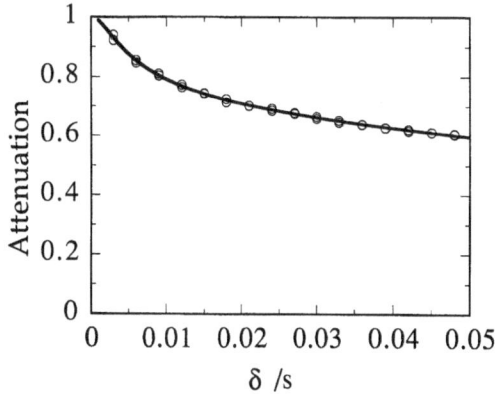

Fig. 3 Echo intensity for the entrapped water in droplets formed in a low-calorie spread containing 60% fat vs. δ. The solid line corresponds to the predictions of Eqs. (4)–(6). The results from the fit are: $d_0 = 0.82$ μm and $\sigma_0 = 0.72$. Figure adapted from [14]

phase in the emulsion droplets, and thus important aspects of the enclosed phase may be studied.

Determination of emulsion droplet radii by means of the NMR PFG method

As pointed out above, the echo attenuation curve for the PFG experiment when applied to molecules entrapped in an emulsion droplet is a signature of the size of the emulsion droplet (cf. Fig. 2). As a consequence, droplet sizes can be determined by means of the PFG experiment.

The NMR sizing method, which was apparently first suggested by Tanner in [4] has been applied to a number of different emulsions ranging from cheese to crude oil emulsions [14–20].

When applied to a real emulsion one has to consider the fact that the emulsion droplets in most cases are polydisperse in size. This effect can be accounted for if the molecules confined to the droplets are in a slow exchange situation, meaning that their lifetime in the droplet must be longer than Δ. For such a case, the echo attenuation is given by

$$E_{poly} = \frac{\int_0^\infty R^3 P(R) E(R) \, dR}{\int_0^\infty R^3 P(R) \, dR}, \tag{5}$$

where $P(R)$ represents the droplet size distribution function and $E(R)$ the echo attenuation according to Eq. (4) (or, within the SGP-approximation, Eq. (3)) for a given value of R. From NMR data it is difficult (although, in principle, not impossible) to determine the actual form of $P(R)$. However, given an analytical expression for $P(R)$ we may determine the parameters of that distribution function. A frequently used form is the log-normal function as

defined in Eq. (6), as it appears to be a reasonable description of the droplet size distribution of many emulsions. In addition, it has only two parameters which makes it convenient for modeling purposes.

$$P(R) = \frac{1}{2R\sigma\sqrt{2\pi}} \exp\left[-\frac{(\ln 2R - \ln d_0)^2}{2\sigma^2} \right]. \tag{6}$$

In Eq. (6), d_0 represents the diameter median and σ a measure of the width of the size distribution.

To illustrate the method and discuss its accuracy we will use as an example some recent results for margarines (or low-calorie spreads) [14]. This system highlights some of the definite advantages of using the NMR method to determine emulsion droplet sizes, since other non-perturbing methods hardly exist for these systems.

Given in Fig. 3 is the echo decay for the water signal of a low calorie spread containing 60% fat. These systems are W/O emulsions and as can be seen the water molecules do experience restricted diffusion (in the representation of Fig. 3, the echo decay for free diffusion would be given by a Gaussian function). Also, given in Fig. 3 is the result of fitting Eqs. (4)–(6) to the data. As is evident, the fit is quite satisfactory and the parameters of the distribution function obtained are given in the figure caption. However, one might wonder how well determined these parameters are, given the fact that the equations describing the echo attenuation are quite complicated. To test this matter further, Monte Carlo error investigations were performed in [14]. Thus, random errors were added to the echo attenuation and the least-squares minimization was repeated 100 times as described previously [21]. A typical result of such a procedure is given in Fig. 4. As can be seen in Fig. 4 the parameters are reasonably well determined, with an uncertainty in R (and σ, data not shown) of about $\pm 15\%$.

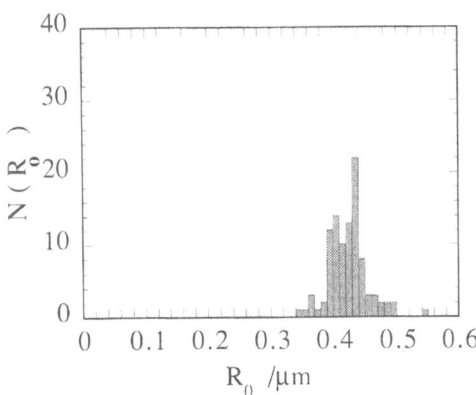

Fig. 4 A Monte Carlo error analysis of the data in Fig. 3. The value of the parameter d_0 in Eq. (20) is $d_0 = 0.82 \pm 0.05 \, \mu$m. Figure adapted from [14]

Creaming or sedimentation of emulsions with droplet sizes above 1 μm causes some experimental difficulty because of the change of the total amount of spins in the NMR active volume of the sample tube during the experiment. This can be accounted for by extra reference measurements with no gradient applied before and after each NMR scan at a particular value of δ. In addition, such reference measurements may provide information on the creaming rate which is a useful characteristic of emulsions. Creaming or sedimentation is not a problem in the study of most food emulsions (such as low-calorie spreads), highly concentrated emulsions or viscous water-in-crude-oil emulsions [15].

Emulsion droplet sizes in the range from 1 up to 50 μm can be measured with rather modest gradient strengths of about 1 T/m. Note that the size determination rests on determining the molecular motion of the dispersed phase, so the method cannot be applied to dispersed phases with low molecular mobility. In practice, oils with self-diffusion coefficients above 10^{-12} m^2 s^{-1} is required for sizing of O/W emulsion. Of course, W/O emulsions with most conceivable continuous media can be sized.

Emulsion droplets below 1 μm can often be characterized by the Brownian motion of the droplet as such (exceptions are concentrated emulsions or other emulsions where the droplets do not diffuse). This is the approach taken in the study of microemulsion droplets, where the diffusion behavior of the solubilized phase is characterized by the droplets (Gaussian) diffusion.

As an example of another application of the method, we show in Fig. 5 the echo signal from the water in a W/O emulsion (panel a), and from the resulting double emulsion obtained when the original W/O emulsion is emulsified in water (panel b) [22]. Also given is the resulting size distri-

butions (panel c). For these complex systems, PFG NMR is very well suited, as few other methods exist that can determine such basic properties as the size distribution.

In conclusion, we summarize the main advantages of the NMR diffusion method as applied to emulsion droplet sizing. It is non-perturbing, requiring no sample manipulation (such as dilution with the continuous phase) and non-destructive which means that the same sample may be investigated many times, which is important if one wants to study long-time stability or the effect of certain additives on the droplet size. It requires small amounts of sample (typically on the order of a few 100 mg). Moreover, the total NMR signal from the dispersed phase in emulsions is usually quite intense because of the large amounts of spins. This fact allows for rapid measurements with a single scan per δ point.

PFG studies of concentrated emulsions

The discussion carried out above, applies to the case where the molecules are confined to the droplets on the time scale of the experiment. This is a reasonable assumption for many emulsions, and it can, in fact, be tested by the NMR diffusion method by varying Δ. However, there are some interesting emulsion systems where this is not always the case. These are the so-called highly concentrated emulsions (often termed high internal phase emulsions) [23, 24], which may contain up to (and in some cases even more than) 99% dispersed phase. Here the droplets are separated by a liquid film which may be very thin (on the order of 100 Å), and which may in some instances be permeable to the dispersed phase.

The case when the lifetime of the dispersed phase in the droplet is of the same order of magnitude as Δ is particularly interesting. Under these conditions, one may in some cases obtain one (or several) peak(s) in the plot of the echo amplitude vs. the parameter q, defined as $q = \gamma g \delta / 2\pi$. This is a surprising result at first sight, as we are accustomed to observe a monotonous decrease of the echo amplitude with q, but it is actually a manifestation of the fact that the diffusion is no longer Gaussian. Such peaks can be rationalized within a formalism related to the one used to treat diffraction effects in scattering methods [25], and the analysis of the data may yield important information regarding not only the size of the droplets but also the permeability of the dispersed phase through the thin films as well as the long-term diffusion behavior of the dispersed phase. We show in Fig. 6 some preliminary results which display such a diffraction-like effect in a concentrated emulsion system. The particular example pertains to a concentrated three-component emulsion based on a nonionic surfactant with composition $C_{12}E_4/C_{10}H_{22}/H_2O$

(1 wt%NaCl), (3/7/90) wt%. The concentrated emulsion was made according to a protocol described in [26].

The data in Fig. 6 are presented with the value of the quantity q on the abscissa. This quantity, which is defined above, has the dimension of inverse length and it is, in fact, related to the scattering vector used in describing scattering experiments. In fact, the inverse of the value of the position of the peak can be related to the center-to-center distance of the droplets. In the example given in Fig. 6 this value is roughly 1.8 μm which is in rough agreement with twice the droplet radii as judged from microscope pictures taken of the emulsion.

In order to analyze the data in more detail, one needs access to a theory of diffusion in these inter-connected systems. One such theory was developed by Callaghan and coworkers [7]. It assumes that the SGP limit described above is valid and it is based on a number of underlying assumptions of which pore equilibration is perhaps the most serious one. This latter assumption implies that an

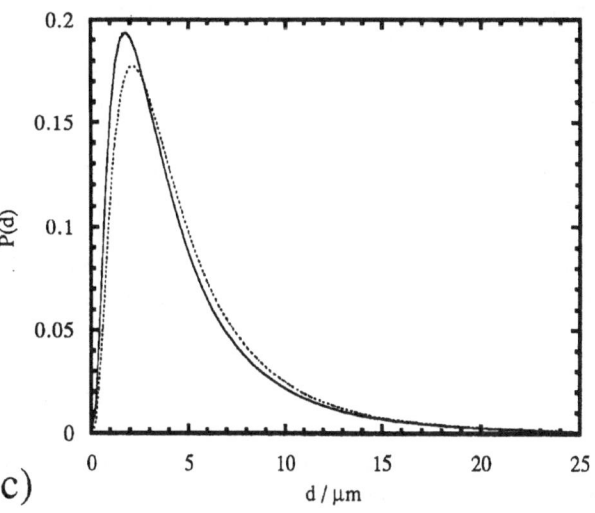

c)

Fig. 5 Echo intensities for water in a W/O emulsion (a), for water in an O/W/O emulsion (b), and the obtained size distribution curves (c) (solid line corresponds to the W/O emulsion while the dashed line corresponds to the W/O/W emulsions). Also, given in (a) and (b) are the results of fitting Eqs. (4)–(6) to the data. The W/O emulsion consists of 40 wt% "oil" solution (consisting of 20 wt% Span 80 dissolved in decane) and 60 wt% water. The W/O/W emulsion consists of 31 wt% of the W/O emulsion, 68 wt% water and 1 wt% Tween 80

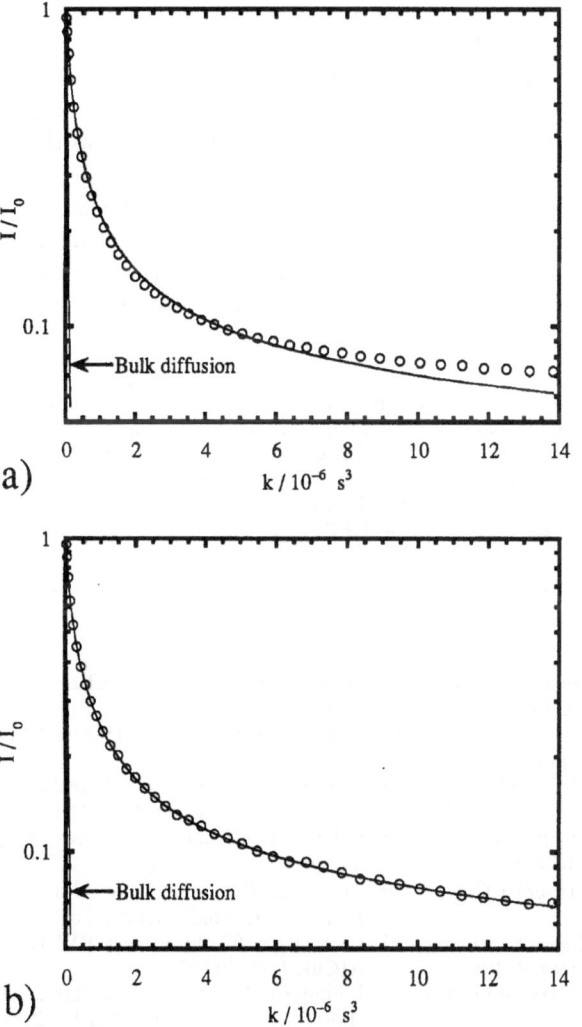

a)

b)

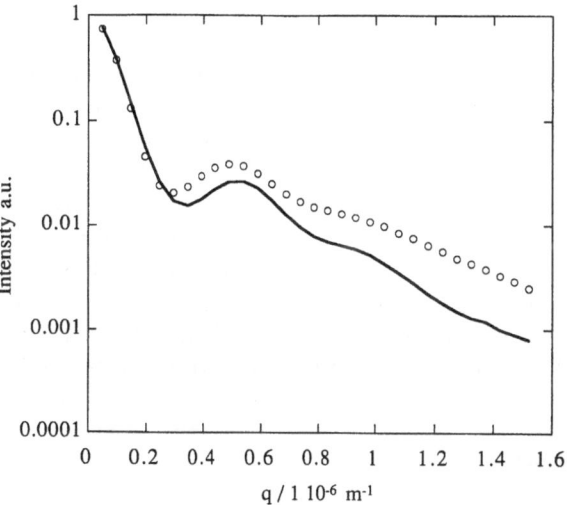

Fig. 6 Echo intensities for water in a concentrated emulsion prepared according to [26]. The composition of the emulsion is: $C_{12}E_4/C_{10}H_{22}/H_2O$ (1 wt% NaCl), (3/7/90) wt%. The solid line is the result of fitting the pore-hopping model [7] to the data

individual molecule in a droplet samples all the positions in the interior of the droplet fully before it migrates to a neighboring droplet. The echo decay for such a case is given by a product of a structure factor for the single pore and a function that depends on the motion of the molecules between the pores. The pore-hopping formalism takes as input the radius of the sphere, the long-time diffusion coefficient, the center-to-center distance between the droplets and, finally, a spread in the center-to-center distance (to account for polydispersity of the droplets). The prediction of the pore-hopping theory for the data in Fig. 6 is included as a solid line. The agreement is not quantitative (the difference most likely owing to problems in defining a relevant structure factor for our system of polydisperse droplets), but the main features of the experimental data is certainly reproduced. The results are: for the pore-to-pore distance 1.2 μm, for the spread in the parameter 0.2 μm and, finally, for the long-time diffusion we obtain $D = 9 \times 10^{-11}$ m^2 s^{-1}. From the latter value one can estimate a value for the life-time of water molecule in a droplet and the value obtained is 3 ms.

A different starting point in the analysis of the data such as the one in Fig. 6 is to make use of Brownian simulations [27]. These are essentially exact within the specified model, although they do suffer from statistical uncertainties. For the present case one lets a particle perform a random walk in a sphere with a semi-permeable boundary. With a given probability the particle is allowed to leave the droplet after which it starts to perform a random walk in a neighboring droplet. We are in the process of applying this model to data, of which those presented in Fig. 6 is a subset. That the approach yields peaks in the echo-decay curves can be seen in Fig. 7, where such simulations have been performed under some different conditions [27]. The simulation scheme yields essentially the same kind of information as the pore-hopping theory. Thus, one obtains the droplet size and the lifetime of a molecule in the droplet (or quantities related to this, such as the permeability of the film separating the droplet).

Clearly, spectra such as the one presented in Fig. 6 can be used to study many aspects of concentrated emulsions of which a few are exemplified above. It can also be used to study the evolution of the droplet size with time (recall that

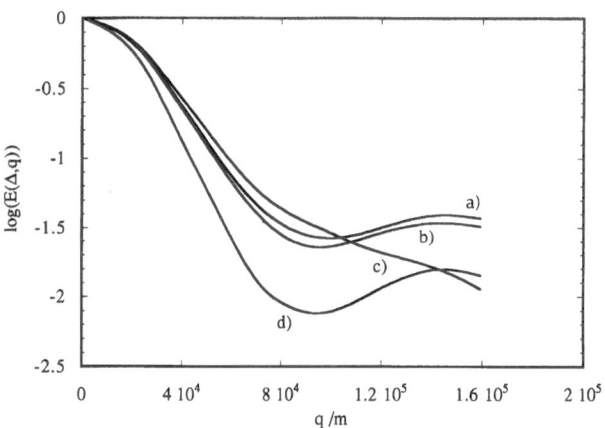

Fig. 7 The common logarithm of the calculated attenuation vs. q for $R = 4 \ \mu$m, $P = 0.032$, and $D_0 = 2.0 \times 10^{-9}$ m^2 s^{-1}. The different curves corresponds to: $\Delta = 100$ ms and $g = 0.16$ T m^{-1} (a), $\Delta = 100$ ms and $g = 0.39$ T m^{-1} (b), $\Delta = 100$ ms and $g = 0.039$ T m^{-1} (c), and $\Delta = 140$ ms and $g = 0.16$ T m^{-1} (d). Figure adapted from [27]

the method is non-perturbing) and also a function of changes in external parameters such as temperature, which is an important variable for the properties of non-ionic surfactant films.

A full account of the concentrated emulsion work presented above is under preparation.

Finally, we note that concentrated emulsions are excellent model systems in the development of the PFG methods as applied to the general class of porous systems, where the method has a great potential in providing relevant and important information.

Acknowledgments It is a pleasure to acknowledge the co-operation with Balin Balinov (Norway), Ramon Pons (Spain) and Björn Håkansson, Per Linse and Bengt Jönsson in Lund. This work has received financial support from the Swedish Board for Industrial and Technical Development (NUTEK) and from the Swedish Natural Science Research Council (NFR).

References

1. Kabalnov A, Wennerström H (1996) Langmuir 12:276
2. Söderman O, Stilbs P (1994) Prog Nucl Mag Reson 26:445
3. Stejskal EO, Tanner JE (1965) J Chem Phys 42:288
4. Tanner JE (1966) Thesis University of Wisconsin
5. Callaghan PT, Coy A (1993) In: Tycko P (ed) PGSE NMR and Molecular Translational Motion in Porous Media. Kluwer Academic Publishers, Dordrecht
6. Callaghan PT (1991) Principles of Nuclear Magnetic Resonance Microscopy. (Clarendon Press, Oxford)
7. Callaghan PT, Coy A, Halpin TPJ, MacGwan D, Packer JK, Zelaya FO (1992) J Chem Phys 97:651
8. Terentjev EM, Callaghan, PT, Warner M (1995) J Chem Phys 102:4619
9. Tanner JE, Stejskal EO (1968) J Chem Phys 49:1768
10. Balinov B, Jönsson B, Linse P, Söderman O (1993) J Magn Reson A 104:17
11. Douglass DC, McCall DW (1958) J Phys Chem 62:1102
12. Neuman CH (1974) J Chem Phys 60:4508
13. Murday JS, Cotts RM (1968) J Chem Phys 48:4938
14. Balinov B, Söderman O, Wärnheim T (1994) J Am Oil Chem Soc 71:513
15. Balinov B, Urdahl O, Söderman O, Sjöblom J (1994) Colloids Surf 82:173
16. Callaghan PT, Jolley KW, Humphrey R (1983) J Colloid Interface Sci 93:521
17. Li X, Cox JC, Flumerfelt RW (1992) AIChE J 38(10):1671
18. Lönnqvist I, Khan A, Söderman O (1991) J Colloid Interface Sci 144(2):401

Progr Colloid Polym Sci (1997) 106:34–41
© Steinkopff Verlag 1997

19. Packer KJ, Rees C (1972) J Colloid Interface Sci 40(2):206
20. Van den Enden JC, Waddington D, Van Aalst H, Van Kralingen CG, Packer KJ (1990) J Colloid Interface Sci 140(1):105
21. Stilbs P, Moseley M (1978) J Magn Reson 31:55
22. Lönnqvist I, Håkansson B, Balinov B, Söderman O (In press) J Colloid Interface Sci
23. Lissant KJ (1966) J Colloid Interface Sci 22:462
24. Princen HM (1983) J Colloid Interface Sci 91:160
25. Callaghan PT, Coy A, MacGowan D, Packer KJ, Zelaya FO (1991) Nature (London) 351(6362):467
26. Pons R, Carrera I, Erra P, Kunieda H, Solans C (1994) Colloids Polym Sci 91:259
27. Balinov B, Linse P, Söderman O (1996) J Colloid Interface Sci 182:539

Progr Colloid Polym Sci (1997) 106:42–48
© Steinkopff Verlag 1997

H. Matsuura

Conformational behavior of nonionic surfactants in the organized phases studied by vibrational spectroscopy

Received: 13 October 1996
Accepted: 4 February 1997

Abstract Conformational behavior of the alkyl chain in α-dodecyl-ω-hydroxytris(oxyethylene) ($C_{12}E_3$) in the aqueous organized phases has been studied by a newly developed technique of C–D stretching vibrational spectroscopy. This method is based on the fact that the wavenumbers of the isolated C–D stretching vibrations are sensitive to the conformation in the vicinity of the C–D bond. The C–D stretching infrared spectra of five selectively monodeuterated species of $C_{12}E_3$ in the lamellar (L_α) and isotropic solution (L_2) phases were analyzed and the fractions of the *trans* conformation around the dodecyl C–C bonds and the oxyethylene-adjoining O–C bond and the fractions of the consecutive *trans* conformations around the two adjoining bonds were evaluated. The conformational change at the phase transition from L_2 to L_α is not significant and only a small increase in the *trans* fraction is observed for the C–C bonds close to the alkyl/oxyethylene interface, implying that the conformational states of the alkyl chain in the L_2 and L_α phases in the vicinity of their boundary are substantially not different. In the L_α phase, when the composition or the temperature approaches the region of the phase separation or transition, the *trans* fractions for the C–C bonds closer to the alkyl/oxyethylene interface and those closer to the chain terminal decrease. These observations indicate that the conformational transformation from *trans* to *gauche* at these chain positions makes the lamellar structure less stable and leads eventually to the structural destruction. The conformational order as evaluated from the fractions of the consecutive *trans* conformation is the highest in the middle of the chain in the L_α and L_2 phases. This vibrational spectroscopic observation, together with the previous NMR observations, indicates that the alkyl/oxyethylene interface is flexible with respect to the conformation and the orientation of the chain.

Key words Nonionic surfactants – organized phases – conformation – vibrational spectroscopy

Prof. H. Matsuura (✉)
Department of Chemistry
Faculty of Science
Hiroshima University
Kagamiyama
Higashi-Hiroshima 739, Japan

Introduction

Amphiphilic molecules self-assemble in water and other solvents to form molecular aggregates with a variety of structures [1–3]. The most typical organized phases of the molecular aggregates are the lamellar phase, the hexagonal phase (normal and reversed) and the cubic phase (normal and reversed), which are liquid-crystalline phases characterized by long-range order. These amphiphilic molecules form in their isotropic solutions the aggregates lacking long-range correlations, such as spherical micelles, cylindrical micelles and bilayers. One of the most interesting classes of amphiphilic molecules is a family of nonionic surfactants consisting of a hydrophilic moiety of the oxyethylene chain and a hydrophobic moiety of the alkyl chain. There have been a great number of studies on the aggregate structures of these surfactants in water [4, 5].

In order to understand the close picture of the organized phases, conformational properties of the molecular chains in the aggregates are fundamentally important [6–8]. No definitive experimental evidence of the conformational relevance to the organized phases has been reported, possibly because of the limited experimental techniques to provide precise information of the conformational state of the molecules in these systems. Vibrational spectroscopy, which implies infrared and Raman spectroscopy, is one of the most powerful techniques for this purpose, because the spectra exhibit a number of bands characteristic of particular conformational states of the molecular chain [9]. One of the practical advantages of this method over the others is its applicability to the molecular systems in any physical states of the substance.

We have used Raman spectroscopy to study the molecular conformation of a series of nonionic surfactants α-n-alkyl-ω-hydroxyoligo(oxyethylene)s $CH_3(CH_2)_{n-1}$-$(OCH_2CH_2)_mOH$ (abbreviated as C_nE_m) with $n = 1$–16 and $m = 1$–8 in the solid state and clarified the relation between the molecular conformation and the chain length [10–13]. Vibrational spectroscopy has also been applied to aqueous solutions of C_nE_m and other related surfactants [14–19]. In most of these studies, the molecular conformation has been discussed in relation to the phase transitions in the surfactant–water systems.

In this paper, we present the experimental results of the conformational behavior of the alkyl chain in $C_{12}E_3$ in the aqueous organized phases [20] and discuss the conformational aspects of the aggregate structure. The $C_{12}E_3$–water system exhibits phase behavior as shown in Fig. 1 [4]. In the phase diagram, the liquid-crystalline lamellar phase, denoted as L_α, is observed in the concentration regions of approximately 45–80 wt% below about 50 °C. In more concentrated regions, the isotropic solution phase L_2 occurs, which is a phase of liquid surfactant containing dissolved water, not fully miscible with water. In lower-concentration regions, the separated phases of $W + L_\alpha$ and $L_2 + L_\alpha$ are observed, where W is a phase of water containing surfactant monomers.

In this work, we utilized a newly developed technique of isolated C–D stretching vibrational spectroscopy. This method using a number of selectively monodeuterated species of the surfactant is capable of determining the conformational state of each of the specified sites of the molecular chain and eventually, after examining all possible monodeuterated species, the conformational state of

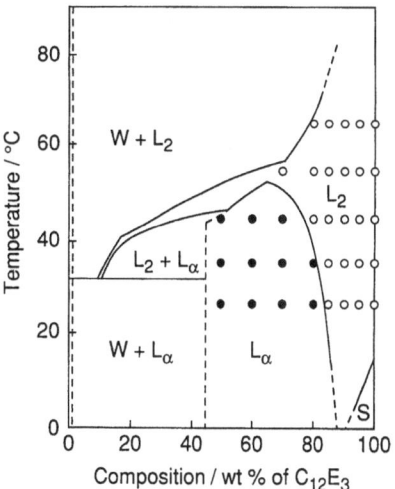

Fig. 1 Phase diagram of the $C_{12}E_3$–water system [4]. The compositions and temperatures at which the spectral measurements were performed are indicated by solid circles for the L_α phase and open circles for the L_2 phase

the chain whole. The present work is the first application of this technique to molecular aggregate systems, aiming at establishing the conformational relevance to the structure of the organized phases. The monodeuterated species of $C_{12}E_3$ we studied are $CH_3(CH_2)_{11-k}CHD(CH_2)_{k-1}$-$(OCH_2CH_2)_3OH$, where $k = 1$ (C_{12}-1-d_1-E_3), $k = 2$ (C_{12}-2-d_1-E_3), $k = 4$ (C_{12}-4-d_1-E_3), $k = 6$ (C_{12}-6-d_1-E_3) and $k = 8$ (C_{12}-8-d_1-E_3).

Method

The conformational analysis by isolated C–D stretching vibrational spectroscopy is based on the fact that the wavenumbers of the stretching vibrations of the isolated C–D bond in the selectively monodeuterated compound, in which deuterium is substituted for just one of the hydrogen atoms in the molecule, are sensitive to the conformation in the vicinity of the C–D bond [21, 22]. The conformation-dependent isolated C–D stretching wavenumbers are correlated directly with the lengths of the C–D bonds. In order to establish this correlation for the alkoxyl chain, the lengths of the C–D bonds and the wavenumbers and infrared absorption intensities of the isolated C–D stretching vibrations were calculated by the ab initio molecular orbital (MO) method on small model compounds for $C_{12}E_3$.

On the basis of the spectral interpretation of the C–D stretching bands of the model compounds, the infrared spectra of the C–D stretching vibrations of the selectively monodeuterated species of $C_{12}E_3$ were analyzed. The populations of the respective conformational states of the C–C and O–C bonds in the alkoxyl chain of $C_{12}E_3$ were evaluated from the observed C–D stretching infrared spectra. Through appropriate relations existing among the pertinent quantities, the fractions, f_{xy}, of the conformations xy for the W–X–CHD–Y–Z structure were derived, where $x, y = T$ for *trans*, G^+ for *gauche*$^+$ or G^- for *gauche*$^-$. The fractions f_x, which represent the conformational populations for the respective single bonds, were obtained from the pertinent f_{xy} values. In the discussions to be given below, the *trans* fraction f_T and the consecutive *trans* fraction f_{TT} are used. Detailed procedure of the conformational analysis by isolated C–D stretching vibrational spectroscopy has been described elsewhere [20].

Experimental and calculations

Five selectively monodeuterated species of $C_{12}E_3$ were synthesized from pertinent selectively monodeuterated 1-chloro- or 1-bromododecane and triethylene glycol by the Williamson method. The infrared spectra were measured on aqueous solutions of these monodeuterated species with compositions of 50, 60, 70, 80, 85, 90, 95 and 100 wt% at 26, 36, 45, 54 and 65 °C. The compositions and temperatures at which the spectral measurements were performed

are indicated on the phase diagram in Fig. 1. The observed spectral profiles of the C–D stretching vibrations were analyzed by fitting with resolved Lorentzian components.

Ab initio MO calculations were carried out on hexane and butyl methyl ether with the GUSSIAN 92 program [23] by the restricted Hartree–Fock method using the 6-31G** basis set.

Results and discussion

The *trans* fractions f_T for the dodecyl chain O–C_1–C_2– ··· –C_{11}–C_{12} in the $C_{12}E_3$ molecule were evaluated from the isolated C–D stretching vibrations of the five selectively monodeuterated species. The analysis of C_{12}-1-d_1-E_3 gave f_T for the O–C_1 and C_1–C_2 bonds and the analysis of C_{12}-2-d_1-E_3 gave f_T for the C_1–C_2 and C_2–C_3 bonds. Each of the spectral analyses of C_{12}-4-d_1-E_3, C_{12}-6-d_1-E_3 and C_{12}-8-d_1-E_3 gave the *trans* fraction f_T that represents average values for the two adjoining bonds C_3–C_4–C_5, C_5–C_6–C_7 and C_7–C_8–C_9 bonds, respectively. The results of the *trans* fractions f_T for the dodecyl chain in $C_{12}E_3$ are displayed in Fig. 2, showing the dependence of f_T on the composition of the $C_{12}E_3$–water system at different temperatures.

Fig. 2 Composition dependence of the *trans* fractions f_T for the respective bonds at different temperatures. Solid and open symbols are used, respectively, for the L_α and L_2 phases

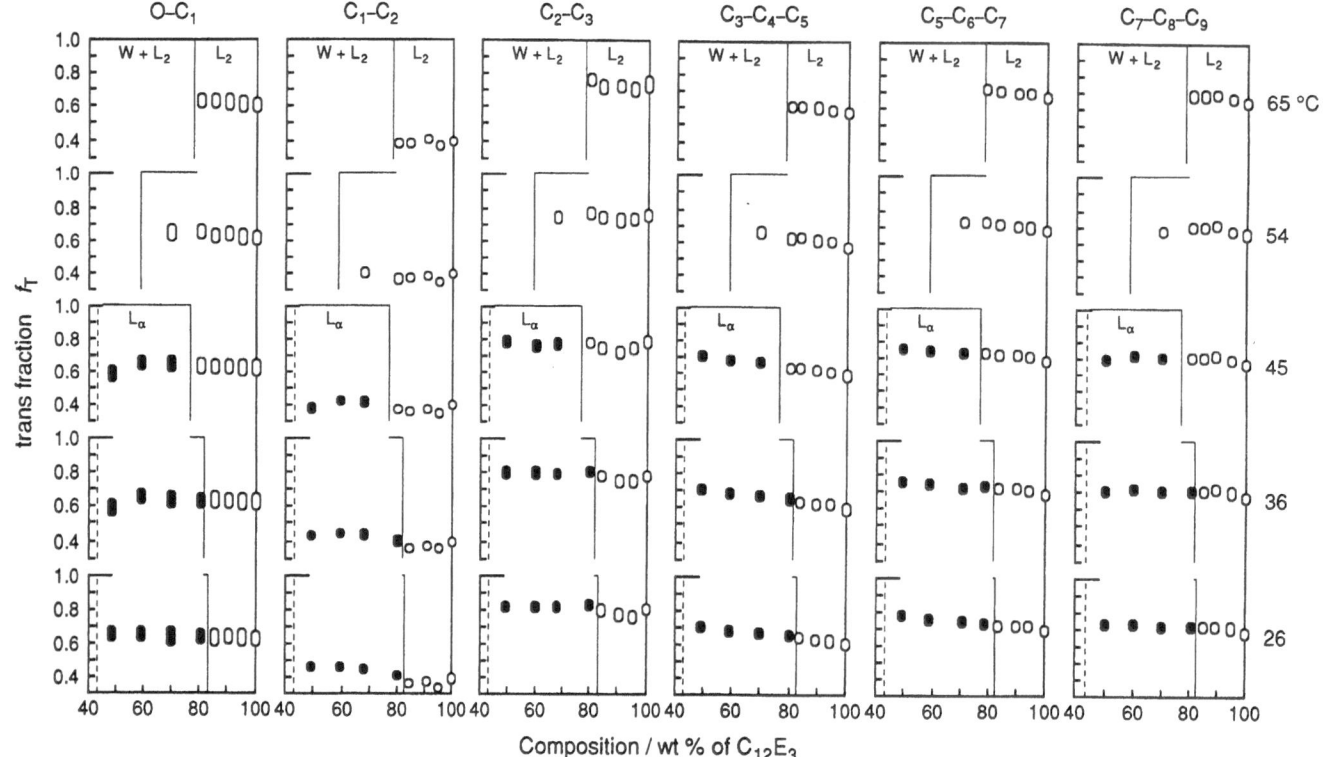

Progr Colloid Polym Sci (1997) 106:42–48
© Steinkopff Verlag 1997

Chain conformation and phase behavior

The conformational change of the dodecyl chain at the phase transition from L_2 to L_α with increasing water content is not significant and only a small increase in the *trans* fraction is observed for the C_1–C_2 and C_2–C_3 bonds at lower temperatures. This implies that only the conformational states of the C–C bonds close to the alkyl/oxyethylene interface change to be slightly more *trans*-populated, so that the alkyl chain may better be fitted into the lamellar structure. The L_2/L_α phase transition with changing temperature gives no substantial conformational changes. In addition, the conformation profiles, to be discussed later, are almost continuous across the phase boundary. It can be stated therefore that the conformational states of the alkyl chain in the L_2 and L_α phases in the vicinity of their boundary are substantially not different.

The conformational changes of the C–C and O–C bonds in the L_2 phase with increasing water content or with increasing temperature depend on the position of the bond in the chain. These characteristic conformational changes should be correlated with the formation of varying aggregate structures in this phase.

In the L_α phase, some of the bonds in the dodecyl chain exhibit notable conformational behavior with changing composition or temperature. On increasing temperature to approach the L_2 phase, the *trans* fractions for the O–C_1 and C_1–C_2 bonds decrease significantly, particularly, at lower surfactant concentrations. On the other hand, the decrease of the surfactant concentration toward the phase separation into $W + L_\alpha$ or $L_2 + L_\alpha$ gives a decrease in the *trans* fraction for these bonds. These observations indicate that the conformational transformation from *trans* to *gauche* at the bonds closer to the alkyl/oxyethylene interface makes the lamellar structure less stable and leads eventually to the phase separation or transition.

The conformational behavior of the C_7–C_8–C_9 bonds located closer to the chain terminal is also peculiar in the L_α phase in that their conformational changes with the composition and temperature are similar, though their magnitudes are smaller, to those observed for the bonds closer to the alkyl/oxyethylene interface. This implies that the conformational change from *trans* to *gauche* at the part closer to the chain terminal is also responsible for the destruction of the lamellar structure.

The molecular conformation in other C_nE_m–water systems has been studied by Gaufrès et al. [16, 17] by means of Raman spectroscopy. They have concluded on the C_8E_5–water and $C_{12}E_8$–water systems that there is no substantial discontinuity in the conformational state of the molecular chain at the L_1/H_1 and L_1/I_1 phase transitions. This conclusion is consistent with the present results for the $C_{12}E_3$–water system, although they were looking at unresolved overall conformational state of the molecular chain.

Conformation profiles and order parameters

Figure 3 shows the single *trans* conformation profiles, i.e. the plots of the *trans* fractions f_T as a function of the chain position for various compositions and temperatures and Fig. 4 shows the consecutive *trans* conformation profiles, i.e. the plots of the fractions, f_{TT}, of the consecutive *trans* conformations around the two adjoining bonds. In Fig. 3, the f_T values calculated with the rotational-isomeric state model [24] are also shown in comparison with the observed results for 100 wt%. These figures show that the conformation profiles are not significantly different between the L_α and L_2 phases. The single *trans* fraction f_T is the highest (0.7–0.8) at the C_2–C_3 bond and is the lowest (0.35–0.5) at the C_1–C_2 bond among the C–C bonds studied. The observed consecutive *trans* fractions f_{TT} in the L_α and L_2 phases show their maximum at the C_5–C_6–C_7 bonds which are located just in the middle of the dodecyl chain.

A previous study of X-ray measurements of the $C_{12}E_3$–water and $C_{12}E_4$–water systems in the L_α phase has indicated that these surfactants give half-bilayer thicknesses considerably less than expected from the all-*trans* dodecyl chain [25]. This observation is consistent with the present finding that the dodecyl chain in this phase contains a considerable amount of the *gauche* C–C bonds. These experimental results demonstrate that the alkyl chains in the lamellar structure are significantly disordered and are conformationally almost liquid like.

Since the disorder of the alkyl chain originates from the conformational transformation from the *trans* state to the *gauche* state around the C–C bonds, the population of the consecutive *trans* conformation is correlated more closely to the order of the chain than the population of the single *trans* conformation. It may be reasonable then to use the f_{TT} values to represent the conformational order of the alkyl chain. An important finding is that, although the long-range order in the system should be greatly different between the L_α and L_2 phases, the consecutive *trans* conformation profiles are essentially the same in the two phases and there is no distinct discontinuity of the profiles at the phase transition.

Figure 4 shows that the conformational order in the region of the L_α phase closer to the phase separation into $W + L_\alpha$ or $L_2 + L_\alpha$ decreases more rapidly toward the chain terminal. This is the same observation as made in the composition dependence of f_T (Fig. 2), indicating that the conformational change from *trans* to *gauche* near

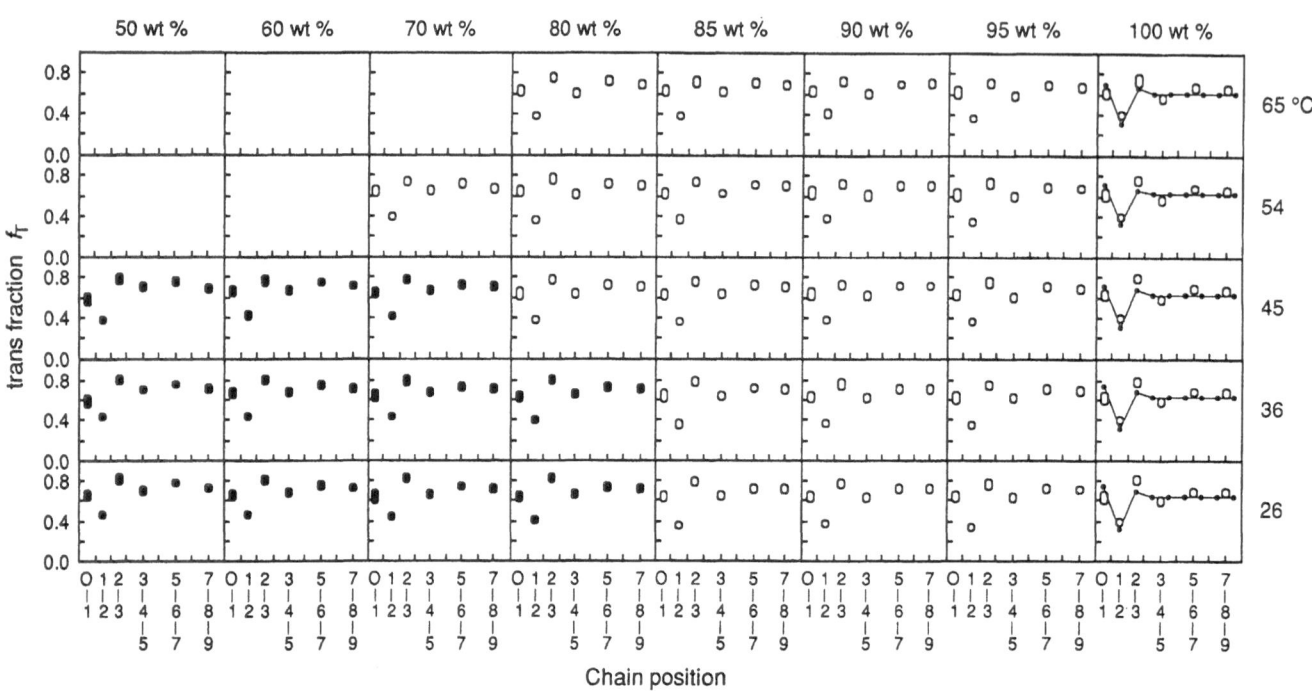

Fig. 3 Positional dependence of the *trans* fractions f_T at different compositions and temperatures. Solid and open symbols are used, respectively, for the L_α and L_2 phases. Carbon atoms are indicated by their numbers. The f_T values calculated with the rotational-isomeric state model [24] are shown by dots in comparison with the observed results for 100 wt%

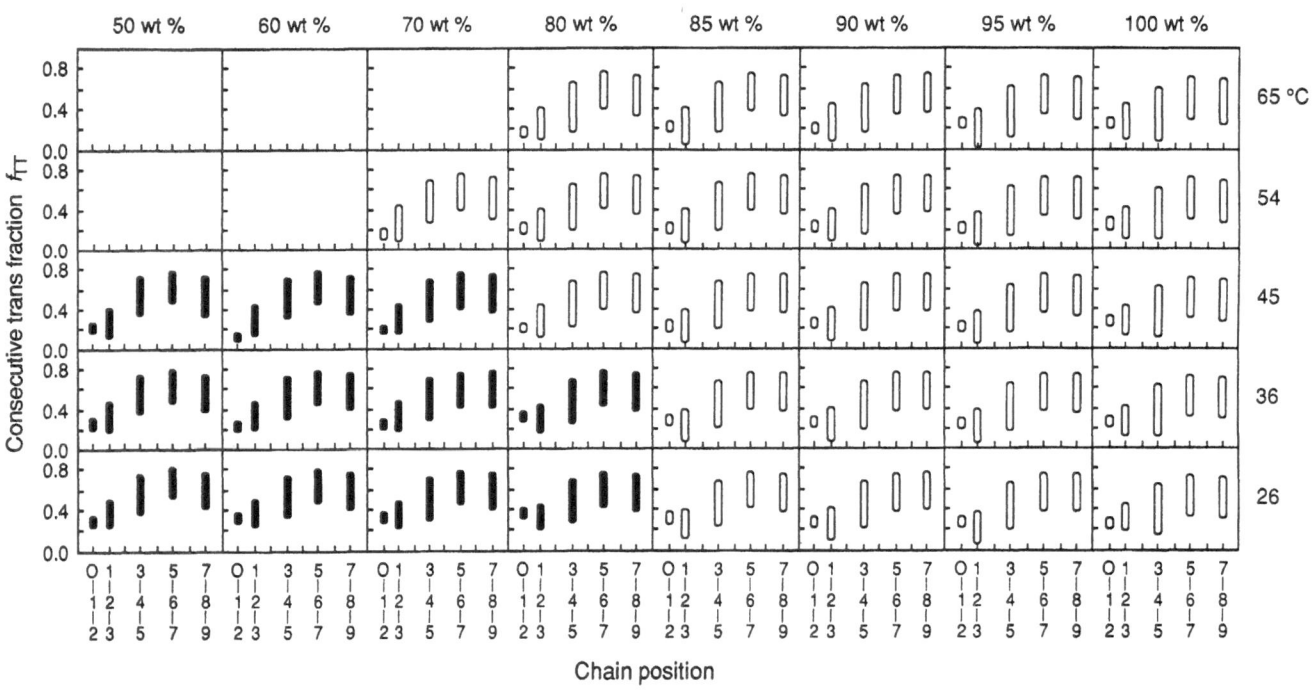

Fig. 4 Positional dependence of the consecutive *trans* fractions f_{TT} at different compositions and temperatures. Solid and open symbols are used, respectively, for the L_α and L_2 phases. Carbon atoms are indicated by their numbers

the chain terminal is responsible for the unstability of the lamellar structure. Another important finding is that the consecutive *trans* conformation profiles show the highest conformational order in the middle of the alkyl chain in the L_α and L_2 phases.

The conformational order determined by infrared spectroscopy may be compared with the order parameters from NMR spectroscopy. Ward et al. [26] and Schnepp and Schmidt [27] have obtained the order parameters from the quadrupolar splittings in the ^2H NMR spectra for the L_α phase of the $C_{12}E_4$–water system and for the same phase of the $C_{12}E_6C_1$–water system, respectively. Their results indicate that the maximum ordering of the alkyl chain occurs at the C_3 to C_5 positions for $C_{12}E_4$ and at the C_2 to C_4 positions for $C_{12}E_6C_1$ and that the order decreases progressively toward the chain terminal.

Ahlnäs et al. [28], on the other hand, have studied ^{13}C NMR relaxation of $C_{12}E_4$ in the neat liquid and of $C_{12}E_5$ in the neat liquid and in the L_1 phase and determined the order parameters in these phases. They obtained the results that the order is maximum at the C_3 to C_6 positions for either of the neat liquid and the L_1 phase. The very low order was again observed at the end of the alkyl chain.

The present results of the consecutive *trans* conformation profiles in the L_α and L_2 phases are consistent with the order parameter profiles obtained from the NMR spectra. The important observation on the conformational behavior and the order parameters for the C_nE_m–water systems is that the chain order is the highest in the middle of the alkyl chain but not at the hydrophobic/hydrophilic interface as observed for the L_α phase of ionic surfactant systems [29]. For the nonionic surfactant systems, the observed behavior of the chain conformation and ordering is consistent with the flexible structure of the alkyl/oxyethylene inferface.

Conclusion

The present analysis by C–D stretching infrared spectroscopy has clarified several conformational properties of the alkyl chain in the $C_{12}E_3$–water system. The conformational change at the phase transition from L_2 to L_α is not significant and only a small increase in the *trans* fraction is observed for the C–C bonds close to the alkyl/oxyethylene interface. This implies that the conformational states of the alkyl chain in the L_2 and L_α phases in the vicinity of their boundary are substantially not different. In the L_α phase, when the composition or the temperature approaches the region of the phase separation or transition, the *trans* fractions for the C–C bonds closer to the alkyl/oxyethylene interface and those closer to the chain terminal decrease. These observations indicate that the conformational transformation from *trans* to *gauche* at these chain positions makes the lamellar structure less stable and leads eventually to the structural destruction.

The conformational order as evaluated from the fractions of the consecutive *trans* conformation is the highest in the middle of the chain in the L_α and L_2 phases. This vibrational spectroscopic observation, together with the NMR observations, indicates that the alkyl/oxyethylene interface is flexible with respect to the conformation and the orientation of the chain. The flexibility at the interface is probably one of the important factors to characterize the aggregate structures of the C_nE_m–water systems.

References

1. Tanford C (1980) The Hydrophobic Effect: Formation of Micelles and Biological Membranes, 2nd ed. Wiley, New York
2. Degiorgio V, Corti M (eds) (1985) Physics of Amphiphiles: Micelles, Vesicles and Microemulsions. North-Holland, Amsterdam
3. Tiddy GJT (1985) In: Eicke HF (ed) Modern Trends of Colloid Science in Chemistry and Biology. Birkhäuser, Basel, pp 148–183
4. Mitchell DJ, Tiddy GJT, Waring L, Bostock T, McDonald MP (1983) J Chem Soc Faraday Trans 1 79:975–1000
5. Degiorgio V (1985) In: Degiorgio V, Corti M (eds) Physics of Amphiphiles: Micelles, Vesicles and Microemulsions. North-Holland, Amsterdam, pp 303–335
6. Tiddy GJT (1980) Phys Rep 57:1–46
7. Gruen DWR (1985) Progr Colloid Polym Sci 70:6–16
8. Ben-Shaul A, Gelbart WM (1985) Annu Rev Phys Chem 36:179–211
9. Painter PC, Coleman MM, Koenig JL (1982) The Theory of Vibrational Spectroscopy and Its Application to Polymeric Materials. Wiley, New York
10. Matsuura H, Fukuhara K (1987) J Phys Chem 91:6139–6148
11. Matsuura H, Fukuhara K, Masatoki S, Sakakibara M (1991) J Am Chem Soc 113:1193–1202
12. Matsuura H (1991) Trends Phys Chem 1:89–109
13. Masatoki S, Fukuhara K, Matsuura H (1993) J Chem Soc Faraday Trans 89:4079–4084
14. Kalyanasundaram K, Thomas JK (1976) J Phys Chem 80:1462–1473
15. Cooney RP, Barraclough CG, Healy TW (1983) J Phys Chem 87:1868–1873
16. Gaufrès R, Bribes JL, Sportouch S, Ammour J, Maillols J (1988) J Raman Spectrosc 19:149–153
17. Gaufrès R, Sportouch S, Ammour JE, Maillols J (1990) J Phys Chem 94:4635–4639
18. Matsuura H, Fukuhara K, Takashima K, Sakakibara M (1991) J Phys Chem 95:10800–10810
19. Nickolov ZS, Earnshaw JC (1995) J Mol Struct 348:273–276
20. Masatoki S, Ohno K, Yoshida H, Matsuura H (1996) J Phys Chem 100:8487–8498

21. Ohno K, Takagi Y, Matsuura H (1993) J Phys Chem 97:5530–5534
22. Ohno K, Abe H, Masatoki S, Yoshida H, Matsuura H (1996) J Phys Chem 100:12674–12679
23. Frisch MJ, Trucks GW, Head-Gordon M, Gill PMW, Wong MW, Foresman JB, Johnson BG, Schlegel HB, Robb MA, Replogle ES, Gomperts R, Andres JL, Raghavachari K, Binkley JS, Gonzalez C, Martin RL, Fox DJ, Defrees DJ, Baker J, Stewart JJP, Pople JA (1992) GAUSSIAN 92, Revision F.3. Gaussian Inc, Pittsburgh, PA
24. Flory PJ (1989) Statistical Mechanics of Chain Molecules. Hanser, Munich
25. Carvell M, Hall DG, Lyle IG, Tiddy GJT (1986) Faraday Discuss Chem Soc 81:223–237
26. Ward AJI, Ku H, Phillippi MA, Marie C (1988) Mol Cryst Liq Cryst 154:55–60
27. Schnepp W, Schmidt C (1993) Ber Bunsen-Ges Phys Chem 97:1399–1402
28. Ahlnäs T, Karlström G, Lindman B (1987) J Phys Chem 91:4030–4036
29. Mely B, Charvolin J, Keller P (1975) Chem Phys Lipids 15:161–173

Progr Colloid Polym Sci (1997) 106:49–51
© Steinkopff Verlag 1997

K. Shigeta
M. Suzuki
H. Kunieda

Phase behavior of polyoxyethylene oleyl ether in water

Received: 13 October 1996
Accepted: 6 March 1997

K. Shigeta · Dr. H. Kunieda (✉)
Division of Artificial Environment Systems
Graduate School of Engineering
Yokohama National University
Tokiwadai 79-5
Hodogaya-ku, Yokohama 240, Japan

M. Suzuki
Oleochemical Research Laboratory
NOF Corporation
Ohama-cho 1-56
Amagasaki 660, Japan

Abstract Phase behavior of penta-, deca- and eicosa-oxyethylene oleyl ether in water were investigated. Differently from ordinary oleyl surfactants, the lipophilic chains of the present surfactants consist of pure oleyl group. The basic feature of the phase behavior extremely resembles that of pure polyoxyethylene hexadecyl ether systems although the solid present regions in the oleyl surfactant systems are very narrow due to the low melting point of oleyl group.

Key words Polyoxyethylene oleyl ether – liquid crystal – phase behavior of nonionic surfactant – small-angle X-ray scattering

Introduction

Various kinds of self-organizing structures are formed in polyoxyethylene-type nonionic surfactant/water systems depending on temperature, surfactant concentration, surfactant structures, etc. [1, 2]. In previous studies on the basic phase behavior of nonionic surfactants, dodecyl chain surfactants have been mainly used. In longer chain nonionic surfactant systems, more phases appear and they are more important for practical applications, although the phase behavior have not been extensively studied. Polyoxyethylene oleyl ethers are biocompatible surfactants used for cosmetics, medicines, etc. The purity of commercially available oleic acid (cis-9-octadecenoic acid) is usually around 60% and the rest contains more than 50 kinds of other fatty acids [3]. Since conventional commercial polyoxyethylene oleyl ethers are synthesized from the low-purity oleic acid, it is difficult to figure out the basic phase behavior of the nonionic surfactants in water.

Recently, polyoxyethylene oleyl ethers were made of high-purity oleyl alcohol (99.7%) synthesized from high-purity oleic acid [3]. In this paper, the effect of hydrophilic chain length on the phase behavior and the structures of liquid crystals in polyoxyethylene oleyl ether/water system were investigated by means of phase study and small-angle X-ray scattering (SAXS).

Materials

Polyoxyethylene oleyl ethers, POE(5), POE(10) and POE(20) were kindly supplied by NOF CO., LTD. POE(5), POE(10), and POE(20) were synthesized from high-purity oleyl alcohol (99.7%). The mean number of oxyethylene unit, which was determined by analyzing end hydroxyl group [4], are 5.1 for POE(5), 10.7 for POE(10) and 19.2 for POE(20), respectively. The trace of oleyl alcohol was detected by high-performance liquid chromatography, but the distribution of number of oxyethylene unit(n) in POE(n) is narrow. Doubly distilled water was used.

Methods

Phase diagram

Various amounts of water and surfactant were sealed in ampoules. A series of sample solutions were heated at about 1 °C/min in a thermostat. If necessary, the samples were kept at constant temperature to observe phase separation. The phase change was detected by direct visual observation with polarizers. The type of liquid crystals were determined by polarizing microscopy and small-angle X-ray scattering.

Small-angle, X-ray scattering (SAXS)

Interlayer spacing was measured using SAXS, performed on a small-angle scattering goniometer with an 18 kW Rigaku Denki rotating anode goniometer (RINT-2500) at about 25 °C. The samples of liquid crystals were lapped by polyethylene terephthalate film (Mylar seal method), because the samples were very viscous and X-ray capillary cannot be used as a sample cell.

Results and discussion

The lipophilic group of polyoxyethylene oleyl ethers (POEs) used in the present study is almost pure oleyl group because POEs were synthesized from high-purity oleyl alcohol (99.7%). On the other hand, their oxyethylene chains have distribution. If a system contains a large amount of oil, one has to consider the effect of the hydrophilic-chain distribution on the phase behavior, because the solubility of each surfactant of the mixture in oil is largely different [5]. However, critical micelle concentrations of POEs are considered to be very low [6], the effect of the distribution of the hydrophilic chains on the phase behavior may be small except in a very dilute region.

The phase diagrams of POE(5)/water, POE(10)/water and POE(20)/water systems are shown in Figs. 1–3. Hexagonal and lamellar liquid crystalline phases were distinguished by polarizing microscopy and SAXS peaks. The ratios of the interlayer spacings from first, second and third peaks are $1 : 1/\sqrt{3} : 1/2$ for hexagonal type (H_1) and $1 : \frac{1}{2} : \frac{1}{3}$ for lamellar type (L_α), respectively [7]. Judging from analogy of polyoxyethylene hexadecyl ether systems, D_2, I_1, V_1 and V_2 phases are considered as sponge phase, discontinuous cubic phase, bicontinuous cubic phase, reverse bicontinuous cubic phase, respectively.

In the POE(5)/water system, L_α was observed as liquid crystal. The liquid crystal intrudes into the cloud point

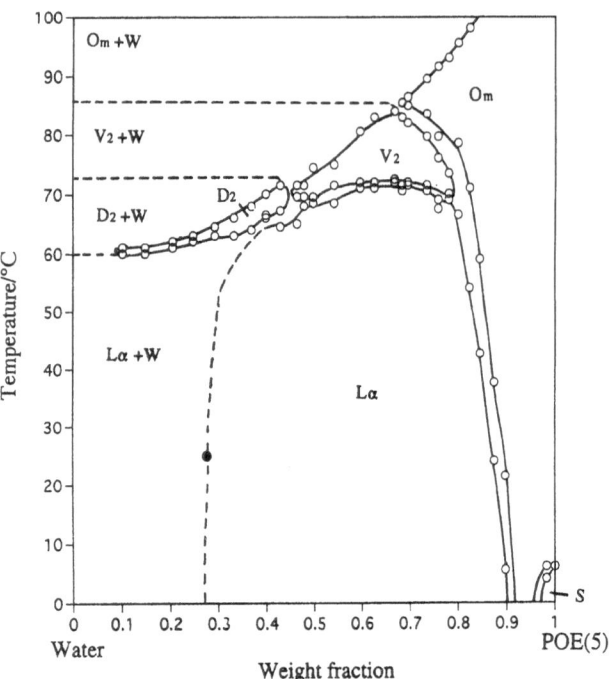

Fig. 1 Phase diagram of water/POE(5) system as a function of temperature. Wm: aqueous micellar solution phase, Om: reverse micellar solution phase, L_α: lamellar liquid crystalline phase, V_2: reverse bicontinuous cubic phase, D_2: isotropic bicontinuous phase, W: excess water phase, S: solid-present region. The filled circle was determined by SAXS

curve and an isotropic fluid phase (D_2) appears. This D_2 phase is considered to be so-called sponge phase, often denoted by L_3 phase [1, 2, 8].

Although it is not easy to determine the phase boundary of single L_α phase in the dilute region, the boundary (filled circle) was determined by SAXS where the interlayer spacing becomes constant. The viscous isotropic phase (V_2) appears above the L_α phase. This phase is considered to be a bicontinuous cubic phase because of the similarity of the phase behavior to a tetraoxyethylene hexadecyl ether ($C_{16}EO_4$)/water system [1]. In the $C_{16}EO_4$ system, however, the Krafft temperature is observed.

In the POE(10)/water system, the cloud temperature is 75.5 °C. The type of liquid crystals changes from discontinuous cubic phase (I_1) to L_α phase via H_1 phase and bicontinuous cubic phase (V_1) by increasing the weight fraction of POE(10). The general feature of the POE(10) phase diagram resembles that of the $C_{16}EO_8$ system [1]. In the POE(20)/water system, a cloud point was not observed at least until 100 °C. The type of liquid crystals changes from I_1 phase to H_1 phase with increasing weight fraction of POE(20). The SAXS peak ratio of the I_1 phase was $1 : 1/\sqrt{2} : 1/\sqrt{3}$ and the structure is considered to be simple or body-centered cubic phase [9].

Progr Colloid Polym Sci (1997) 106:49–51
© Steinkopff Verlag 1997

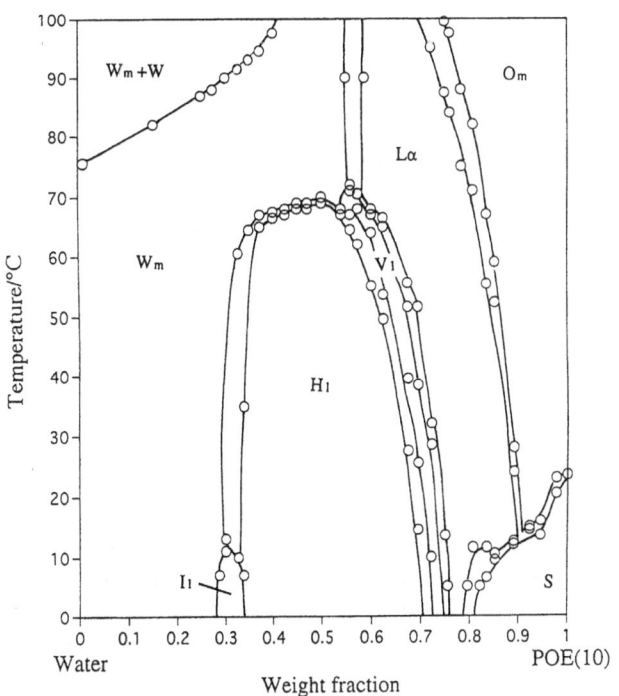

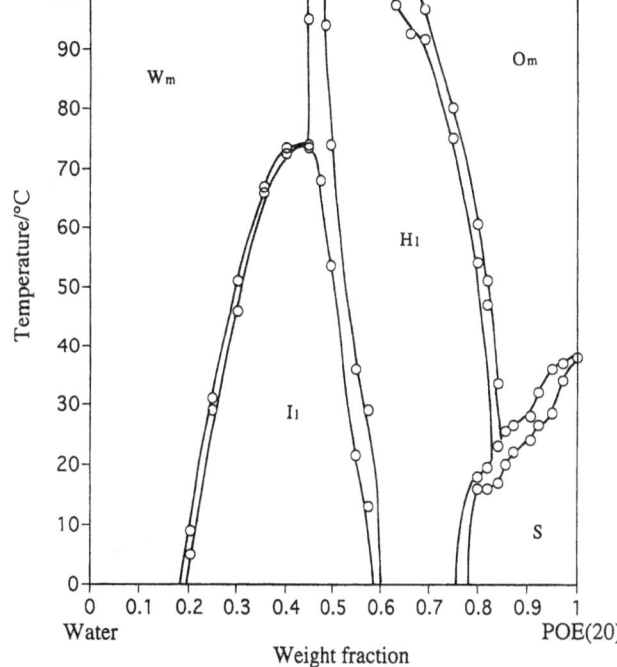

Fig. 2 Phase diagram of water/POE(10) system as a function of temperature. H_1: hexagonal liquid crystalline phase, I_1: discontinuous cubic phase. V_1: bicontinuous cubic phase

Fig. 3 Phase diagram of water/POE(20) system as a function of temperature

The feature of the series of the present phase diagrams in the POE/water systems is similar to those in the $C_{16}EO_n$/water systems [1], because the phase diagrams of POE(5) and POE(10) systems are extremely similar to those of $C_{16}EO_4$ and $C_{16}EO_8$ systems, respectively. Since the carbon number of the oleyl group is 18, the lipophilic nature of the oleyl group is not very much different from that of C_{18} straight hydrocarbon chain. On the other hand, the solid-present region is very small and the Krafft point phenomenon was not observed even in a short hydrophilic-chain POE system because the melting temperature of POE are rather low due to the double bond in the lipophilic chain.

References

1. Mitchel DJ, Tiddy GJT, Waring L, Bostock T, McDonald MP (1983) J Chem Soc Faraday Trans I 79:975
2. Laughlin RG (1994) The Aqueous Phase Behavior of Surfactants, Ch. 8, Academic Press, London
3. Suzuki M (1995) Fragrance J 4:105 (in Japanese)
4. Standard Methods for Analysis of Fats, Oils and Related Materials (1996) J Oil Chem Soc No. 2.3.6.2
5. Kunieda H, Yamagata M (1993) Langmuir 9:3345
6. Rosen M (1989) Surfactants and Interfacial Phenomena, Ch. 3, Wiley, New York
7. Fontell K (1974) In: Gray GW, Winsor PA (eds) Liquid Crystals and Plastic Crystals. Wiley, London
8. Anderson D, Wennerstroem H, Olsson U (1989) J Phys Chem 93:4243
9. Fontell K, Mandell L, Ekwall P (1968) Acta Chemica Scandinavica 22:3209

Progr Colloid Polym Sci (1997) 106:52–56
© Steinkopff Verlag 1997

T. Imae
M.-P. Krafft
F. Giulieri
T. Matsumoto
T. Tada

Fibril–vesicle transition and their structures – investigation by microscopy and small-angle scattering

Received: 13 October 1996
Accepted: 24 March 1997

Dr. T. Imae (⊠)
Department of Chemistry
Faculty of Science
Nagoya University
Nagoya 464, Japan

M.-P. Krafft · F. Giulieri
Laboratoire de Chimie Moléculaire
Unité Associée au CNRS n° 426
Université de Nice-Sophia Antipolis
06108 Nice Cedex 2, France

T. Matsumoto · T. Tada
Division of Material Chemistry
Faculty of Engineering
Kyoto University
Kyoto 606, Japan

Abstract The formation of fibrous assemblies by fluorinated amphiphiles and the temperature-dependent fibril–vesicle transition were investigated by using microscopy and small-angle scattering. The results were compared with that of fibrous assemblies by nonfluorinated amphiphiles. Sodium (F-octylethyl) nonylmethylene O-6-phosphoglucoside (F-Glu) formed hollow tubules in water at room temperature ($\sim 25\,°C$). Tubules transformed into vesicles above $60\,°C$. Similar tubule–vesicle transition for hydrogenated-analog, sodium decylnonylmethylene O-6-phosphoglucoside (H-Glu), was observed below room temperature. The mixture of [2-(F-octyl)penthyl] dimorpholinophosphate (F8C5DMP) and [2-(F-decyl)ethyl] dimorpholinophosphate (F10C2DMP) in water constructed U- or V-shape assemblies, which changed to vesicles at $60\,°C$.

Key words Vesicle – fibril – cryo-TEM – small-angle X-ray scattering – fluorinated amphiphile

Introduction

It has been known that various types of supramolecular assemblies are constructed in the medium by noncovalent self-organization of small molecules. Many workers reported assembly structures and solution properties [1]. Rod-like micelles which are one group of linearly extended supramolecular assemblies are formed by the hydrophobic interaction between long alkyl chains of amphiphiles. Rod-like micelles are transformed from spherical micelles [2]. In a few cases, rod-like micelles change to vesicles when cinnamic acid is added to a solution of hexadecyl-dimethylamine oxide [3]. The spontaneous transition from polymer-like micelles to vesicles was observed in lecithin–bile salt solutions [4].

Fibrous chains are another group of linear assemblies and were found in gel-like solutions of amphiphiles [5–7], steroids [8], oligobipiridine metal complexes [9], and cyclic peptide [10]. These compounds reveal specific intermolecular interaction such as hydrogen bonding, π–π stacking, and so on, besides hydrophobic interaction. Among supramolecular fibers, helical assemblies are characteristic for molecules with chiral hydrophilic head groups such as amino acids, nucleosides, and sugars [1]. It is also known that some of the fibrous assembly systems display fibril–vesicle transition depending on temperature [11–13].

Progr Colloid Polym Sci (1997) 106: 52–56
© Steinkopff Verlag 1997

The formation of supramolecular fibers were mainly confirmed by the transmission electron microscopic (TEM) and the dark-field optical microscopic observation. On the other hand, a few investigations were carried out using techniques of NMR [6], X-ray diffraction [9, 14, 15], small-angle neutron scattering [8, 16, 17], and small-angle X-ray scattering (SAXS) [16, 17].

In this paper, we examined the formation of different kinds of fibrous, linear assemblies from fluorinated amphiphiles by using microscopy and small-angle scattering. We discuss the structure of fibrous chains and the fibril–vesicle transition.

Experimental section

Sodium (F-octylethyl)nonylmethylene *O*-6-phosphoglucoside (F-Glu), sodium decylnonylmethylene *O*-6-phosphoglucoside (H-Glu), [2-(F-octyl)penthyl]dimorpholinophosphate (F8C5DMP), and [2-(F-decyl)ethyl] dimorpholinophosphate (F10C2DMP) were same samples as previously synthesized and used [18–20].

SAXS was measured at 25 °C by a 6 m point focussing SAXS camera at the High-Intensity X-ray Laboratory in Kyoto University. TEM observation was carried out on a Hitachi H-800 electron microscope. A Hitachi H5001-C cold stage was used for the cryo method. Freeze–fracture replica film was prepared by using a balzers BAF 400 freeze–fracture device.

Results and discussion

Figure 1 shows cryo-TEM photographs for aqueous 3% solution of hydrogenated/fluorinated phosphoglucolipid, F-Glu, at various temperatures. F-Glu in water formed very long tubules with hollow inside at room temperature (~25 °C) (Fig. 1a). The diameters were almost uniform. SAXS data (Fig. 2a) fitted to the theoretical curve for a hollow tubule with parameters of external and internal diameters, 280 and 70 Å, respectively. There was a Bragg peak which can be assigned to the reflection from lamellar layers, suggesting the formation of concentric lamellar layers in the cross section of a hollow tubule.

When aqueous F-Glu solution was heated up to more than 60 °C, tubules changed to vesicles (Fig. 1c). The tubule–vesicle transition temperature must be 50–60 °C, because tubules and vesicles coexisted at that temperature (Fig. 1b). It should be noticed that tubules transferred to vesicles through plate (sheet) or tubule expansion. In accord with tubule–vesicle transition, the characteristic Bragg peak in SAXS diminished, as seen in Fig. 2b, indic-

(a)

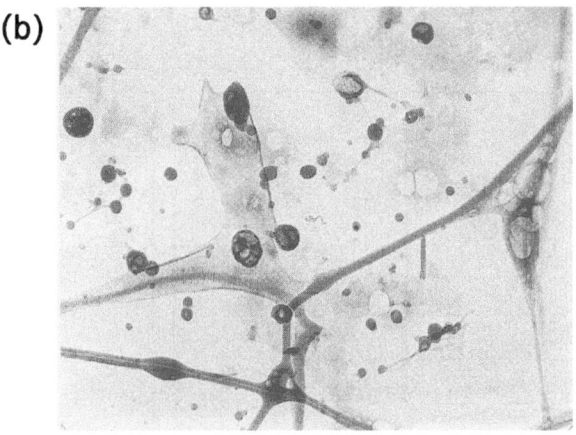

(b)

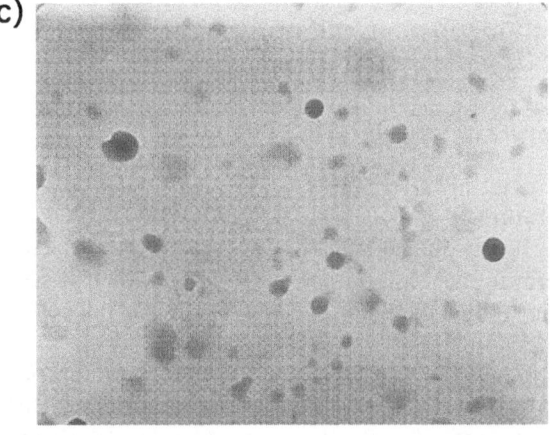

(c)

100nm

Fig. 1 Cryo-TEM photographs of an aqueous 3% F-Glu solution at various temperatures: (a) ~25 °C; (b) 50–60 °C; (c) 60–70 °C

ating the disappearance of multilamellar (concentric lamellar) arrangement of bilayers in vesicles.

Hydrogenated phosphoglucolipid, H-Glu, in water constructed unilamellar vesicles and tubules at room

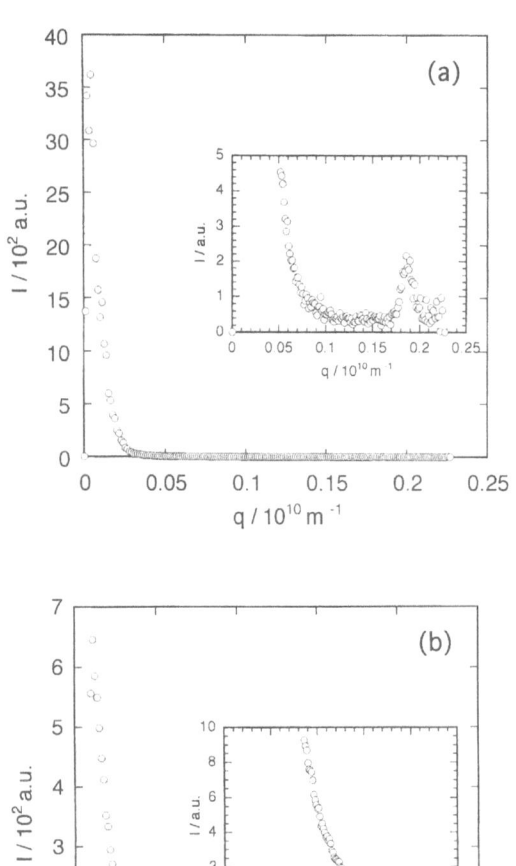

Fig. 2 SAXS data for an aqueous 3% F-Glu solution at different temperatures: (a) 25 °C; (b) 60 °C

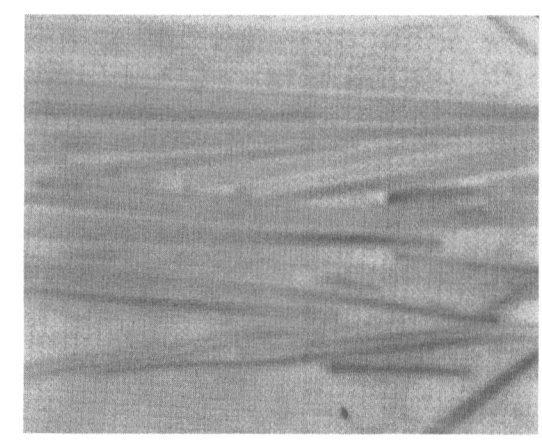

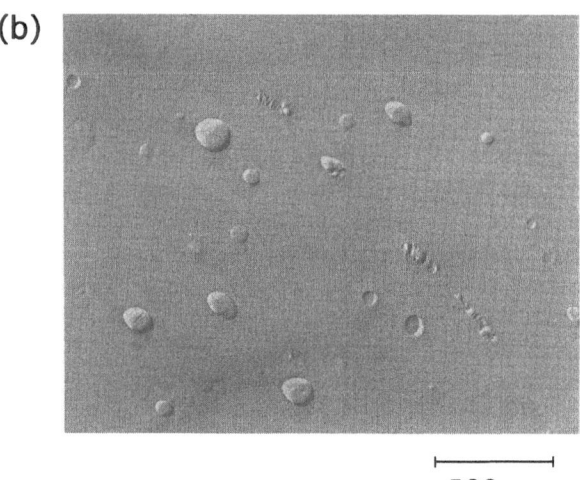

Fig. 3 TEM photographs of an aqueous 3% H-Glu solution at different temperatures: (a) ~4 °C; cryo-TEM; (b) ~25 °C; freeze–fracture TEM

temperature (~25 °C) and ~4 °C, respectively, as seen in Fig. 3. It may be noticed that tubule–vesicle transition temperature was low in comparison with that of F-Glu and that tubule diameter of 300–600 nm was thicker than that of F-Glu.

Although F8C5DMP molecules can form tubules, tubules transformed into a lamellar sheet structure within a short time. On the other hand, the mixture of F8C5DMP with F10C2DMP constructed more stable tubules with V- or U-shape at room temperature (~25 °C) (Fig. 4a). The structure was assumed to be cigar-like roll of a multilamellar sheet. Then a Bragg peak, which is seen in Fig. 5a, may correspond to the reflection from lamellar layers. The Bragg peak disappeared (Fig. 5b), when a solution was heated at 60 °C, suggesting the disappearance of multilamellar structure in molecular assembly. It was supported

by TEM (Fig. 4b) that vesicles without multilamellar layers existed at 60 °C, and the tubule–vesicle transition had to be at 25–40 °C.

These results can be compared with the fibers of single-chain amphiphiles with amino acid head groups, N-dodecanoyl-β-alanine (C_{12}Ala) and N-acyl-L-aspartic acids (C_nAsp). C_{12}Ala molecules in water at pH below 6 were associated into nonhelical, cylindrical fibers without hole with circular cross section of 360–380 Å diameter at room temperature [7]. It was confirmed from the characteristic Bragg spacing in SAXS spectrum that cylindrical C_{12}Ala fibers consisted of concentric multilamellar layers [17]. The cylinder–vesicle transition was also observed for molecular assemblies of C_{12}Ala [13].

On the other hand, C_nAsp in water of medium pH at lower temperature formed fibrous molecular assemblies, which were the double strand of helical bilayer strands

(a)

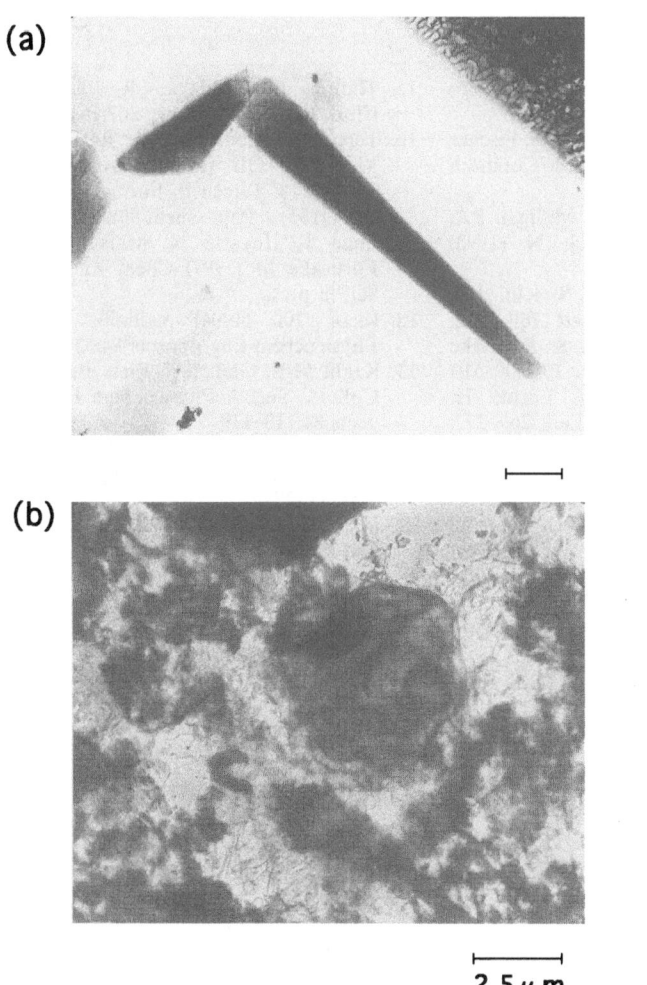

(b)

2.5 μm

Fig. 4 Cryo-TEM photographs of an aqueous 3% solution of F8C5DMP and F10C2DMP mixture at different temperatures: (a) ~25 °C; (b) ~60 °C

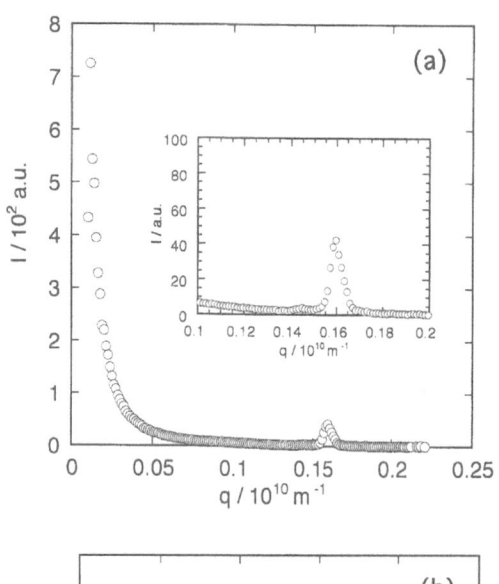

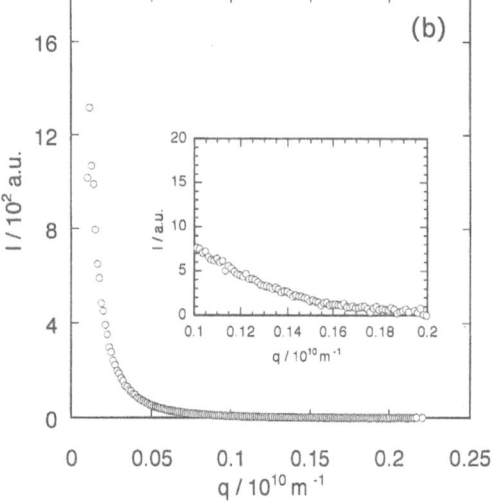

Fig. 5 SAXS data for an aqueous 3% solution of F8C5DMP and F10C2DMP mixture at different temperatures. (a) 25 °C; (b) 60 °C

with unit chain of a 50–60 Å diameter [7, 17]. Unit chains, where linear bilayers twist, formed a double strand with helical sense of ~650 Å pitch and with the distance between strands of ~230 Å. Vesicles were constructed, when fibrous solutions of C_nAsp were heated up above Krafft temperatures, which increased with alkyl chain length from 12 °C of $C_{12}Asp$ to 52 °C of $C_{18}Asp$.

The systems described above formed linearly extended, fibrous self-assemblies, although the morphological structures depended on amphiphiles. The molecular arrangement in fibers were maintained by the formation of concentric multilamellar bilayer, besides hydrophobic interaction and hydrogen bonding, except C_nAsp fibers with chiral character. The assemblies underwent temperature-dependent fibril–vesicle transition. The fibril–vesicle transition resulted from the rearrangement of bilayers. Generally, multilamellar layers were diminished in ves-

icles. Transition temperature was higher for fluorinated amphiphile, F-Glu, than for hydrogenated amphiphile, H-Glu. The packing parameter defining the structure of self-assemblies [21] was lower for F-Glu than for H-Glu, supported by the larger surface curvature of F-Glu hollow tubules than of H-Glu vesicles, when compared at room temperature. The transition temperature was higher for F-Glu with double alkyl chains than for the mixture of F8C5DMP and F10C2DMP with a single alkyl chain, suggesting higher Krafft point of amphiphiles with double alkyl chains.

Acknowledgments This work was supported by the Cosmetology Research Foundation (No. JA-95-10). Authors are grateful to Mr. T. Iwamoto and Mr. K. Funayama of Nagoya University for their help in TEM measurement and SAXS analysis.

References

1. Fuhrhop JH, Koning J (1994) Membranes and Molecular Assemblies: The Synkinetic Approach. Roy Soc Chem, Cambridge, UK
2. Imae T, Ikeda S (1989) In: Mittal KL (ed) Surfactants in Solution. Vol. 7. Plenum Press, New York, pp 455–472
3. Imae T, Kakitani M, Kato M, Furusaka M (1996) J Phys Chem 100: 20051–20055
4. Egelhaaf SU, Schurtenberger P (1994) 98:8560–8573; Pederson JS, Egelhaaf SU, Schurtenberger P (1995) 99: 1299–1305
5. Tachibana T, Kambara H (1965) J Am Chem Soc 87:3015–3016
6. Fuhrhop JH, Schnieder P, Rosenberg J, Boekema E (1987) J Amer Chem Soc 109:3387–3390; Fuhrhop JH, Svenson S, Boettcher C, Rossler E, Vieth HM (1990) J Amer Chem Soc 112:4307–4312
7. Imae T, Takahashi Y, Muramatsu H (1992) J Amer Chem Soc 114:3414–3419
8. Terech P, Volino F, Ramasseul R (1985) J Phys 46:895–903
9. Lehn JM, Mascal M, DeCian A, Fischer J (1990) J Chem Soc, Chem Commun 479–481
10. Ghadiri MR, Granja JR, Milligan RA, Mcree DE, Khazanovich N (1993) Nature 366:324–327
11. Nakashima N, Asakuma S, Kim JM, Kunitake T (1984) Chem Lett 1709–1712; Nakashima N, Asakuma S, Kunitake T (1985) J Amer Chem Soc 107:509–510
12. Yanagawa H, Ogawa Y, Furuta H, Tsuno K (1988) Chem Lett 269–272; (1989) J Amer Chem Soc 111:4567–4570
13. Imae T, Kidoaki S (1995) J Jpn Oil Chem Soc 44:301–308
14. Hanabusa K, Miki T, Taguchi Y, Koyama T, Shirai H (1993) J Chem Soc, Chem Commun 1382–1384; Hanabusa K, Matsumoto Y, Miki T, Koyama T, Shirai H (1994) J Chem Soc, Chem Commun 1401–1402
15. Thomas BN, Safinya CR, Plano RJ, Clark HA (1995) Science 267:1635–1638
16. Terech P, Rodriguez V, Barnes JD, McKenna GB (1994) Langmuir 10: 3406–3418; Terech P, Furman I, Weiss RG (1995) J Phys Chem 99:9558–9566
17. Imae T, Hayashi N, Matsumoto T, Furusaka M (1997) Chem Eng Symp Ser, in press
18. Riess JG (1994) Colloids Surf A Physicochem Eng Aspects 84:33–48
19. Krafft M-P, Giulieri F, Riess JG (1994) Colloids Surf A Physicochem Eng Aspects 84:113–119
20. Krafft M-P, Giulieri F (1994) Colloids Surf A Physicochem Eng Aspects 84:121–127
21. Israelachvili JN (1985) Intermolecular and Surface Forces. Academic Press, London

Progr Colloid Polym Sci (1997) 106:57–60
© Steinkopff Verlag 1997

T. Kato
N. Taguchi
D. Nozu

Structure and dynamics of networks formed in concentrated solutions of nonionic surfactants

Received: 13 October 1996
Accepted: 7 March 1997

Dr. T. Kato (✉) · N. Taguchi · D. Nozu
Department of Chemistry
Faculty of Science
Tokyo Metropolitan University
Minamiohsawa, Hachioji
Tokyo 192-03, Japan

Abstract Dynamic light scattering has been measured on semidilute solutions of nonionic surfactant ($C_{16}E_7$) where formation of networks of worm-like micelles is suggested from our previous studies. In addition to the diffusion mode, we have found the so-called slow mode in time correlation function of the scattered field. By comparing concentration and temperature dependence of the correlation time of the slow mode with that of the surfactant self-diffusion coefficient, dynamics of networks are discussed.

Key words Dynamic light scattering – self-diffusion – micelle – nonionic surfactant – lyotropic liquid crystal

Introduction

In recent years, considerable effort has been devoted to clarify the structure and dynamics of network formed by worm-like micelles [1–4]. In the previous studies [5–8], we have measured light scattering, small-angle X-ray scattering (SAXS), viscosity, and pulsed-gradient spin echo (PGSE) on semidilute solutions of nonionic surfactants $C_{16}E_7$, $C_{14}E_6$, and $C_{14}E_7$ (C_nE_m represents a chemical formula $C_nH_{2n+1}(OC_2H_4)_mOH$). Light scattering results suggest that worm-like micelles are entangled above about $T_c - 15$ K where T_c is the lower critical solution temperature (about 35 °C for $C_{16}E_7$). From the PGSE measurements, it has been found that surfactant molecules migrate to other micelles when micelles are entangled. We have shown also that concentration dependence of surfactant self-diffusion coefficient obtained from PGSE can be explained by a simple model, where intermicellar migration of surfactant molecules at entangled point is taking into account. As the temperature increases further, the results of PGSE and SAXS suggest that cross-linking (or branching) of the worm-like micelles occurs.

In the present study, we have measured dynamic light scattering in the $C_{16}E_7$ system, paying attention to the so-called slow mode in order to discuss dynamical aspects of entanglement from the viewpoint of structural relaxation. In addition, phase behaviors have been investigated in more detail than in our previous study [7, 8].

Experimental section

$C_{16}E_7$ was purchased from Nikko Chemicals, Inc. in crystalline form and used without further purification. Deuterium oxide purchased from ISOTEC, Inc. (99.9%) was used after being degassed by bubbling of nitrogen to avoid oxidation of the ethylene oxide group of surfactants.

Phase diagram was determined by using polarizing microscope (OLYMPUS BHSP) with METLLER FP82HT Hot Stage, ^{2}HNMR, and SAXS. Measurements of ^{2}HNMR and SAXS were made on a JEOL JNM-EX400 Fourier transform NMR spectrometer and an apparatus (MAC Science) constructed from an X-ray generator (SRA MXP18, 18kW), incident monochromator (W/Si multilayer crystal), Kratky slit, and imaging plate (DIP 200).

Dynamic light scattering measurements were made with a monomode-fiber compact goniometer system (ALV) and an argon-ion laser (Spectra Physics, Stabilite 2016).

58

T. Kato et al.
Structure and dynamics of networks of wormlike micelles

Results and discussion

Phase behaviors and dynamic light scattering

Figure 1 shows a partial phase diagram of a $C_{16}E_7$–D_2O system determined in the present study. In the region indicated as N_c, the ^2HNMR spectra and the observation by polarizing microscope indicates the existence of an anisotropic phase while the SAXS spectra give broader peaks than those in the hexagonal phase. Tiddy and coworkers have found a nematic phase in $C_{16}E_6$ [9] and $C_{16}E_8$ [10] systems. These results suggest that the region N_c may be identified as the nematic phase. However, it is difficult to draw a conclusion at the present stage.

In Fig. 2, examples of time correlation functions of the scattered field are illustrated at different scattering angles. In our previous paper [6], it has been reported, based on viscosity measurements that entanglement of worm-like micelles occurs above about 0.7 wt% at 35 °C. In the present study, we have found that the relaxation is bimodal and that the correlation time τ_s of the slow relaxation is independent of scattering angle at concentrations where micelles are entangled. Such a slow mode has been observed for solutions of cetyltrimethylammonium bromide in the presence of organic/inorganic salt where entanglement of worm-like micelles are expected [11–13]. In these studies, it has been shown from experimental results that τ_s is nearly equal to the rheological relaxation time.

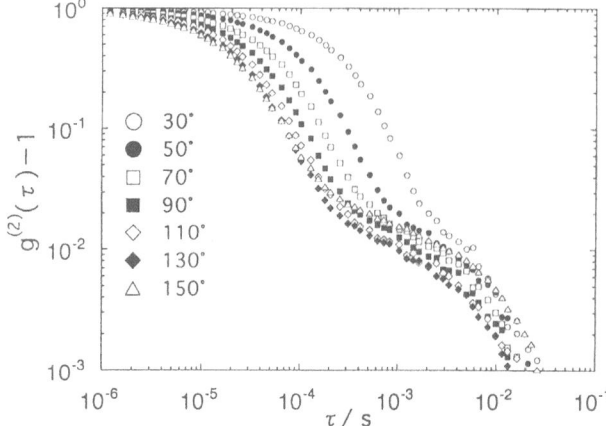

Fig. 2 Examples of time correlation function of the scattered field at 6.4 wt% and 35 °C. Scattering angles are indicated in the figure

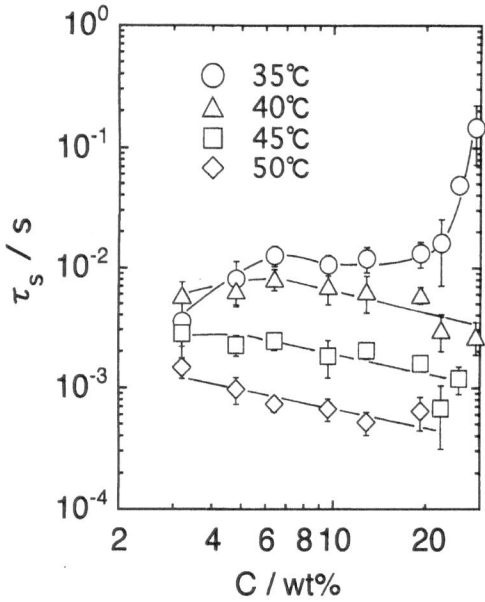

Fig. 3 Double-logarithmic plots of slow relaxation time vs. surfactant concentration

Figure 3 shows double-logarithmic plots of τ_s against surfactant concentration. At 35 °C, τ_s slightly increases in the lower concentration range and then, becomes almost constant. When the concentration exceeds 20 wt%, τ_s increases rapidly, suggesting pre-transition from micellar to nematic phase. Analyses of small angle X-ray and neutron-scattering curves in this range are in progress and will be reported elsewhere [14].

Above 40 °C, τ_s decreases with increasing concentration. From the temperature dependence of τ_s, the

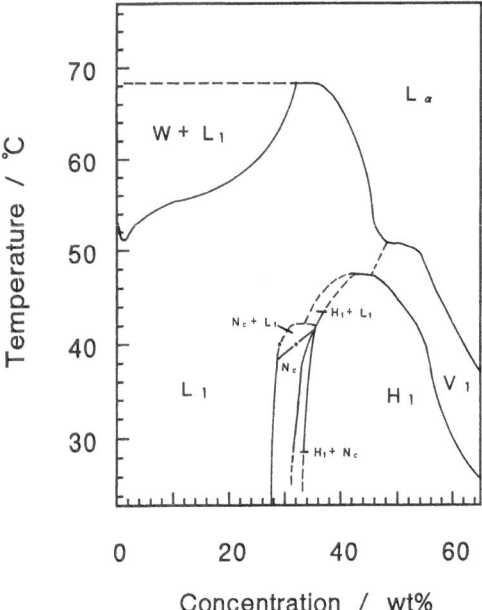

Fig. 1 Phase diagram of the $C_{16}E_7$–D_2O system. L_1, micellar solution; $W + L_1$, co-existing isotropic phases; H_1, hexagonal phase; V_1, cubic phase; L_α, lamellar phase. The phase marked "N_c" may be a nematic phase (see the text)

activation energy is calculated to be about 150–200 kJ mol^{-1}. Such a large activation energy cannot be explained by the reptation of micelles alone. These results suggest that at higher concentrations, τ_s is dominated by the kinetic processes such as fusion and dissociation of micelles.

Comparison with surfactant self-diffusion coefficient

In Fig. 4, observed self-diffusion coefficients of $C_{16}E_7$ (D) are plotted against concentration. It can be seen from the figure that the self-diffusion coefficient increases with increasing concentration above a certain concentration. At the same time, the temperature dependence becomes very large; the activation energy is about 160 kJ mol^{-1} at about 20 wt%. When the concentration exceeds the boundary between micellar and liquid crystal phases, the self-diffusion coefficient increases abruptly and the temperature dependence becomes small.

Cates and his coworkers [15, 16] have proposed a "living polymer" theory for stress relaxation and self-diffusion taking into account the "scission and recombination", "end interchange", and "bond interchange" in addition to the reptation. However, the results in Figs. 3 and 4 are not consistent with the prediction of their theory. In order to explain the self-diffusion results, we have proposed a diffusion model considering intermicellar migration at entanglement points [5, 6]. According to this model, the self-diffusion coefficient can be expressed as

$$D = D_M + \langle d \rangle^2/(6\tau_{mig}),\tag{1}$$

where D_M is the diffusion coefficient of the center of mass of worm-like micelles, $\langle d \rangle$ is the mean distance between the centers of mass of adjacent micelles, and τ_{mig} is the time

during which a surfactant molecule diffuses in a micelle along its contour.

Figures 3 and 4 demonstrate strong correlation between D and τ_s in the concentration range 2–20 wt%. In fact, the product $D\tau_s$ is almost constant in this range. Such a correlation can be explained if we consider that surfactant molecules migrate to another micelles through transient cross-links which may relax the stress of networks [1–3, 17]. In other words, intermicellar migration occurs only when the transient cross-links are formed. Then the residence time τ_{mig} may be replaced by τ_s. From the observed D and τ_s values, we have obtained $\langle d \rangle$ in Eq. (1), assuming that D_M can be neglected in this range. The calculated $\langle d \rangle$ values are around 200 nm, almost independent of concentration and temperature.

From its definition, $\langle d \rangle$ is related to the surfactant concentration (c) and the aggregation number of micelles (N) as

$$\langle d \rangle \propto (N/c)^{-1/3}.\tag{2}$$

In dilute solutions, N is proportional to $c^{1/2}$ after a threshold concentration [18, 19]. If this relation holds up to semidilute region, we have

$$\langle d \rangle \propto c^{-1/6}.\tag{3}$$

Recently, Schurtenberger et al. [20] have proposed a relation

$$N \propto c\tag{4}$$

is semidilute solutions of $C_{16}E_6$ from a detailed analysis of their light scattering data. In this case, $\langle d \rangle$ is expected to be independent of concentration. Both Eqs. (1) and (2) suggest that the $\langle d \rangle$ depends on concentration only slightly, which is consistent with the present results.

Equation (1) holds true only when the surfactant molecules can diffuse over the "end-to-end" distance during τ_s in the same order as or larger than $\langle d \rangle$. This relation has been confirmed as follows. First, we obtained the one-dimensional lateral diffusion coefficient (D_0) along the axis of the rod forming the cubic phase from the self-diffusion coefficient in the cubic phase using the equation proposed by Anderson and Wennerström [7, 21]. Then the diffusion length along the contour of micelles, $L_D(\tau_s)$, was calculated from D_0 and τ_s using the relation

$$L_D(\tau_s) = (2D_0\tau_s)^{1/2}.\tag{5}$$

On the other hand, the persistence length of worm-like micelles (ℓ_p) was estimated to be about 35 nm from static light scattering data in the dilute region [22] using the equation of Polod et al. [23]. These L_D and ℓ_p values were utilized to calculate the "end-to-end" diffusion distance $R(\tau_s)$ ($=[L_D(\tau_s)\ell_p]^{1/2}$) of about 200 nm, which is nearly equal to the $\langle d \rangle$ value obtained from Eq. (1).

Fig. 4 Double-logarithmic plots of self-diffusion coefficient of $C_{16}E_7$ vs. surfactant concentration [5–7]. The vertical-dashed lines indicate phase boundaries

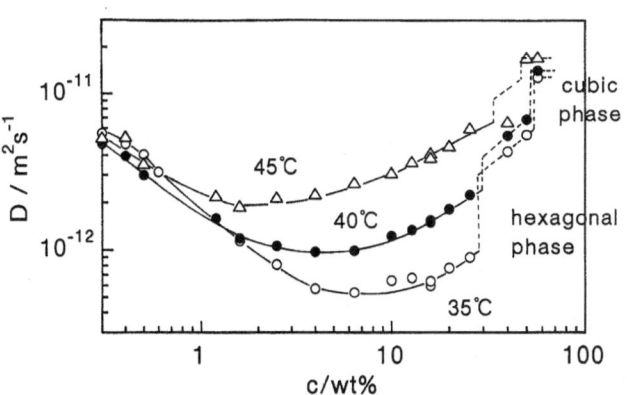

60

T. Kato et al.
Structure and dynamics of networks of wormlike micelles

In the previous study [7], we have shown that when the temperature exceeds about 55 °C, the self-diffusion coefficient approaches the value expected from the data in the cubic phase where the lateral diffusion is dominant. In this temperature range, the slow mode cannot be distinguished from the fast mode because τ_s decreases rapidly with increasing temperature, suggesting that the $R(\tau_s)$ becomes smaller than $\langle d \rangle$. Then it is possible that the self-diffusion coefficient is dominated by lateral diffusion rather than τ_s.

Acknowledgement The author would like to express his sincere thanks to Prof. Gordon J. Tiddy of University of Salford for his helpful comments on the phase behaviors of nonionic surfactants.

References

1. Hoffmann H (1994) In: Herb CA, Prud'homme RK (eds) ACS Symposium Series. vol 578, American Chemical Society, Washington DC, pp 2–31
2. Lequeux F, Candau SJ (1994) In: Herb CA, Prud'homme RK (eds) ACS Symposium Series. vol 578, American Chemical Society, Washington DC, pp 51–62
3. Khatory A, Lequeux F, Kern F, Candau SJ (1993) Langmuir 9:1456–1464
4. Drye TJ, Cates ME (1992) J Chem Phys 96:1367–1375
5. Kato T, Terao T, Tsukada M, Seimiya T (1993) J Phys Chem 97:3910–3917
6. Kato T, Terao T, Seimiya T (1994) Langmuir 10:4468–4474
7. Kato T, Taguchi N, Terao T, Seimiya T (1995) Langmuir 11:4661–4664
8. Kato T (1996) Progr Colloid Polym Sci 100:15–18
9. Funari SS, Holmes MC, Tiddy GJT (1994) J Phys Chem 98:3015–3023
10. Corcoran J, Fuller S, Rahman A, Shinde N, Tiddy GJT, Attard GS (1992) J Mater Chem 2:695–702
11. Brown W, Johansson K, Almgren M (1989) 93:5888–5894
12. Nemoto N, Kuwahara M, Yao M, Osaki K (1995) Langmuir 11:30–36
13. Buhler E, Munchi JP, Candau SJ (1995) J Phys II France 5:765–787
14. Kato T, Taguch N, Imai M, Yoshida H, to be published
15. Cates ME (1987) Macromolecules 20:2289–2296
16. Turner MS, Marques C, Cates ME (1993) 9:695–701
17. Shikata T, Hirata H, Kotaka T (1988) Langmuir 4:354–359
18. Kato T, Kanada M, Seimiya T (1995) Langmuir 11:1867–1869
19. Kato T, Kanada M, Seimiya T (1996) J Colloid Interface Sci 181:149–158
20. Schurtenberger P, Cavaco C, Tiberg F, Regev O (1996) Langmuir 12:2894–2899
21. Anderson DM, Wennerström H (1990) J Phys Chem 94:8683–8694
22. Nozu D, Kato T, unpublished data
23. Yamakawa H (1971) Modern Theory of Polymer Solutions. Harper & Row, New York, pp 56

Progr Colloid Polym Sci (1997) 106:61–63
© Steinkopff Verlag 1997

H. Iijima
T. Kato
T. Seimiya
H. Yoshida
M. Imai

Micellar growth in dilute solutions of cesium perfluorooctanoate studied by small-angle X-ray and neutron scattering

Received: 13 October 1996
Accepted: 18 March 1997

H. Iijima · Dr. T. Kato (✉) · T. Seimiya
Department of Chemistry
Faculty of Science
Tokyo Metropolitan University
Minamiohsawa, Hachioji
Tokyo 192-03, Japan

H. Yoshida
Department of Industrial Chemistry
Faculty of Engineering
Tokyo Metropolitan University
Minamiohsawa, Hachioji
Tokyo 192-03, Japan

M. Imai
Institute for Solid State Physics
University of Tokyo
Tokai
Ibaraki 319-11, Japan

Abstract Small-angle X-ray and neutron scattering (SAXS and SANS, respectively) have been measured in H_2O solutions of cesium perfluorooctanoate (CsPFO) in the concentration range 65–500 mM (below the phase boundary between micellar and discotic nematic phases). The data have been analyzed with a double-layered ellipsoid form factor combined with the rescaled mean-spherical approximation for the intermicellar structure factor. The scattering curves in the lower-concentration range are fitted well with oblate ellipsoid rather than prolate ellipsoid. Both the aggregation number and the degree of counterion binding increase with increasing surfactant concentration, which is consistent with the concentration dependences of the ^{133}Cs and ^{19}F NMR chemical shifts.

Key words Micelle – small-angle scattering – fluorinated surfactant – disk-like micelle

Introduction

It is well known that micelles grow if required conditions (surfactant concentration, temperature, concentration of added salt, and so on) are satisfied. Large amount of data suggest that micelles grow along one dimension (rod) rather than two dimensions (disk). To the authors' knowledge, there has been no report which clearly shows the two-dimensional micellar growth on the basis of small-angle scattering data except for the recent study on mixed surfactant systems [1]. On the other hand, Boden et al. have found that a discotic–nematic phase is formed in aqueous solutions of cesium prefluorooctanoate (CsPFO) [2–4]. They have suggested also the existence of disk-like micelles in an isotropic phase near the discotic–nematic phase from analysis of the position of the SAXS peak. However, the two-dimensional micellar growth has not yet been established at least in the lower-concentration range.

In the present study, we have measured SAXS and SANS in dilute solutions of CsPFO and analyzed observed scattering curves to elucidate whether two-dimensional micellar growth occurs or not.

Experimental section

Materials

Cesium perfluorooctanoate were prepared by neutralizing perfluorooctanoic acid with cesium hydroxide in acetone. The salts obtained were purified by recrystallization from

62

H. Iijima et al.
Micellar growth in cesium perfluorooctanoate by SAXS and SANS

mixed solvent of hexane and ethanol (1 : 1) and dried under vacuum.

Methods

The measurements of small-angle neutron scattering (SANS) and small-angle X-ray scattering (SAXS) were carried out using the SANS-U at the Institute for Solid State Physics of University of Tokyo, and BL10C at National Laboratory for High-Energy Physics, respectively. The SANS and SAXS data were analyzed by using the double-layered ellipsoid model for micellar particle structure factor $P(Q)$ and the rescaled mean-spherical approximation (RMSA) for interparticle structure factor $S(Q)$ [5–8].

Results and discussion

Small-angle scattering

Figure 1 shows an observed SAXS curve of CsPFO in H_2O (100 mM) and least-squares fits for oblate ellipsoid (solid line), prolate ellipsoid (dashed line), and sphere (dash–dot line). It can be seen from the figure that the oblate ellipsoid model can explain the experimental results better than the prolate ellipsoid and the sphere model. Similar results have been obtained for other concentrations, which supports the two-dimensional micellar growth. The results of these fits are summarized in Tables 1 and 2.

It should be noted that the polydispersed spheres give similar scattering curves as those for monodispersed ellipsoids [5, 6]. So, the observed scattering curves may be explained by polydispersed spheres, although we did not perform calculations for polydispersed spheres. In this case, however, the average radius of micellar core should be much larger than the extended length of hydrophobic chain in the higher-concentration range as can be expected from Table 2.

Figure 2 shows concentration dependence of the aggregation number (m) and the degree of counter-ion binding (β) obtained for the oblate ellipsoid. This figure demonstrates that micelles grow with increasing surfactant concentration. At the same time, the degree of counter-ion binding increases, which may be characteristic of two-dimensional growth because the surface area per head group decreases with increasing aggregation number more rapidly than in one-dimensional growth [9].

Table 1 Results of analysis of SAXS data on the basis of the oblate and prolate ellipsoid. Core minor semiaxis was fixed to be 12.5 Å

Model	C/mol dm^{-3}	m	β	Axial ratio of core
Oblate Ellipsoid	0.101	69 ± 2	0.59 ± 0.02	1.70 ± 0.02
	0.152	81 ± 1	0.63 ± 0.02	1.84 ± 0.02
	0.219	100 ± 1	0.72 ± 0.01	2.05 ± 0.01
Prolate Ellipsoid	0.101	59 ± 2	0.70 ± 0.02	2.47 ± 0.07
	0.152	68 ± 1	0.76 ± 0.02	2.86 ± 0.06
	0.219	89 ± 1	0.86 ± 0.01	3.73 ± 0.04

Table 2 Results of analysis of SAXS data on the basis of the sphere

C/mol dm^{-3}	m	β	Radius of core/Å
0.101	71 ± 2	0.58 ± 0.03	18.0 ± 0.2
0.152	84 ± 2	0.61 ± 0.03	19.0 ± 0.2
0.219	103 ± 3	0.71 ± 0.04	20.3 ± 0.2

Fig. 1 SAXS scattering curve of CsPFO in H_2O (100 mM). The circles represent experimental values. Solid, dashed, and dash–dot lines are fitted curves for oblate, prolate, and sphere, respectively

Fig. 2 Concentration dependence of the degree of counter-ion binding (β) and the aggregation number (m) obtained from the analysis of SAXS (filled symbols) and SANS (open symbols)

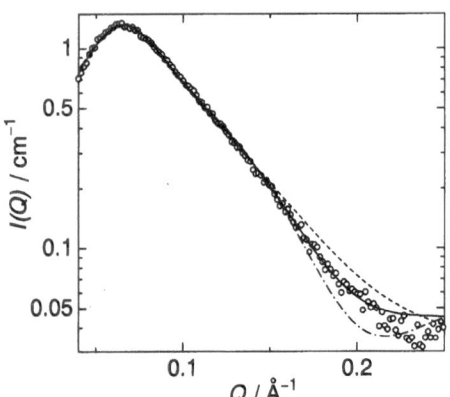

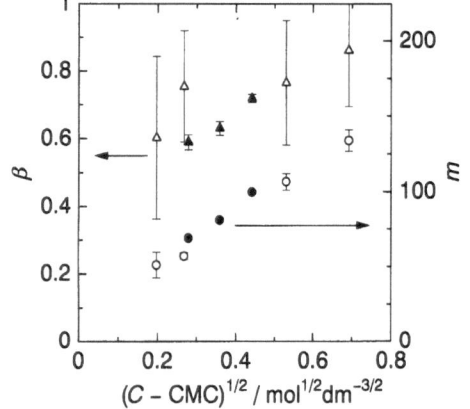

^{133}Cs NMR

NMR data were analyzed by a two-state model where bound (attached to a micelle) and free (dispersed in bulk phase) states are considered. In this model, the observed chemical shift δ_{obs} is expressed as

$$\delta_{obs} = \frac{C_{free}\delta_{free} + C_{bound}\delta_{bound}}{C}, \qquad (1)$$

where C is the surfactant concentration, C_{free} and C_{bound} the concentration of cesium ion in the free and bound state, respectively, and δ_{free} and δ_{bound} the corresponding chemical shifts. If the concentration of surfactant monomers is equal to the CMC, Eq. (1) can be rewritten as

$$\delta_{obs} - \delta_{free} = (1 - CMC/C)\beta(\delta_{bound} - \delta_{free}), \qquad (2)$$

where β is the degree of counter-ion binding. When $\beta(\delta_{bound} - \delta_{free})$ is independent of concentration, $\delta_{obs} - \delta_{free}$ is proportional to the reciprocal surfactant concentration. In this system, however, the plot of the ^{133}Cs chemical shift against reciprocal surfactant concentration does not give a straight line [10]. This indicates that $\beta(\delta_{bound} - \delta_{free})$ depends on the surfactant concentration. So, we have obtained $\beta(\delta_{bound} - \delta_{free})$ from

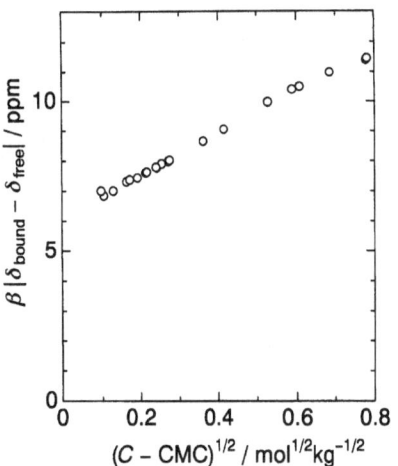

Fig. 3 Concentration dependence of $\beta(\delta_{bound} - \delta_{free})$ obtained from the observed ^{133}Cs NMR chemical shift (see the text)

$\delta_{obs} - \delta_{free}$ at each concentration by using Eq. (2). The results are shown in Fig. 3. This figure demonstrates that β is also proportional to $(C - CMC)^{1/2}$ ($\delta_{bound} - \delta_{free}$ is expected to be independent of concentration), which supports the small-angle scattering results.

References

1. Lin TL, Liu C-C, Roberts MF, Chen SH (1991) J Phys Chem 95:6020–6027
2. Boden N, Corne SA, Jolley KW (1987) J Phys Chem 91:4092–4105
3. Boden N, Jolley KW, Smith MH (1993) J Phys Chem 97:7678–7690
4. Boden N (1994) In: Gelbert WM, Ben-Shaul A, Roux D (eds) Micelles, Membranes, Microemulsions, and Monolayers. Springer, Berlin, pp 153–217
5. Kotlarchyk M, Chen SH (1983) J Chem Phys 79:2461–2469
6. Chen SH, Lin TL (1987) In: Price DL, Sköld K (eds) Methods of Experimental Physics. Vol. 23 Part B Ch 16, Academic Press, New York
7. Hayter JB, Penfold J (1981) Mol Phys 42:109–118
8. Hansen JP, Hayter JB (1982) Mol Phys 46:651–656
9. Tanford C (1980) In: The Hydrophobic Effect: Formation of Micelles and Biological Membranes. Wiley, New York, pp 54–55
10. Iijima H, Koyama S, Fujio K, Uzu Y (to be published)

Progr Colloid Polym Sci (1997) 106:64–69
© Steinkopff Verlag 1997

M. Imai
T. Kato
D. Schneider

Fluctuations of cubic network observed in non-ionic surfactant system

Received: 13 October 1996
Accepted: 24 March 1997

Dr. M. Imai (✉)
Institute for Solid State Physics
University of Tokyo
Tokai, Ibaraki 319-11, Japan

T. Kato
Faculty of Science
Tokyo Metropolitan University
Minamiohsawa, Hachioji
Tokyo 192-03, Japan

D. Schneider
Biology Department
Brookhaven National Laboratory
Upton, New York 11973-5000, USA

Abstract We have investigated the stability of a cubic lattice having a periodic-minimal surface in a non-ionic surfactant system using a time-resolved small-angle neutron-scattering technique. The cubic lattice transformed from a hexagonal phase showed a polycrystalline scattering pattern and strong intensity fluctuation as a function of time. A characteristic correlation time of the intensity fluctuation obtained by time correlation function analysis was about 150 min. In the case of the cubic lattice from a lamellar phase the diffraction peaks showed mono-crystalline-like pattern and no intensity fluctuation. We consider that the fluctuations arise from the polycrystalline cubic network trapped in a metastable state. In other words, the cubic phase from the hexagonal phase is in an incubation time of *finite*-to-*infinite* transition of the cubic network.

Key words Cubic phase – infinite periodic minimal surface – phase transition – incubation time – surfactant

Introduction

Amphiphilic molecules, such as surfactants, will aggregate spontaneously so as to avoid hydrocarbon/water contacts in aqueous solution. The resultant interfaces between the surfactants and water show a variety of curved surfaces [1, 2]. In general, the curved surfaces can be classified into three categories using a Gaussian curvature K; (i) ellipsoid having $K > 0$, (ii) paraboloid having $K = 0$, (iii) hyperboloid having $K < 0$. For surfactant–water systems, spherical or ellipsoidal micelles are observed in the low-concentration region and cylinder (hexagonal phase) or sheet (lamellar phase) assemblies classified into the paraboloid are observed in the high-concentrated region.

In 1960, Luzzati et al. [3, 4] found the so-called cubic phase between the hexagonal and lamellar phases in a lipid–water system. The cubic structure attributed to space group I_{a3d} consists of rods of finite length that join three by three to form two three-dimensional networks, mutually interwoven and unconnected. This bicontinuous cubic phase is considered to be a gyroid which is one of the infinite periodic minimal surfaces (IPMS) having negative Gaussian curvature [5]. After that, similar bicontinuous cubic phases were found in many systems such as surfactants [6], copolymer [7] and plasmidic compounds [7].

The phase diagram of such curved surfaces in the amphiphilic system has been studied by Huse and Liebler [8] on the basis of the elastic energy of the surface. The phase behavior is determined by the balance of the surface tension, the bending rigidity and the saddle-splay modulus of the curved interface and the IPMS structure appears when the saddle-splay modulus increased to a some critical value. Mathematically, more than 30 species of curved surfaces have been reported as IPMS [9] and, experimentally, several types of IPMS structures have been found in cubic phases of lipid–water systems [10]. Concerning the formation mechanism of the cubic network, Rançon and Charvolin [6] found the epitaxial relationships between reticular planes of these three phases in the

$C_{12}E_6$–H_2O system. The relationship can be summarized as follows:

Hexagonal Cubic   Lamellar
Cylinder axis — [1 1 1] direction
(1 0) plane — (2 1 1) plane — lamellar plane

The morphology transitions take place keeping the fixed orientation of each geometry. Clerc and coresearchers [11] investigated the kinetics of the morphology transition for the $C_{12}E_6$/H_2O system using synchrotron X-ray diffraction with 50 ms time resolution and explained the observed transition on the basis of the nucleation and growth mechanism. However, how does the cubic domain having complicate network structure develop into the large domains? The domain coarsening of the cellular pattern has been widely investigated theoretically and experimentally [12]. In many systems the average domain size $\langle R(t) \rangle$ at time t obeys the following power law at the long-time limit:

$$\langle R(t) \rangle \sim t^v, \tag{1}$$

where v is the growth exponent which specifies the universality class of the domain growth. In this study, we have investigated the dynamics of polycrystalline cubic phase, namely, the finite-size cubic network system using a time-resolved small-angle neutron-scattering technique.

Experimental

A nonionic surfactant $C_{16}E_7$ ($CH_3(CH_2)_{15}(OCH_2CH_2)_7$ OH) was used for the amphiphilic molecule because the phase behavior of the $C_{16}E_7$–D_2O system has been investigated in detail [13]. $C_{16}E_7$ was purchased from Nikko Chemicals, Inc., and used without further purification and deuterium oxide purchased from ISOTEC, Inc. (99.9%) was used after being degassed by bubbling of nitrogen to avoid oxidation of the ethylene oxide group of surfactant. Samples containing 49 wt% of D_2O sealed in a glass vial were annealed 3 h at about 55 °C and then held at room temperature for 21 h. This heat treatment was repeated after one week for homogenization of the sample and the homogeneity of the sample was checked by a polarizing microscope observation. The sample in the vial was stirred just before transfer to the SANS cell. In order to obtain monocrystalline phases, Clerc and coresearchers [14] have used rectangular glass-capillary tubes with a thickness not exceeding 0.2 mm, while Rançon and Charvolin [6] have used rotating glass-capillary tubes having 1.5 mm diameter. In this study we adopted the quartz cell having 10 (width) × 50 (height) × 2 mm (thickness) dimension to obtain polycrystalline phases of the $C_{16}E_7$–D_2O system.

SANS measurements were performed using the H9B instrument [15] at HFBR of Brookhaven National Laboratory and the SANS-U instrument [16] of the Institute for Solid State Physics, University of Tokyo at JRR-3M of the Japan Atomic Energy Research Institute. The samples were heated from room temperature (hexagonal phase) to the cubic phase with a rate of 1 °C/min and then annealed isothermally for 2000 min at 47 °C (± 0.1 °C). During this annealing process, we performed *in situ* time-resolved SANS measurements. The scattering patterns were recorded on 2D-detectors, having 128 × 128 pixels of 5 × 5 mm² resolution. The observed scattering vector Q ($Q = 4\pi \sin \theta / \lambda$) range 0.03–0.3 Å^{-1}. The measurement time resolution is 5 min for H9B and 10 min for SANS-U. After the isothermal annealing, the samples were heated to the lamellar phase and then cooled to the cubic phase to examine the scattering behavior of the cubic phase transformed from the lamellar phase. We repeated a series of these experiments to check the reproducibility.

Results and discussions

The phase behavior of the 51 wt% $C_{16}E_7$–D_2O system is shown in Fig. 1a with corresponding scattering patterns for each phases in bird's-eye-view expression (Figs. 1b–e). The phase diagram was determined by polarized optical microscope observations, small-angle X-ray scattering (SAXS) measurements and ^2H NMR measurements [13]. The symmetry of each phase could be confirmed using characteristic higher-order reflections in SAXS profiles. At 34 °C corresponding to the hexagonal phase, the scattering pattern shows a Debye ring, indicating randomly oriented polycrystalline state (Fig. 1b). The peak position, Q_{max} is 0.105 Å^{-1} and this diffraction pattern did not depend on temperature in the hexagonal phase. At the transition temperature from hexagonal to the cubic phase, the diffraction pattern changed to a speckle ring as shown in Fig. 1c but the diffraction position was almost the same as that in the hexagonal phase. The hexagonal-to-cubic transition could be verified by the disappearance of $\sqrt{3}$ hexagonal reflection. In the lamellar phase, the speckle pattern transforms to the Debye ring again (Fig. 1d), keeping their diffraction position. On the other hand, when the sample was cooled from the lamellar to the cubic phases, the numerous diffraction peaks on the Debye ring converged to some strong peak as shown in Fig. 1e. The scale of the ordinate in Fig. 1e is contracted by seven times than in Fig. 1c. The agreement of peak position was observed in this transition again.

The diffraction peaks from hexagonal, cubic and lamellar phases shown in Fig. 1 are attributed to (1 0), (2 1 1) and lamellar planes, respectively. This epitaxial relationship in the $C_{16}H_7$–D_2O system indicates that the adjacent phases are strongly correlated and the cubic lattice grows with

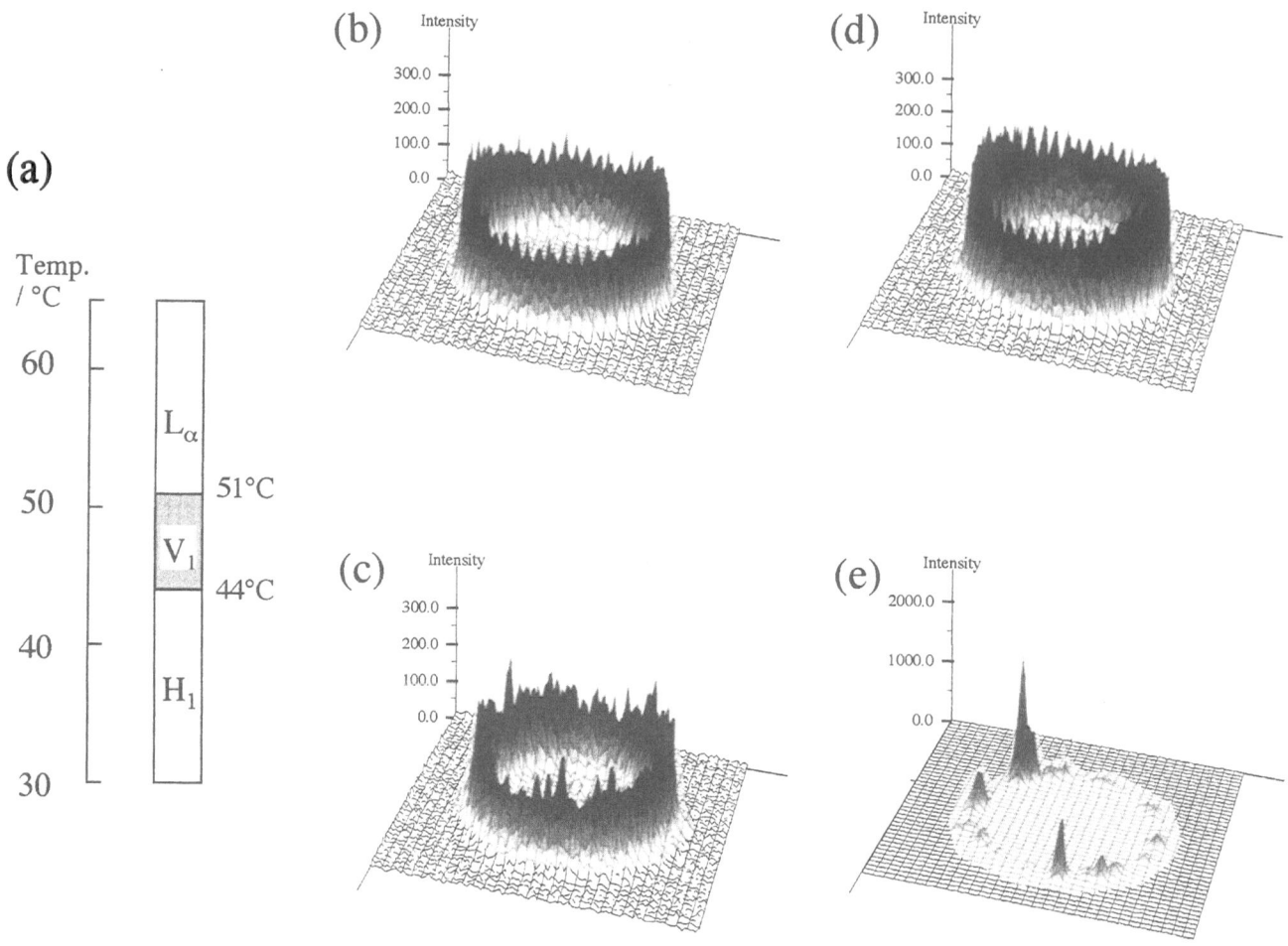

Fig. 1 (a) Phase diagram for the 51% $C_{16}E_7$–D_2O system, H_1, V_1 and L_α designate hexagonal, cubic and lamellar phases, respectively. The corresponding SANS patterns are shown in (b) hexagonal, (c) cubic from hexagonal, (d) lamellar and (e) cubic from lamellar phases

a fixed orientation with respect to the hexagonal or lamellar lattice. However, starting from polycrystalline hexagonal lattice, we obtained not the monocrystal cubic lattice or the IPMS but the polycrystalline cubic lattice. In this case, the continuity of the periodic minimal surface is broken at the boundary of the cubic domains and the system is considered to be trapped in some local potential minimum states. On the other hand, the cubic phase transformed from the lamellar phase showed the monocrystalline-like pattern, although the starting lamellar phase has also a polycrystalline state. Then we consider that the ideal stable state is not polycrystalline or finite cubic network, but monocrystalline or infinite cubic network. The morphology change from hexagonal to cubic phases occurs on the basis of the reorganization of the cylinders, whereas that from lamellar phase proceeds through channels in the lamellar plates. This difference probably makes the polycrystalline cubic phase observed for hexagonal/cubic

transition. Then, it is worthwhile to investigate the dynamics of the polycrystalline cubic lattice trapped in the metastable states.

The time evolution of SANS patterns with 5 min resolution for the cubic phase from the hexagonal phase is shown in Fig. 2. Adjacent scattering patterns corresponding to 5 min interval shows good memory of the last pattern. With the elapse of time some diffraction peaks lose their intensity and new peaks appear at other positions on the ring. The diffraction intensity at a fixed pixel is plotted as a function of time as shown in Fig. 3a. The statistical errors are indicated by the $\pm 1\sigma$ error bars. The intensity fluctuations of up to 150% relative amplitude are evident at the 47 °C cubic phase. These are much larger in amplitude than the noise due to counting statics, and there is a clear correlation between neighboring points. This indicates that the cubic network domains fluctuate with several ten minutes time scale.

Progr Colloid Polym Sci (1997) 106:64–69
© Steinkopff Verlag 1997

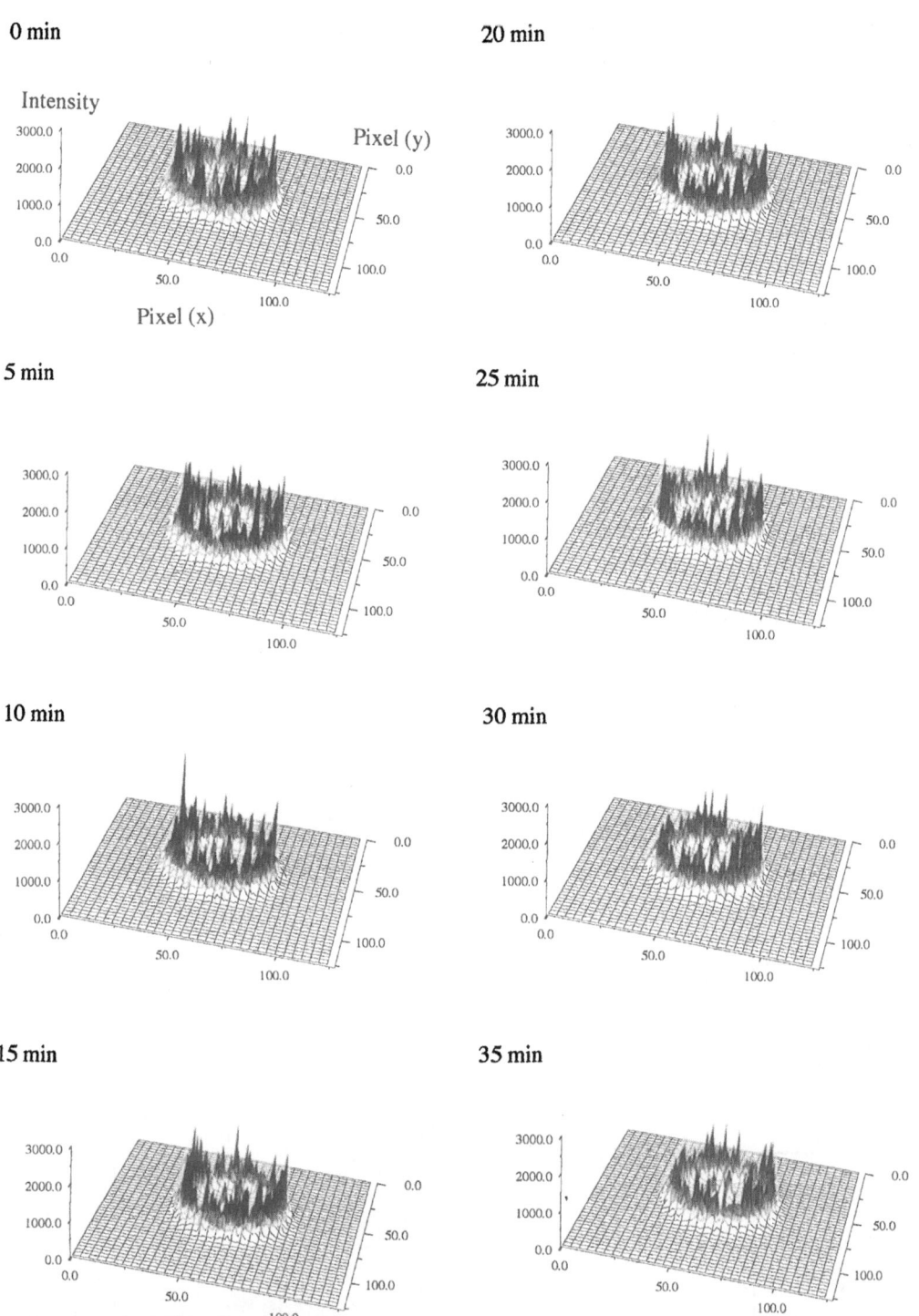

Fig. 2 Time-resolved SANS pattern of the 51% $C_{16}E_7$–D_2O system for the cubic phase transformed from the hexagonal phase at 47 °C. The time resolution is 5 min

The fluctuations can be characterized by introducing the normalized time correlation function

$$g(\tau) = \frac{\langle [I(t) - \langle I \rangle][I(t + \tau) - \langle I \rangle] \rangle}{\langle [I(t) - \langle I \rangle]^2 \rangle}, \qquad (2)$$

where $\langle \; \rangle$ represents a time average and $I(t)$ is the detected neutron counts at time t. We calculated the time correlation function using time-resolved SANS data of 1500 min and with 10 min time resolution and then averaged the autocorrelation function obtained in all azimuthal angle

68

M. Imai et al.
Fluctuations of cubic network

having $Q = 0.105\,\text{Å}^{-1}$ (241 pixels), in order to obtain a sufficient ensemble of the system. The calculated time correlation function is shown in Fig. 4. A single-exponent relaxation function describes the observed time correlation function very well. The fitting function gives characteristic decay time of about 150 min. This corresponds to the mean lifetime of the cubic domains transformed from the hexagonal phase and this value is fairly slow compared with the time constant for the formation of the cubic lattice, i.e., order of seconds or less than 100 ms [11]. However, in the case of the cubic domains from the lamellar phase, most of the reflections did not fluctuate as shown in Fig. 3b.

One candidate to explain this fluctuation is the cellular pattern growth [12] found in coarsening grain aggregates and soap froths, which means that each cubic domain favors the development of a large crystal. During the cellular pattern growth, some domains disappear the other domains with self-similarity, which may be the cause of the fluctuation. According to the growth model [17], the average volume of cells $\bar{V}(t)$ is expressed by $\bar{V}(t) \sim t^{3/2}$, and the diffraction intensity distribution of the cubic phase should be shifted to the high-intensity side and the distribution functions are scaled with the average intensity. The

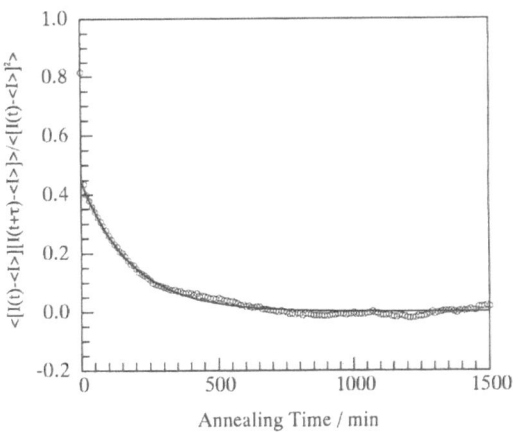

Fig. 4 Autocorrelation function of the diffraction intensity from the cubic phase transformed from hexagonal phase at 47 °C. The solid curve indicates the single-exponent decay function

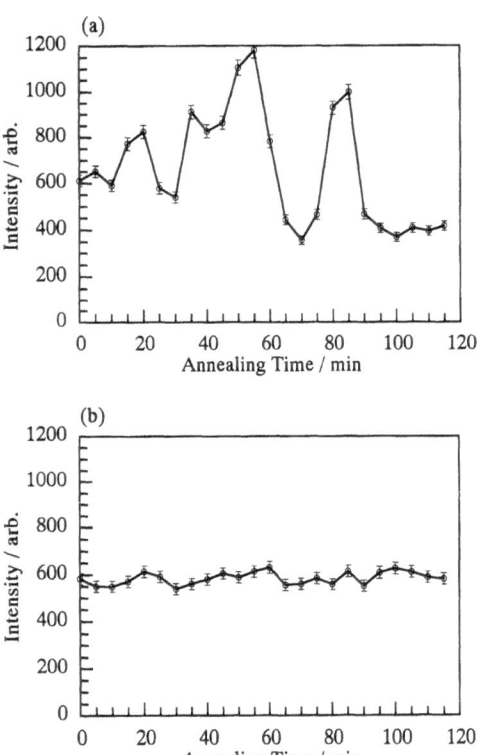

Fig. 3 Typical examples of time dependence of the diffraction intensity of cubic lattice recorded at a fixed pixel: (a) The sample was heated from the hexagonal phase and hold at 47 °C. (b) The sample was cooled from the lamellar phase and hold at 47 °C

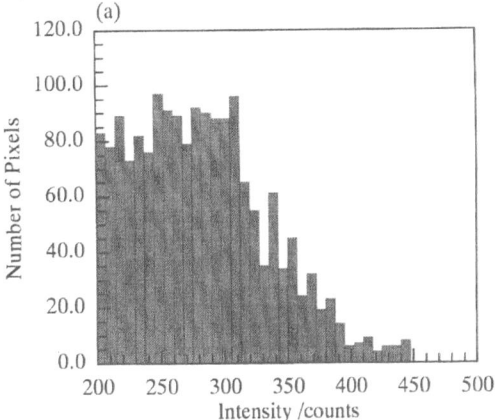

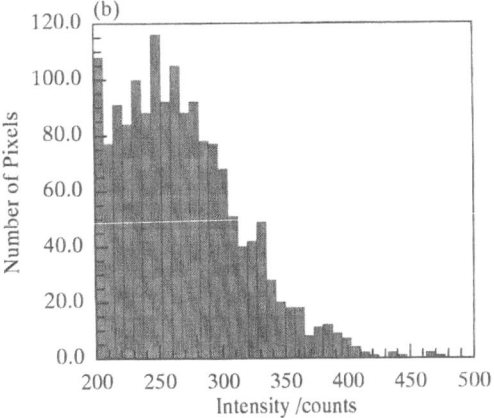

Fig. 5 Intensity distribution of the diffraction ring of the cubic phase: (a) $t = 0$ min; (b) $t = 800$ min

observed intensity distribution of the diffraction ring at annealing time $t = 0$ min was compared with that at $t = 800$ min as shown in Fig. 5. We could not detect significant shift of the distribution which is expected from

Progr Colloid Polym Sci (1997) 106:64–69
© Steinkopff Verlag 1997

the cellular pattern growth. This means that the average domain size of the polycrystalline cubic phase does not develop and that the system fluctuates around an equilibrium state.

At present we consider that the cubic phase transformed from the hexagonal phase is considered to be trapped in the metastable state and the energy barrier between the finite and infinite cubic network is fairly high compared with the thermal fluctuation at 47 °C. Then the coarsening of the cubic domains is suppressed in the polycrystalline cubic state. In other words, the observed cubic network from the hexagonal phase is probably in an incubation time of the nucleation growth of cubic network and the incubation time is fairly long compared with the observed time scale.

Acknowledgments We would like to express our thanks to Dr. T. Kawakatsu, Department of Physics, Tokyo Metropolitan University for helpful discussions. This work is supported by a Grant-in-Aid for Scientific Research on Priority Areas "Cooperative Phenomena in Complex Liquids" from the Ministry of Education, Science and Culture of Japan (No. 07236103). The present work was also supported by the US–Japan Cooperative Research Program on Neutron Scattering operated by the United States Department of Energy and the Ministry of Education, Science and Culture (MONBUSHO of Japan).

References

1. Cates ME (1993) Philos Trans Roy Soc London 344:339
2. Gompper G, Schick M (1994) In: Domb C, Lebowitz J (eds) Phase Transitions and Critical Phenomena. Vol. 16, Academic Press, London
3. Luzati V, Mustacchi H, Skoulios AE, Husson F (1960) Acta Cryst 13:660
4. Luzzati V, Tardieu A, Gulik-Krzywicki T, Rivas E, Reiss-Husson F (1968) Nature 220:485
5. Scriven LE (1976) Nature 263:123
6. Rançon Y, Charvolin J (1988) J Phys Chem 92:2646
7. Duois-Violette E, Pansu B (eds) (1990) International Workshop on Geometry and Interfaces. J Phys (France) Colloq 51:C7 (Suppl 23)
8. Huse DA, Leibler S (1988) J Phys (France) 49:605
9. Fischer W, Koch E (1989) Acta Cryst A 45:726
10. Luzzati V, Vargas R, Mariani P, Gulik A, Delacroix H (1993) J Mol Biol 229:540
11. Clerc M, Laggner P, Levelut AM, Rapp G (1995) J Phys II France 5:901
12. Okuzono T, Kawasaki K (1994) Trends Stat Phys 1:65
13. Kato T, Taguchi N, Terao T, Seimiya T (1995) Langmuir 11:4661
14. Clerc M, Levelut AM, Sadoc JF (1991) J Phys II (France) 1:1263
15. (1992) HFBR Handbook. Brookhaven National Laboratory, New York, USA
16. Ito Y, Imai M, Takahashi S (1995) Physica B 213&214:889
17. Nagai T, Ohta S, Kawasaki K, Okuzono T (1990) Phase Transit 28:177

Progr Colloid Polym Sci (1997) 106:70–74
© Steinkopff Verlag 1997

Y. Murata
Y. Maeda
H. Kuniyasu
T. Yamaguchi
M. Tanaka
R. Shimozawa
H. Wakita
Y. Tabata

Effects of temperature and chain length on the properties of long chain polyoxyethylene dodecyl ethers in aqueous solution

Received: 20 December 1996
Accepted: 31 March 1997

Dr. Y. Murata (✉) · Y. Maeda
H. Kuniyasu · T. Yamaguchi · M. Tanaka
R. Shimozawa · H. Wakita
Department of Chemistry
Fukuoka University
Nanakuma, Jonan-ku
Fukuoka 814-80, Japan

Y. Tabata
Nikko Chemicals Co. Ltd.
1-4-8 Nihonbashi-Bakurocho
Chuoku, Tokyo 103, Japan

Abstract The micelle formation and adsorption phenomena at the air–water interface of four long-chain polyoxyethylene dodecyl ethers, $C_{12}H_{25}O(C_2H_5O)_n H$ abbreviated as $C_{12}E_n$ ($n = 6, 7, 10, 20, 30,$ and 40), which have newly been synthesized to have more narrow polyoxyethylene chain-length distribution than previously used, have been investigated in aqueous solution by surface tension measurements over a concentration range of the surfactants ($10^{-5}–10^{-3}$ m) at different temperatures (25, 35, 40, and 45 °C). The critical micelle concentration (CMC), the area (A) per surfactant molecule, and the standard Gibbs free-energy change of micellization (ΔG_m) were determined from the surface-tension data of the solutions and the Gibbs adsorption equation at the various temperatures. It has been found that the CMC values of $C_{12}E_n$ increase with increasing chain length of polyoxyethylene but decrease with increasing temperature. The area per molecule A increases with elevating temperature, their increment from 25 °C to 45 °C being 3.3, 12.2, 20.5, and 36.9 Å^2 for the $C_{12}E_n$ ($n = 10, 20, 30,$ and 40), respectively. Application of a scaling rule in polymer science to parameters A and ΔG_m has suggested that a coil–globule transition takes place between 35 and 45 °C both at the micellar surface and at the air–water interface for the $C_{12}E_{30}$ and $C_{12}E_{40}$ around 40 °C.

Key words Scaling theory – micelle formation – polyoyethylene dodecyl ethers – coil-globule transition – surface tension

Introduction

Nonionic surfactants of hydrophilic polyoxyethylene chain and hydrophobic alkyl chain, $C_mH_{2m+1}O$ $(CH_2CH_2O)_n H$ abbreviated as C_mE_n, have widely been investigated and many physico-chemical data are compiled. However, most of C_mE_n so far investigated, are limited to short polyoxyethylene chains ($n \leq 8$) [1–6]. There are some papers on the surfactants of long polyoxyethylene chains ($n \geq 8$) [1, 7–10]. The CMC and area per molecule for these surfactants reported in the literature are not consistent with each other. One of the reasons for this inconsistency may be that the C_mE_n surfactants ($n \geq 8$) are mixtures of compounds having different polyoxyethylene chain lengths.

In this paper, $C_{12}E_n$ surfactants with long polyoxyethylene chains ($n = 10, 20, 30,$ and 40) have been newly synthesized; their surface tension in aqueous solution has been measured. The CMC and the area per molecule at the air–water interface are calculated and discussed together with those for $C_{12}E_n$ ($n \leq 8$).

Progr Colloid Polym Sci (1997) 106:70–74
© Steinkopff Verlag 1997

Experimental

Materials

Surfactants $C_{12}E_6$, $C_{12}E_7$, $C_{12}E_{10}$, $C_{12}E_{20}$, $C_{12}E_{30}$, and $C_{12}E_{40}$ were obtained from Nikko Chemicals, Tokyo, and used without further purification. The high purity of $C_{12}E_6$ and $C_{12}E_7$ was confirmed by the absence at any detectable minimum in the curves of measured surface tension data vs. the surfactant concentration [5]. The chain-length distribution of $C_{12}E_{10}$ sample was checked by gas chromatography and found to be narrower than that of previously used samples [1,7–10]. A similar check for $C_{12}E_{20}$, $C_{12}E_{30}$, and $C_{12}E_{40}$ was not possible because of their high boiling points. All sample solutions were prepared by dissolving the surfactants into triple distilled water.

Surface-tension measurements

The surface tension of surfactant solutions was measured as a function of bulk surfactant molal concentration (mol/kg H_2O) with a Wilhelmy-plate-type tension meter (Kyowa CBVP). The accuracy of the surface tension measurements is ± 0.1 mN/m. The equilibrium surface tension was determined as the value obtained when the surface tension became constant within 0.1 mN/m for 10 min. Near CMC, about 3–5 h were usually required to reach the equilibrium. All measurements were carried out in a thermostated device maintained at a constant temperature of 25, 35, 40, and 45 °C. All glassware and Wilhelmy glass plate were washed in an aqueous solution of chromic acid and sulfuric acid, rinsed in triple distilled water and dried before each measurement.

Results and discussion

Effect of polyoxyethylene chain length on CMC

The surface-tension data measured for the aqueous $C_{12}E_n$ solutions at 35, 40 and 45 °C are shown in Fig. 1. As is seen in the figure, a minimum in the measured surface-tension curves becomes larger as the increasing polyoxyethylene chain length, showing that the present $C_{12}E_n$ ($n = 10, 20, 30$, and 40) still contain impurities with different polyoxyethylene chain lengths. The CMC of the surfactants was determined as a concentration indicated with arrows in Fig. 1. The CMC, the value of constant surface-tension γ, and the area A per molecules obtained for the $C_{12}E_n$ ($n = 10, 20, 30$, and 40) are summarized in Table 1. The CMC values increase with increasing length of polyoxyethylene chain, not considerably at $n \simeq 40$. Figure 2 shows the plot of log CMC vs. log n at 25 °C. It is apparent that two straight lines cross each other around $n = 10$. This result suggests that two different mechanisms are responsible for the micelle formation of $C_{12}E_n$ ($n = 6, 7$, and 10) and of $C_{12}E_n$ ($n = 10, 20, 30$ and 40). Some polymer-like behavior of the long polyoxyethylene chain may become significant in micelle formation for the latter surfactants. Another reason for this inflection might be ascribed to the distribution of the polyoxyethylene chain length in the present surfactant ($n > 10$). A final conclusion on this point should not be made until purer surfactants ($n > 10$) are synthesized.

Temperature effect on CMC and ΔG_m

The values of CMC at various temperatures are given in Table 1. The CMC values for each surfactant decrease with

Fig. 1 Surface tension vs. logarithm concentration of $C_{12}E_{10}$ (○), $C_{12}E_{20}$ (△), $C_{12}E_{30}$ (□), and $C_{12}E_{40}$ (◇) at 35, 40, and 40 °C

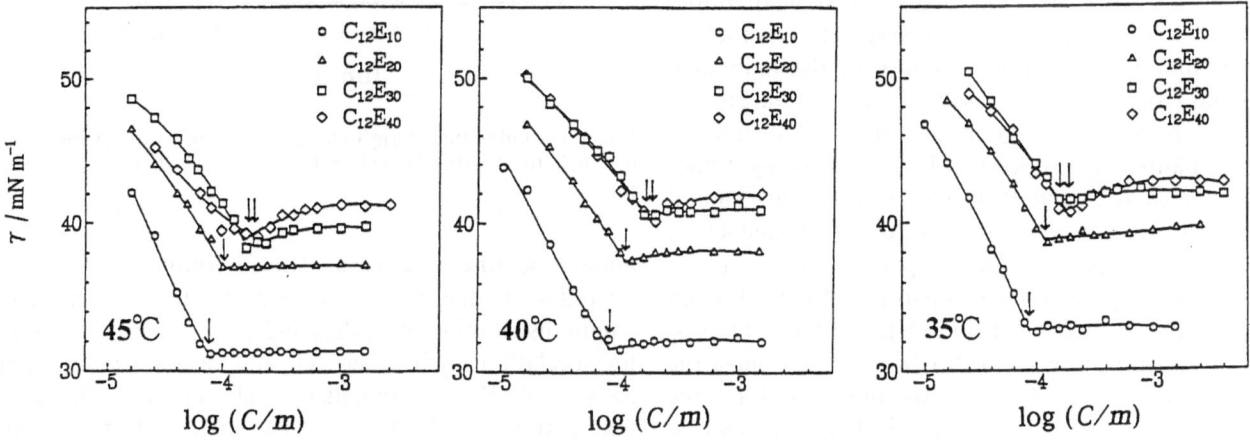

Table 1 Surface properties and *CMC* of the dodecyl polyoxyethylene surfactants $C_{12}E_n$, ($n = 10, 20, 30$ and 40) in aqueous solution at various temperatures. The estimated uncertainties in *CMC*, γ, and A are ± 0.05 mM, ± 0.1 mN/m, and ± 0.3 Å² respectively

Surfactant	t [°C]	*CMC* [mN]	γ [mN/m]	A [Å²/mol]
$C_{12}E_{10}$	25	0.0887	34.2	58.3
	35	0.0847	32.9	59.2
	40	0.0818	32.1	62.1
	45	0.0743	31.4	61.6
$C_{12}E_{20}$	25	0.174	40.7	77.2
	35	0.117	39.5	88.1
	40	0.11	38.1	88.5
	45	0.101	37.2	89.4
$C_{12}E_{23}$[a]	25	0.06		
	55	0.3		
$C_{12}E_{28}$[b]	15	0.124		
	25	0.079		
	40	0.042		
	55	0.024		
$C_{12}E_{30}$	25	0.236	43.7	87.5
	25	0.08[a]		
	35	0.162	41.8	94.2
	40	0.164	40.8	106
	45	0.164	39.6	108
	55	0.04[a]		
$C_{12}E_{40}$	25	0.279	44.9	85.1
	35	0.18	42.7	96.9
	40	0.175	41.8	112
	45	0.171	41.2	122

[a] Ref. [1].
[b] Ref. [7].

increasing temperature, but those for $C_{12}E_{10}$ do in a different manner from those of the other surfactants.

In order to further discuss the temperature effect by a scaling theory in polymer science, the quantity of $\log CMC$ was converted to the standard Gibbs free-energy change of micellization ΔG_m by eq. (1).

$$\Delta G_m = 2.303 \, RT \log X_{CMC} \qquad (1)$$

ΔG_m may be calculated by choosing for the standard initial state of the nonmicellar surfactant species a hypothetical state at unit mole fraction x, but with the individual surfactant molecules behave as at infinite dilution, and for the standard final state, the micelle itself. The values of ΔG_m are plotted against $\log n$ in Fig. 3. The straight lines were obtained at 25 and 35 °C; their slopes were 4.98 and 3.33, respectively. At higher temperatures 40 °C and 45 °C, however, the values of ΔG_m were separated into two parts: the short polyoxyethylene chain part ($n = 10$ and 20) and the long polyoxyethylene chain one ($n = 30$ and 40). For the short-chain part the slope (2.62) of ΔG_m is almost the same as that (3.33) at 35 °C; on the other hand, for the long-chain part the slope of ΔG_m is 1.18, the smallest

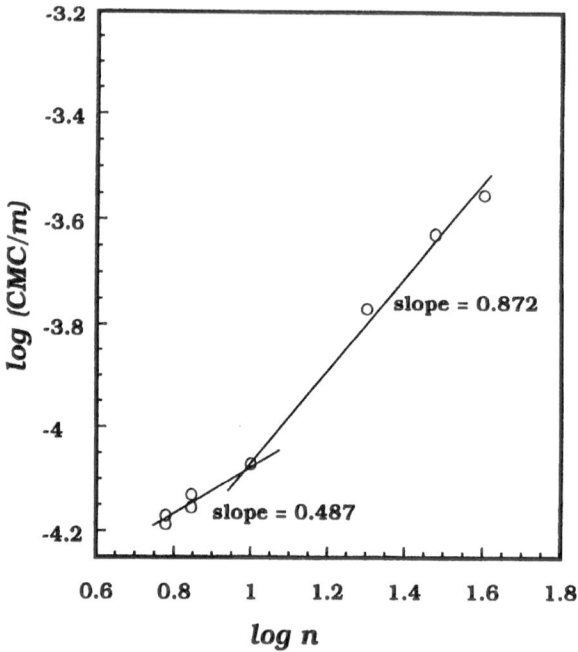

Fig. 2 $\log CMC$ vs. $\log n$ for the $C_{12}E_n$ ($n = 6, 7, 10, 20, 30,$ and 40) at 25 °C

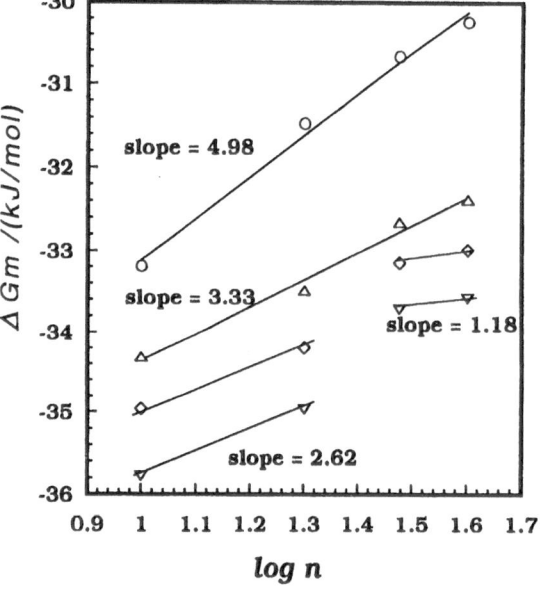

Fig. 3 The Gibbs free energy of micellization ΔG_m vs. $\log n$ for $C_{12}E_n$ ($n = 6, 7, 10, 20, 30$ and 40) at 25 °C (○) 35 °C (△), 40 °C (□) and 45 °C (◇)

among the other three cases. This discontinuity in ΔG_m at 40 and 45 °C does not probably originate from the chain length distribution of $C_{12}E_{30}$ and $C_{12}E_{40}$ because a good linear relationship between ΔG_m and $\log n$ was obtained at 35 °C and 25 °C. Fujimatsu et al. reported the theta temperature of a $C_{12}E_{9.3}$ micellar solution to be around

Progr Colloid Polym Sci (1997) 106:70–74
© Steinkopff Verlag 1997

45 °C from light scattering measurement and free-boundary experiment over a temperature range from 30 °C to 80 °C [11]. Since this theta temperature of 45 °C is near the transition temperature found in this study, a kind of coil–globule transition [12, 13] may occur in the long polyoxyethylene chain region at the present micellar surface.

Temperature effect on area per molecule

In the previous section we concluded that a coil–globule-like transition occurs for the micelles of long polyoxyethylene surfactants between 35 and 40 °C. A similar transition may take place at the air–water interface. From the slope of the curves of the surface-tension γ vs. $\log C$ below the CMC and the Gibbs equation, the surface concentration Γ and the area per molecule A $(= 1/\Gamma)$ were calculated at each temperature. The results are given in Table 1 and the values are plotted against temperature in Fig. 4. The area per molecule A for all surfactants increase with increasing temperature; the increment of area with temperature becomes larger and larger with lengthening polyoxyethylene chain. In the figure a similar transition in A is observed between 35 and 40 °C for $C_{12}E_{30}$ and $C_{12}E_{40}$ as found for ΔG_m in Fig. 3, implying the coil–globule transition also occurs in the long polyoxyethylene chain at the air–water interface.

In the literature, the area per molecule was reported by using a Wilhelmy tension meter for $C_{12}E_{14}$, $C_{12}E_{23}$, and $C_{12}E_{30}$ [1] and for $C_{12}E_{28}$ [7]. Contrary to the present results, the reported values in ref. [1] were constant or decreased only by 2 Å^2 in spite of lengthening polyoxyethylene chain when the temperature was elevated from 25 °C to 55 °C. They explained the decreasing tendency of the area with temperature in terms of dehydration of the polyoxyethylene chain at elevated temperature. However, our recent small-angle neutron-scattering study on $C_{12}E_{40}$ micelles at 25 °C has revealed that the polyoxyethylene layer contains only 5.9 hydrated water molecules per $C_{12}E_{40}$ surfactant molecule; thus, the dehydration effect does not seem significant at the air–water interface for the $C_{12}E_n$ $(n = 23–30)$. Furthermore, the CMC value for $C_{12}E_{28}$ was reported to be 0.079 mM at 25 °C, which corresponds to the values for $C_{12}E_n$ $(n < 10)$ in the present study and in ref. [1]. Therefore, the A values reported in refs. [1, 7] seem less accurate for the surfactants with long polyoxyethylene chains; this might be due to that the surfactants measured contain impurities with different polyoxyethylene chains.

The area per molecule for $C_{16}E_n$ $(n = 17, 32, 47,$ and 63) were also measured with a du Nouy tension meter with a platinum ring [8]. The values increases by 4–6 Å^2 when the temperature was raised from 25 °C to 40 °C; this trend is consistent with the present findings, but their increments with temperature are much lower than those obtained in

Fig. 4 Area per molecule A vs. temperature curves of $C_{12}E_{10}$ (○), $C_{12}E_{20}$ (△), $C_{12}E_{30}$ (◇), and $C_{12}E_{40}$ (▽)

Fig. 5 Logarithm radius of micelle R vs. $\log n$ for the $C_{12}E_n$ $(n = 6, 7, 10, 20, 30$ and 40) at 25 °C (○) 35 °C (△), 40 °C (◇) and 45 °C (▽)

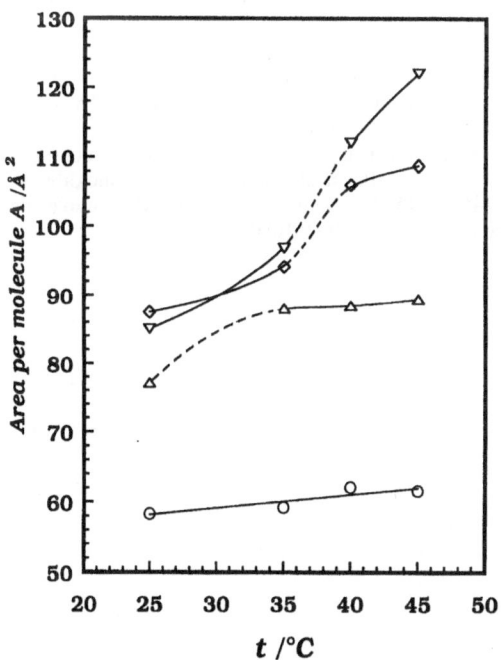

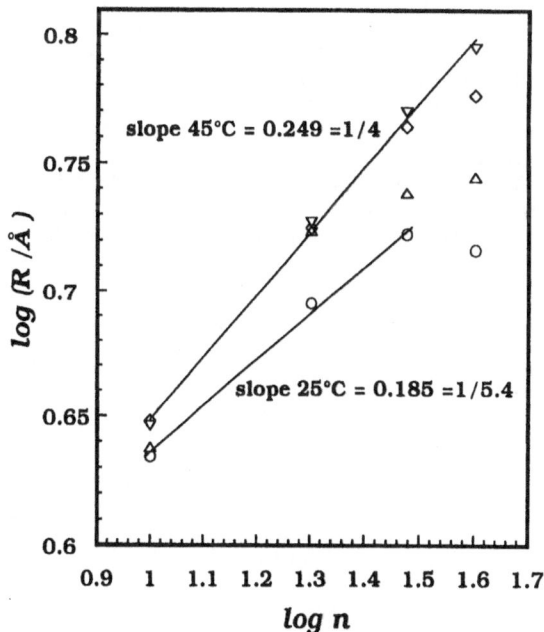

the present study, in particular, for the surfactants with $n = 30$ and 40. This disagreement in A may be ascribed to the different methods employed in the surface tension measurements.

It is well known that the end-to-end distance or the radius of gyration for linear polymers has a good correlation with the degree of polymerization by the scaling theory. In the scaling theory, the power law of the coil–globule transition below the theta temperature for polymers shows the power of n changing from $\frac{1}{6}$ to $\frac{1}{2}$. The coil globule transition is observed in a gradual change (crossover) in the radius for short polymers, but it occurs as the first-order phase transition when polymers become very long [14].

By assuming that the area per molecule A is equal to πR^2, where R is the radius of a disk containing one polyoxyethylene chain at the air–water interface, the value of $\log R$ are plotted against $\log n$ in Fig. 5. The linear relationship between $\log R$ and $\log n$ was obtained at $40\,^\circ C$ and $45\,^\circ C$. The power of n for polymers in a coil state was $\frac{1}{2}$, different from the present value ($\frac{1}{4}$) at $40\,^\circ C$ and $45\,^\circ C$. At $25\,^\circ C$ another linear relationship was obtained, except for $C_{12}E_{40}$, and the power of n obtained was $1/5.4$ instead of $\frac{1}{6}$ at $25\,^\circ C$. Thus, for the transition of the polyoxyethylene chain at the adsorption layer, the power of n changed from $1/5.4$ to $\frac{1}{4}$. This transition probably corresponds to the conformational transition of the polyoxyethylene chain, because the scaling rule holds in this system and the power of n increases from the low-temperature state to the high-temperature state as observed for the coil–globule transition of bulk solution of polymer. The reasons for smaller n obtained in this work than expected for polymers may be as follows: (i) one end of the polyoxyethylene chain is fixed to the air–water interface by a dodecyl group; (ii) the segment of the polyoxyethylene chain is difficult to extrude to air, therefore they are more restricted than in bulk solution. The effect of a fixed terminal end of the polymer has already been discussed by de Gennes [15]. As is seen in Fig. 5 the polyoxyethylene chains of all surfactants are in their expanded state at $45\,^\circ C$. The expanded state will be transformed to more condensed state with decreasing the temperature to $25\,^\circ C$. Our small-angle neutron-scattering experiment on the $C_{12}E_{40}$ micelles has confirmed that this condensed state of the polyoxyethylene chain at the micellar surface contains only 5.9 hydrated water per $C_{12}E_{40}$ molecule.

Acknowledgments We express our sincere thanks to Prof. Jacob Israelachvili of University of California, Santa Barbara for valuable suggestions for Fig. 2. We are also grateful to Mikuo Akimaru (Nikko Chemicals CO., Ltd.), who contributed his technical assistance to the accomplishment of the experimental program.

References

1. Schick MJ (1962) J Colloid Sci 17: 801–813
2. Schick MJ (1963) J Phys Chem 67: 1796–1799
3. Crook EH, Fordyce DB, Tebbi GF (1963) J Phys Chem 67:1987–1994
4. Ohoba N, Takahashi A (1968) Chem Phys Appl Prat Ag Surface CR Congress Int Deterg 5th 2:481–1991
5. Schick MJ (1987) In: Nonionic Surfactants–Physical Chemistry. Marcel Dekker, New York, and the references therein
6. Nikas YJ, Puvvada S, Blankschtein D (1992) Langmuir 8:2680–2689
7. Schott H (1969) J Pharm Sci 58: 1521–1524
8. Barry BW, El Eini DID (1975) J Colloid Interface Sci 54:339–347
9. Ogino K, Kakihara T, Uchiyama H, Abe M (1988) J Amer Oil Chem Soc 65:405–347
10. Reddy NK, Foster A, Styring MG, Booth C (1990) J Colloid Interface Sci 134:588–592
11. Fujimatsu H, Takagi K, Matsuda H, Kuroiwa S (1983) J Colloid Interface Sci 94:237–242
12. Sun ST, Nishio I, Swislow G, Tanaka T (1980) J Chem Phys 73:5971–5975
13. Chu B, Ying Q, Grosberg AY (1995) Macromolecules 28:180–189
14. Tanaka F (1994) In: Introduction to Physical Polymer Science. Shokabo
15. De Gennes PG (1980) Macromolecules 13:1069–1080

Progr Colloid Polym Sci (1997) 106:75–78
© Steinkopff Verlag 1997

S. Komura
H. Kodama

Microemulsions under steady shear flow

Received: 13 October 1996
Accepted: 7 March 1997

Dr. S. Komura (✉)
Department of Mechanical System
Engineering
Kyushu Institute of Technology
Iizuka 820, Japan

Dr. H. Kodama
Graduate School of Science and Technology
Kobe University
Kobe 657, Japan

Abstract Dynamic response of microemulsions to shear deformation on the basis of two-order-parameter time dependent Ginzburg–Landau model is investigated by means of cell dynamical system approach. Time evolution of anisotropic factor and excess shear stress under steady shear flow is studied by changing shear rate and total amount of surfactant. As the surfactant concentration is increased, overshoot peak height of the anisotropic factor increases whereas that of the excess shear stress is almost unchanged.

Key words Microemulsions – rheology – shear flow – Ginzburg–Landau model

Microemulsions being mixture of oil, water and surfactant are known to exhibit various interesting mesoscopic structures depending on the temperature or the composition [1]. For relatively low concentration of surfactant and if the volumes of oil and water are not very different, microemulsions form a bicontinuous structure where a multiply connected randomly oriented monolayer of surfactants separate oil-rich and water-rich subvolumes with a mesoscopic length scale (10 ~ 100 nm).

When one applies an external flow to the microemulsion system, its mechanical response is deeply affected by its internal structure. Using a single-order-parameter time-dependent Ginzburg–Landau (TDGL) model, Mundy et al. have investigated rheological properties of microemulsions theoretically [2]. In their model, the order parameter represents the concentration difference between oil and water, and the presence of surfactants is taken into account through the surface tension parameter. Their work has been extended by Pätzold and Dawson, and it was shown that the microemulsions behave in an essentially non-Newtonian manner [3].

In this proceeding, we investigate the rheological properties of microemulsions using a two-order-parameter TDGL model. The two-order-parameter Ginzburg–Landau free energy for microemulsions has already been proposed by Laradji and his coworkers to study their microphase separation dynamics [4]. However, the proposed free energy is not applicable to the rheological study, since the pattern does not deform even in the presence of a flow [5]. This is related to the fact that their free energy is unbounded from below for the configurations with large surfactant concentration at the oil/water interfaces [1]. Thus, we use a different free energy which has been proposed in our previous paper as an improved model [5].

Let $\psi(\mathbf{r})$ describe the local concentration difference between oil and water, and $\rho(\mathbf{r})$ the local surfactant concentration. What we have assumed in our model is that (i) the profiles of ψ and ρ at oil/water interfaces do not depend on the average values of ψ and ρ (denoted hereafter as $\bar{\psi}$ and $\bar{\rho}$, respectively) and that (ii) the coarse-graining dynamics of ψ based on the free energy becomes slow when the amplitude of ρ at the interface takes a certain saturated value. One of the minimum models which reproduce the properties of microemulsions is [5]

$$F = \int d\mathbf{r}\,[w(\nabla^2\psi)^2 + d(\nabla\psi)^2 - a\psi^2 + u\psi^4$$

$$+ e\rho^2(\rho - \rho_s)^2 - s\rho(\nabla\psi)^2]\,, \tag{1}$$

where w, d, a, u, e, ρ_s and s are positive constants. The first term $w(\nabla^2\psi)^2$ with positive w prevents the model from becoming unbounded, whereas the double-minimum potential $e\rho^2(\rho - \rho_s)^2$ guarantees that ρ locally takes either 0 or ρ_s. The last term $-s\rho(\nabla\psi)^2$ favors the surfactants to sit at the oil/water interfaces [4].

For the evolution of $\psi(\mathbf{r}, t)$ and $\rho(\mathbf{r}, t)$, we assume the standard TDGL equations with a macroscopic flow $\mathbf{v}$. Since both ψ and ρ are conserved quantities, TDGL

equations are given by

$$\frac{\partial\psi}{\partial t} + \nabla\cdot(\mathbf{v}\psi) = M_\psi\nabla^2\frac{\delta F}{\delta\psi} + \eta_\psi(\mathbf{r}, t)\,,$$

$$\frac{\partial\rho}{\partial t} + \nabla\cdot(\mathbf{v}\rho) = M_\rho\nabla^2\frac{\delta F}{\delta\rho} + \eta_\rho(\mathbf{r}, t)\,. \tag{2}$$

Here M_ψ and M_ρ are transport coefficients, η_ψ and η_ρ represent the thermal noise which satisfy the fluctuation–dissipation theorem $\langle\eta_{\psi(\rho)}(\mathbf{r}, t)\eta_{\psi(\rho)}(\mathbf{r}', t')\rangle = -2k_B T M_{\psi(\rho)} \times \nabla^2\delta(\mathbf{r} - \mathbf{r}')\delta(t - t')$, where k_B is the Boltzmann constant and T the temperature. As regards the macroscopic flow in Eq. (2), we consider a simple shear flow $v_x(\mathbf{r}) = \dot{\gamma}y$, $v_y = v_z = 0$, where the shear rate $\dot{\gamma}$ is the time derivative of the strain γ. In our work, we have entirely ignored the

Fig. 1 Time evolution of ψ (left) and ρ (right) for (a) $\bar{\rho} = 0.1$, $\dot{\gamma} = 2 \times 10^{-4}$ ($\dot{\gamma}\tau_0 = 0.1$) and (b) $\bar{\rho} = 0.1$, $\dot{\gamma} = 2 \times 10^{-3}$ ($\dot{\gamma}\tau_0 = 1$)

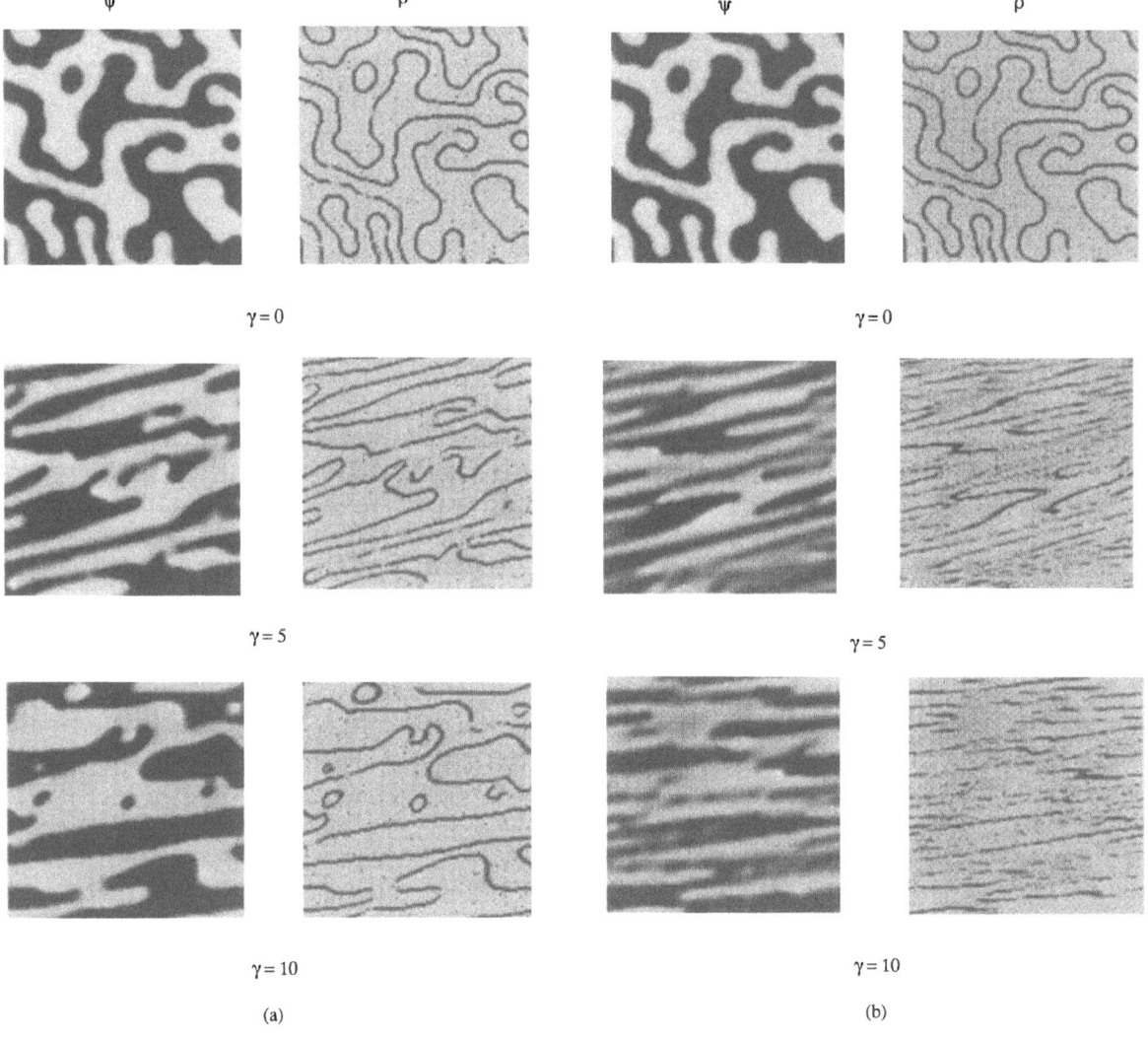

$$\psi \qquad\qquad \rho \qquad\qquad\qquad \psi \qquad\qquad \rho$$

$$\gamma = 0 \qquad\qquad\qquad\qquad\qquad \gamma = 0$$

$$\gamma = 5 \qquad\qquad\qquad\qquad\qquad \gamma = 5$$

$$\gamma = 10 \qquad\qquad\qquad\qquad\qquad \gamma = 10$$

$$(a) \qquad\qquad\qquad\qquad\qquad (b)$$

Progr Colloid Polym Sci (1997) 106: 75–78
© Steinkopff Verlag 1997

hydrodynamic interactions which might play an important role in microemulsions.

We solved the above time-evolution equations numerically by using the "cell dynamical system" approach proposed by Oono et al. [6]. The details of the numerical method have been mentioned in our previous paper [5]. The simulation conducted here is essentially equivalent to solving Eq. (2) with the parameter values fixed as $w = 0.1$, $d = 0.25$, $a = 0.3$, $u = 0.18$, $e = 0.125$, $\rho_s = 1$, $s = 0.25$ and $M_\psi = M_\rho = 0.05$. The noise terms are treated as a random noise with an amplitude 0.02. We fixed $\bar{\psi}$ as $\bar{\psi} = 0$, whereas $\bar{\rho}$ has been changed as $\bar{\rho} = 0.1, 0.2, 0.3$ and 0.4. Here, we restricted ourselves to a two-dimensional system. The system size is chosen as 128×128, and the sheared-periodic boundary condition is imposed both for ψ and ρ. We first started from random uniform distributions of ψ and ρ in the range $[\bar{\psi} - 0.01, \bar{\psi} + 0.01]$ and $[\bar{\rho} - 0.01, \bar{\rho} + 0.01]$, respectively. We relaxed them according to Eq. (2) without any shear flow up to $t = 5 \times 10^5$, and then applied the shear flow with a constant shear rate $\dot{\gamma} = 2 \times 10^{-4}$, 5×10^{-4}, 1×10^{-3} and 2×10^{-3}. Since we have not included any hydrodynamic interactions, our model lacks a bare viscous time scale against which we can compare these shear rates. Nevertheless, we can choose the model intrinsic time scale τ_0 as the inverse of the initial growth rate of the most unstable mode, which is estimated as $\tau_0 \approx 5.0 \times 10^2$ for the above parameter values. The investigated shear ranges from weak to medium shear ($\dot{\gamma}\tau_0 = 0.1 \sim 1$).

Typical time evolutions of ψ and ρ are shown in Fig. 1 for (a) $\bar{\rho} = 0.1$, $\dot{\gamma} = 2 \times 10^{-4}$ ($\dot{\gamma}\tau_0 = 0.1$) and (b) $\bar{\rho} = 0.1$, $\dot{\gamma} = 2 \times 10^{-3}$ ($\dot{\gamma}\tau_0 = 1$). By changing $\bar{\rho}$ and $\dot{\gamma}$, we found the following general behaviors. When the shear rate is small (Fig. 1a), surfactants move under the flow keeping themselves attached to the oil/water interfaces. The total amount of the interface does not seem to change appreciably during the deformation. On the other hand, when the shear rate is large (Fig. 1b), some surfactants are blown off the interfaces by the shear. The coagulation and break-up processes take place as has been observed in the spinoidal decomposition under steady shear flow [7], and the total amount of the interface increases.

Given the evolving patterns, we have evaluated the anisotropic factor defined by [2, 3]

$$Q_{xy} = -2d\left\langle \frac{\partial \psi}{\partial x} \frac{\partial \psi}{\partial y} \right\rangle - 4w\left\langle \frac{\partial^2 \psi}{\partial x \partial y} \nabla^2 \psi \right\rangle, \qquad (3)$$

where $\langle \cdots \rangle$ denotes the average over the total volume. Although this quantity essentially represents the xy-component of the macroscopic excess stress tensor in the case of one-order-parameter model, this is not the case in the present model since there should be a contribution to the

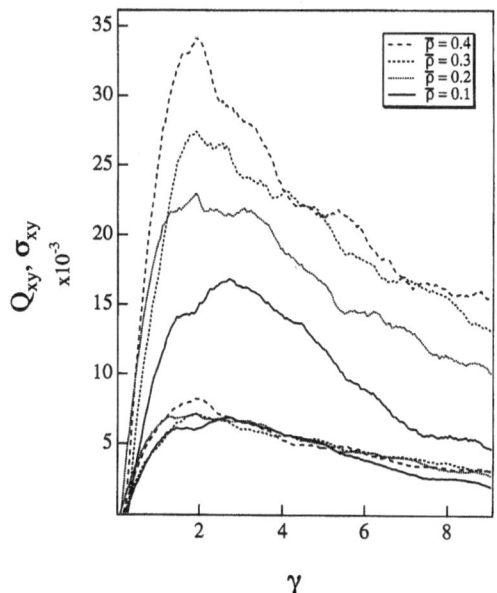

Fig. 2 The anisotropic factor Q_{xy} (upper 4 curves, see Eq. (3)) and the shear stress σ_{xy} (lower 4 curves, see Eq. (4)) as a function of the shear strain $\gamma = \dot{\gamma}t$ for fixed $\dot{\gamma} = 2 \times 10^{-4}$ ($\dot{\gamma}\tau_0 = 0.1$) and $\bar{\rho} = 0.1, 0.2, 0.3, 0.4$

stress due to the non-local coupling term in Eq. (1). Nevertheless, the problem of stress division between ψ and ρ is theoretically not yet clear. Instead, we propose here the following quantity:

$$\sigma_{xy} = -2\left\langle (d - s\rho)\frac{\partial \psi}{\partial x} \frac{\partial \psi}{\partial y} \right\rangle - 4w\left\langle \frac{\partial^2 \psi}{\partial x \partial y} \nabla^2 \psi \right\rangle, \qquad (4)$$

which is assumed to express the excess shear stress. Figure 2 shows the plots of Q_{xy} and σ_{xy} as a function of the shear strain γ for several values of $\bar{\rho}$. Here the shear rate is fixed as $\dot{\gamma} = 2 \times 10^{-4}$ ($\dot{\gamma}\tau_0 = 0.1$). In the time region where Q_{xy} and σ_{xy} increase the domains are elongated, whereas these quantities start to decrease once the burst and the recombination take place. We observed that the strain giving the peak position of Q_{xy} and σ_{xy} is almost constant throughout the present simulation ($\gamma \approx 2$). On the other hand, the peak height of Q_{xy} is larger than that of σ_{xy} as a whole. $\bar{\rho}$ dependencies of the peak height of Q_{xy} and σ_{xy} are also different; the peak height of Q_{xy} increases linearly with $\bar{\rho}$, while that of σ_{xy} is almost independent of $\bar{\rho}$. However, a clear shear rate dependence of the peak height of σ_{xy} could be observed as in ref. [7].

Details of the present work have been published elsewhere [8].

This work is supported by the Ministry of Education, Science and Culture, Japan (Grant-in-Aid for Scientific Research Nos. 08226233, 08740324 and 0657).

References

1. Gompper G, Schick M (1994) Self-Assembling Amphiphilic Systems. Academic Press, New York
2. Mundy CJ, Levin Y, Dawson K (1992) J Chem Phys 97:7695–7698
3. Pätzold G, Dawson K (1996) J Chem Phys 104:5932–5941
4. Laradji M, Guo H, Grant M, Zuckermann MJ (1991) J Phys A 24:L629–L635; (1992) J Phys Condens Matter 4: 6715–6728
5. Komura S, Kodama H (1997) Phys Rev E 55:1722–1727
6. Oono Y, Puri S (1987) Phys Rev Lett 58: 836–839; (1988) Phys Rev A 38:434–453
7. Ohta T, Nozaki H, Doi M (1990) Phys Lett A 145:304; (1990) J Chem Phys 93:2664–2675
8. Kodama H, Komura S (1997) J Phys II France 7:7–14

Progr Colloid Polym Sci (1997) 106:79–82
© Steinkopff Verlag 1997

R. Kawai-Hirai
Y. Matsumoto
M. Hakoda
M. Oya

Interparticle interaction in aerosol-OT microemulsion system studied by dynamic light scattering and viscosity

Received: 13 October 1996
Accepted: 14 March 1997

Dr. R. Kawai-Hirai (✉)
Meiwa Women's Junior College
Maebashi
Gunma 371, Japan

Y. Matsumoto · M. Hakoda · M. Oya
Faculty of Engineering
Gunma University
Kiryu, Gunma 386, Japan

Abstract Aerosol–OT water-in-oil microemulsions in isooctane was studied. By using dynamic light scattering and viscosity we measured two coefficients relating to hydrodynamic interparticle interaction, that is, α from DLS and Huggins coefficient k_H from viscometry. The W_0 ($=[H_2O]/[AOT]$) dependence of these coefficients in the range of $W_0 = 5$–15 showed a critically different tendency compared to that in the rest of the range of W_0 measured ($W_0 = 15$–50). The present results are in good agreement with our previous evidences which show the existence of an oligomeric phase in the W_0 range 5–15 using small-angle X-ray scattering.

Key words w/o microemulsion – AOT – interparticle interaction – dynamic light scattering – viscosity

Introduction

Microemulsion systems have been attractive scientific concerns from various viewpoints [1–3]. The water-in-oil microemulsions where sodium bis(2-ethylhexyl)sulfosuccinate (Aerosol–OT(AOT)) stabilizes water in 2,2,4-trimethylpentane (isooctane) are known to form a transparent dispersion of discrete water droplets with a surfactant layer in a continuous oil phase. In our previous papers using synchrotron radiation small-angle X-ray scattering (SAXS) [4, 5], we clarified the structure of a biological buffer/AOT/isooctane system depending on water–surfactant ratio W_0 ($=[H_2O]/[AOT]$), AOT concentration and temperature. We reported in the above papers that in a high W_0 range ($16 < W_0 < 50$) the formation of the AOT microemulsions satisfies a linear relation between the micellar radius and the W_0 value, which is described by a geometrical model considering the molar volume of water and the surface area per AOT molecule at the interface. This observed linear relation is mostly in agreement with other previous results [6–8]. However, we also found a serious deviation from such a linearity in a low W_0 range ($0 < W_0 < 16$), and we showed that such a deviation is attributable to a transient oligomerization of the microemulsions, namely that with increasing water content the oligomerization of microemulsions proceeds successively from an oligomeric phase ($0 < W_0 < 16$) to a monomeric phase ($16 < W_0$) through a mesophase.

In this report, we will clarify the relation between the previous structural evidences and the hydrodynamic parameters of AOT microemulsions obtained by using dynamic light-scattering (DLS) method and viscometry.

Experimental

Materials

AOT was purchased from Nacalai Tesque Inc. and isooctane from Wako Pure Chemical Industries Ltd. These were used without further purification. Other reagents with special grade were used. Water was purified by a Millipore system. The w/o microemulsions with different W_0 values were prepared by injection method. The mixtures were stored at 30 °C for at least 20 h prior to the measurements. Analyses of the water content in the mixtures were performed by the Karl Fischer method using a Mitsubishi Kasei CA-05 moisture meter.

Dynamic light scattering and viscosity measurements

Dynamic light scattering measurements were performed using DLS-7000 dynamic light scattering photometer (Otsuka Electronics Co.) with 75 mW argon laser operating at 488 nm. Sample solutions were filtered through Millipore 0.22 μm filters. The data were analyzed using the program with DLS-7000. Viscosity measurements were carried out by using an Ubbelohde viscometer. Densities were measured using a standard pyknometer. All the above measurements were performed at 30.0 ± 0.1 °C.

Results and discussion

Dynamic light scattering

DLS measurements give us the collective diffusion coefficient D and the corresponding correlation length l with a relation defined as

$$l = kT/6\pi\eta D , \qquad (1)$$

where k is the Boltzmann constant, η is the solvent viscosity. The collective diffusion coefficient D means a self-diffusion coefficient for a dilute monodisperse system (for small solute particle volume fractions) and a mutual-diffusion coefficient for a condensed monodisperse system (for higher solute particle volume fractions), and the correlation length l corresponds to the hydrodynamic radius at the infinite-dilution limit. Under the present experimental conditions the correlation lengths measured were independent of the scattering angle ($30° < \theta < 150°$) for every W_0 value at 30 °C, then we fixed the scattering angle at 90°. Figure 1a shows the W_0 dependence of the correlation lengths of water/AOT/isooctane solutions with different AOT concentrations. It is evident that the correlation lengths increase with increasing both W_0 value and AOT concentration. As is well known under linear interaction theory, the collective-diffusion coefficient D displays the concentration dependence as presented by

$$D = D_0(1 + \alpha\phi) , \qquad (2)$$

where D_0 is the infinite-dilution limiting value of D, ϕ the volume fraction of AOT and water in the solution, and the α provides an index of interparticle interactions quantitatively [9–11]. Thus, the observed values of l in Fig. 1a

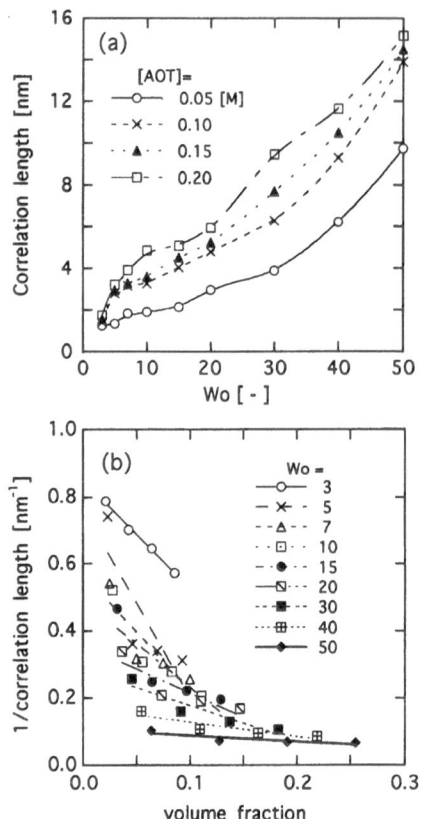

Fig. 1 (a) W_0 ($=$[H$_2$O]/[AOT]) dependence of correlation length of water/AOT/isooctane system at various AOT concentrations at 30 °C. (b) ϕ (the volume fraction of AOT and water in the solution) dependence of the inverse of correlation length at various W_0 values at 30 °C. AOT concentration and W_0 value were varied from 0.05 to 0.20 M and from 3 to 50, respectively. In Fig. 1b, the linear lines were obtained by least-squares fit based on Eq. (2)

includes some contribution from interparticle interactions, and can be equated with particle hydrodynamic radii at the infinite-dilution limit. We showed previously that the size and shape of the microemulsions in the similar systems mostly do not depend on AOT concentration [4]. Then, by using Eq. (2) we can estimate both values of the hydrodynamic radius r_H and the coefficient α, as shown in Fig. 1b. Figure 2a shows the W_0 dependence of the obtained hydrodynamic radii estimated from Fig. 1b. Assuming the linear relation between r_H (nm) and W_0, the data in Fig. 2a is mostly described by

$$r_H = 0.16 + 0.16 W_0 . \qquad (3)$$

It seems that the value of the slope is in good agreement with the value estimated from the simple geometrical model considering the molar volume of water and the surface area per AOT molecule at the interface [12–14], however, the relation between r_H and W_0 does not satisfy a simple linearity in the whole W_0 range measured.

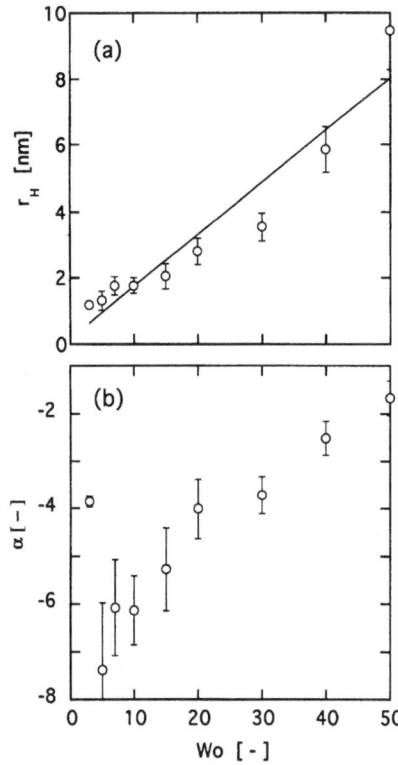

Fig. 2 Hydrodynamic radius r_H and interparticle interaction coefficient α of water/AOT/isooctane system at 30 °C as functions of W_0: (a) r_H; (b) α. The values of r_H and α were estimated from Fig. 1b

Figure 2b shows the variation of α depending on W_0 at 30 °C. In the presence of repulsive interactions between hard spheres α is thought to be about 1.5, while in the presence of attractive interactions it is presumed to be a negative value [15]. The α values obtained were negative in the whole W_0 range, suggesting that an attractive interparticle interaction exists in the present AOT/isooctane system. The α values obtained are smaller than that obtained in AOT/n-hexane [16], suggesting that an attractive interaction between particles in the present system is relatively stronger compared to that in AOT/n-hexane system. Moreover, in Fig. 2b we can well recognize that the W_0 dependence of the α value shows a large negative value at $5 < W_0 < 15$ within the experimental error, indicating that the strength of the attractive interaction would become stronger in this range. Alternatively, in the W_0 region from 5 to 15 the collision probability between microdroplets probably increases, resulting in the increase of the oligomers of the microemulsions. This picture agrees very well with our previous SAXS results that show a transient oligomerization phase of microemulsions in the W_0 range below 12 for the similar AOT system as used here.

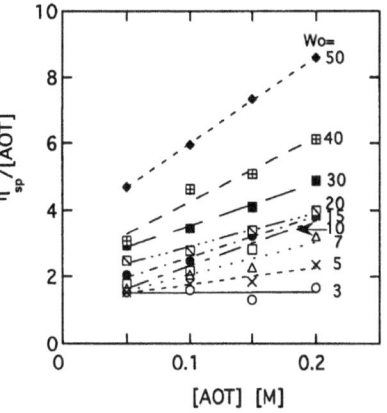

Fig. 3 η_{sp}/[AOT] of water/AOT/isooctane system as a function of AOT concentration at 30 °C. AOT concentration and W_0 value were varied as in Fig. 1. The linear lines were obtained by least-squares fit based on Eq. (5)

Viscosity measurements

Figure 3 shows the plots of η_{sp}/[AOT] against [AOT] for various W_0 values, where η_{sp} is the specific viscosity given by

$$\eta_{sp} = (\eta - \eta_{solvent})/\eta_{solvent} . \tag{4}$$

The plots for each W_0 in Fig. 3 were fitted by

$$\eta_{sp}/[AOT] = a + b[AOT] \tag{5}$$

to obtain the coefficients of a and b. On the other hand, η_{sp} is usually given by a virial-type expansion using the volume fraction Φ of the suspended particles as

$$\eta_{sp} = [\eta]\Phi + k_H[\eta]^2\Phi^2 + \cdots , \tag{6}$$

where $[\eta]$ is the intrinsic viscosity and k_H the Huggins coefficient indicating the contribution of the hydrodynamic interactions among particles to the intrinsic viscosity. In the case of hard spheres the k_H is known to range from 0.7 to 0.8. In this system the Φ of AOT microemulsions can be expressed as

$$\Phi = [AOT](V_{AOT} + W_0 V_{H_2O} + \Delta V) , \tag{7}$$

where V_{H_2O} is the partial volume of H_2O, V_{AOT} the partial molar volume of AOT molecules, and ΔV an additional volume by the solvation. By using Eqs. (5)–(7), the following relation is derived as

$$k_H = b/a^2 . \tag{8}$$

By using Eq. (8) we can estimate k_H values in the W_0 range of 3–50, as shown in Fig. 4. Evidently, we can recognize a strong dependence of k_H on W_0, and cannot describe this dependence based on an assumption of a simple hard-sphere potential between solute colloidal spheres. The k_H variation depending on W_0 suggests the solvation state of the solute particles and the presence of a long-range

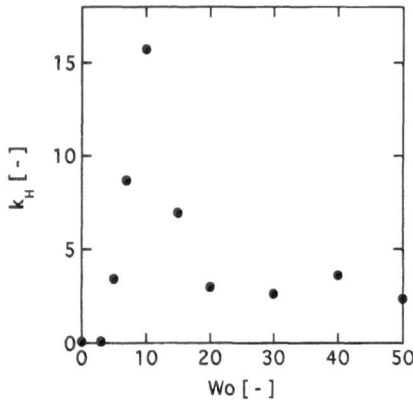

Fig. 4 Huggins coefficient k_H of water/AOT/isooctane system as a function of W_0 at 30 °C. The values of k_H were estimated by Eq. (8) using the linear relations in Fig. 3

interaction among them. Moreover, the k_H takes an evident maximum value in the W_0 range of 7–15, which suggests that the state of microemulsions in the W_0 range of 7–15 is critically different from that in the rest region of W_0 ($W_0 = 0$–7 and 15–50). These results are well consistent with the above results of the DLS measurements.

Conclusions

The present DLS analysis clearly shows the presence of an attractive interparticle interaction in the H_2O/AOT/

isooctane microemulsions depending on the W_0 value. The viscosity measurements suggest the long-range interactions among the solute particles also depend on the W_0 value. Both interactions, suggested by the DLS and viscosity measurements, appear most evidently in the W_0 region from 7 to 15. Then we can conclude that in the W_0 range from 7 to 15 there is a clearly different state of dispersion, showing an attractive intermicellar interaction in the H_2O/AOT/isooctane w/o microemulsions. This situation is very different from that observed in the other W_0 region studied well in the previous reports. In addition, the present results agree very well with our previous SAXS results showing the presence of an oligomeric phase around this W_0 region. We should mention a little difference between samples in the present study and in the previous SAXS study [4, 5]. We used pure water and biological buffer in the present study and previous SAXS studies, respectively. However, in the same W_0 region of 5–15 which we named in the previous report as an oligomeric-phase region, we have also observed the evidence of the presence of attractive interparticle interactions. Therefore, there is no doubt of the existence of the mesophase, and the region of this mesophase would be independent of the presence of buffer salt (Hepes) injected in aqueous solution.

Acknowledgments We thank Professor K. Nakamura (The University of Tokyo) for the use of the DLS-7000 and also thank Mr. S. Mii for assisting in DLS measurements.

References

1. Fendler JH, Fendler EJ (1975) Catalysis in Micellar and Macromolecular Systems. Academic Press, New York
2. Luisi PL, Straub BE (eds) (1984) Reversed Micelles. Plenum Press, New York
3. Pileni MP (ed) (1989) Structure and Reactivity in Reversed Micelles. Elsevier, Amsterdam
4. Hirai M, Kawai-Hirai R, Yabuki S, Takizawa T, Hirai T, Kobayashi K, Amemiya Y, Oya M (1995) J Phys Chem 99:6652–6660
5. Hirai M, Kawai-Hirai R, Takizawa T, Yabuki S, Nakamura K, Kobayashi K, Amemiya Y, Oya M (1995) J Chem Soc Faraday Trans 91:1081–1089
6. Zulauf M, Eicke HF (1979) J Phys Chem 83:480–486

7. Day RA, Robinson BH, Clarke JHR, Doherty JV (1979) J Chem Soc Faraday Trans I 75:132–139
8. Pileni MP, Zemb T, Petit C (1985) Chem Phys Lett 118:414–420
9. Felderhof BU (1978) J Phys A 11:929–937
10. Cazabat AM, Langevin D (1980) In: Degiorgio V, Corti M, Giglio M (eds) Light Scattering in Liquids and Macromolecular Solutions. Plenum Press, New York, pp 139–155
11. Cazabat AM, Langevin D (1993) In: Brown W (ed) Dynamic Light Scattering. Clarendon Press, Oxford, pp 555–591

12. Israelachvili JN, Mitchell DJ, Ninham BW (1976) J Chem Soc Faraday Trans II 72:1525–1568
13. Israelachvili JN, Mitchell DJ, Ninham BW (1977) Biochim Biophys Acta 470:185–201
14. Mitchell DJ, Ninham BW (1981) J Chem Soc Faraday Trans II 77:601–619
15. Caszabat AM, Langevin D (1981) J Chem Phys 74:3148–3158
16. Nicholson JF, Clarke JHR (1983) In: Mittal K (ed) Surfactants in Solution. Plenum Press, New York, pp 1663–1674

Progr Colloid Polym Sci (1997) 106:83–85
© Steinkopff Verlag 1997

Y. Terada
H. Maeda
T. Odagaki

Effects of hydrogen bonding on the surface energy of ionic-neutral dodecyldimethylamine oxide micelles

Received: 13 October 1996
Accepted: 4 February 1997

Dr. Y. Terada (✉) · T. Odagaki
Department of Physics
Faculty of Science
Kyushu University
Fukuoka 812-81, Japan

H. Maeda
Department of Chemistry
Faculty of Science
Kyushu University
Fukuoka 812-81, Japan

Abstract We calculate the surface energy of a micelle consisting of neutral and ionized dodecyl-dimethylamine oxides (DAOs) by Monte Carlo simulation based on a lattice model of the micelle. We obtain the dependence of the surface energy on the fraction, α_M, of ionized DDAO and show that the hydrogen bond between DDAOs plays a significant role in determining the properties of micelles. In particular, a conclusion is drawn that a DDAO molecule supports two hydrogen bonds. We also discuss the effects of the coexisting salt and temperature.

Key words Dodecyldimethylamine oxide – cmc – surface energy – hydrogen bond – mixed micelle

Dodecyldimethylamine oxide (DDAO) is a surfactant with an amine-oxide functional group, which has been used in many practical applications. A neutral DDAO molecule (n-DDAO) can be ionized by a hydrogen ion H^+,

$$C_{12}H_{25}N(CH_3)_2O + H^+ \rightleftharpoons C_{12}H_{25}N(CH_3)_2OH^+$$

and one can control the fraction, α_M, of ionized DDAO (i-DDAO) by the pH. Recent systematic experimental studies [1] for a low-concentration DDAO solution have revealed that the critical micelle concentration (CMC) and the surface free energy of the solution depend strongly on α_M and take a minimum as a function of α_M and the aggregation number of a micelle takes a maximum at the same values of α_M. The studies also found that DDAO micelles are stabilized by adding salts to the solution. These properties cannot be explained quantitatively by the existing theories [2].

In this paper, we study the surface energy of a micelle consisting of n-DDAO and i-DDAO by Monte Carlo (MC) simulation and investigate the effects of hydrogen bonding in micelle formation. DDAO micelles consist of a few hundred molecules. We prepare a 16×16 square lattice with periodic boundary conditions and consider the lattice sites as the head of the DDAO molecules. (Although tails of molecules are considered to be on one side of the surface, we do not include the tails explicitly in the following discussion.) We initiate our Monte Carlo simulation by attaching H^+ ions randomly to the lattice sites with a fixed concentration α_M. We assume that an H^+ ion ionizes the amine-oxide group and a hydrogen atom (H atom) is located on one of four bonds leaving the site and that there are no hydrogen bonds (H-bonds) in the initial state of MC simulation. In each step of the succeeding MC simulation, we choose randomly an H atom and move it by the standard Metropolis algorithm. As the final state of each MC step, we move the H atom to one of six adjacent bonds and allow it to form or not to form an H-bond. We also allow the H atom to stay on the same bond (attached to the adjacent site or remain on the same site), making an H-bond if it does not have or breaking an H-bond if it has already. If it is moved to bonds associatd with the neighboring site, we transfer the charge as well. We also take account of the fact that (1) a single bond cannot accommodate more than one H atom and (2) the number of H-bonds formed by a DDAO cannot be more than two.

We calculate the additional surface energy $\Delta E(\alpha_M)$ due to hydrogen bonding and the Coulombic repulsion (we employ the screened Coulomb interaction), which is the measure of the stability of a micelle at α_M compared to those at $\alpha_M = 0$. We can ignore the dipole interaction energy between n-DDAO heads, because it was estimated as much smaller compared to other two energies. In our model system, we can identify three different types of H-bond, depending on the number of H-bonds supported by DDAOs at the end of the H-bond; for type A, both DDAOs support only one H atom under consideration; for type B, both DDAOs support two H-bonds; for type C, one DDAO suports only one H-bond and the other accompanies an H atom (H-bonded or not) besides the H-bond. We assign different energies h_A, h_B and $\frac{1}{2}(h_A + h_B)$ to type A, type B and type C H-bond, respectively.

We show in Fig. 1 the hydrogen bonding energy, the Coulombic energy and the sum of them per DDAO molecule as a function of α_M at room temperature $T_R = 25\,^\circ\text{C}$, salt concentration $C_S = 0.20\,\text{mol}\,\text{dm}^{-3}$ and $h_B/|h_A| = -1.0$ (namely, types A, B and C H-bonds stabilize equally the system). It is clear from this figure that the total energy shows a minimum at a certain value, α_M^*, of α_M. In order to see the effect of DDAO supporting two H-bonds, we show the energy when $h_B/|h_A| = 100$ (that is type B H-bonds are energetically unfavorable) in the inset. Although the total energy for this case also shows a minimum, the overall qualitative behavior does not agree with the experiments in that the total energy shows a rapid increase near $\alpha_M = 1.0$. Comparing our results with experiments for the surface energy of pure n-DDAO and pure i-DDAO, we conclude that a DDAO molecule can support two H-bonds with little difference in energies of types A and B H-bonds [3] and the second H-bonds play a significant role in determining the properties of micelles. This fact, in turn, indicates that H-bonds will form a chain-like cluster structure. We should discuss the hydrogen bonds in our model system briefly in relation to the surface curvature. By using the stick-and-ball model to construct the DDAO chains, we can conclude that the H-bonds chain of more than five DDAOs are hardly formed in two dimensions because of the steric hindrance due to the functional groups. However, our simulation shows that the long H-bonds can be formed. This is due to the fact that our lattice model ignores the steric hindrance. In the actual micelles, the effect of the steric hindrance may be reduced, because the curvature of the micelles surface and the ragged nature of the surface will allow the three-dimensional packing of molecules. Therefore, the assumption of the present model simulation that the long H-bond chains can be formed will be valid in the actual DDAO micelles.

We can understand the α_M dependences of the energies in Fig. 1 as follows. The Coulombic energy will be in

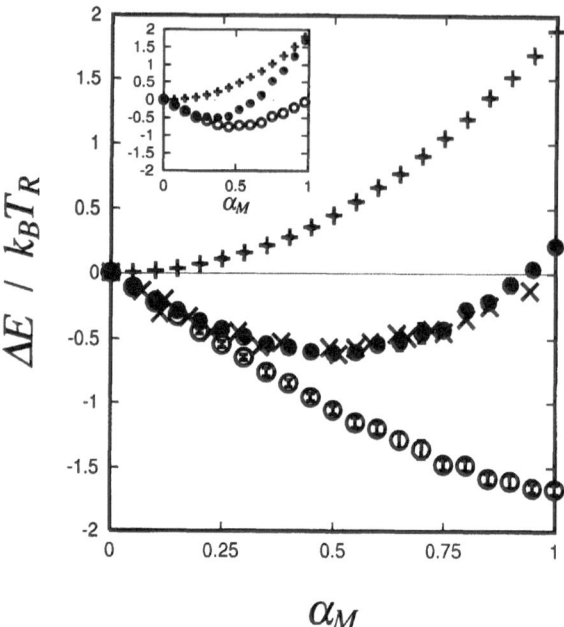

Fig. 1 The hydrogen bonding energies (open circles), and Coulomb energy (pluses) and the total surface energies (closed circles) as the function of α_M when $h_B/|h_A| = -1.0$. ($T = 25\,^\circ\text{C}$, $C_S = 0.20\,\text{mol}\,\text{dm}^{-3}$.) The surface energies when $h_B/|h_A| = 100$ is also shown in the inset. Crosses are reported experimental data [1]

proportion to α_M^2, because it is determined mainly by the number of pairs of i-DDAOs. On the other hand, the hydrogen bonding energy decreases in proportion to α_M when α_M is small, because an H-bond can be formed if an H^+ ion exists. When α_M is close to 1, the slope gets smaller due to the topological constraint that a DDAO cannot support more than two H-bonds. The total energy shows a minimum as a function of α_M because of the competition of the Coulombic energy and hydrogen bonding energy.

We have also studied the effects of coexisting salt and temperature. We can see from our simulation for various C_S that the coexisting salt stabilizes the micelle, which agrees qualitatively with experiments [1]. In particular, we find that the C_S is larger than a critical value $C_S^* \cong 0.22\,\text{mol}\,\text{dm}^{-3}$, the system with $\alpha_M = 1.0$, i.e. pure i-DDAO solution is more stable than n-DDAO solution. This behavior can well be understood by assuming that it is possible for a DDAO to support two H-bonds. We also calculated the surface energy for various temperatures T and found that α_M^* is a decreasing function of T and $\Delta E(\alpha_M^*)$ is an increasing function of T. We expect these results can be tested by experiment.

Acknowledgement This work was partially supported by Grants from the Ministry of Education, Science and Culture of Japan.

Progr Colloid Polym Sci (1997) 106:83–85
© Steinkopff Verlag 1997

References

1. Kaimoto H, Shoho K, Sasaki S, Maeda H (1994) J Phys Chem 98:10243–10248; Maeda H, Muroi S, Ishii M, Kaimoto H, Nakahara T, Motomura K (1995) J Colloid Interface Sci 175:497–505; Maeda H (1996) Colloids Surfaces A 109:263–271

2. Mille M (1980) J Colloid Interface Sci 81:169–179; Holland PM, Rubingh DN (1983) J Phys Chem 87:1984–1990

3. Terada Y, Maeda H, Odagaki T, preprint

Progr Colloid Polym Sci (1997) 106:86–90
© Steinkopff Verlag 1997

Pressure-induced phase transition from disordered microemulsion to lamellar structure in a water/ AOT/*n*-decane system

M. Nagao
H. Seto
S. Komura
T. Takeda
M. Hikosaka

Received: 13 December 1996
Accepted: 24 March 1997

Mr. M. Nagao (✉)
Graduate School of Biosphere Science
Hiroshima University
1-7-1 Kagamiyama
Higashihiroshima 739, Japan

H. Seto · S. Komura · T. Takeda
M. Hikosaka
Faculty of Integrated Arts and Sciences
Hiroshima University
Higashihiroshima 739, Japan

Abstract Pressure effect on the structure of a 3-component microemulsion AOT/water/*n*-decane was investigated by means of small-angle neutron scattering. The measured pressure range was from 0.1 (1 atm) to 83.5 MPa. The present sample was at the composition of the molar ratio of [water]/[AOT] = 40.8 and the volume fraction of both water and AOT against the whole volume 0.6. A phase transition from disordered microemulsion to ordered lamellar structure was observed at about 50 MPa through a coexisting region. The pressure dependence of the repeat distance and the correlation length in both high- and low-pressure phases was calculated from the scattering profiles by fitting to Teubner and Strey's formula and/or a single Gaussian. The low-pressure phase became more disordered with increasing pressure and then the phase transition to the ordered phase occurred.

Key words Neutron scattering – microemulsion – lamellar structure – first-order phase transition – high pressure

Introduction

A mixture of water, oil and amphiphile is known to form various structures under different external conditions, for example, oil-in-water or water-in-oil droplet structure, disordered bicontinuous structure, ordered lamellar structure and so on. Many researchers have directed their attentions to these structural formations and phase transitions among these structures [1, 2]. Their attentions have been focused on the formation of new structures of mesoscopic scale, the self-assembling mechanism of the systems, and interactions among different molecules.

Phase transitions among these structures may be observed by changing an external condition such as compositions of the constituents, temperature or pressure. Replacing water with brine, changing the chain length of oil or the kind of amphiphile will also affect the phase behavior of these systems. So far, in a case that the volume fraction between oil and water is almost the same, a phase transition between a disordered microemulsion and a lamellar structure was investigated as a function of either concentration or temperature. With increasing amphiphilic volume fraction ϕ_S at a given water-to-oil ratio, the disordered microemulsion structure tends to transform into an ordered lamellar phase [1]. In this case, the phase transition is understood in terms of increasing interfaces with increasing ϕ_S. With increasing temperature at a given composition, the same phase transition occurs [3]. This is interpreted in terms of the change of the hydrophile–lypophile balance of the membranes with increasing temperature.

A mixture consisting of water, oil, and AOT (dioctyl-sulfosuccinate sodium salt) is well known to form a stabilized microemulsion at room temperature without cosurfactant, and has been well investigated by many researchers. Chen et al. [4] investigated the effect of

Progr Colloid Polym Sci (1997) 106:86–90
© Steinkopff Verlag 1997

salinity in the AOT/brine/n-decane system. They indicated that the bicontinuous structure does not appear without salt at room temperature and that the water-in-oil droplet is dominant at less amphiphile composition in this system. They discussed the difference of scattering profiles between a dense droplet structure and a bicontinuous structure by fitting to Teubner and Strey's [5] model [6]. They found that the bicontinuous structure was a less-ordered phase than the dense droplet in the one-phase microemulsion. They also indicated that increasing salinity plays the same role as decreasing temperature in structural formation, because the hydrophile–lypophile balance of membrane, which determines the structure, changes with temperature and salinity and further the change with temperature is in the opposite direction compared to that with the salinity. Cametti et al. [7] investigated the temperature dependence of electrical conductivity in the AOT/water/decane system. From their results, it is clear that the conformation of this system is water-in-oil droplet microemulsion at room temperature. They showed that with increasing temperature the percolation transition occurs, in other words the system tends to form continuous water region, and then transforms into the ordered lamellar structure.

The pressure as well as the temperature is expected to influence the phase behavior of fluids, because these two are equally important physical parameters in free energy. However, the effect of pressure is not just the same as that of temperature. The effect of pressure on microemulsions has been investigated by many researchers recently. Eastoe et al. [8] indicated that the radius of droplets was not affected by the hydrostatic pressure for the ionic surfactant system, while it decreased with increasing pressure for the non-ionic surfactant system. They concluded that pressure effects on the structure is due to the surfactant type. Beckman and Smith [9] investigated the pressure effect on the structure of the water/propylene/anionic surfactant system at room temperature. They indicated that increasing pressure decreased the balance of the individual cohesive energy, which plays roughly the same role as mean curvature. In other words, an increasing pressure caused a change in the same direction as decreasing temperature. On the other hand, an increasing pressure is in the same direction as an increasing temperature on the phase behavior of the AOT system [10]. Quite recently, Saïdi et al. [10] have observed a cloud point at around 20 MPa at the composition of [water]/[AOT] = 30 and $\phi \sim 0.6$ for the AOT/H$_2$O/undecane system, where ϕ is the volume fraction of both water and AOT against the whole volume. They suggested that the higher pressure phase would have a lamellar structure; however, it is not yet confirmed experimentally.

In this paper we present an experimental evidence of the pressure-induced phase transition from disordered microemulsion to ordered lamellar structure in AOT/D$_2$O/n-decane system by means of small-angle neutron scattering (SANS) measurements. With increasing pressure the lower-pressure disordered structure became more disordered and subsequently an ordered lamellar structure appeared. The lamellar structure existed as a single phase and it was shown that the effect of increasing pressure is not just the same as that of increasing temperature in this AOT system.

Experimental

A combination of neutron scattering and a high-pressure technique has a lot of advantage for the investigation of the self-assembling mechanism in complex fluids systems. From a SANS experiment, we may determine the spatial correlation of semi-microscale structures. Therefore, we have developed a high-pressure cell for SANS such that the structural evolution can be measured as a function of both temperature and pressure [11]. Two sapphire windows faced each other in order to pass the neutron beam and keep the sample inside. The pressure is applied by a hand pump through a 1:1 free piston and measured with a 6 in. Heise Gauge. An accuracy of the pressure was ± 0.5 MPa. In order to control temperature, the cell is covered by an aluminum-cast heater block with two Pt resisters put inside the cell as temperature sensors. We control the temperature using a personal computer within ± 0.01 K during a measurement with a thermal shield in order to suppress temperature fluctuations. A photograph of this high-pressure cell without the thermal shield is shown in Fig. 1.

The composition of the mixture was in the ratio of 3 g of AOT, 5 ml of heavy water, and 5 ml of n-decane, which can be expressed as [water]/[AOT] = 40.8 and $\phi = 0.6$. This composition of mixture is known to form a dense water-in-oil droplet microemulsion at room temperature and ambient pressure [6, 7]. 99% purity AOT was purchased from Fluka, 99.9% heavy water from Isotech Inc. co., and 99% n-decane from Katayama Chemical Company.

SANS experiments were performed at the SANS-U spectrometer in JRR-3M at JAERI, Tokai [12]. A wavelength of 0.7 nm for incident cold neutron beam was used with its wavelength resolution of about 10%. A two-dimensional position-sensitive proportional counter was placed at 4 m for 0.1–30 MPa and 2 m for 30–83.5 MPa from the sample position, in order to cover the momentum transfer range of $0.08 \leq Q \leq 0.72$ (nm^{-1}) for 4 m position and $0.15 \leq Q \leq 1.45$ (nm^{-1}) for 2 m position, respectively. The pressure was increased from ambient pressure ($P = 0.1$ MPa) to 83.5 MPa with a step of about 10 MPa and its temperature was kept at $T = 297.4$ K. All the

Fig. 1 A photograph of the high-pressure cell

obtained data were radially integrated and normalized to an absolute intensity by a Lupolen standard.

Results and discussion

Figure 2 shows scattering profiles at various pressures. A peak at lower $Q(\sim 0.48$ nm$^{-1})$ represents the correlation peak of the disordered microemulsion phase. With increasing pressure, the peak shifted a little to higher Q up to about 22 MPa, moved oppositely to lower Q between 22 and 50 MPa, and finally vanished above 50 MPa. Above 30 MPa, another sharp peak became distinct at $Q \sim 0.80$ nm^{-1} and its intensity grew up to about 50 MPa. The shape of this peak did not change above 50 MPa. At 83.5 MPa a slight shoulder appeared at $Q \sim 0.40$ nm^{-1} on the diffuse baseline at low-Q region.

In order to analyze these scattering profiles, we used the Teubner and Strey's formula [5] for the lower-pressure phase. They proposed the following expression to explain the structure of microemulsion systems,

$$I(Q) = \frac{1}{A + BQ^2 + CQ^4} \quad (1)$$

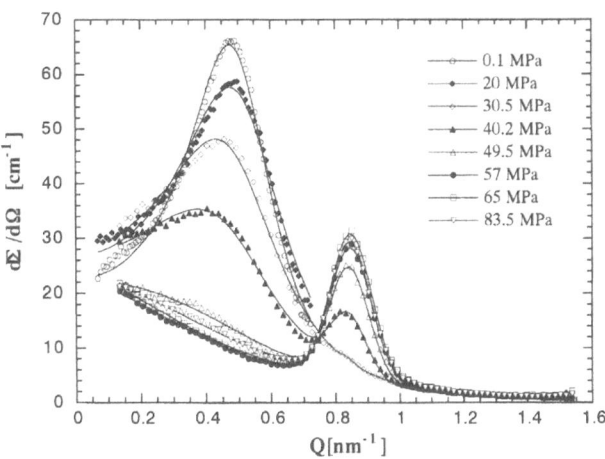

Fig. 2 SANS profiles at various pressures are shown. The peaks at lower Q are from the disordered microemulsion phase. Other peaks at higher Q superimposed on diffuse baselines are from the higher pressure phase. The solid lines indicate the fitted functions (see text)

on the basis of the order parameter expansion in the phenomenological Ginzburg–Landau free-energy functional. The parameters in this equation A and C must be positive. For positive B, Eq. (1) represents a scattering function decreasing monotonically with increasing Q. On the other hand, Eq. (1) has a peak at finite Q for negative B, which is the case of disordered microemulsion structure. From these coefficients A, B and C, the repeat distance d_1 and the correlation length ξ_1 at the lower-pressure phase are expressed [5] as

$$d_1 = 2\pi \left[\frac{1}{2} \left(\sqrt{\frac{A}{C}} - \frac{B}{2C} \right) \right]^{-1/2}, \quad (2)$$

$$\xi_1 = \left[\frac{1}{2} \left(\sqrt{\frac{A}{C}} + \frac{B}{2C} \right) \right]^{-1/2}. \quad (3)$$

The observed profiles in the lower-pressure phase were fitted to Eq. (1) to obtain d_1 and ξ_1. We assumed that the scattering profiles in the higher-pressure phase yielded a Gaussian peak superimposed on a diffuse baseline expressed by Eq. (1). The fitting function to the scattering profiles in the higher-pressure phase was chosen as

$$I(Q) = \frac{1}{A + BQ^2 + CQ^4} + D \exp \left\{ -\frac{1}{2} \left(\frac{Q - Q_0}{\sigma} \right)^2 \right\}, \quad (4)$$

where Q_0 corresponds to the center of a peak, from which the repeat distance of layers is expressed by $d_h = 2\pi/Q_0$. From the peak width σ we calculated the correlation length of the layer displacement $\xi_h = 2\pi/\sigma$. For the coexisting region the scattering profiles were expressed also

Table 1 Fitting parameters obtained and repeat distances d, correlation lengths ξ, and the disorder parameter $1/k\xi$ calculated for both disordered microemulsion (subscript l) and lamellar (subscript h) phases are shown. Thin lines separates the coexistence region from both the microemulsion and the lamellar phase

Pressure [MPa]	A [10^5 nm]	B [10^5 nm^3]	C [10^6 nm^5]	D [cm^{-1}]	d_l [nm]	d_h [nm]	ξ_l [nm]	ξ_h [nm]	$1/k_l\xi_l$	$1/k_h\xi_h$ [10^{-2}]
0.1	4.41	−25.8	5.77	—	12.6	—	6.14	—	0.323	—
10.0	4.18	−22.1	4.86	—	12.3	—	5.56	—	0.352	—
20.0	3.72	−17.8	3.97	—	12.2	—	4.93	—	0.394	—
30.5	3.42	−14.5	3.64	0.00145	12.5	7.76	4.31	384	0.463	0.322
40.2	3.69	−12.3	4.36	0.146	13.5	7.49	3.65	114	0.588	1.05
49.5	4.45	6.92	2.48	0.311	16.7	7.46	1.88	102	1.41	1.16
54.7	4.22	19.2	0.941	0.394	—	7.43	—	97.6	—	1.21
57.0	4.58	23.9	0.597	0.395	—	7.42	—	96.7	—	1.22
65.0	4.44	22.3	0.606	0.432	—	7.41	—	96.9	—	1.22
83.5	4.42	15.3	1.21	0.394	—	7.42	—	99.0	—	1.19

by Eq. (4). From the structural parameters d and ξ we measured the disorder character of the system. For the lower- and higher-pressure phases, characteristic wavelengths $k_l = 2\pi/d_l$ and $k_h = 2\pi/d_h$ are defined, respectively. From the fitting we obtained the parameters A, B, C, the structural parameters d_l, ξ_l, d_h, ξ_h for all the pressure measured, $1/k_l\xi_l$ for the lower-pressure phase, and D and $1/k_h\xi_h$ for the higher-pressure phase as summarized in Table 1.

In Fig. 3 the pressure variation of d_l, d_h and ξ_l, ξ_h are plotted. The vertical dashed lines separate the coexisting region from both the lower-pressure phase and the higher-pressure phase. In the lower-pressure region d_l decreased a little with increasing pressure. Because the oil is more compressible than the other components, the mean distance between droplet decreases with increasing pressure. In the coexisting region, d_l increased steeply and the lower-pressure phase vanished at about 50 MPa. The repeat distance d_h, which is much smaller than d_l, remained almost the same with increasing pressure above 50 MPa, however, it showed a tendency to decrease a little with applying pressure between 22 and 50 MPa. These evidences suggested that an interaction between layers in the lamellar structure is stronger than that in the disordered microemulsion structure. The correlation length ξ_l for the lower-pressure phase decreased gradually with increasing pressure and ξ_h for the higher-pressure phase was more than an order of magnitude larger than ξ_l. Therefore, we can conclude that the higher-pressure phase is an ordered structure. Chen et al. proposed a quantity $1/k\xi$ as a disorder parameter, and it is proportional to the polydispersity of a domain size [6]. Thus, the larger value of $1/k\xi$ corresponds to the more disordered structure. They also proposed that the system forms a bicontinuous structure when the parameter is larger than 0.446 and in the

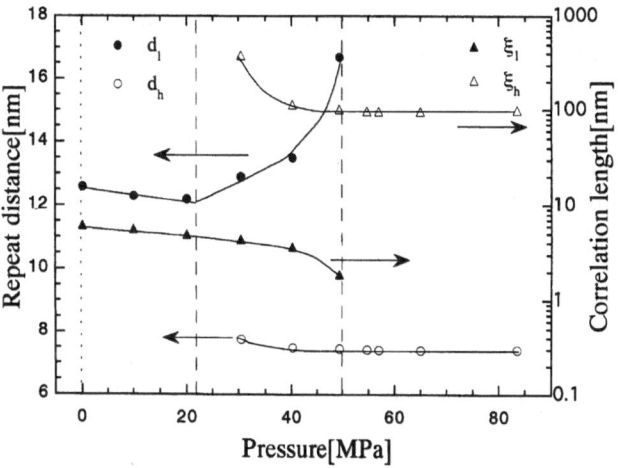

Fig. 3 Pressure dependence of the repeat distances d_l, d_h and correlation lengths ξ_l, ξ_h in both disordered microemulsion and lamellar phases is shown. The ξ_l and ξ_h are plotted on the logarithmic scale as indicated on the right vertical axis. Lines are guides to the eye

opposite case it forms a dense-droplet structure. In Fig. 4 the pressure dependence of the disorder parameter $1/k_l\xi_l$ is shown. The horizontal line indicates the value of 0.446. $1/k_l\xi_l$ for the lower-pressure phase becomes larger with increasing pressure, namely the system becomes more disordered. Above 30 MPa $1/k_l\xi_l$ becomes larger than 0.446, and the structure might be the bicontinuous. We consider that the structure at lower pressure is the dense droplet and the bicontinuous structure and the lamellar structure coexist at intermediate pressure. $1/k_h\xi_h$ for the higher-pressure phase is much smaller than $1/k_l\xi_l$. Therefore, the structure at the higher pressure is a highly ordered one.

Since the difference between d_l and d_h is very large, it is possible that the higher pressure phase is two-phased, the oil-rich phase and the lamellar phase. However, we also

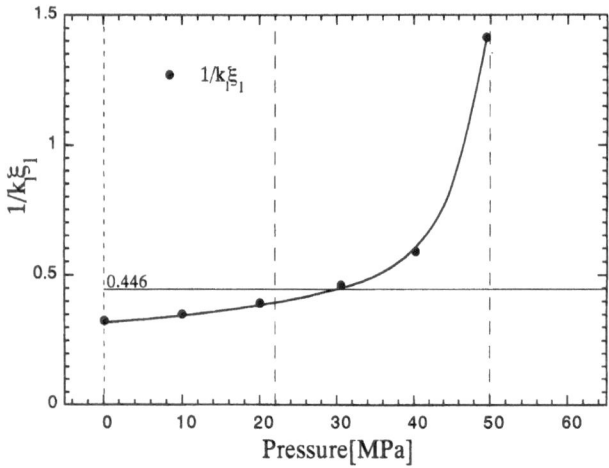

Fig. 4 Pressure dependence of the disorder parameter $1/k_1\xi_1$ for the lower-pressure phase is shown. The lower-pressure phase became more disordered with increasing pressure. The horizontal straight line indicates the value of $1/k_1\xi_1 = 0.446$ which is proposed by Chen et al. [6] in order to distinguish the bicontinuous from the droplet structure (see text). The line is a guide to the eye

observed the temperature-induced phase transition for the same system by means of small-angle X-ray scattering and confirmed that the higher-temperature phase is the single-lamellar phase, where the repeat distances are almost the same as that of the higher-pressure phase. Thus, it is likely that the ordered lamellar phase at high pressure exists as a single phase. This result confirmed the hypothesis suggested by Saïdi et al. that the hydrostatic pressure induces a lamellar phase [10]. This phase transition indicates that the pressure plays a similar role as temperature to induce a phase transition, however the effects on its structure is not exactly the same. The mean repeat distance in the

disordered microemulsion phase was affected significantly by pressure, while the mean repeat distance in lamellar structure did not change. At the coexisting region, with increasing temperature the lamellar repeat distance increases a little, while it decreases with increasing pressure. Increase of the mean repeat distance with increasing temperature might be caused by increasing repulsions between membranes, while decrease of the mean repeat distance with increasing pressure might be caused by the forcible suppressions of membranes. Therefore, the mechanisms that form lamellar structures with increasing temperature and pressure should be different.

Conclusions

We observed a pressure-induced phase transition from disordered microemulsion to ordered lamellar structure by means of small-angle neutron scattering. The structure at lower-pressure phase, a dense water-in-oil droplet structure, becomes more disordered with increasing pressure. Within the coexistence region, a new ordered lamellar phase grows and the remaining microemulsion region becomes more disordered, namely the bicontinuous structure. The high-pressure lamellar structure is confirmed to be highly ordered and exists as a single phase.

Acknowledgments We would like to appreciate Prof. Y. Matsushita and Dr. M. Imai at University of Tokyo for helping the SANS experiments. The SANS experiments were done under the approval of the Neutron Scattering Program Advisory Committee (Proposal No. 96-107). One of the authors (H.S.) was supported from a Grant-in-Aid for Scientific Research (No. 07740329) from the Japanese Ministry of Education, Science and Culture.

References

1. Safran SA (1994) In: Gelbart WM, Ben-Shaul A, Roux D (eds) Micelles, Membranes, Microemulsions, and Monolayers. Springer-Verlag, New York, pp 451–474
2. Komura S, Seto H, Takeda T, Nagao M, Ito Y, Imai M (1996) J Chem Phys 105: 3264–3277
3. Kahlweit M, Strey R, Haase D, Kunieda H, Schmeling T, Faulhaber B, Borkovec M, Eicke HF, Busse G, Eggers F, Funck Th, Richmann H, Magid L, Söderman O, Stilbs P, Winkler J, Dittrich A, Jahn W (1987) J Colloid Interface Sci 118: 436–453
4. Chen SH, Chang SL, Strey R (1990) J Chem Phys 93:1907–1918
5. Teubner M, Strey R (1987) J Chem Phys 87:3195–3200
6. Chen SH, Chang SL, Strey R (1990) Progr Colloid Polym Sci 81:30–35; Chen SH, Chang SL, Strey R, Samseth J, Mortensen K (1991) J Phys Chem 95:7427–7432
7. Cametti C, Codastefano P, Tartaglia P, Rouch J, Chen SH (1990) Phys Rev Lett 64:1461–1464
8. Eastoe J, Steytler DC, Robinson BH, Heenan RK (1994) J Chem Soc Faraday Trans 90:3121–3127
9. Beckman EJ, Smith RD (1991) J Phys Chem 95:3253–3257
10. Saïdi Z, Daridon JL, Boned C (1995) J Phys D: Appl Phys 28:2108–2112
11. Takeno T, Nagao M, Nakayama Y, Hasegawa H, Hashimoto T, Seto H, Imai M (1997) Polymer Journal 29: 931–939
12. Ito Y, Imai M, Takahashi S (1995) Physica B 213&214:889–891

Progr Colloid Polym Sci (1997) 106:91–97
© Steinkopff Verlag 1997

S.K. Ghosh
S. Komura
J. Matsuba
H. Seto
T. Takeda
M. Hikosaka

Structural changes and interaction parameters in amphiphilic system C$_{12}$E$_5$/water/n-octane

Received: 13 October 1996
Accepted: 17 April 1997

Dr. S.K. Ghosh (✉) · J. Matsuba
Graduate School of Biosphere Sciences
Hiroshima University
1-7-1 Kagamiyama
Higashi-Hiroshima 739, Japan
E-mail: ghosh@minerva.ias.hiroshima-u.ac.jp

S. Komura · H. Seto · T. Takeda
M. Hikosaka
Faculty of Integrated Arts and Sciences
Hiroshima University
1-7-1 Kagamiyama
Higashi-Hiroshima 739, Japan

Abstract A ternary amphiphilic system consisting of water, n-octane and nonionic amphiphile C$_{12}$E$_5$ has been studied by means of small-angle X-ray scattering in order to understand the differences of structures and interaction parameters among the low-temperature microemulsion (LTM), middle-temperature lamellar (MTL) and high-temperature microemulsion (HTM) phases. Scattering spectra for microemulsion phases are analyzed by the phenomenological scattering formula of Teubner and Strey (TS), whereas the Ornstein–Zernike scattering formula was combined with the TS formula for the lamellar phase. From the fitting parameters we have determined structural parameters such as the repeat distance D, correlation length ξ and $2\pi\xi/D$ for both the lamellar and the two microemulsion phases. Observed results suggest that the LTM phase is more ordered than the HTM phase. Thermal fluctuations in the HTM phase may make the structure less ordered than that in the LTM phase. Again a finite value of $k\xi$ suggests the existence of an imperfectly ordered MTL phase. Microscopic interaction parameters were determined in terms of Gompper and Schick's (GS) microscopic spin-lattice theory for a suitable lattice parameter $d = 15$ Å. Observed interaction parameters have reasonable agreement with GS's prediction. The inequality condition $|L| > J/4$, suggests strong amphiphilic interaction for all three phases.

Key words Interaction parameter – lamella – microemulsion – structure parameter – spin-lattice model

Introduction

Amphiphile molecules have a hydrophilic polar head and hydrophobic nonpolar tails. In an aqueous–oil–amphiphile ternary system, almost all amphiphiles reside on the oil–water interface and form a variety of structures depending on the temperature and concentration of the components [1, 2]. These structures include macroscopically isotropic disordered phases, such as, micelles, vesicles, microemulsion of droplet type or bicontinuous type or lyotropic ordered phases, such as hexagonal array of cylinders, lamella, etc.

In the past decade, structures in microemulsion systems have been studied by means of small-angle X-ray scattering (SAXS) [3] and small-angle neutron scattering (SANS) [4]. Several theoretical models have emerged to explain novel aspects of amphiphilic systems. These are distinguished into three different approaches, (i) the phenomenological Ginzburg–Landau theory proposed by Teubner and Strey [5], (ii) the microscopic spin-lattice theory proposed by Gompper and Schick [6, 7] and

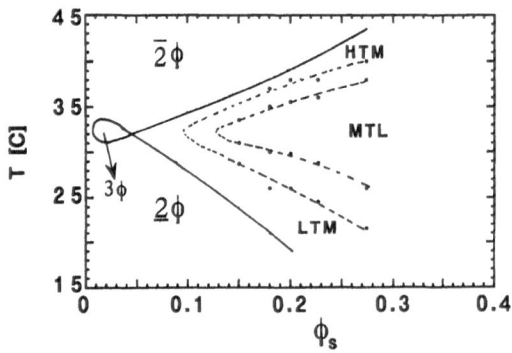

Fig. 1 Phase behavior of $C_{12}E_5$/water/n-octane system for equal volume fraction of water and oil. LTM, MTL and HTM are the low-temperature microemulsion, middle-temperature lamellar and high-temperature microemulsion phase, respectively. The area between the broken lines is the region where the lamellar and microemulsion phases coexist. It is drawn according to X-ray small-angle experiment results

(iii) the theory of bending elasticity of membranes devised by Milner et al. [8]. These studies have revealed many important aspects of microemulsion of both droplet and bicontinuous type.

In this study we have investigated the structural and interaction parameters of ternary water/octane/$C_{12}E_5$ system by means of SAXS. Phase behavior of this system was studied by Kahlweit et al. [9]. This system shows interesting phase behavior (Fig. 1). One can study the structures of low-temperature microemulsion (LTM) phase, middle-temperature lamellar (MTL) phase and high-temperature microemulsion (HTM) phase by changing temperature only, provided that the sample contains approximately more than 12 wt% of surfactant at equal volume fraction of water and oil. Bodet et al. have clarified the structural evolution of this system by means of pulsed-field gradient spin-echo NMR, quasi-elastic light scattering and freeze-fracture transmission electron microscopy [10]. Local structure of the bilayer and monolayer of the same system was also studied by Strey et al. [11]. Recently, we have studied the mechanism of the phase transition [12].

Our main purpose in this paper is to clarify the essential differences of structures and interaction parameters among middle-temperature lamellar (MTL) phase and the two microemulsion, LTM and HTM, phases. We have determined the structural parameters by phenomenological model analysis and the microscopic interaction parameters by microscopic spin-lattice model analysis, respectively.

Experiment

99.7% pure $C_{12}E_5$ was purchased from Tokyo Chemical Company, 99% n-octane from Aldrich Chemical Co. and

highly pure water was used from our laboratory. These materials were mixed without further purification. After mixing all ingredients in a sealed bottle, temperature was increased to 80 °C in a water bath. The bottles were then shaken in a vibrator for 10 min. We prepared five samples, where the volume fraction of oil ϕ was kept constant at $\phi = 0.50$ and the volume fraction of amphiphile ϕ_S was varied at $\phi_S = 0.150, 0.180, 0.200, 0.226$ and 0.274. Here $\phi \equiv v_o/(v_o + v_w)$ and $\phi_S \equiv v_S/(v_w + v_o + v_S)$, respectively. v_o, v_w and v_S represent the volume of oil, water and amphiphile, respectively.

Small-angle X-ray scattering experiments were performed by a small-angle camera using Cu K_α characteristic line from a rotating anode X-ray generator (Rigaku, RU-200) at the power of 50 kV and 150 mA. We used a one-dimensional position-sensitive proportional counter (MAC science, PSPC-5) to measure the scattered beam. Temperature was controlled by a personal computer within an accuracy of ± 0.01 °C.

Data analysis

In order to determine the structural parameters and the interactions from the small-angle scattering experiments, scattering curves were analyzed in terms of both the phenomenological model of Teubner–Strey [5] and the microscopic spin-lattice model of Gompper and Schick [6].

Phenomenological Ginzburg–Landau model

Scattering intensity $I(q)$ in phenomenological model [5] is expressed in the form

$$I(q) = \frac{1}{A + Bq^2 + Cq^4} + I_b , \tag{1}$$

where $B < 0$, $A > 0$ and $C > 0$ and I_b represents a constant background intensity. This model can successfully explain the scattering spectra for microemulsion phases, but it cannot explain the additional scattering spectra at low q region that appear in the lamellar phase. For the scattering spectra in the lamellar phase. Ornstein–Zernike scattering formula

$$I_1(q) = \frac{1}{D_1 + Eq^2} \tag{2}$$

is added to Eq. (1), in order to account for the background scattering at lower q on which a scattering peak is superimposed. From this model one can calculate the structural parameters, such as the average repeat distance between two water or oil domains D and the spatial correlation length ξ, which are given by

$$D = 2\pi\left[\frac{1}{2}\left(\sqrt{\frac{A}{C}} - \frac{B}{2C}\right)\right]^{-1/2} \tag{3a}$$

and

$$\xi = \left[\frac{1}{2}\left(\sqrt{\frac{A}{C}} + \frac{B}{2C}\right)\right]^{-1/2}. \tag{3b}$$

There is another structure parameter $k\ (\equiv 2\pi/D)$ that gives an important quantity $k\xi$, which characterizes the structural order. In addition, it is possible to calculate the correlation length $\xi_1 = \sqrt{E/D_1}$ from Eq. (2), which is specific to the undulation of monolayers in lamellar phase.

Microscopic spin-lattice model

A microscopic lattice model is proposed for microemulsion system by Gompper and Schick [6, 7]. They choose a Hamiltonian H, which incorporates a statistical variable P_i^α on each site of the lattice. The variable has a value, $P_i^\alpha = 1$, if there is a α molecule at site i, otherwise $P_i^\alpha = 0$. Starting from such a Hamiltonian, Gompper and Schick calculated scattering function for a particular amphiphilic system in which the role of water and oil is symmetric or "balanced". In a balanced system, the volume fractions of water and oil are equal, and water and oil are equally favored by the amphiphiles, i.e., hydro- and lipophilicity of the amphiphiles are equal.

The scattering function for a "balanced" system with bulk scattering contrast of water over oil in the (1 1 1) direction of the scattering vector q is approximately given by

$$S(q) = \frac{1}{(1-\phi_s)^{-1} - 6(J/T)\cos(qd) + 6(|L|/T)\phi_s\cos(2qd)}, \tag{4}$$

where, $d = l/\sqrt{3}$ and l is the size of the lattice cell, which is unknown except that it is in the order of the molecular length of the surfactant. J is the interaction parameter that represents the segregation between water and oil. L, which is always negative, is the strength of amphiphilic interaction. It represents the reduced energy when an amphiphile lies between a water and an oil molecule, whereas increased energy when the amphiphile lies between two water or two oil molecules.

Gompper and Schick's scattering formula can be derived from Eq. (4) in the form of

$$I^{GS}(q) = \frac{1}{a + b\cos(qd) + c\cos(2qd)}, \tag{5}$$

which in turn can be transformed into Teubner and Strey's scattering formula (1) by expanding the *cosine* function for small q, so that $qd \ll 1$. The coefficients of the two models are related by [13]

$$a = A + \frac{15}{6d^2}B + \frac{6}{d^4}C,$$

$$b = -\frac{8}{3d^2}B - \frac{8}{d^4}C,$$

$$c = \frac{1}{6d^2}B + \frac{2}{d^4}C. \tag{6}$$

Now, the interaction parameters can be equated from Eq. (4), and are given by

$$\frac{J}{T} = -\frac{b}{6a(1-\phi_s)} \tag{7a}$$

$$\frac{|L|}{T} = \frac{c}{6a\phi_s(1-\phi_s)}, \tag{7b}$$

$$\phi_s^c\left(\equiv \frac{J}{4|L|}\right) = -\frac{b\phi_s}{4c}. \tag{7c}$$

From the fitting of Teubner–Strey's scattering formula to the experimental scattering curve, the coefficients A, B and C can be determined. The coefficients of Gompper and Schick's scattering formula, a, b and c are then calculated from Eq. (6) using the proper value of d, and subsequently the interaction parameters J, $|L|$ and ϕ_s^c are determined by Eq. (7).

The structure function, $S(q)$ has an extremum for q when

$$[J - 4|L|\phi_s\cos(qd)]\sin(qd) = 0. \tag{8}$$

There are two possibilities. For weak amphiphilic interactions, $|L| < J/4$, $S(q)$ has a maximum at $q = 0$ for all ϕ_s. For strong amphiphilic interactions, $|L| > J/4$, the structure function $S(q)$ has a maximum at $q = 0$ for amphiphile concentrations ϕ_s less than the value ϕ_s^c, whereas $S(q)$ has a maximum at nonzero wave vectors $(q_m \neq 0)$ for $\phi_s > \phi_s^c$, which satisfy

$$q_m d = \cos^{-1}(\phi_s^c/\phi_s). \tag{9}$$

Results

We have recorded SAXS spectra from the LTM, MTL and HTM phases of the water/n-octane/$C_{12}E_5$ system. We observe a single Bragg peak in SAXS for the LTM, HTM

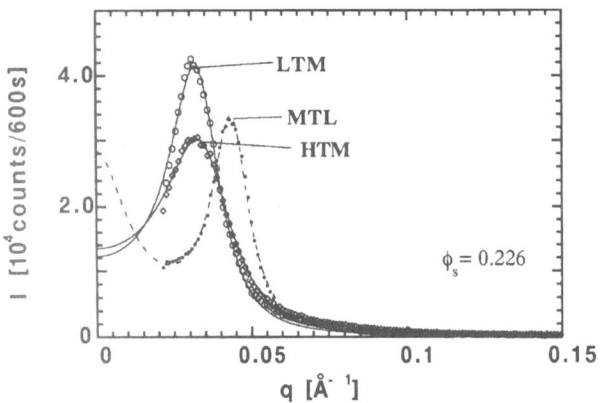

Fig. 2 Scattering intensity for $\phi_S = 0.226$ is plotted as a function of q. Solid and broken lines are the fitting curves by Eq. (1) for LTM and HTM phases. The broken line is the fitting curve by Eq. $(1+2)$ for MTL phase

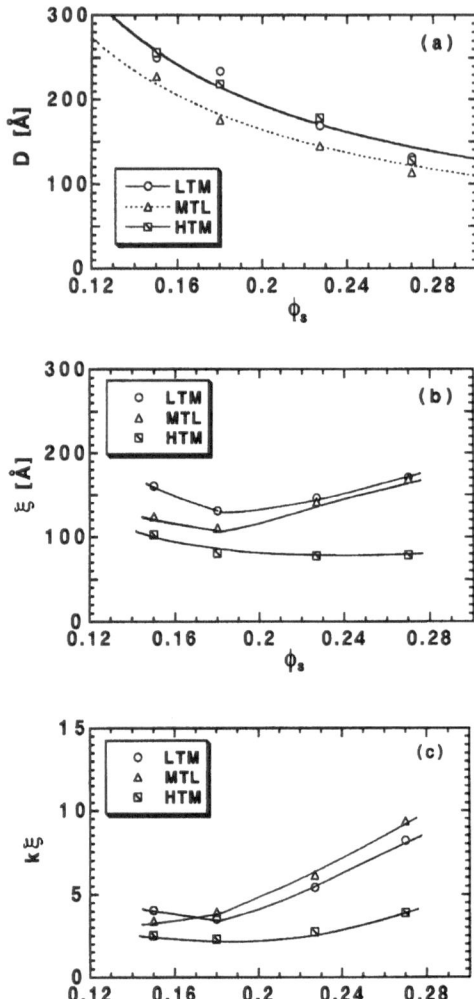

Fig. 3 (a) D as a function of ϕ_S. Solid line is the fitting curve of Eq. (10a) and dotted line is the fitting curve of Eq. (10b). (b) ξ and (c) $k\xi$ are plotted as a function of ϕ_S. Solid lines are guides to the eye

and MTL phases, respectively. However, there is an intensive scattering at lower q in the MTL phase. Teubner–Strey's (TS) scattering formula (1) is then fitted with experimental scattering curves for the LTM and HTM phases. In order to fit the scattering intensity in the MTL phase, Eqs. (2) and (1) were combined. A few selected fitting curves for the sample $\phi_S = 0.226$ are shown in Fig. 2. We observed that, in the microemulsion (both LTM and HTM) phase the T–S formula (1) cannot fit the experimental results well in the high q region. From the fitting parameters A, B and C we have determined the structure parameters such as average repeat distance D, spatial correlation length ξ and dimensionless parameter $k\xi$ for all phases. These parameters are plotted as a function of ϕ_S in Fig. 3. From the fitting of the scattering profile in lamellar phase another correlation parameter $\xi_1 = \sqrt{E/D_1}$ is calculated. ξ_1 has a value of the order ~ 680 Å, which is much larger than the spatial correlation length ξ.

In the next step, we have determined microscopic interaction parameters. We adopted an approximation, which was successfully used in calculating the interaction parameters of the water/AOT/decane system [13]. We calculated the parameters a, b and c of Gompper and Schick's scattering formula (5) in terms of the fitting parameters A, B and C of Teubner–Strey's scattering formula (1) for variable values of d and subsequently the interaction parameters J/T, $|L|/T$ and ϕ_S^c from Eq. (7). Figure 4 is an example of the result of J/T, $|L|/T$ and ϕ_S^c as a function of d for the LTM phase. In order to determine the interaction parameters we need a proper value of lattice parameter d, which is related to the molecular length of the surfactant. We have chosen $d = 15$ Å for all three phases, which was found to be suitable in this system for the reason described

in next section. The interaction parameters J/T, $|L|/T$ and ϕ_S^c are then determined, and are plotted in Fig. 5 as a function of the surfactant volume fraction ϕ_S. The dimensionless parameter $q_m d$ is determined from Eq. (9) and is plotted in Fig. 6 as a function of $\phi_S/\bar{\phi}_S^c$, where $\bar{\phi}_S^c$ is the average of ϕ_S^c for different samples.

Discussion

We discuss here the implications of the results, in order to clarify the differences of structures and interaction parameters among the LTM, MTL and HTM phases.

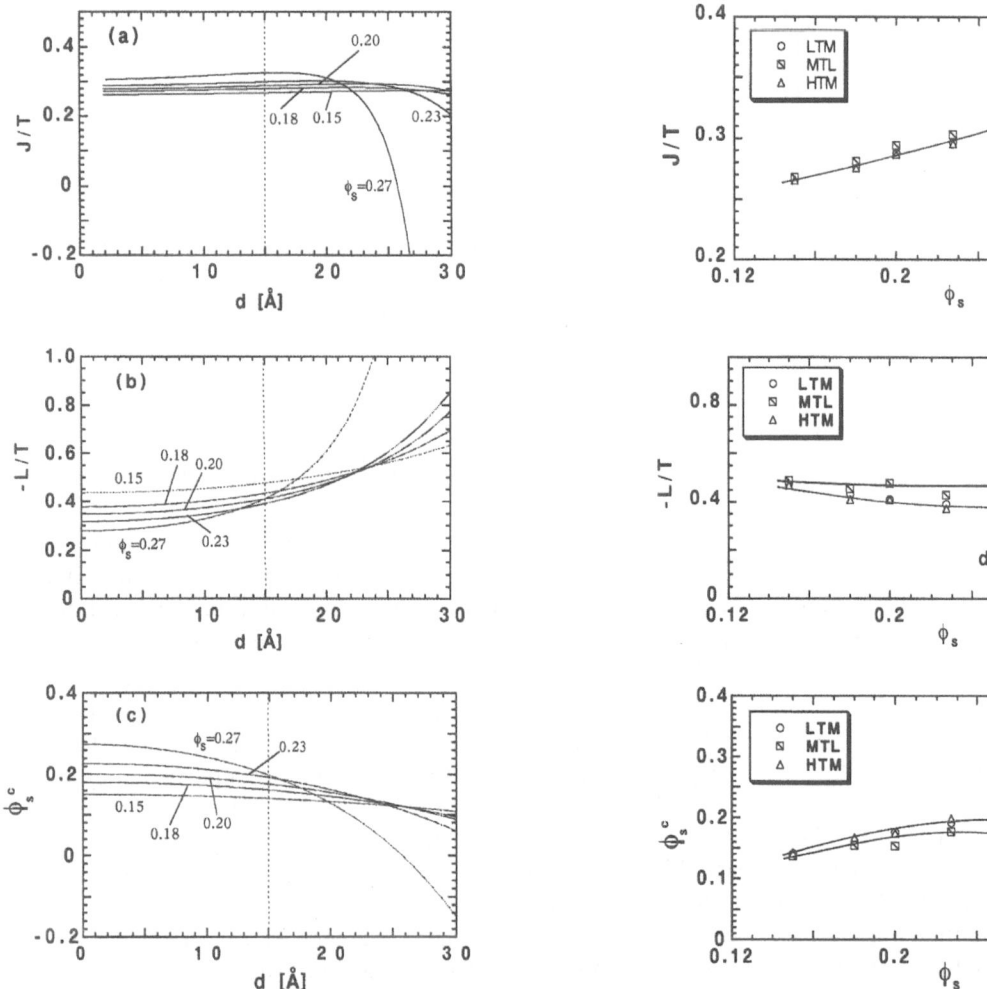

Fig. 4 Interaction parameters (a) J/T, (b) $-L/T$ and (c) $\phi_s^\S$ as a function of d for LTM phase. These are calculated by Eq. (7). The dotted lines indicates the value for $d=15\,\text{Å}$, which we chose to calculate the interaction parameters of Fig. 5

Fig. 5 Interaction parameters (a) J/T, (b) $-L/T$ and (c) ϕ_s^c as a function of ϕ_S. Solid lines are guides to the eye

Structure parameters

Figure 3 shows a systematic variation of D, ξ and $k\xi$ as a function ϕ_S in LTM, MTL and HTM phases. Results indicate that average repeat distance D decreases upon increasing surfactant volume fraction almost in the same way (Fig. 3a) in both LTM and HTM phases, whereas D has the smaller value in the MTL phase. The repeat distance D is calculated by a random mixing approximation [14] for the microemulsion phases and by a simple geometrical argument for the lamellar phase. D's are given by

$$D_m = 2l_m Z\phi(1-\phi)/\phi_S \,, \tag{10a}$$

$$D_l = 2l_l/\phi_S \,, \tag{10b}$$

where Eq. (10a) is for the microemulsion (LTM and HTM) phase and (10b) is for the lamellar (MTL) phase. The calculated D_m and D_l are shown in Fig. 3a by solid and dotted lines, respectively. Z is the coordination number, which is 6 for a cubic lattice. l_m and l_l are the effective molecular lengths of surfactants in the microemulsion and lamellar phases, respectively. We found $l_m = 12.89\,\text{Å}$ and $l_l = 16.39\,\text{Å}$. In order to compare these values, we calculated l_m using the same model (10) from the data measured by Magid et al. in a $D_2O/octane/C_{12}E_5$ system [15]. The value of l_m at equal oil–water volume fraction was found to be $12.4\,\text{Å}$ for the LTM and $12.8\,\text{Å}$ for the HTM phase at $\phi_S = 0.144$, in close agreement with our experimental rersults. However, the values are surprisingly small, because the length for the polyoxyethylene chain of a $C_{12}E_5$ molecule is $32.0\,\text{Å}$ in zigzag configuration and $24.8\,\text{Å}$ in meander configuration [16]. In another paper,

96

S.K. Ghosh et al.
Structural changes and interaction parameters in amphiphilic system

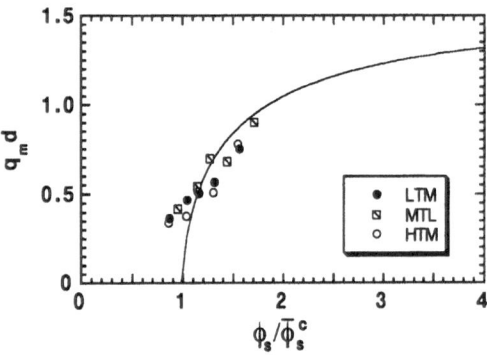

Fig. 6 Dimensionless parameter $q_\mathrm{m}d$ as a function of normalized amphiphilic concentration, $\phi_\mathrm{S}/\bar{\phi}_\mathrm{S}^\mathrm{c}$. Solid line represents the relation $q_\mathrm{m}d = \cos^{-1}(\bar{\phi}_\mathrm{S}^\mathrm{c}/\phi_\mathrm{S})$

we determined the thickness of a $C_{12}E_5$ monolayer by means of SANS using isotopic substitution of water and n-octane [17]. We found that the thickness was approximately 21 Å for LTM and HTM phases, whereas 31.2 Å for the MTL phase. These indicate that we cannot determine molecular length correctly from Eq. (10). Considering the zigzag configuration, meander configuration and our experimental results, we chose a value $d = 15$ Å, which corresponds to a molecular length $l = \sqrt{3}d = 26$ Å.

Almost a decade ago Bodet et al. concluded that LTM has a droplet structure and the HTM phase has a bicontinuous structure [10]. Although the microsructures are different, we observed that the average repeat distance D between two water (or oil) domains are almost equal in the LTM and HTM phases. Correlation length ξ has an increasing tendency (Fig. 3b) with ϕ_S in LTM and MTL phases, while in the HTM phase ξ has a lower value, decreasing slowly with ϕ_S. Dimensionless parameter $k\xi$, which characterizes the structure, increases (Fig. 3c) with ϕ_S in both the LTM and MTL phases while it remains almost constant in the HTM phase. A higher value of $k\xi$ in the LTM phase with respect to the HTM phase implies that the LTM has a better order than the HTM. Thermal fluctuations in the HTM phase may make the structure less ordered than that in the LTM phase. For an ideal lamellar phase $k\xi$ should be infinitely large [7], which indicates that the MTL phase we studied may not be a perfect lamellar phase or it may have a lower order. A low-ordered MTL phase might be constructed by the randomly oriented domains of lamellar liquid crystal.

Interaction parameters

Experimental results are analyzed by the Gompper and Schick's scatering function (5), which is estimated for

a "balanced" system (at equal hydro- and lipophilicity) with bulk-scattering contrast of water against oil in the (1 1 1) direction of the scattering vector q. In order to determine the fitting parameters correctly, one must choose the lattice parameter d properly. The interaction parameters are calculated as a function of d for LTM phase (Fig. 4). The self-consistency of the lattice model suggests that the interaction parameters should be constant in the same microemulsion system for different concentration and the value of d should be consistent with the molecular length of the surfactants, such as $d = l/\sqrt{3}$. We chose $d = 15$ Å, which was found to be suitable for this system, although the condition $qd \ll 1$ is not valid for large q.

Figure 5 shows that an approximate constancy of the interaction parameters J/T, L/T and $\phi_\mathrm{S}^\mathrm{c}$ as a function of ϕ_S with some deviations from it. For example J/T has almost the same values in all three phases, but increasing slowly with ϕ_S. The estimated value of the interaction parameters from our calculation are roughly $J/T \approx 0.29$, $L/T \approx -0.45$ and $\phi_\mathrm{S}^\mathrm{c} \approx 0.17$ for $d = 15$ Å. Hence, $|L| > J/4$, which is satisfied, indicating a strong amphiphilic interaction. Gompper and Schick [6] estimated $J/T = 0.23$ and $L/T = -0.79$ for a particular system with $d = 17.32$ Å. Komura et al. [13] found $J/T \approx L/T \approx 0.35$ and $\phi_\mathrm{S}^\mathrm{c} \approx 0.24$ for $d = 8.42$ Å in an AOT system. The observed interaction parameters are reasonable in view of the earlier works.

Dimensionless parameter $q_\mathrm{m}d$ is calculated for LTM, MTL and HTM phases (Fig. 6) as a function of normalized amphiphilic concentration, $\phi_\mathrm{S}/\bar{\phi}_\mathrm{S}^\mathrm{c}$. Solid lines are the theoretical prediction by Eq. (9) where the value of $\phi_\mathrm{S}^\mathrm{c}$ is replaced by the averaged one $\bar{\phi}_\mathrm{S}^\mathrm{c}$ for different samples. Results are in reasonable agreement with the theoretical relation (9), indicating that the choice of d is consistent for this system.

Conclusions

To quest the differences of the structures and interaction parameters among LTM, MTL and HTM phases, we observed that the average repeat distance D in a lamellar (MTL) phase is smaller than that of both LTM and HTM phases. D is almost equal in LTM and HTM phases, although the microstructure is different. Correlation length ξ for a HTM phase is lower than that of LTM and MTL phases. ξ is almost constant with respect to ϕ_S in the HTM phase, whereas it is increasing in both the LTM and MTL phases. From the behavior of $k\xi$, we conclude that a finite value of $k\xi$ suggests the existence of an imperfectly ordered MTL phase, whereas the LTM phase is more

Progr Colloid Polym Sci (1997) 106:91–97
© Steinkopff Verlag 1997

ordered than the HTM phase. Thermal fluctuations in the HTM phase may make the structure less ordered than in the LTM phase.

The interaction parameters that were determined for the lattice parameter $d = 15$ Å have reasonable values. The condition $|L| > J/4$, which is satisfied, suggests the existence of strong amphiphilic interaction in all the three phases.

Acknowledgement S.K. Ghosh is thankful to the Japanese Ministry of Education, Science, Sports and Culture (Monbusho) for supporting his graduate study at Hiroshima University, during which this work was done.

References

1. Degiorgio V, Corti M (eds) (1985) Physics of Amphiphiles: Micelles, Vesicles and Microemulsion, North-Holland, Amsterdam (for general review)
2. Gelbart WM, Ben-Shaul A, Roux D (eds) (1994) Micelles, Membranes, Microemulsions and Monolayers. Springer, New York
3. Auvray L, Cotton JP, Ober R, Taupin C (1984) J Physique 45:913–928
4. Kotlarchyk M, Chen SH, Huang JS, Kim MW (1984) Phys Rev A 29:2054–2069; Kotlarchyk M, Chen SH, Huang JS, Kim MW (1984) Phys Rev Lett 53:941–944
5. Teubner M, Strey R (1987) J Chem Phys 87:3195–3200
6. Gompper G, Schick M (1989) Phys Rev Lett 62:1647–1650; Gompper G, Schick M (1990) Phys Rev B 41:9148–9162
7. For a general review see also Gompper G, Schick M (1994) Self-Assembling Amphiphilic Systems, Academic Press, London
8. Milner ST, Safran SA, Andleman D, Cates ME, Roux D (1988) J Phys France 49:1065–1076
9. Kahlweit M, Strey R, Haase D, Kunieda H, Schmeling T, Faulhaber B, Borkovec M, Eicke HF, Busse G, Eggers F, Funck TH, Richmann H, Magid L, Söderman O, Stilbs P, Winkler J, Dittrich A, Jahn W (1987) J Colloid Interface Sci 118:436–453
10. Bodet J-F, Bellare JR, Davis HT, Scriven LE, Miller WG (1988) J Chem Phys 92:1898–1902
11. Strey R, Winkler J, Magid L (1991) J Phys Chem 95:7502–7507
12. Ghosh SK, Matsuba J, Komura S, Seto H, Takeda T, Hikosaka M (1996) submitted to Japanese J App Phys
13. Komura S, Seto H, Takeda T, Nagao M, Itoh Y, Imai M (1996) J Chem Phys 105:3264–3277
14. Jouffroy J, Levinson P, de Gennes PG (1982) J Physique 43:1241–1248
15. Magid L, Butler P, Payne K (1988) J App Cryst 21:832–834
16. Degiorgio V (1985) In: Physics of Amphiphiles: Micelles, Vesicles and Microemulsion. North-Holland, Amsterdam, pp 303–335
17. Ghosh SK, Komura S, Seto H, Matsuba J, Takeda T, Hikosaka M, Imai M (1997) to be published in Prog Colloid Polym Sci (this issue)

Progr Colloid Polym Sci (1997) 106:98–103
© Steinkopff Verlag 1997

Structure functions and interfacial mean curvatures in a ternary amphiphilic system $C_{12}E_5$/water/ *n*-octane

S.K. Ghosh
S. Komura
H. Seto
J. Matsuba
T. Takeda
M. Kokosaka
M. Imai

Received: 13 October 1996
Accepted: 1 April 1997

Dr. S.K. Ghosh (✉) · S. Komura · H. Seto
J. Matsuba · T. Takeda · M. Hikosaka
Graduate School of Biosphere Sciences
Faculty of Integrated Arts and Sciences
Hiroshima University
1-7-1 Kagamiyama
Higashi-Hiroshima 739, Japan
E-mail: ghosh@minerva.ias.hiroshima-u.ac.jp

M. Imai
Institute of Solid State Physics
University of Tokyo
Roppongi
Tokyo 106, Japan

Abstract Partial structure functions and mean curvatures of a ternary amphiphilic system consisting of water, *n*-octane and nonionic amphiphile $C_{12}E_5$ with equal volume fraction of oil and water have been studied by means of small-angle neutron scattering (SANS). Isotopic contrast variation technique was used to obtain suitable scattering contrasts of oil and water. We observed that the cross partial structure functions, that arise from the correlation between water–surfactant or oil–surfactant, change sign when the temperature passes through the hydrophile–lipophile balance (HLB) point. It indicates that the interaction parameter C of the spin-lattice model invert on HLB line. We have calculated mean curvatures considering that water or oil molecules can partially penetrate into the surfactant film. Experimental results show that mean curvatures increase as the temperature increases at a rate of 1.5×10^{-3} ($\mathring{A}^{-1} K^{-1}$).

Key words Amphiphile – microemulsion – lamella – partial structure function – mean curvature

Introduction

Amphiphiles, the representatives of which are soap, surfactant and lipid, have a hydrophilic polar head and lipophilic nonpolar tails. They always remain on the interface between water and oil and form monolayers of surfactants in a water/oil/amphiphile ternary system. This monolayers or interfacial film reduce the surface tension between water and oil domains. In a three-component system the surfactant film exists in various topologically different structures such as micelles, vesicles, bicontinuous microemulsions, hexagonal arrays of cylinders or lamellar structures depending upon the pressure, temperature and the concentration of the components [1, 2]. Microemulsions are thermodynamically stable, isotropic and transparent mixtures of ternary amphiphilic systems. When almost equal volume fractions of water and oil are mixed with a dilute concentration of surfactants, they take an interconnected bicontinuous structure with continuous interfacial film.

In order to investigate the mechanism of the self-assembly, we performed a small-angle neutron-scattering experiment on $C_{12}E_5$/water/*n*-octane system with isotopic contrast variation and determined the partial structure functions and mean curvatures of the interfacial film. Phase diagram of water/*n*-octane/$C_{12}E_5$ system was studied by Kahlweit et al. [3]. This system has, in sequence, a low-temperature microemulsion (LTM) phase, a middle-temperature lamellar (MTL) phase and a high-temperature microemulsion (HTM) phase, when water and oil volume fraction is almost equal with dilute concentration of surfactants (Fig. 1). Thus, one can study all these phases as a function of temperature only. However, the isotopic substitution of hydrogen atoms by deuterium in water and oil may change the transition temperatures by a few degrees. Several authors [4–7] have studied the mean curvatures of the surfactant monolayers using different

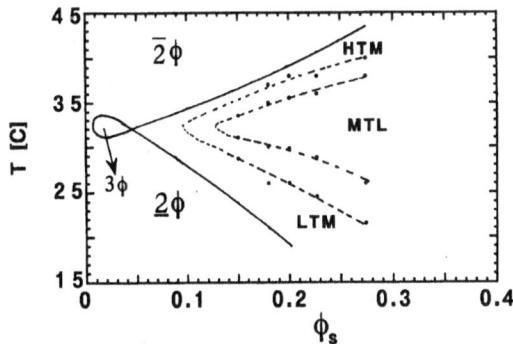

Fig. 1 Phase behavior of $C_{12}E_5$/water/n-octane system for equal volume fraction of water and oil. LTM, MTL and HTM are the low-temperature microemulsion, middle-temperature lamellar and high-temperature microemulsion phase, respectively. The area between the broken lines is the region where lamellar and microemulsion phase coexist. It is drawn according to X-ray small-angle experiment results

methods. The way of measuring mean curvature is improved by an idea that the area of the amphiphile–water interface is larger than that of oil–amphiphile interface if the surfactant film is curved toward oil [6]. Strey and coworkers proposed that oil or water may partially penetrate into the surfactant film [8]. This has lead to an assumption of the smeared nature of the scattering amplitude density instead of a rectangular profile across the amphiphilic layer. Thus, for a diffuse interfacial film the characteristic Porod decay q^{-4} can be modified with an exponential term $\exp(-t^2q^2)$, where $2t$ is an effective thickness of the film.

In this paper we have determined the partial structure functions and explained the nature of different phases as a function of temperature in terms of Gompper and Schick's spin-lattice model [9, 10]. We measured partial structure functions in order to determine mean curvature from it. The mean curvature was measured as a function of temperatures using the method proposed by Lee and Chen [6].

Theory

Partial structure functions

Partial structure functions stem from the correlation between the volume fraction fluctuations of any two components, such as, water–water, water–surfactant, etc., of a three-component system. Partial structure functions $S_{\alpha\beta}(q)$ are defined by

$$S_{\alpha\beta}(\mathbf{q}) = \int \langle \delta\phi_\alpha(0)\delta\phi_\beta(\mathbf{r})\rangle e^{i\mathbf{q}\cdot\mathbf{r}} \, dr \tag{1}$$

where $\delta\phi_\alpha(\mathbf{r}) = \phi_\alpha(\mathbf{r}) - \bar{\phi}_\alpha$ is the fluctuation of αth component volume fraction $\phi_\alpha(\mathbf{r})$ at position $\mathbf{r}$ from its mean value $\bar{\phi}_\alpha$, the bracket means the ensemble average and the subscripts $\alpha = s$, w or o refer to surfactant, water or oil, respectively. Neutron-scattering cross section per unit volume of the sample is given for the oil-contrast variation experiment as [4]

$$I^o(q) = (\varrho_w - \varrho_o)^2 S_{ww}(q) + (\varrho_s - \varrho_o)^2 S_{ss}(q)$$
$$+ 2(\varrho_w - \varrho_o)(\varrho_s - \varrho_o)S_{ws}(q), \tag{2a}$$

and for water-contrast variation experiment as

$$I^w(q) = (\varrho_o - \varrho_w)^2 S_{oo}(q) + (\varrho_s - \varrho_w)^2 S_{ss}(q)$$
$$+ 2(\varrho_o - \varrho_w)(\varrho_s - \varrho_w)S_{os}(q), \tag{2b}$$

where ϱ_s, ϱ_w and ϱ_o refer to the scattering amplitude density of surfactant, water and oil, respectively. The value of ϱ_o is changed while ϱ_s and ϱ_w are kept constant in octane-contrast variation experiment. A similar technique applies for water-contrast variation experiment where the roles of ϱ_w and ϱ_o are interchanged.

In order to understand the behavior of an amphiphile in terms of partial structure functions in a ternary system, we consider the interaction energy parameter C, characterizing the imbalance of amphiphilicity defined by Gompper and Schick [9, 10], which is the difference between the interaction of amphiphile with water and oil. Hence, one would expect $C = 0$ at the hydrophile–lipophile balance (HLB) line, amphiphile favors oil to water for $C > 0$ and the opposite holds for $C < 0$. C has an interesting role in the phase diagram. For equal volume fractions of water and oil ($\delta\phi = \phi_w - \phi_o = 0$) and $C = 0$, the phase diagram is symmetric and the disordered phase is equally separated from the oil-rich or water-rich phase. At the same temperature, a positive value of C makes the oil-rich phase closer to the disordered phase and the symmetry of the phase diagram is destroyed. If C is larger than a certain positive value, the disordered phase becomes accessible from the oil-rich phase and the three-phase coexistence disappears. A negative value of C with the same magnitude plays the same role in the phase diagram except that oil and water are interchanged.

Mean curvatures of interfacial film

The space-averaged mean curvatures $\langle c \rangle$ in a ternary amphiphilic system is related to the water–surfactant interfacial area A_w, the oil–surfactant interfacial area A_0, the interfacial area at the mid-plane of surfactant film A_s and

100
S.K. Ghosh et al.
Structure functions and mean curvatures in a ternary amphiphilic system

film thickness d by [6, 11]

$$A_{\mathrm{w}} = A_{\mathrm{s}}\left(1 - d\langle c \rangle + \frac{d^2}{4}\langle K \rangle\right), \tag{3}$$

$$A_{\mathrm{o}} = A_{\mathrm{s}}\left(1 + d\langle c \rangle + \frac{d^2}{4}\langle K \rangle\right), \tag{4}$$

where $c = \frac{1}{2}((1/r_1) + (1/r_2))$ is the mean curvature of the interfacial film and $K = (1/r_1 r_2)$ the Gaussian curvature, r_1 and r_2 being two principal curvatures. Here, a positive curvature means that surfactant film bends toward water. A_{w}, A_{o}, A_{s} and d can be determined from the modified Porod's law [6, 8] at higher q, such that

$$S_{\mathrm{ww}}(q) = 2\pi \frac{A_{\mathrm{w}}}{V}\frac{1}{q^4}\exp(-t_{\mathrm{w}}^2 q^2) + c_{\mathrm{b}}, \tag{5a}$$

$$S_{\mathrm{oo}}(q) = 2\pi \frac{A_{\mathrm{o}}}{V}\frac{1}{q^4}\exp(-t_{\mathrm{o}}^2 q^2) + c_{\mathrm{b}}, \tag{5b}$$

$$S_{\mathrm{ss}}(q) = 2\pi \phi_{\mathrm{s}}^2 \frac{V}{A_{\mathrm{s}}}\frac{1}{q^2}\exp(-t^2 q^2/2\pi) + c_{\mathrm{b}}, \tag{5c}$$

where $2t_{\mathrm{w}}$ and $2t_{\mathrm{o}}$ are the penetration lengths of the water and oil phases into the film, $d = \phi_{\mathrm{s}}(V/A_{\mathrm{s}})$ the effective thickness of the film, with ϕ_{s} being the volume fraction of surfactant, and c_{b} is the background scattering. This arises from nuclear-spin interaction and the mixing of different isotopes, such as, H_2O and D_2O for water and C_8H_{18} and C_8D_{18} for oil. Some multiple coherent scattering may also contribute to c_{b}. Mean curvature $\langle c \rangle = [(A_{\mathrm{o}} - A_{\mathrm{w}})/(A_{\mathrm{o}} + A_{\mathrm{w}})]/d$ is determined from Eqs. (3) and (4) for negligibly small K.

Experiment

To prepare the samples, $C_{12}E_5$ (99.7% pure, Tokyo Chem. Co.), n-octane (99%, Aldrich Chem. Co.), deutrated n-octane (99 at% Isotec Inc.), D_2O (99.9 at%, Isotec Inc.) and high-pure water (Res. Genetic Co.) were mixed as purchased. Using the isotopic substitutions of water and octane, we prepared a series of samples with different scattering contrasts of octane and water, where surfactant volume fraction was fixed at $\phi_{\mathrm{s}} = v_{\mathrm{s}}/(v_{\mathrm{o}} + v_{\mathrm{w}} + v_{\mathrm{s}}) = 0.2$ and the oil volume fraction against water was kept at $\phi = v_{\mathrm{o}}/(v_{\mathrm{o}} + v_{\mathrm{w}}) = 0.5$. v_{w}, v_{o} and v_{s} are the volume of water, oil and surfactant, respectively. After mixing all ingredients in a sealed bottle, temperature was increased to $80\,^{\circ}\mathrm{C}$ in a water bath. The bottles were then shaken in a vibrator for 10 min. Samples were measured within 24 h after preparation. Samples were filled in the rectangular quartz cells with an inner dimension of $10 \times 40\ \mathrm{mm}^2$ and

an inner thickness of 1.0 mm. Sample cells were sealed by air-tight caps and parafilm. In order to avoid multiple scattering, we mixed H_2O with D_2O and n-octane with deutrated n-octane.

Small-angle neutron scattering was performed by SANS-U. SANS-U is installed at the cold neutron beam port C_{1-2} of JRR-3M at Japan Atomic Energy Research Institute [12]. Neutron wavelength was $\lambda = 7\ \text{Å}$ with a resolution $\Delta\lambda/\lambda = 0.1$. For the counter distance $L = 2\ \mathrm{m}$, it covers the range of scattering vector $0.012 < q < 0.15\ \text{Å}^{-1}$. Temperature was controlled and measured with an accuracy of $\pm 0.1\,^{\circ}\mathrm{C}$. Scattered neutron intensity was measured by a two-dimensional detector with a sensitive area of $64 \times 64\ \mathrm{cm}^2$. Since the detector sensitivity was almost isotropic [12], no tailoring of the observed profile was performed. The observed scattering intensities were corrected for transmission, an empty cell scattering intensity was subtracted and calibrated by a standard measurement from Lupolen to get absolute scattering intensity in the units of (cm^{-1}).

Results and discussion

Partial structure function

We have measured the scattering cross section $I(q)$ and obtained the independent partial structure functions $S_{\mathrm{ww}}(q)$, $S_{\mathrm{oo}}(q)$ and $S_{\mathrm{ss}}(q)$. We have also measured the cross partial structure functions $S_{\mathrm{ws}}(q)$ and $S_{\mathrm{os}}(q)$ independently. Partial structure functions for a few selected temperatures are shown in Fig. 2. Here the temperature $19\,^{\circ}\mathrm{C}$, $21\,^{\circ}\mathrm{C}$ are for LTM, $31\,^{\circ}\mathrm{C}$ for MTL and $38\,^{\circ}\mathrm{C}$, $40\,^{\circ}\mathrm{C}$ for HTM phase. Partial structure functions can be determined from Eq. (2) using the literature values of ϱ. However, to avoid the uncertainty in the calibration of scattering length density due to the penetration effect of water and oil into the surfactant monolayer, we have divided the measured $I(q)$ by the invariant $(\Delta\varrho)^2 = \{1/2\pi^2\phi(1-\phi)\}\int q^2 I(q)\,\mathrm{d}q$, instead of the literature values of $(\Delta\varrho)^2$ [13]. Here, $\Delta\varrho$ is the difference between two scattering amplitude densities. We have subtracted only the incoherent background due to nuclear-spin interaction (using the existing table) from $I(q)$ before calculating the partial structure functions. However, all other possible background scattering, particularly due to the mixing of different isotopes of water and octane, are still remaining in the data. Figure 2a shows that scattering profile is dominated by water–water partial structure functions. Figure 2c shows that $S_{\mathrm{ws}}(q)$ has a peak with positive value at LTM ($21\,^{\circ}\mathrm{C}$) and negative at HTM ($40\,^{\circ}\mathrm{C}$) phase. Thus, as temperature increases, the partial structure functions due to water–surfactant correlation changes from positive to negative value. On the other

Progr Colloid Polym Sci (1997) 106:98–103
© Steinkopff Verlag 1997

Fig. 2 Partial structure functions (a) $S_{ww}(q)$; (b) $S_{oo}(q)$; (c) $S_{ws}(q)$ and (d) $S_{os}(q)$ are plotted as a function of q. LTM, MTL and HTM phases correspond to the temprature 19 °C, 31 °C and 38 °C, respectively

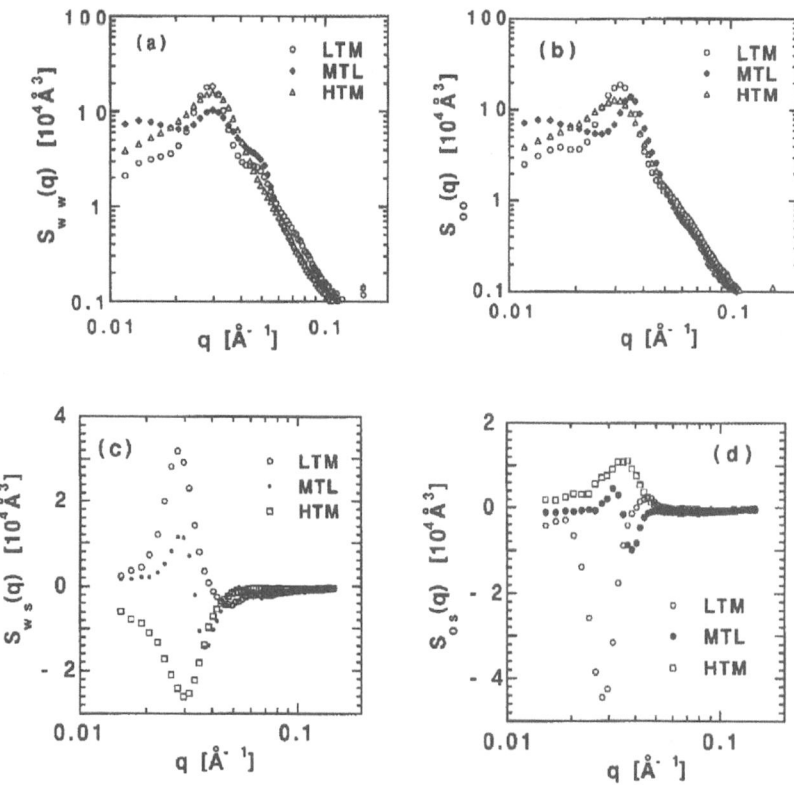

hand, oil–surfactant correlation $S_{os}(q)$ behaves in an opposite way as shown in Fig. 2d.

According to the theory of Gompper and Schick [9], all the partial structure functions $S_{\alpha\beta}(q)$ for ternary systems involving amphiphiles can be expressed by a linear combination of two characteristic functions $S^-(q)$ and $S^+(q)$, corresponding to the two order parameters, oil volume fractions against water ϕ and amphiphile volume fractions ϕ_s. The former function $S^-(q)$ is, in general, one order of magnitude larger than the latter $S^+(q)$, both being positive. Therefore, the contribution of $S^+(q)$ can be neglected in most cases. For particular case where $\phi = 0.5$, the cross partial structure function $S_{ws}(q)$ is nearly proportional to $-CS^-(q)$, whereas $S_{os}(q)$ is to $CS^-(q)$, where C is the imbalance of amphiphilicity toward oil defined as lipophilicity minus hydrophilicity. C is expected to be zero in a symmetric "balanced" system. The behaviors of $S_{ws}(q)$ and $S_{os}(q)$ at 21 °C show that $C < 0$ (Fig. 2c), whereas at 40 °C $C > 0$ (Fig. 2d).

Mean curvature of surfactant film

In order to determine the mean curvature $\langle c \rangle$ of interfacial films, Eq. (5) were fitted to the observed partial structure functions $S_{ww}(q)$, $S_{oo}(q)$, $S_{ss}(q)$ at higher q ($q > 0.08$ Å^{-1}) values, where $q^4 S_{ww}(q)$, $q^4 S_{oo}(q)$ and $q^2 S_{ss}(q)$ were almost linear as a function of q^2. Few selected fitting curves are shown in Fig. 3. From the fitting, the parameters A_w/V, A_o/V and d were determined. A_w/V, A_o/V and d vary with temperature, which is shown in Table 1. We observed that the characteristic thickness parameters t_w, t_o or t have values around 7–9 Å, in agreement with an earlier investigation [8], where they found the characteristic parameter to be 7 Å for a $C_{12}E_5$ monolayer. The average value of the film thickness d was 21 Å for LTM and HTM phases whereas 31.2 Å for MTL phase. The value of d is reasonable if we consider the molecular length of $C_{12}E_5$. The length for polyoxyethylene chain of a $C_{12}E_5$ molecule is 32.0 Å in a zigzag configuration and 24.8 Å in meander configuration [14]. The background c_b, which was determined from the fitting of Eq. (5), arises from the error in absolute cross section. It was found to be ~ 1 (cm^{-1}).

Mean curvature $\langle c \rangle$ is plotted as a function of temperature in Fig. 4. From the fitting of a linear relation $\langle c \rangle = b(T - T_{HLB})$, where T_{HLB} is the hydrophile-lipophile balance temperature, we have determined the characteristic parameter $b = 1.50 \times 10^{-3}$ (Å^{-1} K^{-1}) and $T_{HLB} = 31.9$ °C. Lee and Chen [6] found that $b = 0.86 \times 10^{-3}$ [Å^{-1} K^{-1}] for water/n-octane/$C_{10}E_4$

102

S.K. Ghosh et al.
Structure functions and mean curvatures in a ternary amphiphilic system

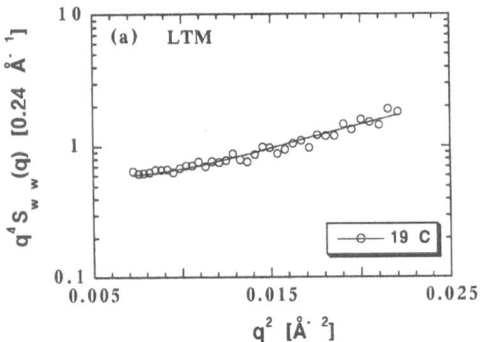

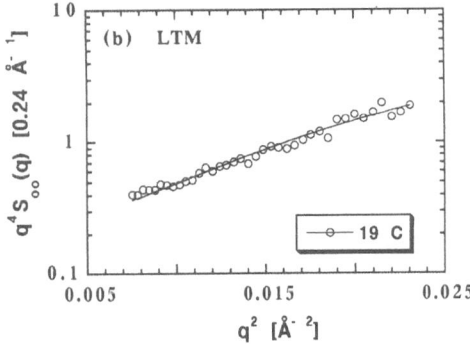

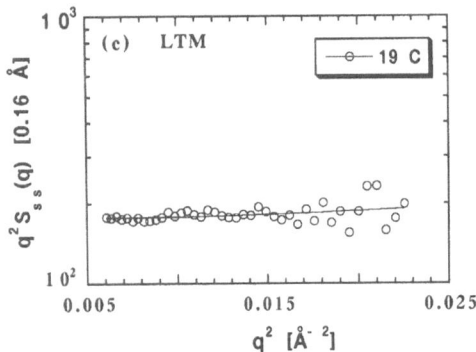

Fig. 3 (a) $q^4 S_{ww}(q)$; (b) $q^4 S_{oo}(q)$ and (c) $q^2 S_{ss}(q)$ are plotted as a function of q^2. Solid lines are the fitting curves of Eq. (5). Temperature 19 °C corresponds to a LTM phase

system, which is supposed to have a weaker amphiphilicity than $C_{12}E_5$. On the other hand, Strey [7] determined the mean curvature in a $C_{12}E_5$ system assuming a droplet type of microemulsion. They measured the mean curvature at lower surfactant concentration ($\phi_s = 0.076$), where the system has in sequence 2Φ phase and 1Φ phase and 2Φ phase with increasing temperature. But lamellar phase does not exist in this region of phase diagram. Their method of calculation was different and the observed value of b was 1.22×10^{-3} ($\text{Å}^{-1} \text{K}^{-1}$). In contrast, our samples have surfactant concentration $\phi_s = 0.2$. At this concentration we have two 2-phase regions and LTM, MTL and HTM

Table 1 A_w/V and A_o/V are the fitting parameters of Eq. (5a) and (5b). Effective thickness of the surfactant film d was calculated from the fitting parameters of Eq. (5c) by the relation $d = \phi_s(V/A_s)$. Mean curvature $\langle c \rangle$ was determined by the relation $\langle c \rangle = [(A_o - A_w)/(A_o + A_w)]/d$

T [°C]	$\frac{A_w}{V}$ [Å^{-1}]	$\frac{A_o}{V}$ [Å^{-1}]	d [Å]	$\langle c \rangle$ [Å^{-1}]
19	0.0304	0.0121	22.01	− 0.0196
21	0.0295	0.0138	21.52	− 0.0169
23	0.0277	0.0156	22.37	− 0.0125
31	0.0207	0.0189	31.19	− 0.0014
38	0.0197	0.0279	19.99	0.0086
40	0.0190	0.0311	19.19	0.0126

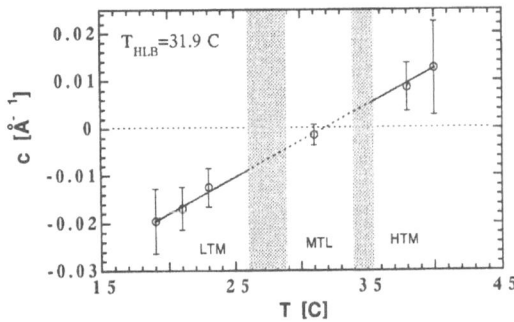

Fig. 4 Mean curvature $\langle c \rangle$ as a function of temperature. The solid line is a fitting to an equation $\langle c \rangle = b(T - T_{HLB})$. The shaded areas are the coexistence regions of both microemulsion and lamellar phase

phases as a function of temperature. We expected that a larger surfactant concentration may induce a bigger value of b. We also expected that the mean curvature in lamellar phase should be zero. Although the observed value of mean curvature in lamellar phase is found to be nonzero, but the zero value of mean curvature is within the limit of the possible error associated with the fitting parameters. We cannot say definitely about lamellar phase, since we investigated only one temperature point in lamellar phase.

In this system the behaviors of mean curvature i.e., $\langle c \rangle < 0$ at lower temperature and $\langle c \rangle > 0$ at higher temperature corresponds to those of the imbalance of amphiphilicity toward oil i.e., $C < 0$ at lower temperature and $C > 0$ at higher temperature. Thus, $\langle c \rangle$ and C correlate to each other in the same sense.

Conclusions

Partial structure functions and mean curvatures in ternary water/n-octane/$C_{12}E_5$ system with equal volume fraction of water and oil were determined by means of small-angle

Progr Colloid Polym Sci (1997) 106:98–103
© Steinkopff Verlag 1997

neutron scattering (SANS). Cross partial structure functions due to the correlation between water–surfactant or oil–surfactant change sign when the temperature passes through the hydrophile–lipophile balance (HLB) point, corresponding to the change of sign of the interaction parameter C, which is the imbalance of the amphiphilicity toward oil. We have determined mean curvatures considering that water or oil molecule can partially penetrate the surfactant film. The observed mean curvatures increase as temperature increases at a rate of 1.5×10^{-3} ($\text{Å}^{-1} \text{K}^{-1}$). A positive value of mean curvature $\langle c \rangle$ corresponds to a positive value of C.

Acknowledgements S.K. Ghosh sincerely acknowledges the financial support provided by the Japanese Ministry of Education, Science, Sports and Culture (Monbusho). The SANS experiment was done under the approval of the Neutron Scattering Program Advisory Committee proposal No. 6748.

References

1. Degiorgio V, Corti M (eds) (1985) Physics of Amphiphiles: Micelles, Vesicles and Microemulsion. North-Holland, Amsterdam
2. Gelbart WM, Ben-Shaul A, Roux D (eds) (1994) Micelles, Membranes, Microemulsions and Monolayers. Springer, New York
3. Kahlweit M, Strey R, Haase D, Kunieda H, Schmeling T, Faulhaber B, Borkovec M, Eicke HF, Busse G, Eggers F, Funck TH, Richmann H, Magid L, Söderman O, Stilbs P, Winkler J, Dittrich A, Jahn W (1987) J Colloid Interface Sci 118: 436–453
4. Auvray L, Cotton J-P, Ober R, Taupin C (1984) J Phys Chem 88:4586–4589
5. Huang JS, Kotalarchyk M (1986) Phys Rev Lett 57:2587–2589
6. Lee DD, Chen S-H (1994) Phys Rev Lett 73:106–109
7. Strey R (1994) Coll Polym Sci 272: 1005–1019
8. Strey R, Winkler J, Magid L (1991) J Phys Chem 95:7502–7507
9. Gompper G, Schick M (1990) Phys Rev B 41:9148–9162; Gompper G, Schick M (1989) Phys Rev Lett 62:1647–1650
10. For a general review, see Gompper G, Schick M (1994) Self-Assembling Amphiphilic Systems. Vol. 16, Academic Press, London
11. Teubner M (1990) J Chem Phys 92: 4501–4507
12. Ito Y, Imai M, Takahashi S (1995) Physica B 213&214:889–891
13. Porod G (1982) In: Glatter O, Kratky O (eds) Small angle X-ray Scattering. Academic Press, London, pp 17–51
14. Degiorgio V (1985) In: Degiorgio V, Corti M (eds) Physics of Amphiphiles: Micelles, Vesicles and Microemulsion. North-Holland, Amsterdam, pp 303–335

Progr Colloid Polym Sci (1997) 106:104–107
© Steinkopff Verlag 1997

H. Seto
G.D. Wignall
R. Triolo
D. Chillura-Martino
S. Komura

Small angle neutron scattering studies of critical phenomena in a three-component microemulsion

Received: 13 October 1996
Accepted: 30 April 1997

Dr. H. Seto (✉) · S. Komura
Faculty of Integrated Arts and Sciences
Hiroshima University
1-7-1 Kagamiyama
Highashihiroshima 739, Japan

G.D. Wignall
Solid State Division
Oak Ridge National Laboratory
P.O. Box 2008 Oak Ridge, TN 37831-6393
USA

R. Triolo · D. Chillura-Martino
Dipartimento de Chimica Fisica
University of Palermo
90123 Palermo, Italy

Abstract Critical density fluctuations of a "water-in-oil" microemulsion consisting of water, benzene, and BHDC (benzyldimethyl-n-hexadecyl ammonium chloride) have been observed near the phase boundary by a small-angle neutron scattering (SANS). The observed profiles were well described by the product of a form factor of spherical droplets and a structure factor, consisting of a term describing the inter-droplet correlations and also an Ornstein–Zernike component describing the droplet density fluctuations. Allowance was also made for the droplet polydispersity, though the width of the distribution turned out to be very small (1–2%). The observed temperature dependence of the osmotic compressibility was fitted using the crossover function proposed by Belyakov et al., and the Ginzburg numbers were obtained of the order 10^{-2}. These results indicate that long-range inter-droplet forces are not significant in this system, which displays upper critical solution temperature (UCST) behavior. In contrast, previous studies of systems displaying with lower critical solution temperature (LCST) behavior [e.g. water, n-decane and AOT (dioctyl sulfosuccinate sodium salt)] indicate that long-range interactions appear to dominate the phase separation behavior.

Key words Small-angle neutron scattering – critical phenomena – crossover – microemulsion

Introduction

The critical behavior of density fluctuations in microemulsions with a droplet structure can be treated analogously to simple fluids, because the radius is virtually constant throughout the phase separation and the droplet density may be regarded as an order parameter. Because of the nature of the droplet systems, its critical behavior is expected to belong to the 3D-Ising universality class. However, the observed critical exponents do not always coincide with the exact values of the 3D-Ising model. In particular, the well-known ternary system (WDA), consisting of an oil-rich mixture of water, n-decane, and AOT (dioctyl sulfosuccinate sodium salt) has been the subject of a range of experimental investigations [1–4]. This system is homogeneous at room temperature and decomposes as the temperature is increased, thus, exhibiting critical phenomena below the decomposition temperature, i.e. the phase diagram show an LCST transition. The various experimental studies performed on this system gave values of the critical exponents, γ and v, which did not coincide with either the 3D-Ising or Fisher's renormalized 3D-Ising values.

Alternatively, Kiselev and his coworkers [5, 6] have proposed an analytical form of the susceptibility, which describes the fluctuations of the order parameter over the whole temperature range in the mixed state of the system,

$$\hat{\tau} = (1 + 2.333\hat{S}_0^{\Delta/\gamma})^{(\gamma-1)/\Delta}[\hat{S}_0^{-1} + (1 + 2.333\hat{S}_0^{\Delta/\gamma})^{-\gamma/\Delta}] , \quad (1)$$

where $\hat{S}_0 = S_0 Gi/C_{MF}$ and $\hat{\tau} = \tau/Gi$ are a renormalized susceptibility and temperature. S_0, Gi, C_{MF}, and $\tau = |T^{-1} - T_C^{-1}|/T_C^{-1}$ represent the susceptibility of the order parameter fluctuation, the Ginzburg number (i.e. the value of reduced temperature separating the mean field and 3D-Ising regions), the critical amplitude of the mean-field approximation and the reduced temperature, respectively. This theory has the advantage that it may be applied quite generally over the whole crossover regime including the scaling limits both very near and very far from the critical point. Seto et al. [7] have recently used this approach to interpret SANS data from WDA microemulsions, and showed that the critical behavior can be explained by a crossover from the mean field to 3D-Ising regimes. The Ginzburg number was found to be of the order of 10^{-3}, which is an order of magnitude smaller than in simple fluids. This finding suggests that the dominant interactions underlying the phase separation are long ranged. In this system, three types of interactions between droplets are known; a hard-core repulsive force, a hydrophobic attraction, and an electrostatic interaction originating in the charge fluctuations and/or the van der Waals-type dipole–dipole interaction [8]. Among these interactions, the electrostatic one should control the phase separation phenomenon, because rather strong attractive forces are required to produce LCST behavior, as the system decomposes as the temperature is raised and the entropy increases. This hypothesis is consistent with the small value of the Ginzburg number because the electrostatic interaction must be long ranged.

In order to clarify the relation between the phase behavior, interactions between droplets, and the Ginzburg number, we have undertaken further SANS studies of critical phenomenon in a different three-component microemulsion system called WBB, consisting of water, benzene, and BHDC (benzyldimethyl-n-hexadecyl ammonium chloride). This system also has a water-in-oil-type droplet structure at room temperature and decomposes with decreasing temperature. Above the (UCST) phase separation point, critical phenomena have been investigated by Beysens and coworkers [9, 10], who obtained the critical indexes, $\gamma = 1.18$ and $\nu = 0.60$, and concluded that their data could be interpreted within the 3D-Ising universality. However, Fisher's renormalized critical exponents were not obtained.

Experimental

The current SANS experiments were performed at 30 m-SANS facility at Oak Ridge National Laboratory with a wavelength of 0.4750 nm and sample–detector distance of 14 and 5 m, which gave a range of momentum transfer

Table 1 Parameters for sample preparation

Sample	D$_2$O volume fraction [%]	M_{water}/M_{BHDC}	c_B
C091	60	2.026	0.91
C092	60	2.026	0.92
C191	50	2.005	0.91
C192	50	2.005	0.92

of $0.047 \leq Q \leq 1.2$ [nm^{-1}]. Quartz cells (1 mm path length) were used to contain the samples, whose temperature was controlled by a circulation bath system with an accuracy of ± 0.2 K. The chemicals, BHDC, benzene, D$_2$O, and H$_2$O were purchased from Tokyo Kasei Co. Ltd., Mallinckrodt Inc., Aldrich Chem. and J.T. Baker Inc. It is not known whether the phase transition may be perturbed by isotope effects, due to H$_2$O with D$_2$O, so in order to minimize this possibility, we prepared four mixtures where the volume fraction of each composition is almost the same as the critical composition for pure H$_2$O, i.e. mass ratios of water to BHDC were 2.026 and 2.005 for the 60% D$_2$O and 50% D$_2$O samples, respectively. The mass fractions of benzene with respect to the total mass (c_B) were 0.91 and 0.92, respectively (see Table 1).

Results and discussion

In Fig. 1, typical scattering profiles $I(Q)$ are depicted. In SANS measurements for the three component microemulsion system, neutrons are sensitive to the differences of scattering amplitude densities among partially deuterated water, non-deuterated benzene, head-groups of surfactants, and hydrocarbon chains of surfactants. Because the neutron-scattering density of hydrocarbon chains is comparable with that of benzene and the layer of head-groups is thin enough, the SANS intensity can be approximated by a model of "water droplets surrounded by benzene background". Therefore, we assumed a scattering function $I(Q)$ which is the product of an intraparticle form factor, $P(Q)$, and an interparticle structure factor, $S(Q)$, with a constant background term I_B in order to fit the observed SANS profiles. Kotlarchyk et al. [11] have already introduced a model to explain SANS profiles from such "water-in-oil" droplet systems. They presented an analytical form of $P(Q)$ for the Schultz size distribution to model possible polydispersity effects. Thus, the fitting parameters are the scattering density difference between the droplet and the solvent ($\Delta\rho$), the mean radius of droplet (R), and width parameter of polydispersity (Z). In their analysis on WDA system, the structure factor $S(Q)$ was assumed to be the

106

H. Seto et al.
SANS studies of critical phenomena in a microemulsion

Table 2 The obtained fit parameters to explain the scattering intensities

Sample	I_B [cm^{-1}]	Z	R [nm]	$\Delta\rho$ [10^{10} cm^{-2}]	S_0^G [cm^{-1}]	Q_0 [nm^{-1}]	σ [nm^{-1}]
C091	0.541	70.1	4.91	2.073	0.161	0.318	0.095
C092	0.565	70.0	4.91	2.074	0.189	0.314	0.108
C191	0.621	69.3	4.56	1.393	0.269	0.319	0.134
C192	0.592	71.8	4.43	1.454	0.237	0.317	0.115

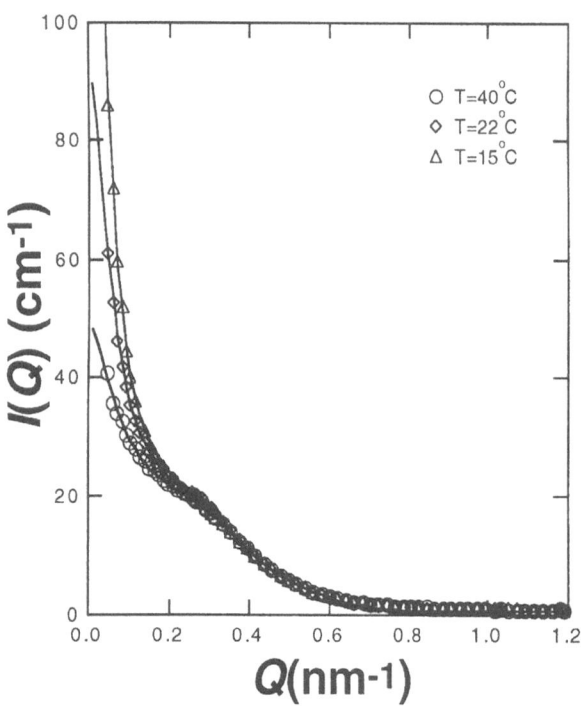

Fig. 1 Observed SANS profiles for the sample C091 and its temperature variation. Solid lines are obtained by the fitting

Table 3 The parameters to fit the crossover function

Sample	T_s [K]	Gi [units of 10^{-2}]	C_{MF} [units of 10^{-2}]
C091	282.8 ± 0.7	0.68 ± 0.48	8.20 ± 0.24
C092	281.7 ± 1.8	1.57 ± 0.16	9.58 ± 1.62
C191	282.8 ± 1.8	5.94 ± 0.82	8.35 ± 4.68
C192	281.7 ± 1.5	3.84 ± 0.38	7.97 ± 2.21

ing the critical point. The temperature variation of the susceptibility for the droplet density fluctuation S_0 was obtained and fitted to the crossover function (Eq. (1)) for the case of 3D-Ising model, (i.e. $\gamma = 1.26$ and $\Delta = 0.51$). The experimental observations were well described by the function and resultant fitting parameters, T_S (in this case, a spinodal temperature was obtained in spite of T_c, because the samples were not at the critical concentration), Gi and C_{MF}, are shown in Table 3. The values of Ginzburg number are of the order of 10^{-2}, which is an order of magnitude greater than in the case of WDA and is of the same order as in simple fluids.

The polydispersity was estimated from the observed values of Z and was very small (i.e. a few percent). The value of Q_0 indicates that an inter-droplet correlation exists with a spacing of about 20 nm. From these results, we conclude that the system consists of relatively monodisperse droplets (radius ~ 5 nm), which scatter uniformly. Therefore, the general features of this system are quite similar to those of a simple fluid and may be expected to exhibit 3D-Ising critical behavior.

In Fig. 2, the renormalized susceptibilites are plotted vs. the renormalized temperature. The data points fall in the intermediate region between the pure-Ising and the crossover regions, in contrast to the WDA system, which stayed in the crossover region [7]. This behavior is reflected in the values of Gi for the WDA and WBB; the former being more mean-field-like, while the latter is more Ising-like. As discussed above, mean-field behavior is expected when long ranged interactions are dominant in the phase separation phenomenon. The LCST-type phase diagram of the WDA system suggests that a temperature-dependent attractive force drives the phase separation behavior. However, the UCST-type transitions (as observed in the

sum of an asymptotic Ornstein–Zernike form

$$S_{OZ}(Q) = 1 + S_0/(1 + \xi^2 Q^2) \qquad (2)$$

and an inter-droplet component $S_{ID}(Q)$ to explain the deviation from unity far away from the critical point. First, we tried to use the hard-sphere model [12] and the mean spherical approximation (MSA) ansatz [13] as the inter-droplet component, however, their contributions were too strong to explain the data. Therefore, a single Gaussian

$$S_{ID}(Q) = S_0^G \exp(-(Q - Q_0)^2/\sigma^2) \qquad (3)$$

was introduced as a small perturbation to the structure factor.

The fitting parameters obtained for each samples are shown in Table 2. Apart from the parameters of $S_{OZ}(Q)$, there was little variation with temperature in one-phase region. Thus, we conclude that the size and shape of droplets do not change in one-phase region in approach-

Progr Colloid Polym Sci (1997) 106:104–107
© Steinkopff Verlag 1997

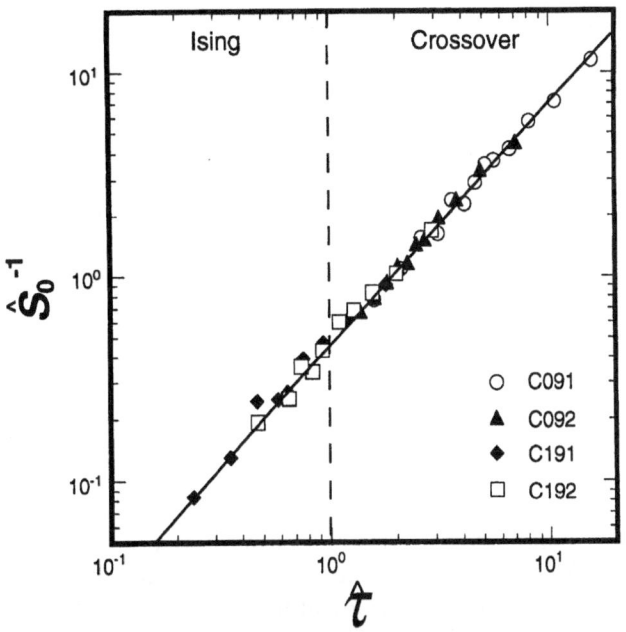

Fig. 2 All the observed temperature dependence of the inverse susceptibility of droplet density fluctuation for the WBB system is shown. The vertical axis indicates the inverse of the renormalized susceptibility and the horizontal the renormalized temperature $\hat{\tau} = \tau/Gi$. The pure Ising region is characterized by the reduced temperature which is smaller than the Ginzburg number i.e. $\hat{\tau} < 1$

WBB system) do not require such an interaction. The present results from the WBB system indicate that the droplets scatter rather uniformly and the Ginzburg number is of the same order as that of simple fluids and are therefore consistent with its phase behavior.

Conclusion

Small-angle neutron-scattering measurements have been carried out on critical scattering from a water-in-oil droplet microemulsion system (WBB) consisting of water, benzene and BHDC. The observed critical divergence of the susceptibility for droplet density fluctuation was explained within the framework of the crossover between mean-field behavior and 3D-Ising. The obtained Ginzburg number was of the order of 10^2, which is the same as that of simple fluids and an order of magnitude more than that of the WDA system. This value was explained by taking the nature of the phase transitions into account.

Acknowledgements The research at Oak Ridge was supported by the Division of Materials Sciences, U.S. Department of Energy under contract No. DE-AC05-96OR22464 with Lockheed Martin Energy Research Corp., and also by the Japan–U.S. Cooperative Research Program on Neutron Scattering.

References

1. Hung JS, Kim MW (1982) Phys Rev Lett 47:1462
2. Kotlarchyk M, Chen SH, Huang JS (1983) Phys Rev A 28:508
3. Honorat P, Roux D, Bellocq AM (1984) J Phys (Paris) Lett 45:L-961
4. Rouch J, Safouane A, Tartaglia P, Chen S-H (1989) J Chem Phys 90:3756
5. Belyakov MY, Kiselev SB (1992) Physica A 190:75
6. Anisimov MA, Kiselev SB, Sengers JV, Tang S (1992) Physica A 188:487
7. Seto H, Schwahn D, Nagao M, Yokoi E, Komura S, Imai M, Mortensen K (1996) Phys Rev E 54:629
8. Tingey JM, Fulton JL, Smith RD (1990) J Phys Chem 94:1997
9. Jayalakshmi Y, Beysens D (1992) Phys Rev A 45:8709
10. Aschauer R, Beysens D (1993) Phys Rev E 47:1850
11. Kotlarchyk M, Chen S-H, Huang JS, Kim MW (1984) Phys Rev A 29:2054; Kotlarchyk M, Chen S-H (1983) J Chem Phys 79:2461
12. Percus JK, Yevik GJ (1958) Phys Rev 110:1
13. Hayter JB, Zulauf M (1982) Colloid Polym Sci 260:1023

Progr Colloid Polym Sci (1997) 106:108–111
© Steinkopff Verlag 1997

K. Miyakawa
F. Sakamoto
H. Akata

Aggregation dynamics of micro-hydrogels dispersed in iso-octane

Received: 13 October 1996
Accepted: 14 April 1997

Dr. K. Miyakawa (✉)
F. Sakamoto · H. Akata
Department of Applied Physics
Faculty of Science
Fukuoka University
Fukuoka 814-80, Japan

Abstract Aggregation dynamics of water-in-oil microemulsion containing polyacrylamide gels have been investigated. Gels were formed only in an aqueous phase of water-in-oil microemulsions composed of a surfactant Aerosol OT, iso-octane, and water. Increasing the temperature, the percolation transition first occurred and then the phase separation occurred. The percolation transition was characterized by a steep increase of the electrical conductivity, which was interpreted in terms of a power law. The phase separation process initiated by a temperature jump was investigated by using a time-resolved light scattering. It was found that the scattering intensity can be scaled with the scaling law similar to that for the ordinary spinodal decomposition if the fractal dimension is taken as the dimensionality of the system.

Key words Microemulsion – gel – aggregation – percolation – spinodal decomposition

Introduction

Microemulsions are macroscopically homogeneous, isotropic, thermodynamically stable mixture of oil, water and surfactant. Water-in-oil (W/O) microemulsions are known to be modified by polymerizing some polymers in the aqueous phase under appropriate conditions of temperature and polymer concentration. Such nanometer-sized microemulsion-based gels, termed nanogels, are of considerable interest from the biophysical and biotechnological points of view. Haering et al. and Quellet et al. have found that the microemulsion-based gelatin gel undergoes a percolation transition on heating, which is marked by a dramatic increase of electrical conductivity [1, 2]. The transition involves the formation of a permanently cross-linked network of gelatin-filled nanodroplets, which is thermally irreversible.

In this study we realize microemulsion-based gels using polyacrylamide. These nanogels have the ability to aggregate reversibly for the change of the temperature. With increasing the temperature, the percolation transition first occurs and then the phase separation occurs. We investigate the process of aggregation leading to the phase separation at a low volume fraction of nanogels. The percolation transition is investigated by measuring the low-frequency electrical conductivity as a function of temperature. For the phase separation, the measurements are made by a temperature jump above the phase separation temperature, using a time-resolved small-angle light scattering.

Experimental section

Materials

The W/O microemulsion was composed of water, iso-octane and sodium bis(2-ethylhexyl) sulfosuccinate (AOT). The pregel solution was prepared by dissolving the following compounds in distilled water, brought up to a final

Progr Colloid Polym Sci (1997) 106:108–111
© Steinkopff Verlag 1997

volume of 10 ml: linear constituents, 1.58 g of acrylamide (AM), 0.167 g of sodium salt of acrylic acid (AA); a cross-linking constituent, 0.083 g of N,N'-methylenebis(acrylamide) (BIS). An aliquot of the pregel solution was added to the mixture of AOT–iso-octane, and then 4.3 mg of ammonium persulfate and 0.035 ml of N,N,N',N'-tetramethylenediamine were added to initiate the polymerization reaction. The solution was vigorously stirred for 1 h at 17 °C until all the components are fully dispersed. The measurements were made after equilibration time of about 12 h to 3 days in order to obtain clear dispersed solutions. A molar ratio of water to surfactant $w = $ [water]/[AOT] was changed in the range 45–60, for the whole range of volume fractions ϕ of nanogels from 0.06 to 0.15. The concentration of polymer constituents c of nanogels ranged from 0.1 to 0.65% (w/v), where the polymer concentration is expressed in the sum of the number of grams of AM, AA, and BIS per 100 ml of sample. The concentration of AOT was always fixed at 0.1 M. Control parameters, c, w, and ϕ are related to hardness, size and number of nanogels, respectively.

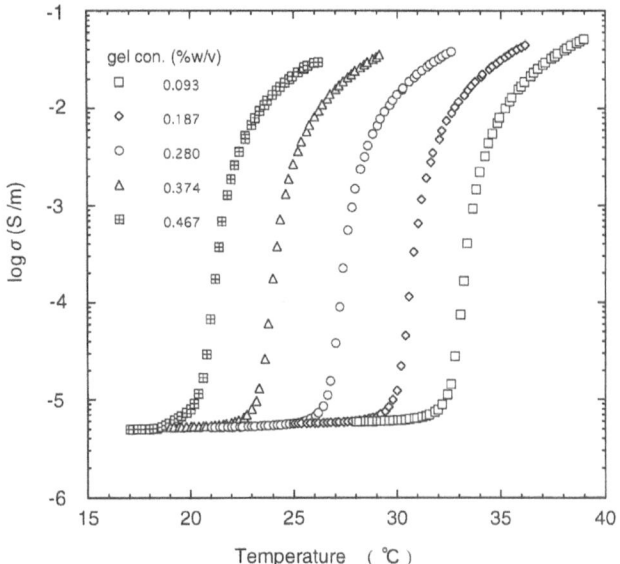

Fig. 1 Electrical conductivity as a function of temperature for various gel concentrations (% (w/v)) at $w = 60$ and $\phi = 0.148$

Electrical conductivity

The electrical conductivity of nanogels dispersed in iso-octane was measured at a frequency of 10 kHz by using a LCR meter (HP4263A). The conductance cell consists of two electrodes plated with platinum black in order to reduce the electrode polarization. The cell constant was 1.0 S cm^{-1}. The temperature of the cell was controlled with an accuracy of ± 0.01 °C.

Phase separation

The time evolution of phase separation was detected by a light scattering apparatus equipped with the linear photodiode array of 512 channels (Hamamatsu Photonics, C4834). This covered the range of wave number $q = (4\pi n/\lambda)\sin(\theta/2)$ of 1–1.7 μm^{-1}, where n, λ and θ are the refractive index of the medium, the wavelength of the incident beam (632.8 nm) and the scattering angle, respectively. To minimize multiple scattering, we used a quartz cell with 1 mm thickness.

Results and discussion

Percolation

The hydrodynamic radius of nanogels was determined by a dynamic light scattering measurement, and was in the range of 10–15 nm depending on w. This value is a little larger than that of AOT–water–iso-octane microemulsion [3]. We measured a low-frequency electrical conductivity σ as a function of temperature T, polymer concentration c, volume fraction ϕ and a molar ratio of water to surfactant w. In Fig. 1, we show the temperature dependence of σ for various polymer concentrations at $w = 60$ and $\phi = 0.148$. On increasing the temperature, the electrical conductivity steeply increases by about 4 orders of magnitude for every case. This is indicative of the appearance of percolation phenomena. The conductivity curves move parallel to lower temperature with an increase of the polymer concentration. This suggests that fundamental features of the transition remains unaffected by the presence of gel in water droplets. A similar result was found for microemulsion-based gelatin gels [2]. The data were analyzed in terms of the power laws,

$$\sigma = A|T_p - T|^{-s} \quad T < T_p,$$
$$\sigma = B|T - T_p|^t \quad T > T_p, \tag{1}$$

where T_p is the percolation temperature, and A and B are constants. In Fig. 2 the scaled conductivities $T_p(\sigma/A)^{-1/s}$ and $T_p(\sigma/B)^{1/t}$ are plotted as a function of $T - T_p$. A nonlinear least-squares fit gives the exponents $s = 1.1 \pm 0.1$ and $t = 1.7 \pm 0.1$. The values of s agrees with that from the dynamic percolation theory [4], while the value of t is slightly smaller than that from the static one [5].

Furthermore, we also investigated the temperature dependence of σ for various volume fractions at $w = 60$ and

110
K. Miyakawa et al.
Aggregation dynamics of micro-hydrogels dispersed in iso-octane

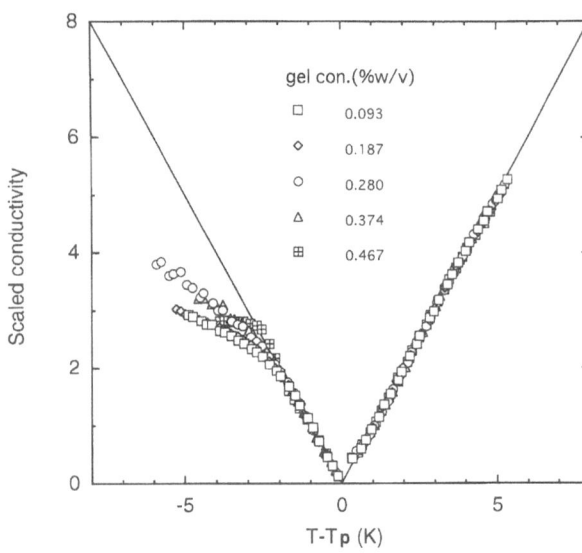

Fig. 2 The scaled conductivities as a function of temperature difference $T - T_{\mathrm{p}}$ for various gel concentrations, which refers to the same run of Fig. 1

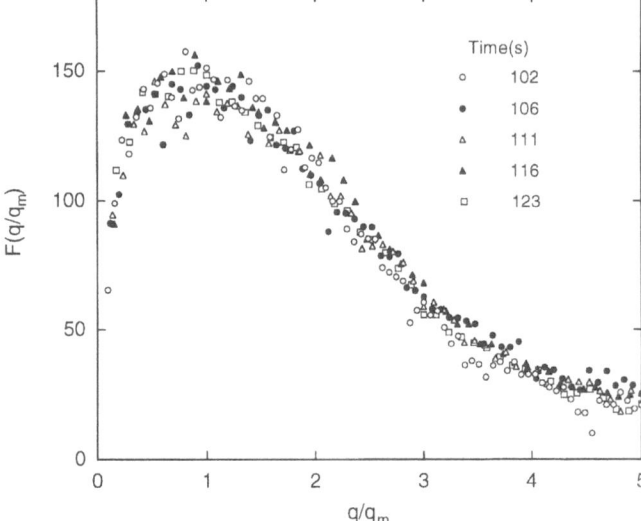

Fig. 3 Scaling function $F(q/q_{\mathrm{m}})$ as a function of q/q_{m} for $w = 60$, $\phi = 0.148$ and $c_{\mathrm{g}} = 0.5\%$ (w/v). The data are for the later stage of the aggregation

$c_{\mathrm{g}} = 0.43\%$ (w/v). The deviation from the power-law behavior appeared below T_{p}, which was more pronounced for lower volume fractions. Such a deviation has been also observed in a AOT–water–decane microemulsion [6]. Thus, the present nanogel system exhibits the percolation behavior very similar to that of simple microemulsions.

Phase separation

When the temperature was further increased above T_{p}, nanogels caused the phase separation at $T = T_{\mathrm{c}}$. We investigated the dynamics of the phase separation as a function of c, ϕ, and w. Here, the composition of the sample was always different from the critical composition. The phase separation was initiated by a thermal jump from a temperature T_{i} of $0.5\,°C$ below T_{c} to a temperature T_{f} of $0.5\,°C$ above T_{c}. Here T_{i} was always higher than the percolation temperature T_{p}. The scattered intensity $I(q,t)$ exhibited a peak at a finite wave vector q. The peak height increased with time, and its position q_{m} shifted to smaller and smaller values. These behaviors are reminiscent of those of spinodal decomposition (SD). However, it should be noted that the time scale greatly differs from that of SD. We tried to scale the intensity distribution $I(q,t)$ with the scaling law of ordinary SD [7],

$$I(x,t) = q_{\mathrm{m}}(t)^{-d}F(x) , \qquad (2)$$

where $x = q/q_{\mathrm{m}}$, d the spatial dimension of the system, and $F(x)$ a time-independent scaling function.

In Fig. 3, the scaling functions $F(x)$ in the later stage are plotted against x. Here we take $d = 1.8$ instead of $d = 3$ for SD. We can see that the data collapse to a single-scaling function. The value $d = 1.8$ corresponds to the fractal dimension of clusters d_{f}. On the other hand, the data in the earlier stage cannot be scaled according to Eq. (2) even if we take $d = d_{\mathrm{f}}$.

Furthermore, we investigated the time behaviors of the peak intensity I_{m} and its position q_{m}. According to the linear theory of SD, q_{m} and I_{m} are expressed as $q_{\mathrm{m}} \sim t^{-\alpha}$ and $I_{\mathrm{m}} \sim t^{\beta}$. The relation between α and β was found to be given by $\beta = 1.7\alpha$, whereas it is given by $\beta = 3\alpha$ for SD. The obtained value of β/α is close to the fractal dimension, as expected.

Thus, the phase separation of present nanogels exhibits features similar to the ordinary spinodal decomposition, except that the fractal dimension is required for d. This surprising similarities to SD may suggest that there exists some underlying common mechanism in the dynamics of both irreversible processes. In the present case, diffusion plays an important role, in contrast to the case of spinodal decomposition. Fractal clusters are formed through the percolation transition, and the clusters themselves diffuse to coalesce into larger clusters and so on. Behaviors described above have been also observed in the aggregation process of dense colloidal solutions [8]. This may come from the similarity between nanogels in the percolated state and dense colloidal solutions.

Progr Colloid Polym Sci (1997) 106:108–111
© Steinkopff Verlag 1997

References

1. Haering G, Luisi PL (1986) J Phys Chem 90:5892–5895
2. Quellet C, Eicke HF, Sager W (1991) J Phys Chem 95:5642–5655
3. Eicke HF, Borkovec M, Das-Gupta B (1989) J Phys Chem 93:314–317
4. Grest GS, Webman I, Safran SA, Bug AL (1986) Phys Rev A 33:2842–2845
5. Derrida B, Stauffer D, Hermann HJ, Vannimenus J (1983) J Phys (France) Lett 44:L701–L706
6. Cametti C, Codastefano P, Tartaglia P, Rouch J, Chen SH (1990) Phys Rev Lett 64:1461–1464
7. Furukawa H (1984) Physica A 123:497–515
8. Carpineti M, Giglio M (1992) Phys Rev Lett 68:3327–3330

Progr Colloid Polym Sci (1997) 106:112–117
© Steinkopff Verlag 1997

M. Monkenbusch
K.D. Goecking
D. Schneiders
D. Richter
L.J. Fetters
J.S. Huang

Dynamical aspects of self-organized (macro) molecular systems investigated by neutron spin-echo spectroscopy

Received: 13 October 1996
Accepted: 4 February 1997

Dr. M. Monkenbusch (✉) · K.D. Goecking
D. Schneiders · D. Richter
Institut für Festkörperforschung
Forschungszentrum Jülich, KFA
D-52425 Jülich, Germany

L.J. Fettes · J.S. Huang
Exxon Corporate Research
and Engeneering Corporation, Annandale
New Jersey, 08801 USA

B. Farago
Institut Laue-Langevin
38042 Grenoble Cedex
France

Abstract Neutron spin-echo (NSE) spectroscopy is a tool to close the gap between light scattering and more conventional neutron spectroscopies. It may also be viewed as an augmentation of small-angle scattering with inelastic, dynamical information.

Using a microemulsion system as example for a system of molecular aggregates and a block-copolymer system forming macromolecular aggregates, it is shown what type of dynamics is observable by NSE-spectroscopy.

The lamellar phase of a SDS–pentanol–water microemulsion allows for the investigation of different layer modes (undulation, peristaltic mode) the dynamics of which are determined by elastic properties of the interface layers and by the friction due to fluid flow between the layers.

Platelet-like aggregates of polyethylene–polyethylenebutylene (PE–PEP) copolymers formed upon cooling a decane solution exhibit a polymeric brush of PEP-hairs on their surface. The fluctuation dynamics of this brush determined by a balance of solvent friction and entropic restoring forces has been observed by NSE spectroscopy. Besides, the direct dynamical information the NSE data allow for the separation of static scattering due to the average structure from the scattering contributions due to mobile fluctuating parts of the sample.

Key words Neutron spin-echo – relaxation dynamics – molecular aggregates – microemulsions – block copolymers

Introduction

Small-angle neutron scattering (SANS) is a unique tool to investigate the structure of mesoscopic molecular systems, e.g. polymeric macromolecules in solution or in the melt or mesoscopic aggregates (micelles, etc.) of small or macromolecules. The substances involved usually contain a considerable amount of hydrogens, thereby allowing the application of the key technique of SANS, contrast variation.

For solutions the variation of the solvent scattering length density allows for the matching of parts of a composite object, aggregate, leaving only scattering signals from the other parts.

However, it being a diffraction method, SANS yields no information on the dynamical properties of the scattering structures. Neutron spin-echo spectroscopy [1] (NSE) may fill this gap. The individual neutron-spin processing in a magnetic field region before and after scattering is used as time keeper to measure the velocity change of the neutron during the scattering. This allows for a broad

Progr Colloid Polym Sci (1997) 106:112–117
© Steinkopff Verlag 1997

incoming neutron velocity band (10–20%) as usual in SANS, yielding reasonable intensity simultaneously with a sensitivity to relative velocity changes in the 10^{-4}–10^{-6} range. The measured signal is proportional to the cosine Fourier transform of the scatterinng function $S(Q, \omega)$, which is, for any reasonable soft matter sample, virtually equal to the intermediate scattering function $S(Q, t) = \langle \rho_Q(0)\rho_{-Q}(t)\rangle$, where $\rho_Q(t)$ denotes the spatial Fourier component of scattering length density at time t. In addition, as in SANS, contrast variation is an additional handle to identify or emphasize different scattering contributions.

The momentum (Q)-range to be observed sets the requirements for the spectral resolution, i.e. the time range to be covered by the observed $S(Q, t)$. Generally, small-angle scattering (low Q) means that the observed distances are large (compared to the respective wavelength) and, usually, also the scattering objects tend to be large. Both lead to a considerable increase of the typical dynamical time constant (10^{0-2} ns for SANS typical Q-values) of these systems compared to the picosecond scale of single-atom movements along interatomic distances. Firstly the spatial displacements needed to change the scattering phase sufficiently are large, and secondly also the moving objects are large and therefore slow.

For all soft matter systems the friction forces dominate any inertial forces as long as the system is viewed with focus on mesoscopic distances. Therefore, the dynamical fluctuation are of the relaxation type. Existing NSE spectrometers cover the range from fractions of a nanosecond to about 10^2 ns. Routinely, 1 ns–20 ns are available at most installation. After a long period, when only one NSE instrument was existing worldwide in the recent years, besides the ancestor IN11, another instrument IN15 [2] has been built at the ILL. The MESS instrument in Saclay [3] has been operational for several years. In Japan,

a more recent NSE spectrometer at JAERI [4] has been set up and this year the NSE instrument at KFA-Jülich [5] has started operation and a sister instrument of it is currently installed at NIST. The above compilation shows that the spin-echo technique is about to become much more available than in the previous years. Therefore, this article is also intended to give examples of the facilities of the method to a broader soft-matter community.

In the next sections results are reported on the dynamics of large aggregates consisting of small surfactant molecules, i.e. lamellar microemulsion phases and of micelles of block-coploymer molecules illustrating the separation of different scattering contributions by their dynamical properties.

Microemulsion dynamics

The quarternary system consisting of the surfactant sodiumdodecylsulfate (SDS), pentanol as cosurfactant, dodecane (= oil) and water or brine exhibits a lamellar phase region in its phase diagram that is stable over a wide temperature range and extends over a broad range of layer distances [6]. The lamellar phases may be oriented [7] to enable experiments with well-defined components of the Q-vector orthogonal ($Q_\perp$) and parallel (Q_z) to the average lamellar plane normal. Selective deuteration of the water and/or the oil components enable to enhance or suppress the scattering contributions of different static or fluctuating structures. Figure 1 shows the 2D SANS pattern of oriented oil-rich brine samples in "full sheet" and "double sheet" contrast. The stacks of alternating layers of water–surfactant–oil–surfactant–water ... yield pseudo-Bragg peaks at positions corresponding to the repeat distance, d. The samples are taken either from an oil–rich or a water-rich phase, thus being composed of a surfactant

Fig. 1 Typical SANS data (PAXE Saclay) from an oriented lamellar SDS phase. A combination of two detector distances (outer field 1 m at $\lambda = 6$ Å, inner field 3 m) is plotted into each data field. The right pattern stems from a sample with "full-sheet contrast", whereas the left pattern is taken from a sample with "double-sheet-contrast". In the latter case, the sheet form factor has a minimum, giving rise to the valley between the inner part of the intensity distribution and the outer bumps around $q_z = 0.2$ Å^{-1}. Besides one pseudo-Bragg peak and a small corresponding ring from residual nonoriented sample, the intensity ridge along $q_\perp = 0$, q_z) is clearly visible. The NSE data shown have been taken on this ridge, of analogous samples, beyond the peak position

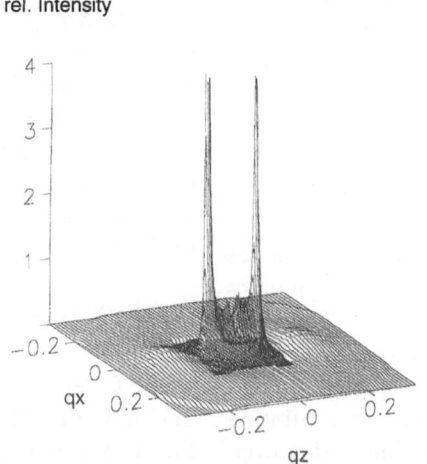
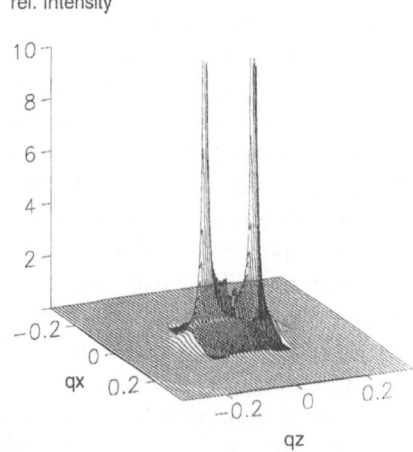

double layer containing the minority component (oil or water) with a total thickness around 20 Å separated by a thicker layer (around 80 Å) of the majority component. The amount of fluctuations in the system determines the width of the peaks as well as the number of observable higher orders [8, 6]. The undulations of the surfactant double layers are the reason for the intensity between the peaks as well as for the power-law divergence-type peaks instead of a δ-functions. The observed scattering intensity depends on the free energy needed to displace the system from its average structure. At ambient temperature the equipartition theorem is applicable, each fluctuation mode is excited according to the ratio $\delta F(q)/kT$, where $\delta F(q)$ denotes the free energy needed to establish a mode q in the system. Therefore, the SANS pattern yields information of the energetics, i.e. elastic moduli, of the system. The basic expression for a layered system with layer displacements u was first given by de Gennes [9] $\langle u_q u_{-q} \rangle = k_B T/(B q_z^2 + K q_\perp^4)$, where B denotes the layer compressibility and $K = \kappa/d$ is the bending elasticity. However for a fluid system – as these microemulsions are – the fluctuation are expected to be of dynamical nature. The elastic forces will lead to the decay of thermally excited modes in the system, the relaxation time being given through the balance of the restoring forces and viscous forces due to flow of water, oil or internal-layer friction connected with the corresponding relaxation motion. The relaxation of these modes is the object of NSE investigations. Contrast-variation techniques allow – at least partly – for a separation of the contributions from qualitatively different modes. Fluctuation intensity that corresponds to undulations of the surfactant double layer is modulated by the scattering form factor of the average double layer. By utilizing a contrast where only the surfactant is hydrogenated ("double-layer contrast") the form factor exhibits a pronounced zero due to interference of the two surfactant-containing surfaces of the double layer (see Fig. 1). Scattering intensity observed near this zero must contain sizeable intensity from shape (thickness) fluctuations of the double layer (peristaltic modes), in contrast to undulation. However, observation of these modes is limited to the region where the average structure form factor is near zero. Figure 2 shows an example exhibiting markedly different dynamics containing a faster component near the double-layer contrast zero compared to the dynamics at the same position observed in full-layer contrast. In the latter case, the nonzero double-layer form factor dominates the intensity and, virtually, only the undulation dynamics is seen.

Hydrodynamic treatments of the relaxation mode structure of such layer systems (see e.g. [10]) yield predictions for **q**-dependent hydrodynamic modes exhibiting corresponding **q**-dependent relaxations times. However,

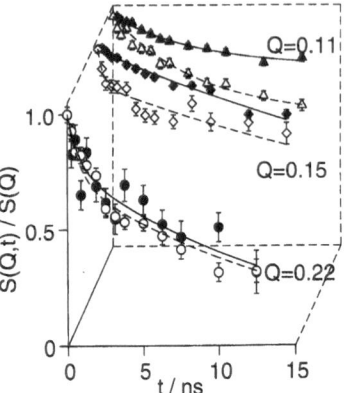

Fig. 2 NSE relaxation spectra corresponding to the scattering patterns shown in Fig. 1. The data at $Q = 0.11\,\text{Å}^{-1}$ and $Q = 0.15\,\text{Å}^{-1}$ taken near the form factor minimum at $Q = 0.13\,\text{Å}^{-1}$ differ markedly from the data taken near the form factor maximum ($Q = 0.22\,\text{Å}^{-1}$) showing a faster contribution due to intra-double-layer fluctuation in the "double-layer contrast"

besides the simplification of thin interfaces for the double layers, unfortunately the relaxation dynamics observed at some $\mathbf{Q} = (Q_\perp, Q_z)$ cannot be mapped directly onto the dynamics of the hydrodynamic mode with wave vector **q**. It turns out – and this is one of the more methodical result of this study – that the observed dynamics contains contributions from many modes with $\mathbf{q} \neq \mathbf{Q}$ as a direct consequence of the large undulatory displacements. As is recognizable in Fig. 1, the main scattering intensity of the lamellar microemulsion phases is located on a ridge along the Q_z-axis, i.e. in the scattering direction perpendicular to the layer planes, which is also the direction where most of the NSE experiments have been performed. As Fig. 2 exhibits, relaxation dynamics on the time scale of a few nanoseconds is observed. On the other hand, a naive picture – mapping **q** directly to **Q** – would lead to very long (infinite) relaxation time, because the corresponding hydrodynamic undulation mode would consist of a large-scale distance variation between the surfactant layers for which transport of oil (water) through the water (oil) layer would be necessary. The fact that a fast dynamic is seen as well as the fact that there is scattering intensity at all off the Bragg position is a consequence of the large undulation amplitudes u_z extending beyond the validity of $\exp(iQu) \approx 1 + iQu$. The basic expression for the scattering intensity in this situation has been formulated by Caille [8]

$$I = |f_0|^2 \sum_n \int d^2\rho \, e^{iq_z nd + iq_\perp \cdot \rho} \, e^{-q_z^2 \frac{1}{(2\pi)^3} \int d^3 q [1 - e^{-iq \cdot r}] \langle u_q u_{-q} \rangle}, \quad (1)$$

where f_0 denotes the form factor of the (double) layer. The 2D SANS data (Fig. 1) could be fitted with a full version of Eq. (1) (involving multidimensional numerical integration)

Progr Colloid Polym Sci (1997) 106:112–117
© Steinkopff Verlag 1997

with addition of a density – density correlation scattering [11] contributing extra intensity only near the central beam. This yields reasonable values for the layer–bending modulus κ and the interlayer compressibility B. The somewhat unexpected observed dynamical behavior can be explained by assuming a simplified mode dispersion which only takes into account the high-$q_\perp$ dependence of the undulation rate (i.e. no relaxing modes on the line $(q_\perp = 0, q_z)$) predicted by hydrodynamic theory [10]: $\Gamma = q_\perp^3 \kappa/4\eta$, η being the fluid viscosity, as time dependence of the correlation function $\langle u_q(0)u_{-q}(t)\rangle = \langle u_q u_{-q}\rangle e^{-\Gamma t}$ that enters Eq. (1), the modification only changes $[1 - e^{-i\mathbf{q}\cdot\mathbf{r}}]$ into $[1 - e^{i\mathbf{q}\cdot\mathbf{r}} \times e^{-\Gamma t}]$ in the argument of the second integral. A full evaluation of the, now time-dependent, expression for the scattering intensity utilizing accuracy-controlled numerical integrations is a major computational task and quite time consuming. However, it could be shown that, by this mechanism, dynamical relaxations in accordance with the observed Q-dependence may be obtained. Besides the other parameters, an upper cutoff q_0 for the Fourier space integrals involved has to be introduced to avoid divergent contributions from impossible modes with wavelength less than the molecular size or other smallest dimensions of the system. Taking the double-layer thickness as natural cutoff yields $q_0 \approx 2\pi/20\,\text{Å} = 0.3\,\text{Å}^{-1}$. Unlike the predicted static scattering intensity, the dynamics results are more sensitive to this value.

Besides, the dynamics may allow for the discrimination of scattering contributions from different modes occuring at the same **Q** value, as has been demonstrated for the inter double-layer fluctuations, i.e., the peristaltic mode (see Fig. 2). The extra information supplied by observation of the dynamics, primarily consist of the frictional coefficients dominated by the solvent viscosities η. Considering the rather complicated procedure leading from the basic physical properties to the observed relaxation dynamics, it is difficult to extract much new information on the friction coefficients from the dynamical measument on only one sample. However, observation of the influence of a variation of the solvent viscosities should yield more insight on the importance of extra (internal) friction contributions. Therefore, experiments on lamellar phases where the water viscosity has been increased by addition of glycerol and measurements with reduced solvent viscosities due to temperature rise have been performed. The composition corresponds to previously used samples without glycerol at ambient temperature. As long as only one viscosity value determines the friction, all relaxation times should scale by the same, Q-independent, factor α. As Fig. 3 shows scaling works for each Q-value, but however, the inset shows that α depends on the selected wave vector, indicating the increasing importance of internal (to the double layer)

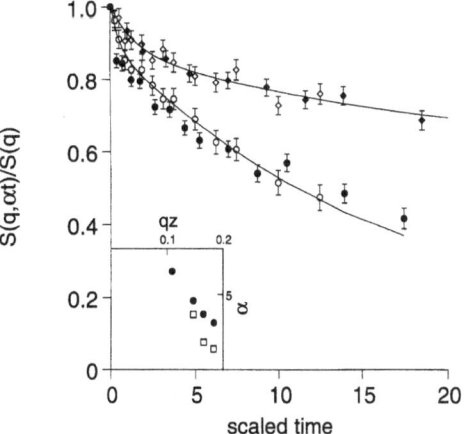

Fig. 3 NSE relaxation curves for two Q-vlaues: 0.146, 0.183 Å^{-1}. The filled symbols correspond to the original composition, whereas the open symbols correspond to the relaxation of an analogous sample the water-phase viscosity which has been increased by a factor of 2.5 by glycerol addition. The time of the glycerol data has been scaled such that they match the original data. The scaling factor α is shown in the inset (open squares) the closed circles correspond to the scaling factor obtained by viscosity variation due to temperature rise

friction contributions with increasing wave vector. Thus, parameter variation shows the potential to be a tool to get physical information on the system even without elaborate mathematical fitting procedures.

Composed aggregates

A second example is the aggregate forming from a solved A-B block-copolymer polyethylene(PE)-polyethylene-propylene(PEP) in decane upon cooling from above 70 °C to ambient temperature. Here, the micellarization is driven by the crystallization of the PE strands, with the consequence that the micellar shape is determined by the habitus of the crystalline PE core, i.e. the aggregates are platelets with a PE core and a PEP brush extending into the solution. These platelets form stacks due to the weak attractive van der Waals attraction of the brushes integrated over a large contact area. The stacks have elongated shapes of about 5000 nm × 500 nm × 500 nm, as may be inferred from the combined results of SANS, SANS after applied shear, light scattering and direct inspection by phase-contrast light microscopy.

The stacking is subject to large paracrystalline distance fluctuations, therefore the stacking structure factor deviates significantly from 1 only for very small values of $(Q < 0.025\,\text{Å}^{-1})$. A detailed account on the structural properties is given in ref. [12].

Spin-echo experiments have been performed in the range $0.037 \leq Q/\text{Å}^{-1} \leq 0.146$ probing density fluctuation inside the brush. As the polymer concentration (volume fraction) in considerable parts of the brush is in the range of about 0.3 the dynamics in this Q-range is expected to be dominated by the collective dynamics of the corresponding semidilute solution [13], rather than by the single-chain relaxations (Zimm dynamics). The brush has a varying density and restrictions due to boundary conditions. If the brush region is considered to have the properties of a semidilute polymer solution, the elastic restoring forces are of entropic nature and result from the osmotic pressure, frictional damping results from the relative motion of solvent and polymer. Within the "blob" model, the elastic $E_{\pi} \propto T/\xi^3$ and frictional $\beta \propto \eta/\xi^2$ coefficients scale with the "blob" size ξ which depends on the polymer concentration $\xi \propto c^{-3/4}$ [14, 15]. The fluctuations have been computed using a model treating the brush as elastic slab anchored at the PE-platelet forming the core with a parabolic density profile in the direction normal to the plane. Details of the model are described in ref. [16]. Due to the finite thickness d of the brush and its grafting on the core fluctuations with $Qd < 1$ are suppressed, the corresponding scattering intensity drops below this Q-value to zero at $Q = 0$, see Fig. 4. The resulting $S(Q, t)$ can only be matched (see Fig. 5) to the spin-echo measurement if, within the accessible time range, an additional static fluctuation contribution proportional to Q^{-v} with $v = 2$–2.5 is added. The elastic and friction parameters emerging from the matching procedure are an elastic modulus in the megapascal range and a friction around 10^{16} N/m^4/s which is within

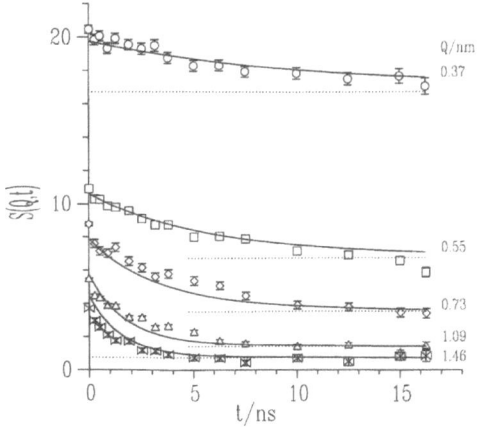

Fig. 5 NSE data taken from PE($M_w = 5000$)-PEB($M_w = 8000$) in decane at ambient temperature. The dotted lines indicate the intensity of the static scattering contribution. The solid lines stem from the scaling of the result of the elastic slab model to the data (minus static component). The scaling factors allow for the determination of the elastic constant and friction

the expectation for a polymeric gel. As shown in Fig. 4, the procedure supplies a method to decompose the fluctuation scattering in a contribution due to thermal fluctuating deformations of the brush and a very slow (static, respectively, frozen) part of fluctuation scattering. The latter may be a consequence of a statistical fluctuation in the grafting point density or the signature of lateral inhomogenieties due do instabilities as predicted in refs. [17, 18] by Monte Carlo or RPA methods. However, such instabilities are predicted to occur only in a poor solvent.

Summary

Two different types of molecular aggregating systems exhibit small-angle scattering with marked dynamical components with a typical time scale in the nanosecond region. The neutron spin-echo technique allows the analysis of this dynamical behavior in terms of the intermediate scattering function. Thereby, different fluctuation modes may be identified and separated by their dynamical behavior. Together with the absolute scattering intensity the spin-echo data yield information on the elastic energy connected with a fluctuation mode and the friction forces connected with the flow in the system to relax to equilibrium.

Acknowledgement Part of the work on that has been performed with the MESS spectrometer and the PAXE and PAXY SANS instruments at the LLB in Saclay has been supported by the CE "Human Capital & Mobility Access to Large Facilities" program under contract no. ERB CHGECT 920001.

Fig. 4 SANS data taken from PE($M_w = 5000$)-PEB($M_w = 8000$) in decane at ambient temperature. The triangles show the scattering in "core-contrast" the circles the scattering in "brush-contrast". The dashed and solid lines are the scattering contribution computed from the average structure. The inset shows the fluctuation scattering range using a linear intensity scale together with the decomposition as obtained by the spin-echo measurement: plain solid line: (computed average structure contribution), open diamonds: dynamical brush fluctuations, solid diamonds: static (frozen) brush fluctuations

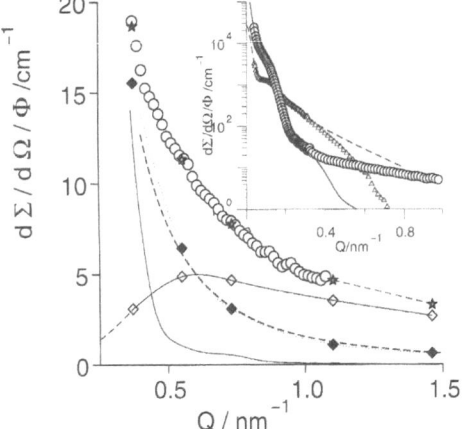

Progr Colloid Polym Sci (1997) 106:112–117
© Steinkopff Verlag 1997

References

1. Mezei F (ed) (1979) Neutron Spin Echo Proc. Lecture Notes in Physics 128. Springer, Berlin
2. The Yellow Book (1994) Institut Laue Langevin
3. Papoular RJ, Millet R, Rosta L, Mezei F (1986) Proc of the Workshop on Neutron Physics, Budapest
4. Takeda T, Komura S, Seto H, Nagai M, Kobayashi H, Ykoi E, Ebisawa T, Tasaki S, Zeyen C, Ito Y, Takahashi S, Yoshizawa H (1995) Phys B 213 & 214: 863
5. Monkenbusch M (1992) Physica B 180 & 181:935
6. Safinya CR, Roux D, Smith GS, Dimon P, Clark NA, Belloqc AM (1986) Phys Rev Lett 57:2718
7. Nallet F, Roux D, Milner ST (1990) J Phys France 51:2333
8. Caille ACR (1972) Acad Sc Paris t 274: 891
9. DeGennes PG (1969) J Phys C 4:65
10. Ramaswamy S, Prost J, Cai W, Lubensky TC (1993) Europhys Lett 23(4): 271–276
11. Nallet F, Roux D, Prost J (1989) J Phys France 50:3147
12. Richter D, Schneiders D, Monkenbusch M, Willner L, Fetters LJ, Huang JS, Lin M, Mortensen K, Farago B (1996) Macromolecules, submitted
13. Ewen B, Richter D, Hayter JB, Lehnen B (1982) J Polym Sci, Polym Lett Ed 20: 233
14. De Gennes PG (1980) Macromolecules 13:1069
15. De Gennes PG (1980) CR Acad Sci Paris Ser II 302:765
16. Monkenbusch M, Schneiders D, Richter D, Farago B, Fetters L, Huang JS (1994) Il Nuovo Cimento 16D:747
17. Lai P, Binder K (1992) J Chem Phys 97: 586
18. Yeung C, Balazs AC, Jasnow D (1993) Macromolecules 26:1914
19. Dubois-Violette E, De Gennes PG (1967) Phys 3:181
20. Monkenbusch M (1992) Physica B 180 & 181:935

Progr Colloid Polym Sci (1997) 106:118–126
© Steinkopff Verlag 1997

FORMATION OF STRUCTURES

T. Hashimoto
H. Jinnai
Y. Nishikawa
T. Koga
M. Takenaka

Sponge-like structures and their Gaussian curvatures in polymer mixtures and microemulsions

Received: 30 November 1996
Accepted: 4 February 1997

T. Hashimoto · H. Jinnai · Y. Nishikawa
T. Koga · M. Takenaka
Hashimoto Polymer Phasing Project
ERATO, JST
15 Morimoto-Cho
Shimogamo Sakyo-ku, Kyoto 606, Japan

Dr. T. Hashimoto (✉)
Department of Polymer Chemistry
Graduate School of Engineering
Kyoto University
Kyoto 606-01, Japan

Abstract Bicontinuous structures formed in the late stage of spinodal decomposition (SD) in polymer mixtures were explored experimentally by time-resolved light scattering (TLS) and laser scanning confocal microscopy (LSCM) and computationally by three-dimensional (3D) simulations based on time-dependent Ginzburg–Landau (TDGL) theory. The 3D structures constructed by LSCM exhibited a sponge-like structure and were found to be statistically identical to those constructed by the computer simulations through equality of their scattering structure factors. Moreover, their structure factors were found to be identical to that obtained by TLS, revealing that the 3D structures truly reflect the structure entities occurring in the polymer mixtures and that the TDGL theory accounts for the phase structures evolving in the late stage SD. Gaussian curvature K and mean curvature H were evaluated from the 3D structures, the results of which were compared with the "scattering-mean-curvature" determined by using the Kirste–Porod theory and with K determined by using a Gaussian random-field theory. The sponge-like structure was found to be strikingly similar to that occurring in an equilibrium microemulsion system at the hydrophile–lipophile balance temperature, though their characteristic length scales are different by two–three orders of magnitude.

Key words Polymer mixtures – microemulsions – spinodal decomposition – sponge-like structure – interface – Gaussian curvature – mean curvature

Introduction

Time evolution of phase-separating morphology (or concentration fluctuations, in general) in binary mixtures has been extensively studied as a research theme on nonlinear and nonequilibrium phenomenon in various fields of science [1]. Especially, a lot of work has been devoted to the spinodal decomposition process (SD) [2], a phase-separation process for the mixtures with thermodynamic instability. Experimental studies, especially time-resolved light scattering (TLS) studies [3], have revealed that the morphology evolved via SD is periodic and bicontinuous, two coexisting and growing domains being continuous in three-dimensional (3D) space. This feature, truly valid for mixtures giving rise to two phases which are nearly isometric, has been also elucidated by computer simulations [4, 5].

We, naturally, begin to ask a more specific question: What kind of bicontinuous structures do we expect to find in the phase-separation process via SD? To answer this question, we have started real-space analysis of the

Progr Colloid Polym Sci (1997) 106:118–126
© Steinkopff Verlag 1997

growing morphology. We take advantage of laser scanning confocal microscopy (LSCM), a special method which enables a 3D scanning of morphology in our specimen [6], and of polymer mixtures as a model system in which *a large structure grows very slowly*, as reviewed in [7], for example. The latter enables *in situ* observation of 3D optical images, because the time required to capture the 3D image is much shorter than the time scale of the structure growth. To answer the above question, we believe it is also important to find out some universal features of our structure when compared with structures in other systems. For this purpose we compare our structure in the late stage SD [3, 7, 8] with the sponge-like structure in microemulsions (μE) at thermal equilibrium and with the 3D structures obtained from computer simulations based on a general time-evolution equation of the time-dependent Ginzburg–Landau (TDGL) equation. In the late stage SD, each phase having an equilibrium composition separates across a well-defined interface, similar to the three-component μE systems of oil/water/surfactant in which most of the surfactant molecules locate at the interfaces between the oil and water phases. In this paper we report highlights of our studies along this line.

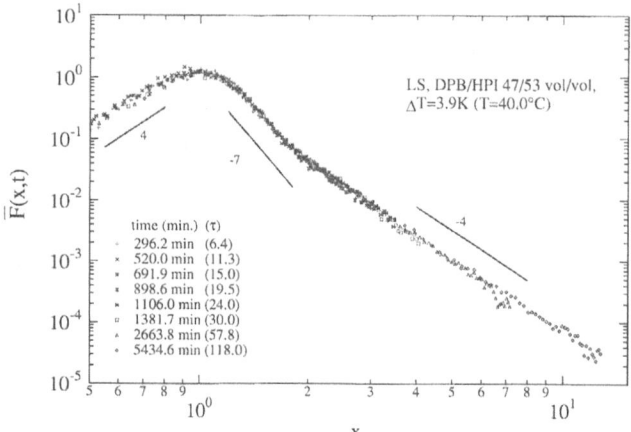

Fig. 1 Scaled structure factor for a near-critical mixture ($\frac{47}{53}$ v/v) deuterated polybutadiane (DPB) and protonated polyisoprene (HPI) obtained from time-resolved light scattering experiments at various times in the late-stage SD at a quench depth $\Delta T \equiv |T - T_s| = 3.9\,°C$ where T_s (36.1 °C) is spinodal temperature and T (= 40 °C) is phase-separation temperature [10]. The time covered corresponds to 6.4–118 in the reduced time τ defined by t/t_c [3, 8] (t_c at 40 °C is 46.6 min)

Universal scaled structure factor in polymer mixtures

In the late stage SD process, a so-called dynamical scaling hypothesis [9] is proposed in that global features of the growing structures can be scaled with a single-length parameter $\Lambda_m(t)$: $\Lambda_m(t)$ characterizes the time evolution of the length scale of the structures and the scaled structure factor $\bar{F}(x)$, universal with time, characterized uniquely a global shape of the structures. Here $\bar{F}(x)$ is defined by

$$\bar{F}(x) \equiv I(q,t)q_m(t)^3 \bigg/ \int_0^\infty I(q,t)q^2 \, \mathrm{d}q \,, \tag{1}$$

$$x \equiv q/q_m(t) \,, \tag{2}$$

where q is magnitude of scattering vector or a wave number for a particular Fourier mode of the structure, q_m is q at the maximum intensity, $I(q,t)$ is the scattering function obtained at time t after onset of SD, and q is defined by

$$q = (4\pi/\lambda)\sin(\theta/2) \tag{3}$$

with λ and θ being wavelength and scattering angle in the sample. The value q_m, interpreted as the wave number for a dominant Fourier mode of the structures, is related to Λ_m by $q_m = 2\pi/\Lambda_m$.

Figure 1 shows typical scaled structure factors obtained for a near-critical mixture of DPB/HPI in the late-stage SD process [10]. Details of the polymers used in this report are summarized in Table 1. As shown in this

figure, the scattering functions $I(q,t)$ obtained at various times in the late-stage SD fall onto a master curve universal with time in the reduced plot. Thus, the universal-scaled structure factor $\bar{F}(x)$ uniquely characterizes the shape of the growing phase-separating structures. It has a maximum and thus, the structure is periodic with periodicity $\Lambda_m(t)$ increasing with t according to a scaling law of t^α ($\alpha \cong 0.8$), revealing that the hydrodynamic interactions [11] play an important role on the growth. It has the following asymptotic behaviors. $\bar{F}(x) \sim x^{+4}$ at $x < 1$ and $\bar{F}(x) \sim x^{-7}$ at $1 < x < 2$, consistent essentially with a scaling function predicted by Furukawa [12]; $\bar{F}(x) \sim x^{-4}$ at $x \geq 4$, revealing that the interfaces of coexisting phases are sharp. It has a higher-order maximum or shoulder at $x \cong 3$. We expect that the structures developed via SD, especially for near-critical mixtures giving rise to nearly isometric two-phase structures, should have a bicontinuous feature in 3D space. However, we do not know yet what kind of morphology corresponds to this universal scaled structure factor.

3D Phase structures determined with LSCM

In LSCM, light emitted from the object is focused on a confocal pinhole which acts as a spatial filter: light from the focal plane passes the pinhole, while light from other planes is effectively suppressed. The pinhole thus ensures that the information only arrives from a particular level of the specimen with a very high resolution (ΔF) along the

Table 1 Characteristics of polymers used

Polymer codes	Polymer	$M_n \times 10^{-4}$ [a]	M_w/M_n [b]	W_{PS} [c] (wt %)	Microstructure [d]			
					Cis-1,4	Trans-1,4	1,2	3,4
DPB	Deuterated polybutadiene	10.3	1.03	—	36	44	20	—
HPI	Polyisoprene	13.6	1.04	—	63	22	—	15
PB	Polybutadiene	5.8	1.2	—	28	56	16	—
PI	Polyisoprene	10.1	1.3	—	70.4	22.1	—	7.5
SBR	Poly(styrene–ran-butadiene)	7.1	1.2	20	31	52	17	—
PB1	Polybutadiene	5.7	1.6	—	37	39	24	—
DPB1	Deuterated polybutadiene	6.1	1.07	—	37.9	46.5	16.5	—
HPI1	Polyisoprene	14.1	1.04	—	62.8	22.3	—	14.9

[a] Number-average molecular weight.
[b] M_w: weight-average molecular weight.
[c] W_{PS}: wt% of styrene monomer in SBR.
[d] Microstructure of butadiene or isoprene unit.

depth direction. Changing the positions of the focal plane in the specimen by vertically moving the specimen makes a series of optically sectioned two-dimensional (2D) images along the optical axis (z-axis) perpendicular to the focal plane (xy plane), which can then be used to construct a 3D image by computationally stacking the images.

LSCM observations were conducted with a 364 nm laser at room temperature by a Carl Zeiss, LSM 410. Sixty 2D optical sections with 0.5 μm increment and $\Delta F \cong 0.65$ μm along the z-axis were taken with an oil-immersed ×40 objective with a numerical aperture (NA) of 1.3. The image in xy plane has an area of 80×80 μm^2, 256×256 pixel^2 and each pixel had 8-bit resolution in the intensity level. The images thus obtained were subjected to the various image processing steps as will be detailed elsewhere [13]. Plate 1 shows a typical 3D LSCM image for a binary mixture of SBR and PB1 containing 50 wt% SBR in the late-stage SD (subjected to SD at 100 °C for 70 h) having $q_m = 0.52$ μm^{-1} and $\Lambda_m = 12.1$ μm [14, 15]. The detailed characteristics of SBR and PB1 [15] are shown in Table 1. The figure represents the interface, one side of which (e.g., PB1-rich phase) is colored in green and the other side (e.g., SBR-rich phase) red. The closest optical slice to the cover slip was 20 μm to avoid surface effects on the structure.

The 3D image clearly reveals: the two phases are bicontinuous in 3D space, the interface is composed of saddle-shaped (or hyperbolic) elements with oppositely signed principal curvatures k_1 (>0) and k_2 (<0) or negative Gaussian curvature $K = k_1 k_2$ (<0) and the two phases are arranged periodically with $\Lambda_m = 12.1$ μm. The 3D image in Plate 1, its solid modeling view [14] and a series of 2D sliced images [16] show a sponge-like structure.

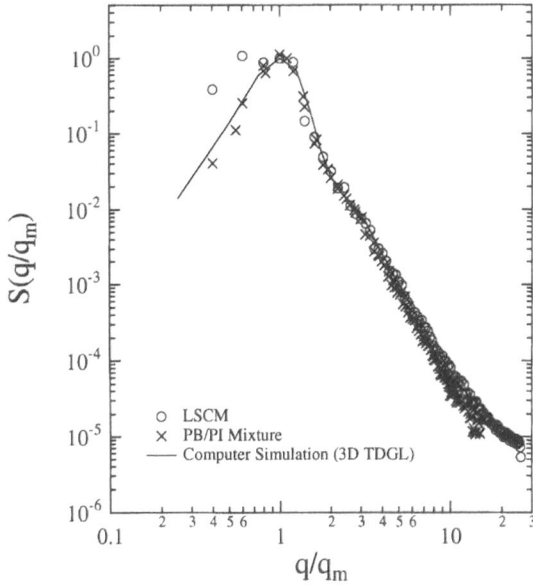

Fig. 2 Comparisons of scaled structure factors $S(q/q_m)$ (proportional to $\bar{F}(x)$ in Eq. (1)) obtained from the 3D LSCM image (unfilled circles), the TLS experiments for the PB/PI $\frac{50}{50}$ wt%/wt% mixture (crosses) and the computer simulations based on the 3D TDGL model (solid line) [14]

In order to confirm whether or not the 3D-phase structure constructed with LSCM truly reflects a real structural entity, we calculated the scattering structure factor from the 3D image [14] and compared it with the structure factor obtained from a TLS experiment. The results are shown in Fig. 2 where the scaled structure factor $S(q/q_m)$ (proportional to $\bar{F}(x)$ in Eq. (1)) obtained from LSCM (unfilled circles) is compared with that from

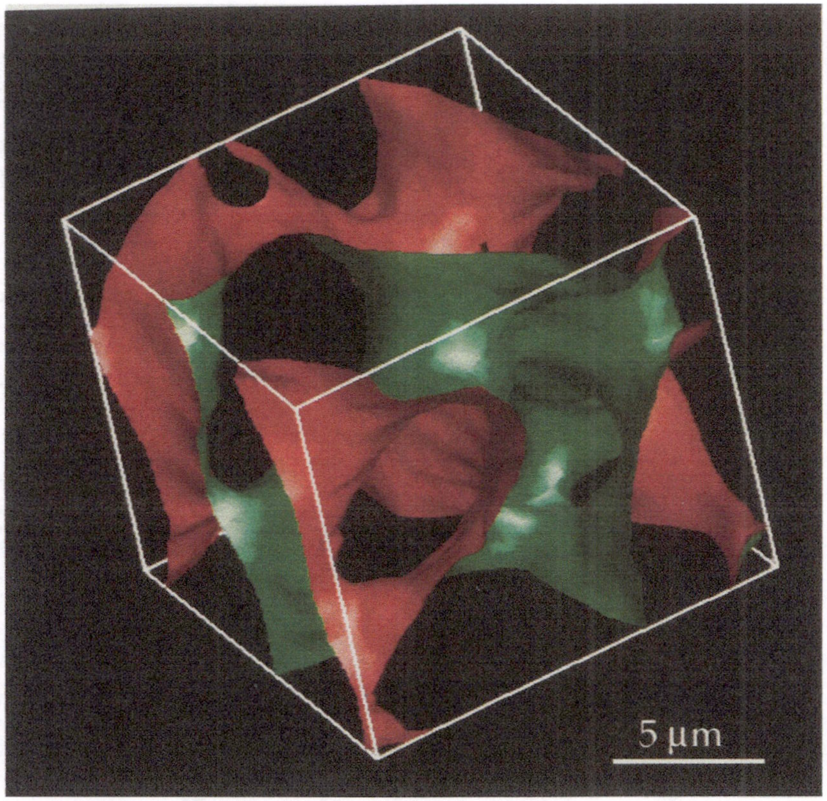

Progr Colloid Polym Sci (1997) 106:118–126
© Steinkopff Verlag 1997

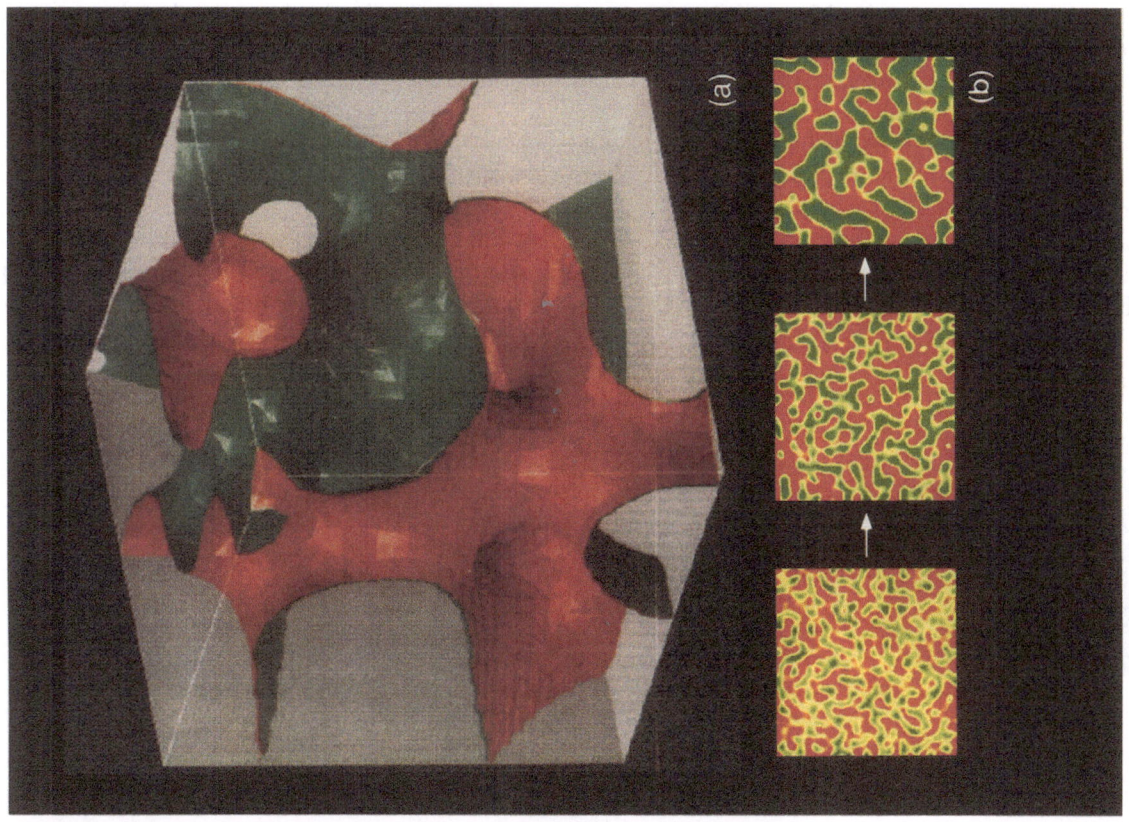

Plate 2 3D image of the interface constructed by the computer simulations with the TDGL model for a binary mixture in the late-stage SD (a), and 2D sliced images showing time evolution of the morphology (b). The edge length of the box in (a) corresponds to about 1.5 Λ_m. The binary mixture has a volume fraction of each component of 0.5

Plate 1 3D LSCM image of the interface for the SBR/PB1 $\frac{50}{50}$ wt%/wt% mixture in the late stage SD having $\Lambda_m = 12.1\ \mu m$

TLS (crosses). Note that $S(q/q_m)$ determined from TLS was obtained for the PB/PI mixture containing 50 wt% PB [17] instead of the SBR/PB1 mixture (see the detailed characteristics of PB and PI in Table 1). This is simply because $S(q/q_m)$ has been studied for a SBR/PB mixture with different compositions [18]. It is quite impressive that two structure factors agree quite well over a large intensity scale, as large as five orders of magnitude, and a large spatial scale, as large as two orders of magnitude. Thus, we conclude that the 3D LSCM image reflects truly the real

structure entity. It should also be noted that the structure factor shown in Fig. 1 for the DPB/HPI mixture is identical to that shown in Fig. 2 for the PB/PI mixture.

3D computer simulation based on TDGL model

The 3D computer simulations were based on the TDGL model, a nonlinear time evolution equation, containing hydrodynamic interactions, the details of which can be

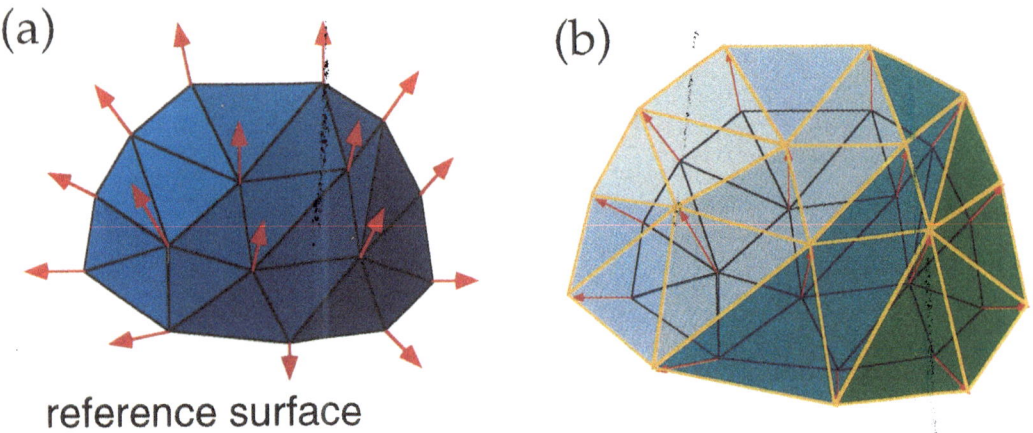

Plate 3 Schematic representation of (a) the interface (blue reference surface) composed of triangles and normal vectors (red arrows) at their vertices and (b) the parallel surface constructed by connecting the normal vectors

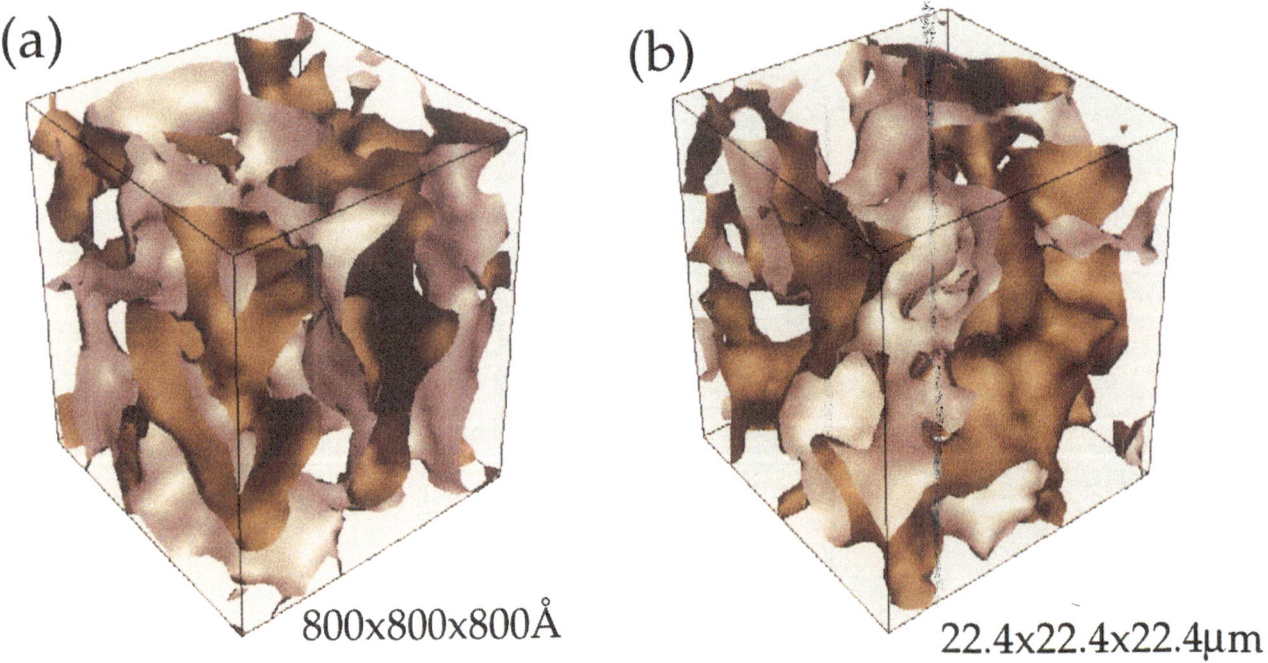

Plate 4 3D image of the interfaces constructed on the basis of Gaussian random-field theory: (a) for the μE system at 22 °C and (b) for the DPB/HPI mixture in the late stage SD [29]

Progr Colloid Polym Sci (1997) 106:118–126
© Steinkopff Verlag 1997

found in Koga and Kawasaki [5]. Detailed comparisons of the results obtained by simulations and scattering experiments are in Koga et al. [19]. Plate 2a shows a 3D image of the interface constructed by the simulations for a binary mixture containing a volume fraction of one component of 0.5 in the late-stage SD, while Plate 2b shows 2D sliced images demonstrating time evolution of morphology. In the simulations the reduced shear viscosity $\tilde{\eta}$ is the only relevant parameter in the TDGL model with the hydrodynamic interaction in the absence of thermal noise, which measures the inverse of the strength of the hydrodynamic interaction. The value of $\tilde{\eta}$ used in the computer simulation is about 0.2. $S(q/q_{\mathrm{m}})$ obtained from the computer simulations is also included in Fig. 2 (solid line). The result nicely agree with $S(q/q_{\mathrm{m}})$ obtained both from the LSCM and TLS experiments. Thus, we conclude that the TDGL model can predict the real structures quite well. In fact, the 3D structure shown in Plate 2a is quite similar to that shown in Plate 1.

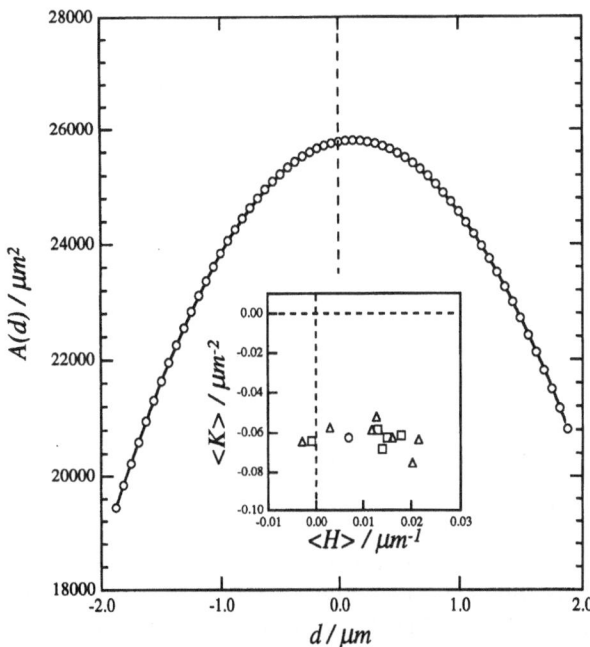

Fig. 3 Area of the parallel surface $A(d)$ as a function of separation distance d from the interface calculated from the 3D image shown in Plate 1 [15]. The inset shows the deviation of the curvatures determined with different size and position of the 3D images; triangles, squares and a circle represent, respectively, the results obtained from the volume of $31.3 \times 31.3 \times 30\ \mu m^3$, $46.9 \times 46.9 \times 30\ \mu m^3$ and $62.5 \times 62.5 \times 30\ \mu m^3$ at the different position of the 3D image

Analysis of Gaussian curvature K and mean curvature H by direct real-space analysis

Now that we have determined the 3D structure, which reflects truly the real structures from global-to-local scale, and that we have obtained a sound physical basis of this structure, we proceed to analyze the structure in terms of interface curvatures, based on differential geometry. We used the "parallel-surface method" to determine the area-averaged mean curvature $\langle H \rangle \equiv \langle (k_1 + k_2)/2 \rangle$ and Gaussian curvature $\langle K \rangle \equiv \langle k_1 k_2 \rangle$

$$A(d) = A_0(1 + 2\langle H \rangle d + \langle K \rangle d^2) , \tag{4}$$

where A_0 is interface area, and $A(d)$ the area of the surface which is parallel to and located at a distance d from the interface. k_1 and k_2 are local principal curvatures.

We used the "Marching cube algorithm" [20] to construct the 3D interface structure in that the interface is composed of triangles, as schematically shown in Plate 3a. To construct the parallel surface at a displacement vector $\mathbf{d}(|\mathbf{d}| = d)$ from the interface, we first calculate normal vectors $\mathbf{n}$ ($|\mathbf{n}| = 1$) at their vertices (red arrows), as detailed elsewhere [21]. Then $\mathbf{d}$ is given by $\mathbf{d} = d\mathbf{n}$, and the parallel surface, composed of yellow triangles as shown in Plate 3b, is constructed by connecting the ends of the displacement vectors. $A(d)$ is then calculated by summing the areas of the triangles.

The results of $A(d)$ obtained from the 3D image shown in Plate 1 are presented in Fig. 3 [15]. The best fit of the data points (open circles) with Eq. (4) (shown by the solid line) yields,

$$A_0/V = 2.2 \times 10^{-1}\ \mu m^{-1} ,$$

$$\langle H \rangle = 7.0 \times 10^{-3}\ \mu m^{-1} ,$$

$$\langle K \rangle = -6.2 \times 10^{-2}\ \mu m^{-2} ,$$

$$[-\langle K \rangle]^{1/2} = 2.5 \times 10^{-1}\ \mu m^{-1} ,$$

where A_0/V is the interface area per unit volume. Thus, we can conclude that the sponge-like structure has a hyperbolic interface having almost zero mean curvature, i.e., $r \equiv \langle H \rangle / [-\langle K \rangle]^{1/2} \cong 0$. The small positive valve of r (ca, 0.03) may reflect portions of bicontinuous domain which tend to burst in the coarsening process.

These results will be discussed later in the following section. In the following two sections we will discuss the curvature analyses based on scattering methods.

Analysis of "scattering mean curvature" R_c^{-1}

Kriste and Porod [22] and Tomita [23] worked out theories concerning the relationship between scattering and surface curvatures of the scattering objects. The asymptotic behavior of density correlation function $\gamma(r)$ at a small length scale r and that of scattering function $I(q)$ at a large q are given by

$$\gamma(r) = 1 - \frac{A_0}{4\phi(1-\phi)V} r \left[1 - \frac{1}{12}\left(\frac{r}{R_c}\right)^2 + \cdots \right], \quad (5)$$

$$I(q) = \frac{2\pi A_0}{\phi(1-\phi)V} q^{-4} \left[1 + \frac{q^{-2}}{R_c^2} + \cdots \right], \quad (6)$$

where ϕ is the volume fraction of one of the phases and R_c^{-1} the "scattering mean curvature" defined by

$$R_c^{-2} \equiv \frac{3}{8} \int \left(k_1^2 + \frac{2}{3} k_1 k_2 + k_2^2 \right) da / \int da \quad (7)$$

$$= (3\langle H^2 \rangle - \langle K \rangle)/2, \quad (8)$$

with da being the area element of the interface. Note that R_c^{-1} depends on both $\langle H^2 \rangle$ and $\langle K \rangle$.

The scattering function $I(q,t)$ at a given time in the late-stage SD was subjected to a Fourier inversion in 3D space to obtain $\gamma(r)$, the results of which are used for a plot shown in Fig. 4 [24]. We can determine R_c^{-1} from the slope and intercept at $r = 0$. In order to compare R_c^{-1} with that determined from the direct real-space analysis, we need to know $\langle H^2 \rangle$. For this purpose, we have developed an algorithm to evaluate distribution function of H and K [25].

We compare the two kinds of curvatures (one determined by the direct real-space analysis and the other by the Kirste–Porod method) in terms of a reduced curvature C_r

$$C_r \equiv \langle R_c^{-2} \rangle^{1/2} / q_m, \quad (9)$$

because they were determined for the different polymer mixtures and different times in the late-stage SD. We denote the value C_r determined from the scattering method (the Kirste–Porod method) by $C_{r,S}$. Since $\langle R_c^{-2} \rangle^{1/2} = 5.21 \times 10^{-1} \ \mu m^{-1}$ (from Fig. 4) and $q_m = 1.0 \ \mu m^{-1}$ for the PB/PI mixture at 55 °C and $t = 85.8$ min, we obtain $C_{r,S} = 0.52$. We denote the value C_r determined from the direct real-space analysis by $C_{r,L}$. Since $\langle K \rangle = -6.2 \times 10^{-2} \ \mu m^{-2}$ and $\langle H^2 \rangle = 2.2 \times 10^{-2} \ \mu m^2$ [25] and $q_m = 0.52 \ \mu m^{-1}$ for the SBR/PB1 mixture subjected to SD at 100 °C for 70 h, we obtain $\langle R_c^{-2} \rangle^{1/2} = 2.53 \times 10^{-1} \ \mu m^{-1}$ and $C_{r,L} = 0.49$. Hence, $C_{r,S}/C_{r,L} = 1.1$, and the reduced curvatures determined by the two methods agree quite well.

Comparison of sponge-like structures in polymer mixtures and microemulsions

The μE systems used were ternary mixtures of heavy water (D_2O), n-octane, and tetraethylene glycol monodecyl ether (C_4E_{10}). Volume fractions of the surfactant, water, and oil were, respectively, 0.13, 0.435, and 0.435. For the particular

Table 2 Comparisons of sponge-like structures in the microemulsion system (μE) at 22 °C and the DPB/HPI mixture (polymer)

Physical quantities	Polymer	μE	Polymer/μE[a]
q_m[c]	$1 \ \mu m^{-1}$	$2 \times 10^{-2} \ \text{Å}^{-1}$	5×10^{-3}
Λ_m[c]	$6.28 \ \mu m$	$3.14 \times 10^2 \ \text{Å}$	2×10^2
$[-\langle K \rangle]^{1/2}$	$1.44 \ \mu m^{-1}$	$1.07 \times 10^{-2} \ \text{Å}^{-1}$	4×10^{-3}
$[-\langle K \rangle]^{1/2}/q_m$	0.44 (0.48[b])	0.54	0.8 (0.9[b])
ξ/Λ_m[d]	1	0.6	1.7

[a] Ratio of a physical quantity for polymer and that of μE.
[b] The value for the SBR/PB1 mixture as determined by direct real-space analysis.
[c] $q_m = 2\pi/\Lambda_m$ where q_m and Λ_m are the characteristic wave number and wavelength, respectively.
[d] Reduced coherence length determined from Eq. (11).

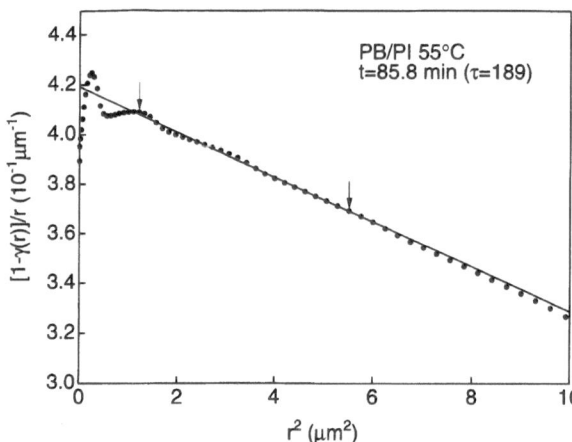

Fig. 4 Plot for estimating the scattering-mean-curvature R_c^{-1} for the PB/PI $\frac{50}{50}$ wt%/wt% mixture in the late-stage SD [24]. The arrows indicate the length scale over which the straight line is drawn: $R_c^{-1} = 2[(3|\text{slope}|)/(\text{Intercept at } r = 0)]^{1/2}$

concentration of surfactant used here, the hydrophile–lipophile balance (HLB) where a microphase separated structure exists in thermal equilibrium spans the temperature range from about 19 °C to 25 °C: the structure consists of interpenetrating domains of water and oil with most of the surfactant molecules sitting at the interface. The microphase separated structures for this μE system was studied by small-angle X-ray scattering (SAXS) [28].

Figure 5 shows the SAXS profile from the μE system described above (equilibrium at 22 °C) together with the TLS profile from the DPB1/HPI1 mixture containing 39 wt% DPB1 at a particular time in the late-stage SD. This mixture also gives nearly isometric phase-separating domains. The solid lines show the best fits of the modified Berk (MB) theory [26, 27] for the experimental profiles (both shown by the data points) [28, 29]. In the MB theory we assume a suitable form of the spectral function $f(k)$

Progr Colloid Polym Sci (1997) 106:118–126
© Steinkopff Verlag 1997

Fig. 5 Comparison of the SAXS profile from the μE equilibrium system at 22 °C with the TLS profile from the DPB1/HPI1 mixture at a particular time in the late-stage SD (both shown by data points [29]). The solid lines are the best fit with the MB theory. The scaled structure factor $\bar{F}(x)$ for the DPB1/HPI1 is identical to that for the DPB/HPI mixture shown in Fig. 1

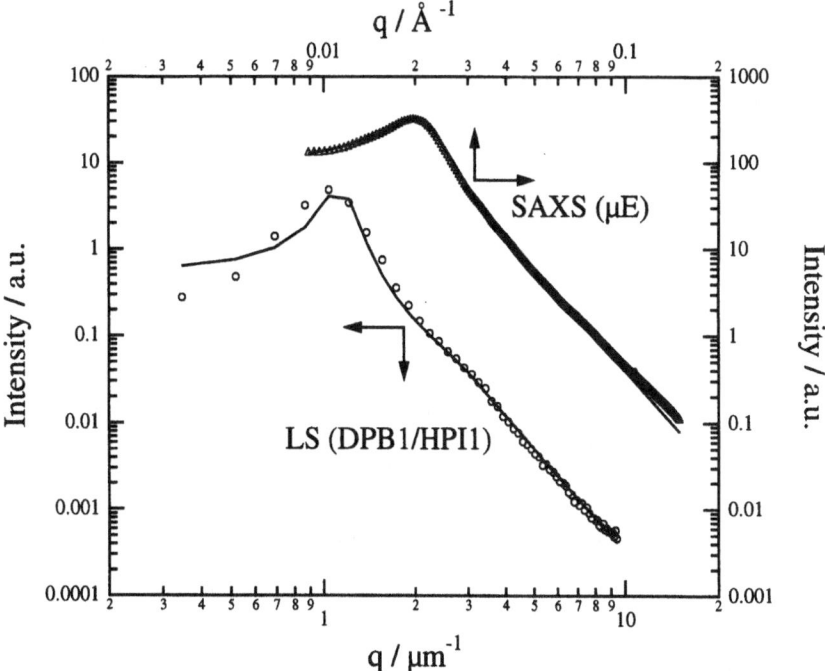

which is an inverse sixth-order polynomial in k and which contains three parameters a, b, and c,

$$f(k) = \frac{8\pi b[a^2 + (b+c)^2]/(2\pi)^3}{(k^2 + c^2)[k^4 - 2(a^2 - b^2)k^2 + (a^2 + b^2)^2]} .$$ (10)

This form is an extension of the well-known structure factor proposed by Teubner and Strey [30] for a bicontinuous μE. The real-space structure as well as the scattering function can be calculated from Eq. (10) [27]. The best fits of the calculated scattering functions for the experimental scattering profile give the three parameters a, b, and c. The parameters thus determined in turn gives important physical parameters such as Λ_m, the coherence length of the local order ξ [31], and $\langle K \rangle$ [29]

$$\Lambda_m \cong 2\pi/a, \quad \xi = 1/b ,$$ (11)

$$\langle K \rangle = \frac{1}{6}\left\{\frac{(a^2 + b^2)^2 + c^2(a-b)(a+3b)}{a^2 + (b-c)^2}\right\} ,$$ (12)

and the corresponding real-space structures are shown in Plate 4 (part a for the μE system and part b for the DPB/HPI system) [29]. In this treatment $\langle H \rangle$ was assumed to be zero.

At a first glance, the two profiles shown in Fig. 5 are very similar, except for absolute length scales (about 100 Å vs. 1 μm). Similarly, the two real-space structures constructed on the basis of the Gaussian random-field theory show a striking resemblance, having sponge-like characteristics. Quantitative comparisons of the two systems were made

on the basis of Eqs. (11) and (12), which are summarized in Table 2. We note the similarity of the two sponge-like structure in the reduced Gaussian curvature $[-\langle K \rangle]^{1/2}/q_m$ on the quantitative base as well. Again, we should note that the values $\langle K \rangle$ thus estimated are negative for the two structures. One important difference in the two structures is discernible in the reduced coherence length ξ/Λ_m. The polymer system has a larger coherence length (a higher degree of the long-range order) than the μE system. This aspect is manifested by the fact that the low q part of $I(q)$ in the polymer system cannot be fitted well by the theoretical scattering profile as shown in Fig. 5. This point deserves future investigations.

Concluding remarks

The real-space analysis has been conducted for binary polymer mixtures giving isometric two-phase structures. The results revealed that the mixtures in the late stage of SD have a sponge-like structure with saddle type or hypabolic interfaces characterized by a negative Gaussian curvature and a nearly zero area-averaged mean curvature ($\langle H \rangle \cong 0$). The almost zero mean curvature, however, does not necessarily mean that the interfaces of the growing nonequilibrium domain structures are characterized approximately by a minimal surface ($H = 0$) [32]. In fact, our preliminary analysis indicates that the distribution of H for our interface is much broader than that for the

126

T. Hashimoto et al.
Sponge-like structures and their Gaussian curvature determination

periodic minimal surfaces of gyroid [25], suggesting deviations of our interfaces from the minimal surface. We elucidated that the sponge-like structure is consistent with the universal scaled structure factors directly determined from TLS and calculated from 3D computer simulations with the TDGL model. We found that the sponge-like structure bears a striking resemblance to that found in the microemulsion system at HLB except for the fact that the polymer systems have a long-range order higher than the microemulsion systems.

Acknowledgements We are grateful to Professors Stephen T. Hyde and Sow-Hsin Chen for their collaboration and stimulating discussions.

References

1. See, for example, Gunton JD, San Miguel M, Sahni PS (1983) In: Domb C, Lebowitz JL (eds) Phase Transitions and Critical Phenomena. Academic Press, New York, pp 269–482; Binder K (1991) In: Cahn RW, Haasen P, Kramer EJ (eds) Materials Science and Technology. Vol 5, Phase Transformations in Materials. Haasen P (Vol ed) VCH, Weinheim, pp 143–212; Komura S, Furukawa H (eds) (1988) Dynamics of Ordering Processes in Condensed Matter. Plenum, New York
2. Cahn JW (1965) J Chem Phys 42:93–99
3. See, for example, Hashimoto T (1988) Phase Transitions 12:47–119
4. Shinozaki A, Oono Y (1993) Phys Rev E 48:2622–2654
5. Koga T, Kawasaki K (1993) Physica A 196:389–415
6. Wilson T (1990) In: Confocal Microscopy. Academic Press, London, pp 1–64
7. Hashimoto T (1993) In: Cahn RW, Haasen P, Kramer EJ (eds) Materials Science and Technology. Vol. 12. Structure and Properties of Polymers. Thomas EL (Vol ed) VCH, Weinheim, pp 251–300
8. Hashimoto T, Itakura M, Hasegawa H (1986) J Chem Phys 85:6118–6128
9. Binder K, Stauffer D (1974) Phys Rev Lett 33:1006–1009
10. Hashimoto T, Jinnai H, Hasegawa H, Han CC (1994) Physica A 204:261–276
11. Kawasaki K, Ohta T (1978) Progr Theor Phys 59:362–374; (1983) Physica A 118:175–190
12. Furukawa H (1989) J Phys Soc Japan 58:216–221
13. Jinnai H, Nishikawa Y, Hashimoto T, in preparation
14. Jinnai H, Nishikawa Y, Koga T, Hashimoto T (1995) Macromolecules 28:4782–4784
15. Jinnai H, Koga T, Nishikawa Y, Hashimoto T, Hyde ST (1997) Phys Rev Lett 78:2248–2251
16. Ribbe A, Jinnai H, Hashimoto T (1996) J Mater Sci 31:5837–5847
17. Hashimoto T, Takenaka M, Jinnai H (1991) J Appl Crystallogr 24:457–466; Takenaka M, Hashimoto T (1992) J Chem Phys 96:6177–6190
18. Takenaka M, Izumitani T, Hashimoto T (1990) J Chem Phys 92:4566–4575
19. Koga T, Kawasaki K, Takenaka M, Hashimoto T (1993) Physica A 198:473–492
20. Lorensen WE, Cline HE (1987) Computer Graphics SIGGRAPH'87 21:163–169
21. Nishikawa Y, Jinnai H, Koga T, Hashimoto T, Hyde ST, submitted to Langmuir
22. Kirste B, Porod G (1962) Kolloid-Z Z Polym 184:1–7
23. Tomita H (1984) Progr Theor Phys 72:656–658
24. Takenaka M, Hashimoto T, to be submitted
25. Nishikawa Y, Koga T, Jinnai H, Hashimoto T, in preparation
26. Berk NF (1987) Phys Rev Lett 58:2718–2721
27. Chen SH, Lee DD, Chang SL (1993) J Mol Struct 296:259–264
28. Chen SH, Lee DD, Kimishima K, Jinnai H, Hashimoto T (1996) Phys Rev E 54:6526–6531
29. Jinnai H, Hashimoto T, Lee DD, Chen SH (1997) Macromolecules 30:130–136
30. Teubner M, Strey R (1987) J Chem Phys 87:3195–3220
31. Chem SH, Chang SL, Strey R (1991) J Appl Crystallogr 24:721–731
32. Hildebrandt S, Tromba A (1985) Mathematics and Optimal Form. Scientific America Library, New York

Progr Colloid Polym Sci (1997) 106:127–130
© Steinkopff Verlag 1997

T. Ohta
M. Nonomura

Formation of micelles and vesicles in copolymer–homopolymer mixtures

Received: 13 October 1996
Accepted: 4 February 1997

Dr. T. Ohta (✉) · M. Nonomura
Department of Physics
Ochanomizu University
Tokyo 112, Japan

Abstract A mixture of A–B type diblock copolymer and homopolymer undergoes macro- and micro-phase separations simultaneously at low temperatures. We have carried out computer simulations for this double-phase separation modeled by a coupled set of equations for the local volume fractions. When the strength of incompatibility between B blocks and homopolymers is sufficiently large, copolymer-rich domains form a concentric pattern with a few layers where A-rich and B-rich domains are arrayed alternatively. When the incompatibility is weak, on the other hand, the stripe domains of A and B blocks tend to be perpendicular to the interface dividing the homopoly-mer-rich and the copolymer-rich domains. The morphological difference is found to affect drastically the kinetics of domain growth. We explore the possibility of vesicle formation such that a bilayer of the copolymer can be formed in the matrix of the homopolymer, which is an interesting feature of the copolymer–homopolymer mixture. We study the statistical properties of a vesicle to calculate the bending and the curvature moduli for the bilayer in the strong segregation limit and in the limit of the negligible membrane width compared to the size of a vesicle.

Key words Micelles – vesicles – copolymer mixture – microphase separation – bending elasticity

Introduction

A copolymer–homopolymer mixture provides us with a variety of domain morphology since both macro- and micro-phase separations take place simultaneously. (Equilibrium property has been studied extensively by a mean-field theory [1].) We consider a mixture of A–B diblock copolymer and C homopolymer assuming a short-range repulsive interaction between A and B monomers and B and C monomers. One may expect a multiple domain structure in a sense that microphase separated domains are developed in a macrophase separated domain [2, 3]. It is also expected that formation of vesicles is also possible in a block copolymer–homopolymer mixture such that a bilayer of copolymer is constituted in the matrix of homopolymers.

In this paper, we investigate the domain morphology in the phase separation of copolymer–homopolymer mixtures. Computer simulations are carried out in two dimensions to explore the kinetics of the double-phase separation. An interfacial approach is applied to study the relative stability of the domain patterns and to evaluate the bending and the curvature moduli of a bilayer membrane.

The structure of this paper is as follows. In the next section, we introduce our model equations and carry out the computer simulations. A morphological transition of multiple domain structures and its kinetics of domain growth are discussed. In Section 3, we investigate the stability of the multiple domain structures. The Helfrich free energy for the membrane of a vesicle is also calculated in the limit of an infinitesimally thin copolymer layer compared to the size of vesicle. Discussions are given in Section 4.

Model and simulations

We consider a mixture of AB-type copolymer and C homopolymer having the molecular weights $N_i (i = A, B, C)$ with $N_A = N_B = N$. It is convenient to choose the independent variables in the combination of the local volume fractions ϕ_A, ϕ_B and ϕ_C as $\psi = \phi_A + \phi_B$ and $\phi = \phi_A - \phi_B$. The variable ψ is the local volume fraction of the AB copolymers and hence it is a measure of the macrophase separation whereas ϕ provides the degree of the microphase separation.

The free energy of this mixture is obtained by generalizing the method in ref. [4] and is given by

$$F[\phi, \psi] = F_S[\phi, \psi] + F_L[\phi, \psi] . \tag{1}$$

The short-range part F_S is written in terms of ϕ and $\eta = \psi - \psi_c$ with ψ_c the volume fraction at a macrophase separation temperature as

$$F_S = \int d\mathbf{r} \left[\frac{c_1}{2} (\nabla \eta)^2 + \frac{c_2}{2} (\nabla \phi)^2 + W(\eta, \phi) \right] , \tag{2}$$

with c_1, $c_2 > 0$ constants. The local part W is given by

$$W(\eta, \phi) = g_0(\eta) + g_1(\phi) + b_1 \eta \phi - \frac{b_2}{2} \eta \phi^2 . \tag{3}$$

The potentials g_0 and g_1 are assumed to be given through the relation $dg_i(x)/dx = -A_i \tanh(x) + x$. In the simulations we put $c_1 = c_2 = 0.5$, $A_0 = 1.3$ and $A_1 = 1.1$.

The long-range part F_L is written as

$$F_L = \frac{\alpha}{2} \int d\mathbf{r} \, d\mathbf{r}' G(\mathbf{r} - \mathbf{r}') \delta\phi(\mathbf{r}) \delta\phi(\mathbf{r}') , \tag{4}$$

where $\alpha \propto (2/N)^2$, $-\nabla^2 G(\mathbf{r} - \mathbf{r}') = \delta(\mathbf{r} - \mathbf{r}')$ and $\delta\phi = \phi - \bar{\phi}$ with $\bar{\phi}$ the spatial average of ϕ. Any bilinear interactions that involve η are not present in F_L when $N_A = N_B$ as we have assumed.

The time-evolution of η and ϕ obeys, respectively,

$$\frac{\partial \eta}{\partial t} = \nabla^2 \frac{\delta F_S}{\delta \eta} , \tag{5}$$

$$\frac{\partial \phi}{\partial t} = \nabla^2 \frac{\delta F_S}{\delta \phi} - \alpha \delta\phi(\mathbf{r}) . \tag{6}$$

For simplicity, we do not consider any dynamical cross-coupling between η and ϕ as well as the possible hydro-dynamic effects.

In the simulations a two-dimensional space is divided into 128×128 square cells with the periodic boundary conditions. The cell size is set to be unity. The initial conditions are the spatially uniform distributions $\eta = \bar{\eta}$, $\phi = \bar{\phi}$ at each cell with small random fluctuations superimposed. Note that $\bar{\phi} = 0$ when $N_A = N_B$. Starting from these initial conditions we solve Eqs. (5) and (6) numerically with the time increment $\Delta t = 1$.

We focus our attention on the domain structures obtained by changing the coupling constant b_1 [5]. The domain patterns at $t = 16\,000$ and $t - 64\,000$ for $\bar{\eta} = -0.2$, $\bar{\phi} = 0$, $\alpha = 0.02$, $b_2 = 0.2$ and $b_1 = 0.05$ are displayed in Fig. 1 while those for $b_1 = 0.03$ with the same values for other parameters are shown in Fig. 2. A copolymer-rich domain where $\eta > 0$ is drawn by gray color whereas a homopolymer-rich domain where $\eta < 0$ is indicated by white color. Regions where ϕ is larger than 0.15 are shown by black color.

It should be noted in Fig. 1 that in the most parts of the copolymer-rich domains, the lamellar stripes tend to be parallel to the interfaces separating copolymer-rich and homopolymer-rich domains so that an onion-ring pattern appears. In Fig. 2, on the other hand, one can see that the striped domains are formed perpendicularly to the interfaces. Comparing these figures, one finds that there is a clear morphological transition. We have examined the b_1-dependence in detail and found that the morphological change occurs at about $b_1 = 0.04$.

One notes that the number of circular domains in Fig. 1 is almost time-independent and hence the domain growth is frozen for the parallel configuration. On the other hand, one can see an ordinary domain growth of polymer blends for the perpendicular configuration in Fig. 2 in such a way that larger domains grow further at the expense of smaller domains. This is a clear evidence that domain morphology affects its kinetics. We have examined the time dependence of the structure functions of η and ϕ to confirm the above kinetic behavior quantitatively [5].

The morphological transition by changing the parameter b_1 described in the above section has been analyzed by calculating the interfacial energy for each domain structure [5]. We have, indeed, found that there is a critical

Progr Colloid Polym Sci (1997) 106:127–130
© Steinkopff Verlag 1997

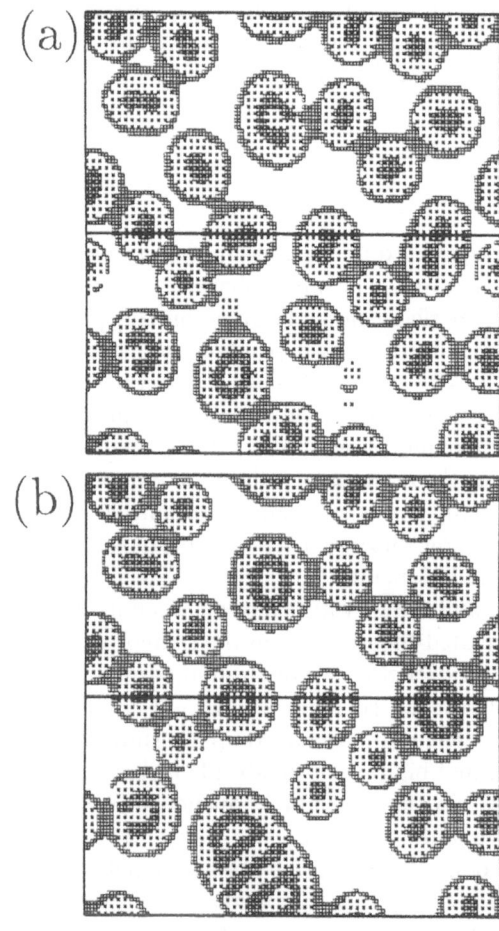

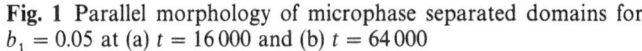

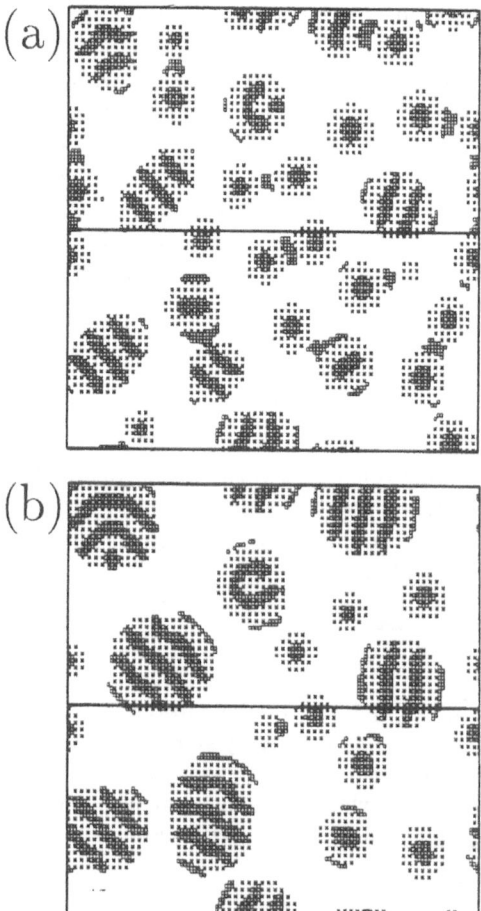

Fig. 1 Parallel morphology of microphase separated domains for $b_1 = 0.05$ at (a) $t = 16\,000$ and (b) $t = 64\,000$

Fig. 2 Perpendicular morphology of microphase separated domains for $b_1 = 0.03$ at (a) $t = 16\,000$ and (b) $t = 64\,000$

value of b_1 given by $b_1^c \sim \alpha^{1/4}$ in the weak segregation regime, above which the interfacial energy of the parallel configuration is smaller than the perpendicular configuration so that the onion-ring domain pattern is energetically more favorable.

Elastic constants of a membrane

In this section, we study the property of a vesicle in the present model free energy. Suppose that there is a stripe bilayer of ABA monomer configuration in a matrix of C homopolymers. Comparing the free energy of this configuration with a multilayer of ABABA configuration, the former free energy is found to be higher than the latter as far as the interfacial energy σ_{AC} between A and C domains is positive. Thus, a bilayer of block copolymer is not a stable structure in a strict sense. However, formation of a bilayer is expected not to be impossible when σ_{AC} is

sufficiently small and in the strong segregation limit with the volume fraction close to $\frac{1}{2}$ where diffusion of block copolymer through the homopolymer-rich domains is almost prohibited.

Under these conditions, we derive the Helfrich Hamiltonian [6] for a bilayer of vesicle membrane. We consider the length scale such that the width of membrane is negligible compared to the size of the vesicle.

The Helfrich Hamiltonian density h is defined by

$$h = \sigma + \kappa_1 H^2 + \kappa_2 K \,, \tag{7}$$

where σ is the interfacial energy of the membrane, H the mean-curvature defined by $1/R_1 + 1/R_2$ with R_1 and R_2 the radii of the principal curvatures whereas $K = 1/(R_1 R_2)$ is the Gaussian curvature. The spontaneous curvature is absent in the present system.

We calculate the bending and the curvature moduli κ_1, κ_2 starting from the long-range part of the free-energy F_L given by (4). One of the simplest way is to consider the free energy of spherical and cylindrical vesicles [7].

First, let us consider a spherical vesicle with ABA layers in the matrix of C homopolymers. The profile of ϕ is assumed to be given by $\phi = 1$ for $r_1 < r < r_2$ and $r_3 < r < r_4$, $\phi = -1$ for $r_2 < r < r_3$ and $\phi = 0$ otherwise, where r is the distance from the center of the spherical vesicle. The conservation of ϕ requires $2(r_4^3 - r_3^3) = 2(r_2^3 - r_1^3) = r_3^3 - r_2^3$. The free energy F_L is given in terms of r_i by

$$F_L = \frac{4\pi\alpha}{6}\left[2r_2^2(r_2^3 - r_1^3) + 2r_3^2(r_3^3 - r_3^3) - \frac{3}{5}(r_4^5 - r_1^5)\right]. \quad (8)$$

We put $r_1 = R - 2\varepsilon - b$, $r_2 = R - \varepsilon$, $r_3 = R + \varepsilon$ and $r_4 = R + 2\varepsilon + a$ and evaluate Eq. (8) in powers of ε/R. The unknown parameters a and b are determined by the conservation condition as a function of ε and R. In this way, one obtains up to $O(R^0)$

$$F_L = 4\pi R^2\alpha\left[\frac{2}{3}\varepsilon^3 + \frac{124}{15}\frac{\varepsilon^5}{R^2}\right]. \quad (9)$$

The first term $2\varepsilon^3/3$ together with the contribution from the short-range part of the free energy gives the interfacial energy of the membrane.

The free energy of a cylindrical vesicle can be evaluated similarly. The result is given up to $O(1/R)$ by

$$F_L = 2\pi R\alpha\left[\frac{2}{3}\varepsilon^3 + \frac{34}{15}\frac{\varepsilon^5}{R^2}\right]. \quad (10)$$

The coefficients of the Helfrich Hamiltonian are readily obtained from Eqs. (9) and (10):

$$\kappa_1 = \alpha\varepsilon^5\frac{34}{15}, \quad (11)$$

$$\kappa_2 = -\alpha\varepsilon^5\frac{4}{5}. \quad (12)$$

Since the curvature modulus is negative, a saddle configuration is energetically unfavorable in this system. If one assumes that the width of the bilayer is proportional to

$N^{2/3}$ as in the mesophase of pure block copolymers [4], these moduli depend on the molecular weight of the copolymer as $\kappa_1, \kappa_2 \sim N^{4/3}$ since $\alpha \sim 1/N^2$.

Discussions

We have investigated domain growth in copolymer–homopolymer mixtures. The microphase separated domains in copolymer-rich region take an onion-ring pattern or a perpendicular morphology depending on the interaction strength b_1. This morphological transition can be understood by calculating the interfacial energies. The difference of domain morphology affects the growth kinetics drastically, i.e., the domain growth in the parallel configuration especially for the average volume fraction $\bar{\eta} = -0.2$ is quite slow once a concentric pattern is formed in a copolymer-rich domain.

The bending and the curvature moduli of a copolymer bilayer are calculated theoretically in the strong segregation limit and in the length scale of an infinitesimal width of membrane. However, so far, we have not found the parameters for vesicle formation in computer simulations. Only interconnected bilayers or quadruple layers are observed transiently in the simulations quenched from the high temperature uniform phase.

Finally, it is pointed out that the present model system can be extended to copolymer–copolymer mixtures. Some of the preliminary results have been published in ref. [8].

The authors are grateful to A. Ito for her collaboration. Thanks are also due to Helmut Brand for valuable discussions during T.O.'s stay at University of Bayreuth under the Japan–Germany cooparative research program supported by Japan Society for the Promotion of Science. This work was also supported by the Grant-in-Aid of Ministry of Education, Science and Culture of Japan.

References

1. Hong KM, Noolandi J (1983) Macromolecules 16:1083
2. Hasegawa H, Hashimoto T (1992) Polymer 33:475
3. Koizumi S, Hasegawa H, Hashimoto T (1994) Macromolecules 27:6532
4. Ohta T, Kawasaki K (1986) Macromolecules 19:2621
5. Ohta T, Ito A (1995) Phys Rev E 52:5250
6. Helfrich W (1973) Z Naturforsch A 28:693
7. Varea C, Robledo A (1995) Physica A 220:33
8. Ohta T, Ito A, Motoyama M (1996) J Phys Condens Matter 8:A65

Progr Colloid Polym Sci (1997) 106:131–135
© Steinkopff Verlag 1997

T. Koga
T. Hashimoto

Computer simulation of ordering processes in block copolymers

Received: 13 October 1996
Accepted: 3 April 1997

Dr. T. Koga (✉)
Hashimoto Polymer Phasing Project
ERATO, JST
15 Morimoto-cho
Shimogamo, Sakyo-ku
Kyoto 606, Japan

T. Hashimoto
Department of Polymer Chemistry
Graduate School of Engineering
Kyoto University
Kyoto 606-01, Japan

Abstract Block copolymer (BCP) systems possess a rich variety of microdomain structures; lamellar, cylindrical, spherical and other bicontinuous structures, which have been extensively studied experimentally and theoretically. On the other hand, there are a few works on the dynamics of ordering processes near the order–disorder transition (ODT) in BCP. Therefore, we investigate the dynamical properties of diblock copolymers by using a computer simulation method. The time-dependent Ginzburg–Landau-type equation with a long-range repulsive interaction is applied for the order-parameter defined as a difference in the local concentration between the two kinds of monomers. This model has three important parameters; strength of repulsive interaction, strength of noise, and block ratio. In the present study, we especially focus on the effects of the noise strength on the lamellar phase near the ODT point in two dimensions and on the formation of the lamellar phase after the system is quenched from the disordered state to the state just below the ODT point. As for the equilibrium phase behavior, we found that the lamellar phase near the ODT point destroyed by noise and fluctuation-induced disordered state is formed, which is different from the disordered phase in the mean-field theory. We also clarified the process of the formation of the lamellar phase after the system is quenched from the disordered state to the state just below the ODT point. The shape of the lamellar grains is consistent with that obtained by the experiments for polystyrene-*block*-polyisoprene copolymers [Hashimoto et al. (1996) Phys. Rev. E 54:5832].

Key words Block copolymer – ordering processes – lamellar phase

Block copolymer (BCP) systems possess a rich variety of microdomain structures; lamellar, cylindrical, spherical and other bicontinuous structures [1, 2]. Such kinds of structures are frequently observed in other systems; microemulsion, Bénard convection, reaction–diffusion systems, etc. [3]. Therefore, BCP is also regarded as a model system to investigate the pattern formation in nature.

The order–disorder transition (ODT) of block copolymers has been extensively investigated [1, 2] since Leibler's mean-field theory [4], where the effects of concentration fluctuations are neglected. Recently, the properties of block copolymers near the ODT have attracted considerable attention and it has become clear that the effects are important near the ODT [4–8]. Fredrickson and Helfand [4] approximately took into account the effects of concentration fluctuations on the ODT and showed that the ODT, which is the second-order phase transition in the mean-field theory, becomes the first-order (weakly first-order) phase transition due to the fluctuation effects.

Recently, the ordering processes after the quench from the disordered state to the ordered state near the ODT have been investigated experimentally [7, 8], where a nearly symmetric polystyrene-*block*-polyisoprene (SI) copolymers was used. It is found by these studies that the peak intensity of the X-ray scattering grows with time after a incubation period of time upon quenching the system to a state below the ODT point but near ODT, indicating that the ordering process is nucleation and growth of lamellar grains. The growth process has also been investigated by transmission electron microscopy (TEM) and an isolated lamellar nucleus after the incubation period of time was clearly observed. It is important to note that the shape of the lamellar nucleus is anisotropic; the dimension parallel to the lamellar normal is larger than that perpendicular to the lamellar normal.

Although a computer simulation method, where the time change of the concentration of BCP is described by the time-dependent Ginzburg–Landau (TDGL) equation with a long-range repulsive interaction [9], has been widely used in many numerical studies to investigate the dynamical properties of BCP [10], the ordering processes near the ODT including the effects of concentration fluctuations have not been systematically studied by such a method. Therefore, in the present paper, we investigate the dynamical properties of BCP near the ODT by using the computer simulation method. Here we especially focus on the formation of the lamellar phase after the system is quenched from the disordered state to the state just below the ODT point. We also present a comparison between results obtained by the computer simulation and the experimental ones mentioned above.

Here we consider an A–B-type diblock copolymer. The local-order parameter $S(\mathbf{r})$ is defined by the difference of the concentration of the two kinds of monomers: $S(\mathbf{r}) \equiv 2c_A(\mathbf{r}) - 1$, where $c_A(\mathbf{r})$ is the local concentration of A monomers. To describe the dynamical behavior of the system, we use the following TDGL equation for the order parameter $S(\mathbf{r})$:

$$\frac{\partial}{\partial t} S(\mathbf{r}, t) = L \nabla^2 \frac{\delta H\{S\}}{\delta S(\mathbf{r})} + \zeta(\mathbf{r}, t), \qquad (1)$$

where L is the Onsager kinetic coefficient, which is assumed to be constant in this study. $\zeta(\mathbf{r}, t)$ is the noise term, which obeys the usual fluctuation–dissipation relation. $H\{S\}$ is the free-energy functional for BCP obtained by the random phase approximation [4, 9]. $H\{S\}$ consists of two parts; the short-range part and the long-range part, the former has the same form as the usual Ginzburg–Landau form and the latter represents the effects of the connection of two kinds of polymers in a diblock copolymer [9]. Since this model is a kind of coarse-grained model, we cannot

argue the microscopic quantities such as the conformation of a polymer chain. On the other hand, the information on the global structures, such as structure of lamellar grains, is easily obtained by computer simulation of the model.

We obtain the following dimensionless equation by substituting the explicit form of $H\{S\}$ into Eq. (1) and by rescaling variables in the equation [11]:

$$\frac{\partial}{\partial t} S(\mathbf{r}, t) = \nabla^2 \left[-\nabla^2 S(\mathbf{r}, t) - S(\mathbf{r}, t) + S(\mathbf{r}, t)^3 \right]$$

$$- \alpha S(\mathbf{r}, t) + \sqrt{B}\zeta(\mathbf{r}, t), \qquad (2)$$

where we assume that the block ratio is $\frac{1}{2}$. $\zeta(\mathbf{r}, t)$ obeys the following relation:

$$\langle \zeta(\mathbf{r}, t)\zeta(\mathbf{r}', t') \rangle = -\nabla^2 \delta(\mathbf{r} - \mathbf{r}')\delta(t - t'). \qquad (3)$$

In Eqs. (2) and (3), we use the same symbols to represent the dimensionless quantities as in the original equation Eq. (1) for simplicity. Since we restrict ourselves to the symmetric case in the present study, where the degrees of polymerization of A and B polymers are the same, there are two independent parameters α and B in Eqs. (2) and (3). α and B is related to the interaction parameter χ and the degree of polymerization N as follows: $\alpha \sim (\chi N)^{-2}$, $B \sim (\chi N)^{-1/2} N^{-1/2}$. In case of $B = 0$, the free energy depends only on α. This means that the phase diagram and the dynamics are determined by the product χN. On the other hand, if χN is fixed, the strength of noise B decreases with increasing N. In case of $B = 0$, the ODT occurs at $\alpha_c = 0.25$, and the system is in the ordered state for $\alpha < \alpha_c$.

To integrate Eqs. (2) and (3), we use the Euler scheme, $\Delta x = 1.0$, $\Delta t = 0.05$. This discretized version of the TDGL equation is regarded as a kind of the cell-dynamic systems [10], which has been used in recent numerical studies on ordering processes in quenched systems. We performed computer simulation runs in two dimensions.

Before we present the results on the dynamics of ordering in BCP, we briefly summarize the effects of noise on the phase behavior. As mentioned above, ODT occurs at $\alpha = \alpha_c = 0.25$ at $B = 0$. We studied the effects of noise on the phase behavior varying B at a fixed value of $\alpha < \alpha_c$, where the lamellar phase is expected to be formed according to the mean-field theory. We find that if B is larger than a certain critical value B_c, the lamellar structure does not form within the computational time. A typical snapshot pattern of $S(\mathbf{r})$ at $B > B_c$ and $\alpha < \alpha_c$ is presented in Fig. 1, where although there is no long-range order, there exist local microdomain-like structures, which are local concentration fluctuations without a clear interface. This is interpreted that the lamellar structure is destroyed by noise. We call this state the fluctuation-induced disordered (D_F) state to distinguish it from the disordered phase in the mean-field sense at $\alpha > \alpha_c$, which is called D_M phase. The value of

Fig. 1 A snapshot pattern of the order parameter $S(\mathbf{r})$ at $t = 10000$ for $\alpha = 0.24$, $B = 0.01$

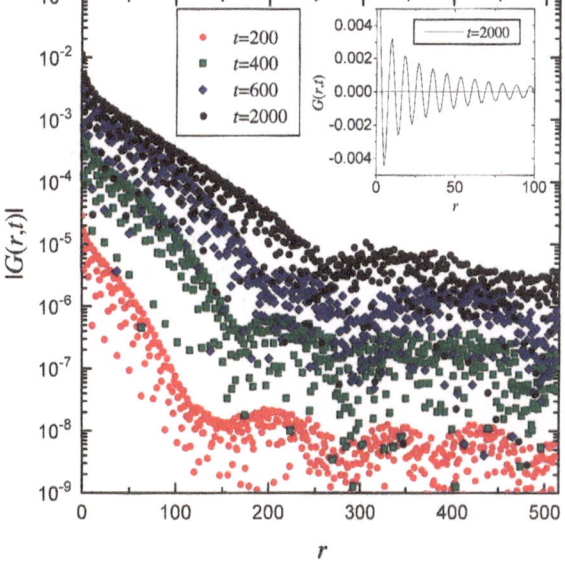

Fig. 2 Time evolution of $|G(r, t)|$ at $t = 200, 400, 600$ and 2000 for $\alpha = 0.24$, $B = 0$. Plot of $G(r, t)$ vs. r at $t = 2000$ is shown in the inset

B_c seems to depend on $\Delta\alpha \equiv \alpha_c - \alpha$: $B_c (\Delta\alpha)$. A similar behavior is observed in case of Swift–Hohenberg equation in two dimensions [12], where the relation between B_c and $\Delta\alpha$ is studied. If we apply the same argument as in ref. [12] to our case, then we obtain $B_c \sim \Delta\alpha$ in the vicinity of α_c.

Next, we study the dynamics of ordering processes near ODT in the absence of noise, $B = 0$. Here we mainly present the results obtained at $\alpha = 0.24$ and the system size is 1024^2. To investigate the formation of the lamellar phase quantitatively, we use the structure function $I_k(t)$ and the pair correlation function $G(r, t)$ of $S(\mathbf{r})$, where k is the wave number. We can characterize the lamellar grain by using the pair-correlation function $G(r, t)$. We present $G(r, t)$ at $t = 2000$ in the inset of Fig. 2, where the lamellar grains fill almost the whole space. As shown in the inset of Fig. 2, $G(r, t)$ sinusoidally oscillates with the period of the spacing of lamellae D, which is about 9 for $\alpha = 0.24$. The amplitude of this oscillation is expected to decay with a larger length scale than D, reflecting the grain structure. A simple functional form of $G(r, t)$ which describes such a situation is as follows: $G(r, t) \sim \cos(q_m r)\exp(-r/\xi_g)$, where $q_m \equiv 2\pi/D$ and ξ_g is the correlation length for the grain structure. In this model, the long-range correlation of $G(r, t)$ decreases as $-r/\xi_g$ in a semi-logarithmic scale. To emphasize such a long-range correlation in $G(r, t)$, we plot the absolute value of $G(r, t)$ in Fig. 2. Here, we are not concerned with the rapid variation in $|G(r, t)|$ reflecting the sinusoidal oscillation with D in $G(r, t)$ as shown in the inset of Fig. 2 but rather asymptotic decays of the peak values of $|G(r, t)|$

with r. We find from Fig. 2 that $|G(r, t)|$ at $t = 2000$ rapidly decreases with r at the small r region ($r \simeq D$) and almost linearly decreases with r at $D < r < r_g(t)$ ($r_g(t) \simeq 250$ at $t = 2000$) and then takes a nearly constant value for $r > r_g(t)$ on the semi-logarithmic scale. Since $r_g(t)$ approximately corresponds to the size of the lamellar grains observed in the real space pattern, the long-range correlation in $|G(r, t)|$ is due to the lamellar grains.

In Fig. 2, the time evolution of $|G(r, t)|$ in the early and intermediate time range are also presented. In the early stage up to $t \simeq 400$, the value of $|G(r, t)|$ becomes large with time as the amplitude of the concentration fluctuations grow. Simultaneously, the correlation with larger length scale than D develops and forms the correlation from grains at $t = 2000$ as mentioned above. It is important to note that there exists large-scale fluctuations even in the early time regime.

To clarify the lamellar grain formation at the early time regime, we also precisely investigate the real space pattern of $S(\mathbf{r})$. We show patterns of $S(\mathbf{r})$ using color figures in Fig. 3 in order to visualize delicate differences in $S(\mathbf{r})$ in the early and intermediate stage of the ordering. Scenario of the formation of the lamellar phase obtained by these analyses is as follows:

(i) Spinodal-like pattern appears from initial disordered state as shown in Fig. 3a at $t = 400$. The typical period of this pattern corresponds to the most unstable mode of the linearized equation of Eq. (2). This process corresponds to the early stage of spinodal decomposition in binary mixtures.

(ii) The region where the value of $S(\mathbf{r})$ is almost equal to its equilibrium value appears and each region looks like a circular microdomain in two dimensions, which is shown as red or blue spot in Fig. 3b. It is important to note that there is long-range correlation between these regions, which looks like a string.

(iii) The local lamellar grains are formed in such a way that the transient circular microdomains connect to each other as shown in Fig. 3c at $t = 1000$. In this case, the

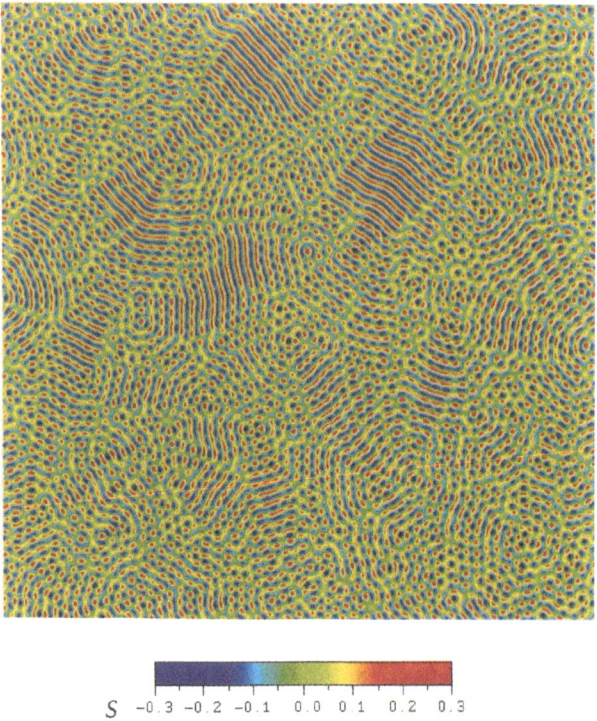

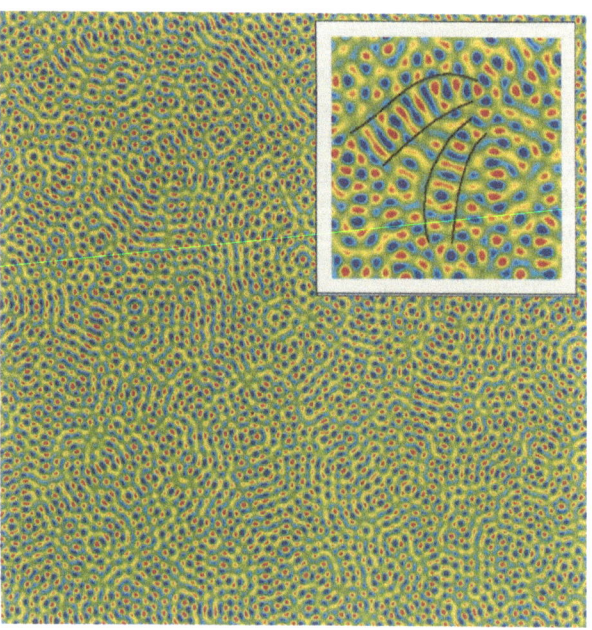

Fig. 3 Snapshot patterns of $S(\mathbf{r})$ at $t = 400$(a), 600(b) and 1000(c) for $\alpha = 0.24$, $B = 0$. The linear dimension of each pattern is 512 in our unit. The inset in part (b) shows a magnitude figure where the correlated regions are marked by black solid lines. The values of $S(\mathbf{r})$ for the patterns in part (a) to (c) are commonly shown with a color representation given beneath the pattern in part (c)

direction along string is almost the same as that of lamellar normal $\mathbf{n}$. As a result, the shape of the lamellar grains are quite anisotropic; the dimension parallel to $\mathbf{n}$ is much larger than that perpendicular to $\mathbf{n}$.

After stage (iii), the local lamellar grains grow with time and fill almost the whole space. As for the structure of the grain boundary, the lamellae at grain boundaries melt and the structure at the grain boundary looks like a hexagonal pattern. If $\Delta\alpha$ is large, such behavior is not seen. Therefore, the structure of the grain boundary also depends on $\Delta\alpha$.

The long-range (string-like) correlation in Fig. 3b is probably related to the correlation of the concentration fluctuations near the ODT point and the initial grain size depends on the correlation length of the concentration fluctuations. To confirm this, we investigate the initial grain size after grains fill the whole system below the ODT point and found that the initial grain size actually depends on $\Delta\alpha$; the initial grain size decrease as $\Delta\alpha$ increases.

In the presence of noise, if $B \ll B_c$, the process of the lamellar phase formation are almost the same as that in the absence of noise. In case that B is slightly smaller than B_c, the concentration fluctuations emerges from the initial disordered state and the state which resembles the D_F

Progr Colloid Polym Sci (1997) 106:131–135
© Steinkopff Verlag 1997

state is formed. In this case, there also exist the strong correlated region as discussed above in stage (ii) in the absence of noise. However, these regions are fluctuating in time and space, which can be regarded as the concentration fluctuations. In this case, the correlation length is anisotropic. From this state, the lamellar grains are gradually formed and grow with time. Since these lamellar grains are fluctuating in time, the growth process is slower than that at $B \ll B_c$.

Recently, the ordering processes in BCP near ODT have been investigated experimentally [7, 8] as mentioned above. The shape of an isolated lamellar nucleus observed after the incubation period of time is quite anisotropic; the dimension parallel to the lamellar normal is larger than that perpendicular to the lamellar normal. This tendency of the shape of the lamellar grain is consistent with that obtained by our computer simulation as shown in Fig. 3c. However, we have not yet clearly obtained the incubation period of time which indicates the nucleation and growth process as in the experiments. For example, the peak height $I_m(t)$ of $I_k(t)$ obtained by the computer simulation

near B_c for $\alpha < \alpha_c$ gradually increase with time after the early stage.

If we use a free-energy functional where the effects of concentration fluctuations are taken into account by a perturbation expansion of the original model $H\{S\}$ near ODT [5, 6], we can reproduce the experimental results indicating the nucelation and growth of lamellar grains mentioned above [8]. Such behavior should be reproduced by using the original model with the noise term. This point is now under investigation and the results together with quantitative details of this study will be described elsewhere.

In conclusion, we investigated the phase behavior and ordering processes in BCP near ODT by using the TDGL-type equation for BCP. As for the equilibrium phase behavior, we found that the lamellar phase near the ODT point destroyed by noise and the fluctuation-induced disordered state is formed. We also clarified that the formation of the lamellar phase after the system is quenched from the disordered state just below the ODT point. The shape of the lamellar grains is consistent with that obtained by the experiments.

References

1. See for example a review article, Hashimoto T (1987) In: Legge NR, Holden GR, Schroeder HE (eds) Thermoplastic Elastomers. Carl Hanser Verlag, Vienna, Ch 12, Section 3 (1st ed) and 2nd ed in press, and references cited therein
2. See, for example, a review article, Bates FS, Fredrickson GH (1990) Annu Rev Phys Chem 41:525, and references cited therein
3. Seul M, Andelman D (1995) Science 267:476–483
4. Leibler L (1980) Macromolecules 13:1602–1617
5. Fredrickson GH, Helfand E (1987) J Chem Phys 87:697–705
6. Hohenberg PC, Swift JB (1995) Phys Rev E 52:1828–1845
7. Hashimoto T, Sakamoto N (1995) Macromolecules 28:4779–4781 and references cited therein
8. Hashimoto T, Sakamoto N, Koga T (1996) Phys Rev E 54:5832–5835
9. Ohta T, Kawasaki K (1986) Macromolecules 19:2621–2632
10. Bahiana M, Oono Y (1990) Phys Rev A 41:6763–6771; Liu F, Goldenfeld N (1989) Phys Rev A 39:4805–4810; Ohta T, Enomoto Y, Harden JL, Doi M (1993) Macromolecules 26:4928–4934; Kawakatsu T (1995) Phys Rev E 50:2856–2862; Ohta T, Ito A (1995) Phys Rev E 52:5250–5260
11. Koga T, Kawasaki K (1993) Physica A 196:389–415
12. Elder KR, Viñals J, Grant M (1992) Phys Rev A 46:7618–7629

Progr Colloid Polym Sci (1997) 106:136–140
© Steinkopff Verlag 1997

S. Shimabayashi
T. Uno
Y. Oouchi
E. Komatsu

Interaction between hydroxypropylcellulose and surfactant and its effect on dispersion stability of kaolinite suspension in an aqueous phase

Received: 13 October 1996
Accepted: 10 April 1997

Dr. S. Shimabayashi (⊠) · T. Uno
Y. Oouchi · E. Komatsu
The University of Tokushima
Faculty of Pharmaceutical Sciences
Sho-machi 1-78-1
Tokushima, Tokushima 770, Japan

Abstract Both sodium dodecylsulfate (SDS) and cetylpyridinium chloride (CPC) formed a complex with hydroxypropylcellulose (HPC) due to hydrophobic interaction. Binding ratio of a surfactant to HPC, reduced viscosity and cloud point of an aqueous solution of HPC were obtained as a function of the surfactant concentration. Intramolecular/ intersegment bridging was speculated at low concentrations of both surfactant and HPC, which resulted in shrinking of the polymer coil. On the other hand, at high concentrations of the surfactant, micelle-like aggregate was formed along the polymer chain p-dimethyl-amino-azobenzene was solubilized in this aggregate. The surface complex was formed on kaolinite particles. Anomalous flocculation of the suspension was observed at a specific mixing ratio of SDS to HPC by virtue of the interparticle bridging by the complex. The flocculation was detected by the measurements of mean diameter of secondary particles and rheological parameters such as Bingham yield values, and apparent and plastic viscosities. The contour lines for these parameters were obtained as a function of the concentrations of SDS and HPC.

Key words Hydroxypropylcellulose – sodium dodecylsulfate – cetyl-pyridinium chloride – hydrophobic interaction – suspension stability

Introduction

There is a lot of literature so far on the study of hydrophobic interaction of ethyl(hydroxyethyl)cellulose (EHEC) with various ionic surfactants in water. The techniques used in these studies are of electromotive force of N-tetradecylpyridinium bromide [1], of thermal gelation with sodium dodecylsulfate (SDS) [2], of Fourier transform NMR in the presence of dodecyltrimethylammonium bromide [3], of phase separation diagram with various kinds of anionic and cationic surfactants [4], and so on. These studies have focused on the complex formation between EHEC and these surfactants and on the concomitant phase separation of the polymer solution.

In the present study, hydrophobic interaction between hydroxypropylcellulose (HPC) and an ionic surfactant in an aqueous phase was discussed. HPC, as well as EHEC, is a nonionic cellulose ether which contains hydrophobic groups in its molecular structure. Therefore, it might be interesting to compare the complex-formation properties of HPC with that of EHEC. The surfactants used here were an anionic surfactant SDS and a cationic one cetyl-pyridinium chloride (CPC). HPC formed a complex with these surfactants, of which cloud point changed with the surfactant concentration in the same manner as that observed in the EHEC-surfactant systems [4]. Effects of the complex on stability of dilute and concentrated kaolinite suspensions were also studied, taking physicochemical properties of the complex into account.

Progr Colloid Polym Sci (1997) 106:136–140
© Steinkopff Verlag 1997

Experimental

Binding ratios of SDS (BHD Ltd.) and CPC (Tokyo Kasei Ltd.) to HPC (M.W. $= 11$–15×10^4, Tokyo Kasei Ltd.) were determined at 30 °C by an equilibrium dialysis method. Cloud point of an HPC solution was observed by the naked eye. Amounts of p-dimethylaminoazobenzene (Nakarai Ltd.) solubilized by the surfactant micelle and the surfactant–polymer complex were determined at 30 °C by colorimetry at $\lambda = 418$ nm. Viscosity was measured by an Ubbelohde-type capillary viscometer, a cone–plate rotary viscometer, or a Brookfield-type rotary viscometer at 25 or 30 °C. Mean diameter of secondary particles of kaolinite in its dilute suspension (1 g/dl) was estimated by a Coulter counter (type TA-II) in 154 mmol/dm³ NaCl at room temperature.

Results and discussion

Binding isotherms of the surfactants on HPC at various concentrations of NaCl are shown in Fig. 1 as a function of a free concentration of the surfactant. The binding ratio was very small at the low-concentration region, while it steeply increased at medium concentrations. The sigmoidal increase suggests that the binding is cooperative, and the cooperativity is accelerated in the presence of NaCl. In the region of the high concentrations, the binding ratio decreased again after attaining a maximum around the critical micellization concentration (cmc). This is because micelle formed in the bulk solution is more stable than that formed together with hydrophobic groups of the polymer chain. It was confirmed that CPC was definitely bound by HPC although it is scarcely bound by polyvinylpyrrolidone [5, 6].

The relationship among the equilibrium concentration of a surfactant, cloud point, solubilized amount of the dye, and reduced viscosity of an HPC solution were studied. Figure 2 shows one of the typical results. Similar trends were observed at any different concentrations of NaCl, even in the presence of SDS as well as in the presence of CPC.

Cloud point and reduced viscosity shown in Fig. 2 increased with an equilibrium concentration of CPC after attaining a minimum at almost the same concentration. This fact means that the polymer coil once shrinks and then swells with an amount of CPC bound to HPC due to the intramolecular/intersegment bridging and due to the following repulsion by the bound CPC, respectively, as shown schematically on the top. Another explanation for the decreases in cloud point and viscosity might be given in terms of the effect of salting-out by an ionic surfactant, irrespective of the concentration of NaCl. This idea might come out from the fact that the binding of the surfactant does not seem to occur at its lower concentrations, while the binding cooperatively occurs at its medium concentrations. We are, however, considering that the binding certainly occurs, even though the binding ratio is very low (see Fig. 1). We prefer the mechanism of the intersegment bridging, as mentioned above, to that of salting-out in explanation of the decreases. The bridging mechanism is convenient and consistent to give an explanation for flocculation/dispersion of kaolinite particles in the presence of HPC and SDS, as mentioned later (Figs. 4 and 5).

Solubilization of the dye started around the concentration of the minima for the cloud point and the reduced viscosity (ca. 0.3 mmol/dm³), although the cmc of CPC in its bulk solution (0.7 mmol/dm³) is far beyond it. That is, micelle-like aggregates are formed along the polymer chain after a short transient period of the "monodisperse" binding of the surfactant (see the middle on the top of Fig. 2) but prior to the formation of micelles in the bulk solution. The concentration where the solubilization starts is almost the same as that where the cooperative binding begins (see Fig. 1B).

Fig. 1 Binding isotherms of SDS (A) and CPC (B) at 30 °C. Added concentration of NaCl/(mmol/dm³) $= 0$ (●); 10 (◐), and 20 (○) for (A) and 0 (○), 1.0 (△), and 1.5 (□) for (B). Initial concentration of HPC/(g/dl) $= 0.2$ for (A) and 0.4 for (B)

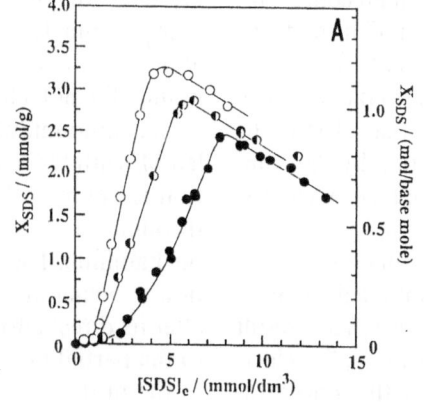

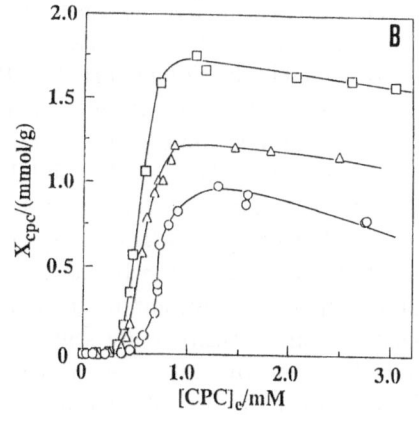

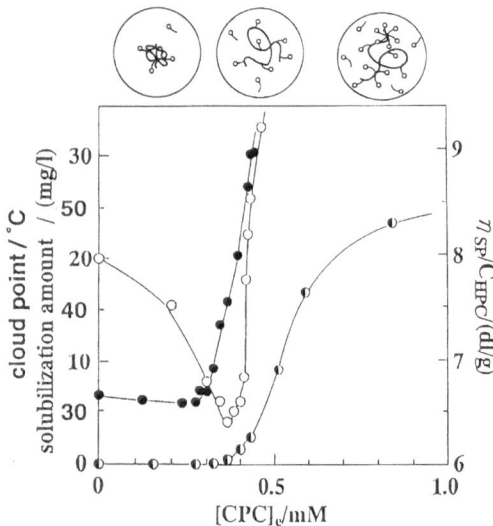

Fig. 2 Cloud point, solubilization amounts, and reduced viscosity as a function of an equilibrium concentration of CPC at 30 °C. The digits of 0, 10, 20, and 30 on the ordinate are for solubilization amount, and those of 30, 40, and 50 for cloud point. An equilibrium concentration on the abscissa was obtained from the binding isotherm shown in Fig. 1. ○: cloud point; ◐: solubilization amount; ●: reduced viscosity [HPC] = 0.2 g/dl; [NaCl] = 1.5 mmol/dm^3

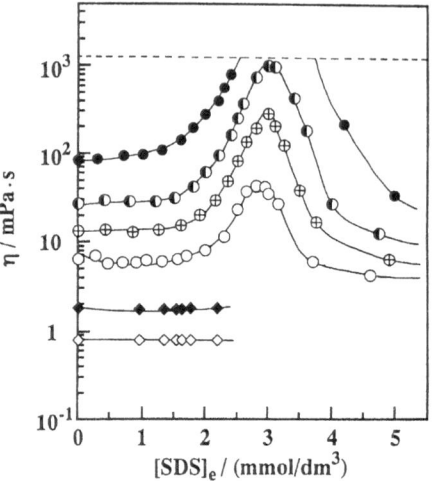

Fig. 3 Solution viscosity as a function of an equilibrium concentration of SDS at 30 °C. An equilibrium concentration on the abscissa was obtained from the binding isotherm shown in Fig. 1. The dotted line on the top shows the upper limit for the measurement by the apparatus. HPC concentration/(g/dl) = 0 (◇), 0.20 (◆), 0.50 (○), 0.75 (⊕), 1.0 (◖), and 1.5 (●), [NaCl] = 0 mmol/dm^3

Viscosity of an HPC solution (0–1.5 g/dl) was measured by a cone–plate-type rotary viscometer as a function of a free concentration of SDS in the absence of NaCl (Fig. 3). Viscosity increased remarkably with an SDS concentration (> 1.5 mmol/dm^3) where the binding ratio is still low. This is owing to the bridging and/or entaglement between HPC molecules (i.e., intermolecular network structure) which are induced through the intermediary of the bound SDS. It should partly bridge to a vacant segment of a neighbor polymer molecule, instead of intramolecular/intersegment bridging shown on the top in Fig. 2, through hydrophobic interaction. The intermolecular bridging is more dominant than the intramolecular one because the polymer concentration is higher in Fig. 3 than in Fig. 2.

The decrease in viscosity was found at high concentrations of SDS after attaining a maximum. Similar tendency was observed in the system of HPC–CPC (data not shown here). The SDS concentration at the maximum (ca. 3 mmol/dm^3) was almost the same as that where the cooperativity (Fig. 1A) and the solubilization by the complex started (data not shown here). That is, when the micelle-like aggregate is formed along the polymer chain, the electrostatic repulsion between the segments concurrently increases and the entanglement is unfolded owing to the high charge density on the segments. These facts result in the break-down of the network structure and, therefore, the decrease in the viscosity. The increase in the concentra-

tion of added SDS contributes to the increases both in the charge density by the bound SDS and in the ionic strength by the free SDS. The increase in the former strengthens the intersegment repulsion while that in the latter, to the contrary, shields it. The former effect is more dominant than the latter one in the HPC–SDS system because the charge density increases with the free concentration of SDS. This fact is in contrast with a polyelectrolyte, in general, where the charge density along the chain is invariable but the repulsion decreases wtih ionic strength.

The effects of SDS and HPC on stability of a kaolinite suspension were studied. Fig. 4A shows contour lines for a mean diameter of the secondary particles in its dilute suspension (1 g/dl) as a function of concentrations of added SDS and HPC. 154 mmol/dm^3 NaCl was used as a dispersing medium, because the measurement by a Coutler counter required an electrolyte solution. In the absence of SDS, the mean diameter decreased after attaining a maximum (7.5 μm) at 7×10^{-4} g/dl HPC, as shown along the abscissa. The concentration of HPC at the maximum did not change even in the presence of SDS by a small amount, where the size decreased and finally levelled off with the SDS concentration. This HPC concentration is optimum for the flocculation which is carried out through the interparticle bridging by HPC adsorbed on the kaolinite. The decrease in size with an SDS concentration is owing in part to the electrostatic dispersing effect after its direct adsorption to the surface and owing in part to the partial breakdown of the bridge by the SDS bound to the bridging polymer HPC.

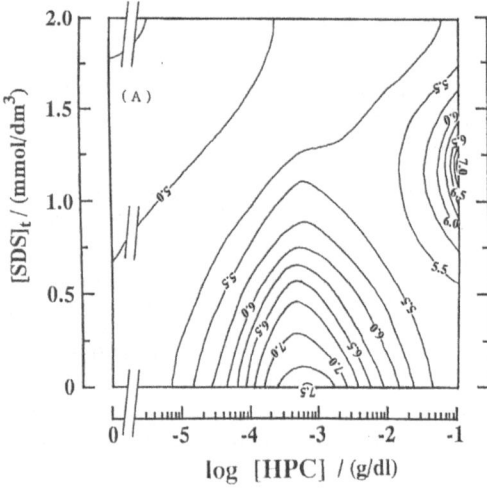

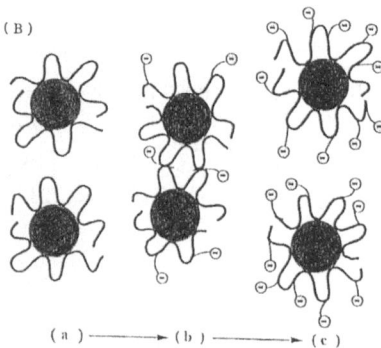

Fig. 4 (A) Contour lines for the mean diameter of the secondary particles in 1 g/dl kaolinite suspension in 154 mmol/dm³ NaCl at a room temperature as a function of the concentrations of added SDS and HPC. The digits on each curve represents the mean diameter in the unit of μm. (B) Schematic illustration of the surface complex of SDS with HPC on the kaolinite particles. SDS concentration increases in the sequence (a)–(c), where there is no SDS in (a). Quoted from Ref. [7] after slight revision

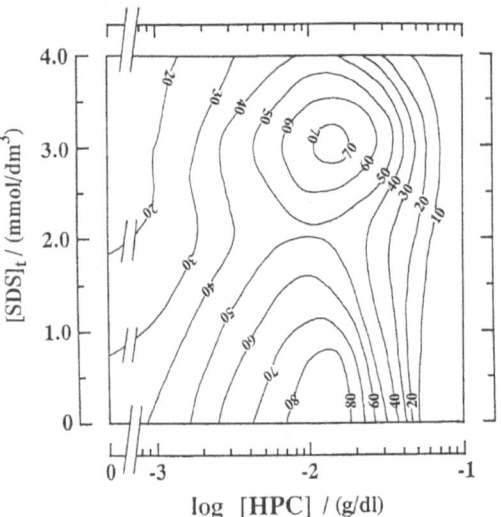

Fig. 5 Contour line of a Bingham-yield value of a concentrated kaolinite suspension (26.7 g/dl-medium) obtained by means of a Brookfield-type rotary viscometer in the presence of 154 mmol/dm³ NaCl at 25 °C as a function of the concentrations of added SDS and HPC. The digit on each curve represents the yield value in an arbitrary unit. Quoted from Ref. [7] after slight revision

The diameter decreased to less than 5.5 μm at 10^{-1} g/dl HPC in the absence of SDS by virtue of the dispersing effect of HPC, as shown schematically in Fig. 4B(a). However, the diameter increased to reach a maximum (7.0 μm at 1.2 mmol/dm³ SDS) when the concentration of HPC was kept constant but that of SDS was increased (see along the ordinate of the right-hand side in Fig. 4A). This effect is explained in terms of the interparticle bridging by the surface complex, as is schematically shown in Fig. 4B(b). This fact was expected from the results discussed above (see Figs. 2 and 3). The surface complex formed under this condition behaved as a flocculating agent, although HPC alone at this concentration (10^{-1} g/dl) behaved as a dispersing agent. On the other hand, in the presence of a high concentration of SDS (2 mmol/dm³, for example) at the same concentration of

HPC, kaolinite particles were dispersed by the surface complex through the electrostatic repulsion (see Fig. 4B(c)). The surface complex in this case behaved as a dispersing agent. Thus, two maxima in the mean diameter were observed in the contour diagram by the different flocculation mechanisms, i.e, bridging effect by HPC alone and that by the surface complex.

Flow curves for the concentrated kaolinite suspension (26.7 g/dl-medium) were obtained by a Brookfield-type rotary viscometer, from which the rheological parameters were estimated. The parameters such as Bingham yield values, apparent and plastic viscosities are the indication for flocculation/dispersion of the concentrated suspension, where their increases mean the interparticle bridging is pronounced (i.e., flocculation/aggregation) while their decreases suggest the interparticle structure is ruined (i.e., dispersion).

Contour lines for Bingham-yield value of the concentrated suspension are shown in Fig. 5 as a function of the concentrations of SDS and HPC. It decreased after attaining amaximum at 10^{-2} g/dl/HPC in the absence of SDS (see along the abscissa). On the other hand, the yield value monotonously decreased with an SDS concentration. That is, SDS alone exhibited a simple dispersing effect (see along the ordinate in left-hand side). However, another maximum was observed at 1.5×10^{-2} g/dl HPC in the presence of 3 mmol/dm³ SDS. Thus, two maxima, one in the absence and the other in the presence of SDS, were again observed. The mechanisms for the appearence of two maxima are the same as those given for the dilute suspension, mentioned above. The optimum concentrations of SDS and HPC are higher than those for the dilute suspension (Fig. 4) naturally because the kaolinite concentration is higher.

In the contour lines for apparent and plastic viscosities of the suspension, the optimum concentrations for the maxima (data not shown here, see in Ref. [7]) were almost the same as those obtained from Fig. 5. This fact means that the flocculation and dispersion of the kaolinite particles in a concentrated suspension reflect correctly in these rheological parameters.

In conclusion (1) an aqueous solution of HPC exhibited a cloud point, which increased after attaining a minimum with a concentration of CPC or SDS. These phenomena were quite similar to those of an aqueous solution of EHEC mixed with a surfactant. (2) Both CPC and SDS formed a complex with HPC through hydrophobic interaction. p-dimethylaminoazobenzene was solubilized in the micelle-like aggregate formed on the polymer chain in a similar manner to that in the micelle formed in the bulk solution. (3) The complex formed on the surface of kaolinite particles affected the stability of the suspension. The mechanisms for flocculation/dispersion of the kaolinite particles were considered, taking the intersegment bridging/repulsion of the HPC molecules in the presence of SDS into consideration. (4) In the present study, the effects of CPC and SDS on the complex formation were quite similar with each other.

References

1. Carlsson A, Lindman B, Watanabe T, Shirahama K (1989) Langmuir 5:1250–1252
2. Carlsson A, Karlstrom G, Lindman B (1990) Colloids Surfaces 47:147–165
3. Carlsson A, Karlstrom G, Lindman B (1989) J Phys Chem 93:3673–3677
4. Karlstrom G, Carlsson A, Lindman B (1990) J Phys Chem 94:5005–5015
5. Nakagaki M, Shimabayashi S (1972) Nippon Kagaku Kaishi 1972:1496–1502
6. Saito S (1987) In: Schick M (ed) Nonionic Surfactants. Marcel Dekker, New York, pp 881–926
7. Shimabayahsi S, Uno T, Nakagaki M (1997) Colloids Surfaces 123–124:283–295

Progr Colloid Polym Sci (1997) 106:141–143
© Steinkopff Verlag 1997

T. Ogawa
S. Isoda
T. Suga
M. Tsujimoto
Y. Hiragi
T. Kobayashi

Analysis by SAXS and cryo-TEM on polymerization with a surfactant

Received: 13 October 1996
Accepted: 6 March 1997

T. Ogawa · Dr. S. Isoda (✉) · T. Suga
M. Tsujimoto · Y. Hiragi · T. Kobayashi
Institute for Chemical Research
Kyoto University
Uji, Kyoto 611, Japan

Abstract The polymerization process in a solution was examined by SAXS and cryo-TEM in the case of chemical polymerization of pyrrole in aqueous solution using sodium cellulose sulfonate (SCS) and $FeCl_3$ as oxidizing agents. Observations with SAXS and cryo-TEM indicated the formation of micelle of SCS in SCS + water system. By adding $FeCl_3$ in the solution, large spherical aggregates were formed as revealed by cryo-TEM, which is considered to be a precursor for the polymer spherical particles produced with a constant diameter.

Key words Small-angle X-ray scattering – polypyrrole – polymerization – low-temperature electron microscopy

Introduction

In order to understand the macroscopic properties of polymers, it is desirable to characterize the polymers from the standpoint of structure and morphology. For example, conductivity of macroscopic polymer material depends not only on its molecular nature but also on its state of aggregation. Accordingly, to control the aggregation so as to improve the properties, detailed knowledge on the polymerization processes, which happen in a solution usually, would be indispensable.

Polypyrroles (PPy) are obtained chemically in an aqueous solution of pyrrole using an oxidizing agent like $FeCl_3$. The shape and size of the polymer particles are, however, not controlled in such a simple chemical reaction, and irregularly shaped PPy is obtained. Wegner et al. have reported that spherical particles of PPy are obtained, when sodium cellulose sulfonate (SCS) is used as an additional oxidizing agent [1]. The diameters of the spherical polymer particles are surprisingly constant in a solution as shown in Fig. 1, and change depending on the concentration of SCS; higher concentration of SCS results in smaller spheres and vice versa. The conduction of these samples follows the variable-range-hopping mechanism, and conductivity depends on the size of the polymer particles [2]. Such a polymerization process of PPy is investigated by SAXS and cryo-TEM in this study, in particular, on the pre-polymerization state in solution.

Experimental

One of the methods to analyze the process of structure formation in a solution could be small-angle X-ray scattering (SAXS). For the present SAXS measurement, synchrotron radiation at Photon Factory in Tsukuba was employed in a range of scattering angles of $2\theta = 0.025$–0.15 rad for $\lambda = 0.149$ nm. The complementary method to analyze the process is transmission electron microscopy (TEM), which reveals the process as images. However, the solution could not be introduced into an electron microscope, so the cryo-TEM technique is necessary. In this study, a new 400 kV TEM (JSFX 4000) with a cryostage of liquid He (cryo-TEM) was used, so as to examine subsequent changes in the structure formation of the polymer formed in aqueous solution, where the objective materials

142

T. Ogawa et al.
SAXS and cryo-TEM on polymerization

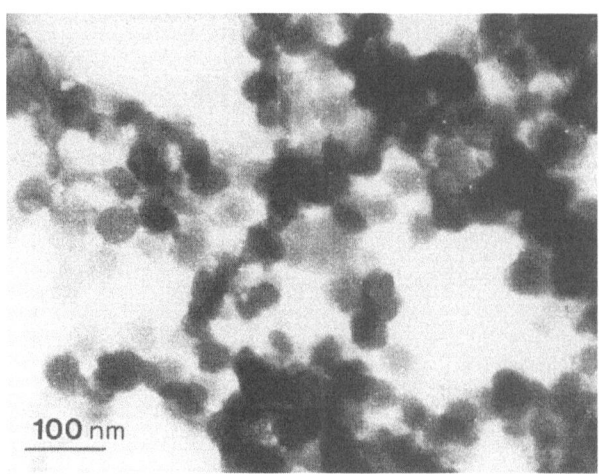

Fig. 1 Conventional TEM image of PPy particles polymerized with SCS. Each PPy has a spherical shape with a constant diameter depending on a concentration of SCS

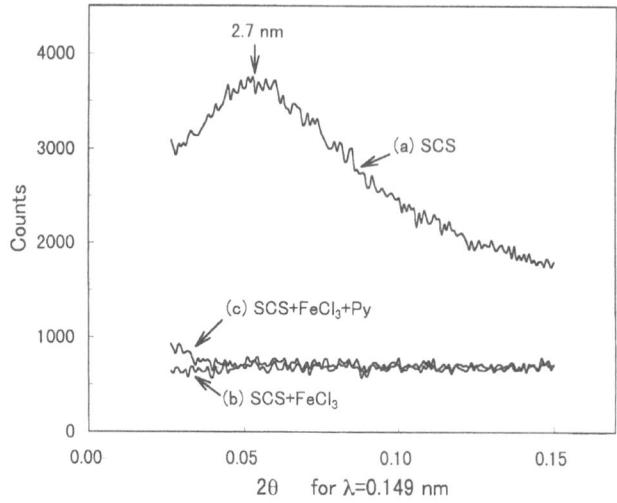

Fig. 2 SAXS curves for (a) SCS + water, (b) SCS + $FeCl_3$ + water and (c) SCS + $FeCl_3$ + pyrrole + water

are embedded in vitreous ice by quenching the solution into liquid propane at $-180\,°C$.

The polymerization in the present study was carried out under the four-components system; pyrrole + $FeCl_3$ + SCS in water, with the concentrations of pyrrole = 0.1 mol/l, $FeCl_3$ = 0.2 mol/l and SCS = 0.57 g/l.

Results and discussion

The polypyrrole synthesized under the present high SCS concentration is regularly shaped spheres with a constant diameter of about 40 nm, which was determined from conventional TEM observation for the dried sample as shown in Fig. 1.

In the SAXS measurement in solution, the following solutions were examined: (1) SCS in water; (2) SCS and $FeCl_3$ in water and; (3) SCS, $FeCl_3$ and pyrrole in water. For the SCS + water system, formation of micelles is concluded, where a structure dimension of 2.7 nm is expected as determined from a peak in the curve (a) in Fig. 2. Of course, it is impossible to draw a conclusion on the shape of object from the SAXS peak. One can only know a corresponding dimension of the object material. By addition of $FeCl_3$ into SCS + water, the peak disappeared as shown by curve (b), which indicates that the structure becomes larger or smaller than the measurable range of SAXS. When pyrrole was introduced successively into the solution, spherical polymer particles are formed in the solution and SAXS exhibits no special features except for a tailing at the smaller angle side (curve (c)), because the particle size is too large to give a peak in the present experimental range of scattering angle. These SAXS data

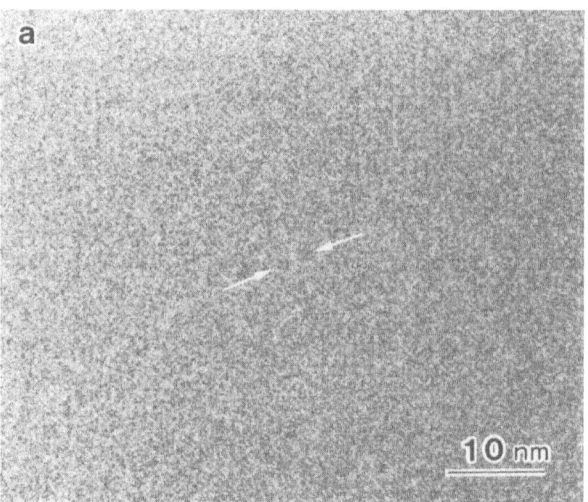

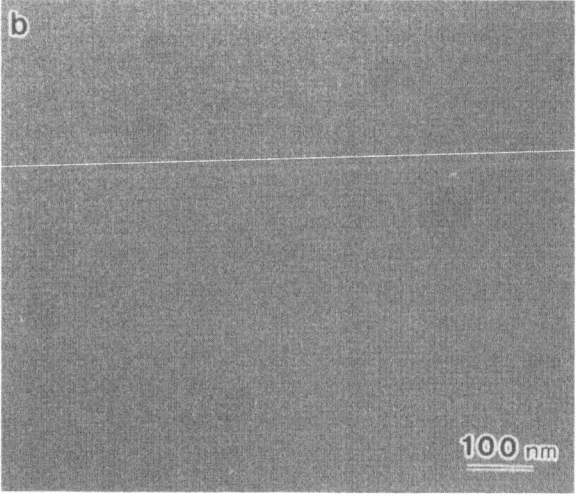

Fig. 3 Images taken with cryo-TEM at 4.2 K for quenched samples of (a) SCS + water and (b) SCS + $FeCl_3$ + water

Progr Colloid Polym Sci (1997) 106:141–143
© Steinkopff Verlag 1997

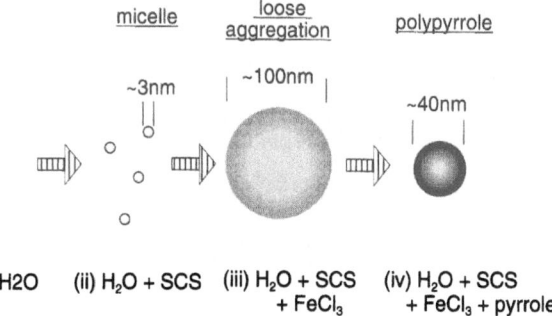

micelle   loose aggregation   polypyrrole

~3nm   ~100nm   ~40nm

(i) H2O   (ii) H₂O + SCS   (iii) H₂O + SCS + FeCl₃   (iv) H₂O + SCS + FeCl₃ + pyrrole

Fig. 4 A structural formation in polymerization of pyrrole with SCS as an additional oxidizing agent

suggest that the polymerization does not happen simply as a continuous growth from the micelles.

Such a structure formation process was examined by cryo-TEM, which can give information on a wider range of object dimension by changing the magnification. The same solutions of (1)–(4) as in the case of SAXS were observed by the cryo-TEM. Figure 3a shows a typical image for the quenched solution of SCS + water, where small faint spherical objects are observed sporadically as indicated by the arrows. They have a diameter of about 3 nm, which corresponds to the peak in SAXS and is considered to be the micelles formed with SCS. Upon addition of $FeCl_3$, however, a much larger spherical object is found in vitreous ice with a diameter of about 100 nm as shown in Fig. 3b, which is too large for the present SAXS measurement. The sphere might be an aggregation of SCS promoted by the addition of $FeCl_3$.

The structure formation process of the PPy in the present chemical reaction is summarized in Fig. 4, where SCS forms micelles in water and changes into larger aggregates by the addition of $FeCl_3$. This aggregation is expected to be a loose texture, and the polymerization starts in the aggregation by addition of pyrrole, which results in smaller tight polymer particles of 40 nm in diameter.

Acknowledgment This work was partly supported by Grant-in-Aid for Scientific Research (B) by The Ministry of Education, Science, Sports and Culture, Japan.

References

1. Ezquerra TA, Mohammadi M, Kremer F, Vilgis T, Wegner G (1988) J Phys C: Solid State Phys 21:927–941

2. Isoda S, Kita Y, Donkai N, Sano Y, Wegner G (1991) Bull Inst Chem Res Kyoto Univ 69:331–341

Progr Colloid Polym Sci (1997) 106:144–146
© Steinkopff Verlag 1997

FORMATION OF STRUCTURES

B. Zhao
H. Li
X. Zhang
R. Zhang
J. Shen
Y. Ozaki

Molecular orientation and structure in ultrathin films of amphiphilic microgel copolymer with the branching of octadecyl groups studied by infrared spectroscopy

Received: 13 October 1996
Accepted: 6 March 1997

B. Zhao · H. Li · X. Zhang · J. Shen
Key Laboratory for Supramolecular
Structure and Spectra
Jilin University
Changchun 130023, P.R. China

R. Zhang
Department of Chemistry
Jilin University
Changchun 130023, P.R. China

Y. Ozaki (✉)
Department of Chemistry
School of Science
Kwansei-Gakuin University
Uegahara, Nishinomiya 662, Japan

Abstract Molecular orientation and structure in five-monolayer Langmuir–Blodgett (LB) films of an amphiphilic polymer consisting of a flexible hydrophilic epichloro-hydrinethylene diamine cross-linking microgels and a number of hydrophobic chains have been studied by infrared (IR) transmission and reflection–absorption (RA) spectroscopy. The downward shifts of the NH and C=O stretching bands show that the amide groups are involved in medium hydrogen bonds. A comparison of the band intensities between the transmission and RA spectra indicates that the hydrocarbon chains are nearly perpendicular to the substrate surface while the C–N bond of the amide group is nearly parallel to the surface. The vibrational frequencies of the two CH_2 stretching bands suggest that the LB films have highly ordered alkyl chains.

Key words IR – ultrathin films – LB films – amphiphilic polymer – microgel

Introduction

Molecular design of polymeric Langmuir–Blodgett (LB) films having both highly ordered structure and structural stability have drawn considerable attention [1–4]. We have been developing a new type of polymeric LB films named "duckweed" and "reversed duckweed" LB films (Fig. 1) [2–4]. The term "duckweed" means that the hydrophobic microgels are able to float on the surface of water; "reversed duckweed" means that the hydrophilic networks extend downward into water and the hydrophobic grafting chains point upward packing away from the surface of water. The amphiphilic polymer employed for the "duckweed" and "reversed duckweed" LB films consists of a flexible hydrophilic epichlorohydrin-ethylenediamine cross-linked network and several hydrophobic stearic acid chains.

In this paper, we report molecular orientation and structure in the "reversed duckweed"-type LB films studied by infrared (IR) transmission and reflection–absorption spectroscopy.

Experimental

The amphiphilic polymer (for simplicity we term it as ES) was synthesized according to the previously reported method [2]. Its number-average molecular weight was 7840. The content of alkyl chains was 75.8%, indicating that one molecule contained 20.8 stearic chains on an average [2].

The LB films were prepared by the previously reported technique [4, 5]. The instruments and sample-handling technique adopted in this work for recording the IR transmission and reflection–absorption spectra were the same as those described previously [5].

Progr Colloid Polym Sci (1997) 106:144–146
© Steinkopff Verlag 1997

(A).

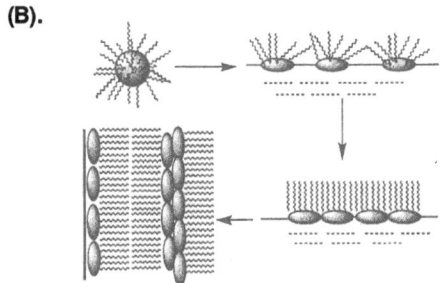

R= -(CH₂)₁₆CH₃

R₁= H, or cross-linked points

n : about 20

(B).

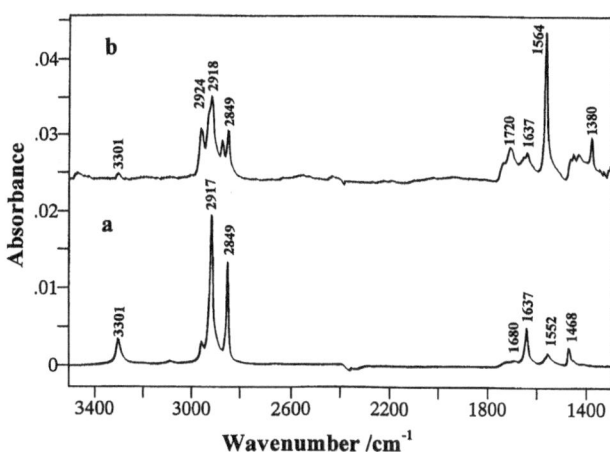

Fig. 1 (A) The structure of the amphiphilic polymer (ES). (B) A schematic model for reversed duckweed LB film

Fig. 2 IR transmission (a) and RA (b) spectra of five-monolayer LB films of ES. The substrates used were CaF_2 plate and gold-evaporated glass slide for IR transmission and RA measurements, respectively

Results and discussion

Figure 2a and b show the IR transmission and reflection–absorption (RA) spectra of five monolayer LB films of ES. The relevant key bands are assigned as: 3301 cm^{-1}, NH stretching; 2917 and 2849 cm^{-1}, CH_2 antisymmetric and symmetric stretching; 1720 and 1680 cm^{-1}, C=O stretching in the –C(=O)–N–C(=O)-type groups; 1638 cm^{-1}, amide I (mono-substituted amide); 1552 cm^{-1}, amide II; 1468 cm^{-1}, CH_2 deformation.

Molecular structure

The vibrational frequencies of the NH stretching and amide II and III modes suggest that the amide groups are in a trans configuration and involved in medium hydrogen bonding [6]. It is well known that the frequencies of the two CH_2 stretching modes are sensitive to the conformation of the hydrocarbon chain; the low frequencies (2918 and 2849 cm^{-1}) of the bands are characteristic of highly ordered alkyl chain [7].

Molecular orientation

Molecular orientation of the hydrocarbon chain and functional groups in a LB film can be investigated by comparison of band intensities in IR transmission and RA spectra [8]. The band intensities of the two CH_2 stretching modes are much stronger in the transmission spectrum than in the RA spectrum. Therefore, it seems that stearic acids in the LB films are nearly perpendicular to the substrate surface. This conclusion is also supported by the results of X-ray analysis [4]. However, special care must be taken here about the intensities of the two CH_2 stretching modes because there are a number of CH_2 groups in the network (approximately 15% of the total CH_2 content). In order to overcome this problem, we are trying to prepare the amphiphilic LB films consisting of deuterated stearic acid.

The NH stretching band appears clearly in the transmission spectrum but is not observed in the RA spectrum. This observation leads us to conclude that the N–H bond is nearly parallel to the substrate surface. The intensity of amide II, which is mainly due to C–N stretching mode [6], is much stronger in the RA spectrum than in the transmission spectrum, indicating that the C–N bond in the amide group is nearly perpendicular to the surface.

In conclusion, the present study has shown that the "reversed duckweed"-type LB film has a highly ordered and well-oriented structure. This conclusion is in good agreement with that obtained by previous studies based upon low-angle X-ray diffraction and polarized IR/attenuated total reflection experiments. Further investigations on the structural characteristics of this amphiphilic polymeric LB film including thermal behavior are now underway and will be reported separately.

146

B. Zhao et al.
IR study of ultrathin films of amphiphilic microgel copolymer

References

1. Ringsdorf H, Laschewsky A, Schmidt G, Schneider J (1987) J Am Chem Soc 109:109–118
2. Yin R, Cha X, Zhang X, Shen JC (1990) Macromolecules 23:5185–5189
3. Li H, Zhang X, Zhang R, Shen J, Zhao B, Xu W (1995) Macromolecules 28: 8178–8181
4. Zhang R, Zhang X, Li H, Zhao B, Shen J (1996) Polym Bull 36:227–232
5. Wang Y, Nichogi K, Terashita S, Iriyama K, Ozaki Y (1996) J Phys Chem 100:368–373
6. Colthup NB, Daly LH, Wiberley SE (eds) (1990) Introduction to Infrared and Raman Spectroscopy, 3rd ed. Academic Press, San Diego, pp 289–325
7. Sapper H, Cameron DG, Mantsch HH (1981) Can J Chem 59:2543–2549
8. Takenaka T, Umemura J (1991) In: Durig JR (ed) Vibrational Spectra and Structure, Vol 19. Elsevier, Amsterdam, p 215

Progr Colloid Polym Sci (1997) 106:147–149
© Steinkopff Verlag 1997

FORMATION OF STRUCTURES

T. Dotera
A. Hatano

Formation of multicontinuous structures in block copolymer melts

Received: 13 October 1996
Accepted: 18 March 1997

Prof. Dr. T. Dotera (✉)
Saitama Study Center
University of the Air
682-2 Nishiki-cho
Omiya 331, Japan

Prof. Dr. A. Hatano
Department of Pure and Applied Sciences
University of Tokyo
Komaba,Meguro-ku
Tokyo 153, Japan

Abstract Using a new lattice polymer simulation method for block copolymers, we have studied a dense polymer system composed of monodisperse linear A–B–C triblock copolymers. We confirm that a tricontinuous structure for linear A–B–C block copolymers observed by Mogi et al. (Macromolecule 25, 5408 (1992) and ibid. 25, 5412 (1992)) could be, indeed, an ordered tricontinuous "double-diamond" (OTDD) structure. In addition, two (symmetric and shifted) OTDD phases and a lamellar phase have been observed in the same simulations. The two OTDD phases are conceivable by shifting one diamond network against the other diamond network.

Key words Block copolymer – microphase separation – double-diamond – computer simulation

Introduction

Much attention has been paid to novel multicontinuous periodic structures of block copolymer systems characterized by mutually penetrated networks: "double-diamond" and "gyroid" morphologies [1–5]. Although these structures are periodic, the structure determination encounters some obstacles: (1) Many complex structures resemble in transmission electron microscopy projection images. For example, double-diamond and gyroid structures give similar wagon-wheel patterns. For this reason, several systems, initially attributed to the ordered bicontinuous "double-diamond" structures (OBDD) [1, 2], are now considered to be different, probably "gyroid" structures [3, 5]. However, there remain some doubts in the final identification for the reasons that follow: (2) Small-angle X-ray and neutron scattering data do not have definitive answers for space groups, because the number of detectable diffraction peaks are few. (3) A possibility of unknown (minimal) surfaces cannot be ruled out [6].

In order to overcome these problems simulation methods are useful; direct structural informations including mesostructures, detailed shape of surfaces, interface roughening, chain conformation and transformations can be obtained: some of them are not obtained by experimental methods mentioned above.

In this paper, we present some simulation results for linear A–B–C triblock copolymers (A–A–A–B–B–B–C–C–C), whose melts are known to produce a tetragonal (not hexagonal) cylindrical structure and an ordered tricontinuous structure [7]. I the case of linear A–B–C triblock copolymers, the multicontinuous structure possesses a *large* portion in both experimental and theoretical [8] phase diagrams, which is different from other systems.

Simulation method

Recently, we have developed a lattice polymer Monte Carlo simulation method named *diagonal bond method* [9]. To enhance the mobility of polymers in melts, we have employed *diagonals* of squares and those of cubes as bonds in addition to edges of squares: bond lengths can be 1, $\sqrt{2}$, or $\sqrt{3}$. The key of the method is that we allow bond crossings while keeping the excluded volume conditions of monomers. The detailed algorithm and the validity and efficiency of the method are described in ref. [9].

We choose a system composed of linear triblock copolymers with $N_A = N_C = 7$, and $N_B = 16$, where N_A is the number of A monomers. The system comprises 1600 polymers and extends over 40 lattice spacings in three directions under periodic boundary conditions. Unit interaction energies are imposed on pairs of different types of monomers and the interaction range is $\sqrt{3}$.

Results

We have prepared a multicontinuous structure through a high temperature quench ($1/k_B T < 0.2$) from a randomized configuration and then annealed at a very low temperature ($1/k_B T \geqslant 0.2$). Figure 1 displays the result; diamond structure composed of A monomers (B–C block copolymers are transparent), while C monomers also form another diamond structure in the void of the figure. Evidently, it is a fourfold coordinated lattice and is not the gyroid morphology, which forms threefold coordinated

Fig. 1 Low-temperature simulation result for linear A–B–C triblock copolymers. Only one diamond structure composed of A monomers is shown. C monomers also form another diamond structure in the center of the void of the figure

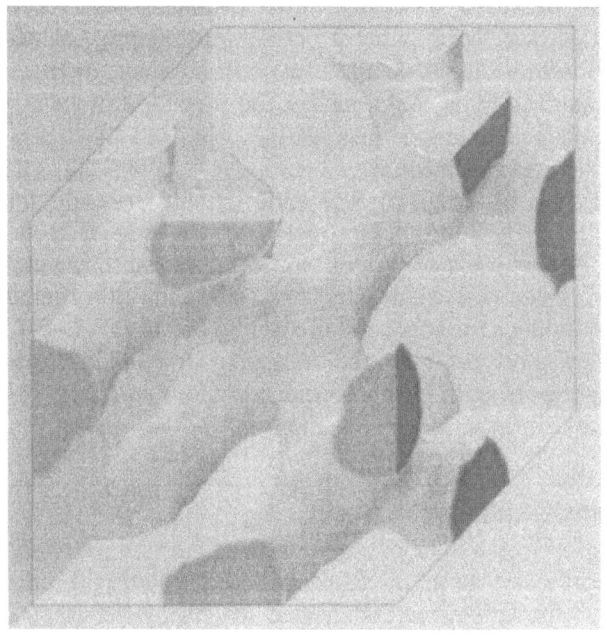

lattices. The figure represents a unit cell of the ordered double-diamond structure.

Suppose two diamond lattices composed of carbon-type atoms and bonds. Set the bond directions of the two lattices to be the same. Fix one of the lattices in space. One can imagine the displacement of the other lattice; there remain degrees of freedom of relative positions of the two lattices. Corresponding to this geometric consideration, we found a phase transition between the low-temperature phase (DD I) and the high-temperature phase (DD II). DD I is a symmetric double-diamond phase: One-diamond lattice occupies the centers of voids of the other lattice. In other words, two lattices are placed at the most symmetric positions. DD II is a shifted double-diamond phase: topologically equivalent but shifted lattices from the symmetric positions; say, one shifts in the x-direction from the other.

This symmetry-lowering transformation upon heating was confirmed by real-space images and by Fourier analysis of the system. We calculated peak intensities $I(j, k, l)$ of the Fourier transform of the A monomers defined by

$$I(j, k, l) = \frac{1}{N_A^2} \left| \sum \exp \frac{2\pi \bar{i}}{L} (jx + ky + lz) \right|^2, \quad (1)$$

where N_A is the number of A monomers and L the size of the system, and the sum is taken over all A monomers. Fig. 2 shows peak intensities $I(j, k, l)$ for two kinds of indices at two temperatures, $1/k_B T = 0.1$ and $1/k_B T = 0.3$.

Fig. 2 Log scale peak intensities $\ln I(j, k, l)$ of the Fourier transform of A monomers at two typical temperatures, corresponding to DD I and DD II phases: an evidence of a preferred direction at a high temperature (DD II). At $1/k_B T = 0.1$, $I(2, 0, 0)$ is stronger than $I(0, 2, 0)$ and $I(0, 0, 2)$ while at $1/k_B T = 0.3$, they are almost equally weak. Similarly, at $1/k_B T = 0.1$ $I(0, 2, 2)$ is stronger than $(2, 0, 2)$ and $(2, 2, 0)$, while at $1/k_B T = 0.3$, they are equally strong

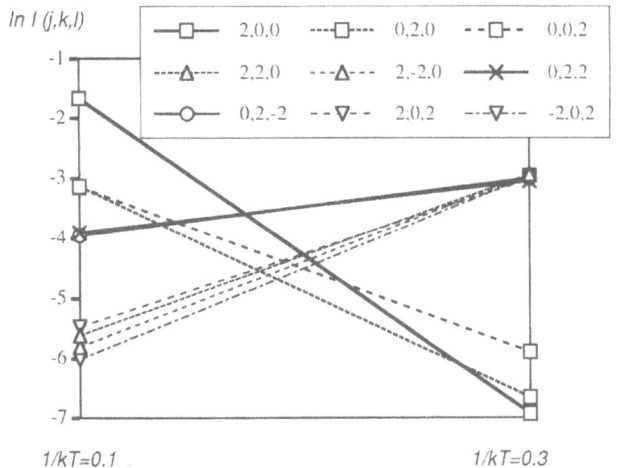

Progr Colloid Polym Sci (1997) 106:147–149
© Steinkopff Verlag 1997

At low temperatures $(0.3 \leq 1/k_B T)$, we have observed $I(2, 2, 0)$ and its permutations are equally strong, while $I(2, 0, 0)$ and its permutations are equally weak; the structure is thus symmetric for (x, y, z) permutations. At high temperatures $(0.06 \leq 1/k_B T \leq 0.14)$, $I(2, 2, 0)$ type is weaker than $I(2, 0, 0)$ type. More interestingly, $I(2, 2, 0)$ is stronger than $I(0, 2, 0)$ and $I(0, 0, 2)$, and also $I(0, 2, 2)$ and $I(0, 2, -2)$ are stronger than their permutations; therefore there is a preferred texture (direction).

We also found a lamellar phase $(ABCB|ABCB| \cdots)$ in the same system by a high-temperature quench of a randomized configuration. The energy of the lamellar phase was higher than that of the diamond phase for low temperatures $(0.19 \leq 1/k_B T \leq 0.25)$, but lower for high temperatures $(0.15 \leq 1/k_B T \leq 0.18)$. The difference of the energies appeared to be very small. We have not found a phase transformation between the double diamond and lamellar phases, because all phases are stable in the simulation time scale. In addition, since systems with different sizes may give different energies, it is not clear whether the lamellar phase is stable or unstable against the diamond phases.

Discussion

The fixed size simulations limit our conclusion. We cannot deny the possibility that the transformation between DD I and DD II is an artifact, because the deformations (shear, or contraction) of the simulation cube cannot be taken into account: e.g. upon heating a copolymer tends to shrink and thus the unit cell of the diamond structure should contract, however, our simulation cannot reproduce this effect. In spite of this limitation, it is still interesting to investigate whether the double-diamond morphology in the experimental systems is shifted or symmetric: this is no trivial problem.

In conclusion, we have demonstrated that linear A–B–C block copolymers can form the double-diamond morphology, but not gyroid. It is evident that the simulation method is useful for investigating morphologies of block copolymer systems. Future studies should include the phase diagram of two-component block copolymers.

Acknowledgements This work was supported by the research grants of the University of the Air.

References

1. Thomas E, Alward D, Kinning D, Martin D, Handlin D, Fetters L (1986) Macromolecule 19:2197
2. Hasegawa H, Tanaka H, Yamasaki K, Hashimoto T (1987) Macromolecule 20:1651
3. Hajduk D, Harper P, Gruner S, Honeker C, Kim G, Thomas E, Fetters L (1994) Macromolecule 27:4063 and Hajduk D, Harper P, Gruner S, Honeker C, Thomas E, Fetters L (1995) Macromolecule 28:2570
4. Olmsted P, Milner S (1994) Phys Rev Lett 72:936 and Milner S, Olmsted P (1997) J Phys II France 7:249
5. Matsen M, Schick M (1994) Phys Rev Lett 72:2660 and Matsen M, Bates F (1996) Macromolecule 29:7641
6. Góźdź W, Hołyst R (1996) Phys Rev Lett 76:2726 and Góźdź W, Hołyst R (1996) Phys Rev E 54:5012
7. Mogi Y, Kotsuji H, Kaneko Y, Mori K, Matsushita Y, Noda I (1992) Macromolecule 25:5408 and Mogi Y, Mori K, Matsushita Y, Noda I (1992) ibid. 25:5412
8. Nakazawa H, Ohta T (1993) Macromolecule 26:5503
9. Dotera T, Hatano A (1996) J Chem Phys 105:8413

Progr Colloid Polym Sci (1997) 106:150–157
© Steinkopff Verlag 1997

A. Onuki
R. Yamamoto
T. Taniguchi

Viscoelastic effects and shear-induced phase separation in polymer solutions

Received: 13 October 1996
Accepted: 4 February 1997

Dr. A. Onuki (✉) · R. Yamamoto
T. Taniguchi
Department of Physics
Kyoto University
Kyoto 606, Japan

Abstract We numerically investigate nonlinear regimes of shear-induced phase separation in entangled polymer solutions. For the purpose a time-dependent Ginzburg–Landau model describing the two-fluid dynamics of polymer and solvent is used. A conformation tensor is introduced as a new dynamic variable to represent chain deformations. Its variations give rise to a large viscoelastic stress. Above the coexistence curve, a dynamical steady state is attained, where fluctuations are enhanced on various spatial scales. At relatively large shear, the elongated polymer-rich regions form a transient network supporting most of the stress. Because such a network is continuously deformed in shear flow, the shear stress and the normal stress difference exhibit large fluctuations. This result explains an early observation of large temporal fluctuations of the normal stress by Lodge.

Key words Polymer solutions – shear flow – phase separation – viscoelasticity

Introduction

In two component viscoelastic fluids the network stress can act on the two components asymmetrically. Then there arises dynamical coupling between stress and diffusion leading to a number of intriguing viscoelastic effects. Most spectacular is shear-induced phase separation in semidilute polymer solutions, in which light scattering can be drastically enhanced even above the coexistence temperature T_{cx} [1–6]. In this system the composition fluctuations give rise to inhomogeneities of the network structure and the applied shear produces stress imbalance resulting in diffusion in the direction of phase separation. Theoretically, this effect was first examined to linear order in the composition fluctuation $\delta\phi$ on the assumption that the polymer stress instantaneously follows $\delta\phi$ [7]. Formal time-dependent Ginzburg–Landau theories have also been developed [8–12], in which a tensor variable **W**, called the conformation tensor, is introduced to represent chain deformations and account for viscoelasticity. In this scheme we set up dynamic equations, in which ϕ and **W** are coupled and the time scale of ϕ needs not to be longer than that of **W**. Note that ϕ changes more rapidly than **W** even at relatively small wave numbers for highly entangled systems. As their first applications, if the dynamic equations are linearized around homogeneous states, they can describe nonexponential decay of the composition fluctuations in dynamic light scattering [13], viscoelastic effects in early stage spinodal decomposition [14], and the fluctuation enhancement in shear in the linear regime [10–13]. However, it has been very difficult to solve the dynamic equations in nonlinear regimes and there is no satisfactory understanding of the physical mechanisms in phase separation with or without shear. The aim of this paper is to investigate the nonlinear regime of shear-induced phase separation by numerically solving the dynamic equations. Viscoelastic effects in spinodal decomposition of deeply quenched polymer solutions are investigated in another paper [15].

Progr Colloid Polym Sci (1997) 106:150–157
© Steinkopff Verlag 1997

Time-dependent Ginzburg–Landau model

We briefly set up a dynamic model of an entangled polymer solution in the semidilute regime, $\phi > \phi_c = N^{-1/2}$ and $\phi \ll 1$, ϕ_c being the critical volume fraction and N being the polymerization index. In terms of the polymer volume fraction ϕ and the conformation tensor $\mathbf{W} = \{W_{ij}\}$, the free energy is given by [10]

$$F\{\phi, \mathbf{W}\} = \int d\mathbf{r} \left[f(\phi) + \frac{1}{2} C(\phi) |\nabla \phi|^2 \right.$$
$$\left. + \frac{1}{4} G(\phi) \sum_{ij} (W_{ij} - \delta_{ij})^2 \right]. \tag{1}$$

Here

$$f(\phi) \cong (k_B T/v_0) \left[\frac{1}{N} \phi \ln \phi + \left(\frac{1}{2} - \chi \right) \phi^2 + \frac{1}{6} \phi^3 \right] \tag{2}$$

is the Flory–Huggins free-energy density [16], where v_0 is the volume of a monomer and χ is the so-called interaction parameter dependent on the temperature. In the second term $C(\phi) \propto 1/\phi$ from the scaling theory. The last term of (1) is the elastic energy of the (transient) network, $G(\phi)$ being the shear modulus. For simplicity, we assume that the deviation of $\mathbf{W}$ from the equilibrium value $\mathbf{I}$ ($=$ the unit tensor) is small, so the elastic free energy is bilinear in $\mathbf{W} - \mathbf{I}$. Because $\mathbf{W}$ represents the network deformation, its motion is determined by the polymer velocity $\mathbf{v}_p$ and its simplest dynamic equation is of the form [10],

$$\frac{\partial}{\partial t} W_{ij} + (\mathbf{v}_p \cdot \nabla) W_{ij} - \sum_k (D_{ik} W_{kj} + W_{ik} D_{jk})$$
$$= - \frac{1}{\tau(\phi)} (W_{ij} - \delta_{ij}) + f_{ij}, \tag{3}$$

where D_{ij} is the gradient tensor of the polymer velocity $D_{ij} = \partial v_{pi}/\partial x_j$. The left-hand side of (3) is called the upper convective time derivative in the rheological literature [17] and $\tau(\phi)$ is the stress relaxation time, very long in the semidilute region [18]. The last term $f_{ij}(\mathbf{r}, t)$ is the random noise characterized by

$$\langle f_{ij}(\mathbf{r}, t) f_{ij}(\mathbf{r}', t') \rangle$$
$$= 2(k_B T/\eta_p)(\delta_{ik} \delta_{jl} + \delta_{il} \delta_{jk}) \delta(\mathbf{r} - \mathbf{r}') \delta(t - t'). \tag{4}$$

From (1)–(3) we may calculate the free-energy changes against infinitesimal motion of the network to obtain the network stress in the form [8–12],

$$\sigma_p = 2\mathbf{W} \cdot (\delta F/\delta \mathbf{W}) = G(\phi) \mathbf{W} \cdot (\mathbf{W} - \mathbf{I}). \tag{5}$$

In particular, in weak, homogeneous, and stationary flow, (3) is solved to give $W_{ij} - \delta_{ij} \cong \tau(D_{ij} + D_{ji})$. In shear flow we require $|W_{xy}| < 1$ for the validity of the bilinear

form of the elastic free energy in (1), which is also the condition of the Newtonian regime $\dot{\gamma}\tau < 1$ in shear flow, where $\dot{\gamma}$ is the shear rate. The solution viscosity in the Newtonian regime is

$$\eta_p = G(\phi) \tau(\phi), \tag{6}$$

which is supposed to be much larger than the solvent viscosity η_0. For rapid motions, on the other hand, our system behaves as a gel and $W_{ij} - \delta_{ij} \cong \partial u_{pi}/\partial x_j + \partial u_{pj}/\partial x_i$, where $\mathbf{u}_p$ is the time integral of $\mathbf{v}_p$ and has the meaning of the displacement of the network.

Note that the solvent velocity $\mathbf{v}_s$ and the polymer velocity $\mathbf{v}_p$ are different when the diffusion is taking place. The volume fraction is convected by $\mathbf{v}_p$ as

$$\frac{\partial}{\partial t} \phi = - \nabla \cdot (\phi \mathbf{v}_p). \tag{7}$$

On the other hand, the average velocity $\mathbf{v} = \phi \mathbf{v}_p + (1 - \phi) \mathbf{v}_s$ obeys the usual hydrodynamic equation,

$$\bar{\rho} \frac{\partial}{\partial t} \mathbf{v} = - \nabla p_1 - \nabla \cdot \mathbf{\Pi} + \eta_0 \nabla^2 \mathbf{v} + \nabla \cdot \boldsymbol{\sigma}_R, \tag{8}$$

where $\bar{\rho}$ is the average mass density and the stress tensor

$$\mathbf{\Pi} = C(\phi)(\nabla \phi)(\nabla \phi) - \boldsymbol{\sigma}_p \tag{9}$$

arises from the gradient term in (1) and the network stress. For simplicity, we are assuming that the mass densities of the pure polymer and solvent are the same and the fluid is incompressible. Then the polymer mass composition and the polymer volume fraction coincide and p_1 in (8) is determined from the condition

$$\nabla \cdot \mathbf{v} = 0. \tag{10}$$

The $\boldsymbol{\sigma}_R$ is the random stress tensor determined by

$$\langle \sigma_{Rij}(\mathbf{r}, t) \sigma_{Rkl}(\mathbf{r}', t') \rangle = 2k_B T \eta_0 (\delta_{ik} \delta_{jl}$$
$$+ \delta_{il} \delta_{jk}) \delta(\mathbf{r} - \mathbf{r}') \delta(t - t'). \tag{11}$$

For slow motions we may use the adiabatic approximation $\partial \mathbf{v}/\partial t = 0$ to obtain

$$\eta_0 \nabla^2 \mathbf{v} = (\nabla \cdot \mathbf{\Pi} - \nabla \cdot \boldsymbol{\sigma}_R)_\perp, \tag{12}$$

where $(\cdots)_\perp$ denotes taking the transverse part.

Furthermore, assuming that the network stress acts on the polymer and not directly on the solvent, we may derive from a two-fluid model.[8–12] the equation of the relative velocity $\mathbf{w} = \mathbf{v}_p - \mathbf{v}_s$,

$$\bar{\rho} \phi \frac{\partial}{\partial t} \mathbf{w} = - \zeta(\phi) \mathbf{w} - \phi \nabla \frac{\delta F}{\delta \phi} + \nabla \cdot \boldsymbol{\sigma}_p$$
$$+ \frac{1}{4} G(\phi) \nabla \sum_{ij} (W_{ij} - \delta_{ij})^2 + \boldsymbol{\theta}, \tag{13}$$

where $\zeta(\phi)$ is the friction coefficient of order $6\pi\eta_0 b^{-2}\phi^2$, b being the monomer size. The last term $\boldsymbol{\theta}(\mathbf{r}, t)$ is the random force characterized by

$$\langle \theta_i(\mathbf{r}, t)\theta_j(\mathbf{r}', t') \rangle = 2k_B T \zeta \delta_{ij}\nabla^2\delta(\mathbf{r} - \mathbf{r}')\delta(t - t') . \tag{14}$$

For slow motions we may set $\partial\mathbf{w}/\partial t = 0$ in (13); then, $\mathbf{w}$ is expressed in terms of ϕ, $\mathbf{W}$, and the random force $\boldsymbol{\theta}$ as

$$\mathbf{w} = \zeta^{-1}\left[-\phi\nabla\frac{\delta F}{\delta\phi} + \nabla\cdot\boldsymbol{\sigma}_p + \frac{1}{4}G(\phi)\nabla\sum_{ij}(W_{ij} - \delta_{ij})^2 + \boldsymbol{\theta} \right] . \tag{15}$$

In the absence of the last two terms, we obtain $\mathbf{w} = [-\phi\nabla(\delta F/\delta\phi) + \nabla\cdot\boldsymbol{\sigma}_p]/\zeta$ which has appeared in the literature [7, 10–14]. Our result shows that stress imbalance $(\nabla\cdot\boldsymbol{\sigma}_p \neq 0)$ gives rise to diffusion, which is called the stress–diffusion coupling. Finally, we remark that our dynamic model ensures the nonnegative definiteness of the heat production rate [10]. Then we can generally prove that the system tends to a homogeneous equilibrium state with $\mathbf{W} - \mathbf{I} = \mathbf{v} = \mathbf{w} = 0$ as $t \to \infty$ in the absence of macroscopic flow field.

Numerical analysis

We numerically integrate (3) and (7) supplemented with $\mathbf{v}_p = \mathbf{v} + (1 - \phi)\mathbf{w}$, and (12) and (15) in two space dimensions on a 128×128 square lattice by applying a shear flow $\langle v_x \rangle = \dot\gamma y$ and $\langle v_y \rangle = 0$, at $t = 0$. In this paper we report results for

$$\langle\phi\rangle/\phi_c = 2, \quad T = T_c \quad \text{or} \quad \chi - 1/2 = N^{-1/2} , \tag{16}$$

where $\langle \cdots \rangle$ is the spatial average. Note that the polymer solution is above the coexistence curve. The coefficient of the gradient free energy in (1) is written as

$$C(\phi) = (k_B T/v_0)C_0/\phi , \tag{17}$$

where $C_0 = b^2/18$ in the random phase approximation [16]. In the equilibrium state of (16), we introduce the thermal correlation length $\xi = (NC_0/5)^{1/2}(\sim$ the gyration radius) and the cooperative diffusion constant

$$D_{co} = \zeta^{-1}\phi(\partial^2 f/\partial\phi^2) = D_1\left[\phi + (1 - 2\chi) + \frac{1}{N\phi}\right], \tag{18}$$

where D_1 is a microscopic diffusion constant of order $k_B T/6\pi\eta_0 b$. We measure space and time in units of $l = (5/3)^{1/2}\xi$ and $\tau_0 = 2.5l^2/D_{co}$. We also set

$$G(\phi) = (k_B T/v_0)\phi^3 , \quad \tau(\phi) = 0.3\tau_0[(\phi/\phi_c)^4 + 1] . \tag{19}$$

The solvent viscosity is taken to be $\eta_0 = (k_B T/v_0)\phi_c^3\tau_0$, which is equivalent to assume $\zeta(\phi) = \eta_0\phi^2/C_0$. Then the Newtonian solution viscosity and the relaxation time are written as $\eta_p(\phi)/\eta_0 = \Phi^3\tau(\phi)/3\tau_0$ and $\tau(\phi) = 0.3(\Phi^4 + 1)$

in terms of

$$\Phi = \phi/\phi_c , \tag{20}$$

which yield $\eta_p/\eta_0 = 13.6$ and $\tau/\tau_0 = 5.1$ in the initial state (16). In our case, the shear modulus $G(\phi)$ is considerably larger and the relaxation time $\tau(\phi)$ is much smaller than in the experiments [4–6] for computational convenience. Furthermore, our functional forms of $G(\phi)$ and $\tau(\phi)$ are not well consistent with the experiment in theta solutions [18] and scaling theories of polymer solutions in theta solvent [19, 20], but we believe that the essential feature is insensitive to the detailed forms of $G(\phi)$ and $\tau(\phi)$.

While there are only small thermal fluctuations in equilibrium, shear can enlarge them on small spatial scales ($\gtrsim$ the mesh size l in the following simulations) in an early stage ($t < \dot\gamma\tau_0$). In later times the fluctuations on various spatial scales appear and the system tends to a strongly fluctuating, dynamical steady state. Hence, random-source terms in the dynamic equations are indispensable at the onset. In our simulations we have therefore added the Gaussian random source terms, $f_{ij}(\mathbf{r}, t)$ in (3), $\nabla\cdot\boldsymbol{\sigma}_R(\mathbf{r}, t)$ in (8) or (12), and $\boldsymbol{\theta}(\mathbf{r}, t)$ in (13) or (15), which are related to $1/\eta$, η_0, and ζ, respectively, to satisfy the fluctuation–dissipation relations. In our model, however, even after the dynamic equations are made dimensionless, these random source terms are still proportional to a common parameter $\varepsilon = [v_0 N^{3/2}/l^d]^{1/2}$, which has not yet been specified, d being spatial dimensionality. The equilibrium distribution in the absence of shear is given by $\exp(-\mathcal{F}/\varepsilon^2)$ where $\mathcal{F}$ is the dimensionless free energy, so ε represents the strength of the thermal fluctuations and our model is self-consistent for arbitrary ε. In this work we set $\varepsilon = 0.1$. The variance

$$\sigma = [\langle(\Phi - \langle\Phi\rangle)^2\rangle]^{1/2} \tag{21}$$

taken over all the lattice points is then equal to 0.038 in thermal equilibrium ($\dot\gamma = 0$).

Furthermore we are interested only in slow disturbances and set $\partial\mathbf{v}/\partial t = \partial\mathbf{w}/\partial t = 0$ in (8) and (13) or use (12) and (15). Then, $\mathbf{v}$ can be expressed in terms of $\boldsymbol{\sigma}_p$ and the random stress tensor. Here we use a new computer method [21], which enables the FFT scheme even in shear flow.

In this paper we mainly show results for

$$\dot\gamma\tau_0 = 0.05 \quad \text{or} \quad \dot\gamma\tau = 0.25 \tag{22}$$

in terms of τ in the initial state (16). Figure 1 displays snapshots of $\Phi(x, y, t) = \phi(x, y, t)/\phi_c$ at $t = 20, 30, 40, 60, 80, 100, 120, 140, 160, 180, 200$, where the darkness represents $(\Phi(x, y, t) - \Phi_{min})/(\Phi_{max} - \Phi_{min})$, $\Phi_{max} \cong 3.6$ and $\Phi_{min} \cong 0.38$ being the maximum and the minimum of $\Phi(x, y, t)$ at these times. In Fig. 2 we show 3D pictures of $\Phi(x, y, t)$ at $t = 20, 30, 40, 60$. We have confirmed $|W_{xy}| \lesssim 0.5$, which is needed to justify the use the bilinear form of the elastic free energy in (1). At short times $t \lesssim 50$

Progr Colloid Polym Sci (1997) 106:150–157
© Steinkopff Verlag 1997

Fig. 1 Time evolution of
$\Phi(x, y, t) = \phi(x, y, t)/\phi_c$ after
application of shear $\dot{\gamma}\tau_0 = 0.05$
at $t = 0$. The numbers below
the figures are the times
measured in units of $\tau_0 =$
$2.5l^2/D_{co}$. The space region in
our simulations is given by
$0 < x, y < 128$, where the space
coordinates are measured in
units of $l = (5/3)^{1/2}\xi$, ξ being
the correlation length in
equilibrium. The x-axis is along
the flow direction and the y-axis
is along the shear direction. At
$t = 30$ and 40 we can see the
fluctuation enhancement in the
abnormal direction $q_x \cong q_y$.

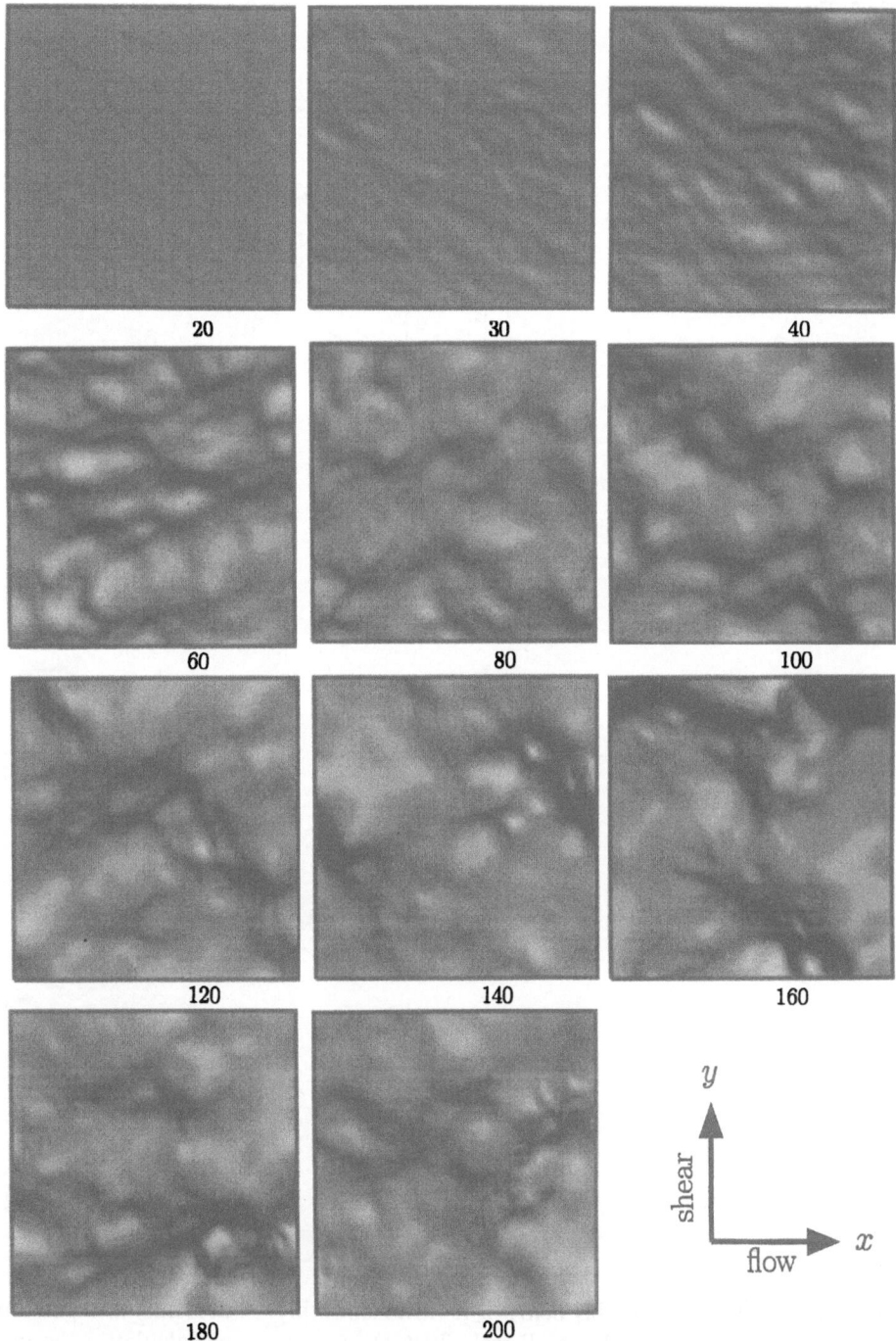

we can see gradual enhancement of the fluctuations elongated in abnormal directions, which are opposite to those in sheared near-critical fluids without elasticity [22, 23]. Note that the previous linear calculations [7, 10–13] have shown that the fluctuations are enhanced in such abnormal directions in steady states. In another paper [24] we will show that for smaller shear, say $\dot{\gamma}\tau_0 = 0.02$, we have patterns similar to that at $t = 40$ in Fig. 1 in steady states.

For our present shear (22), the system tends to a highly nonlinear, dynamical steady state for $t \gtrsim 60$, in which the polymer-rich regions are elongated into long stripes forming a transient network and are continuously deformed by hydrodynamic convection on the time scale of $1/\dot{\gamma}\tau_0 (= 20)$. In addition, looking at Fig. 2, we also notice that in regions where $\Phi(x, y, t) \gtrsim 2$, $\Phi(x, y, t)$ varies irregularly even on the mesh size scale l, whereas in spatial regions in which

Fig. 2 Three-dimensional
pictures of $\Phi(x, y, t)$ for
$\dot{\gamma}\tau_0 = 0.05$, which show initial
fluctuation enhancement in (a),
elongation of the fluctuations in
abnormal directions in (b) and
(c), and fully enhanced
fluctuations in (d)

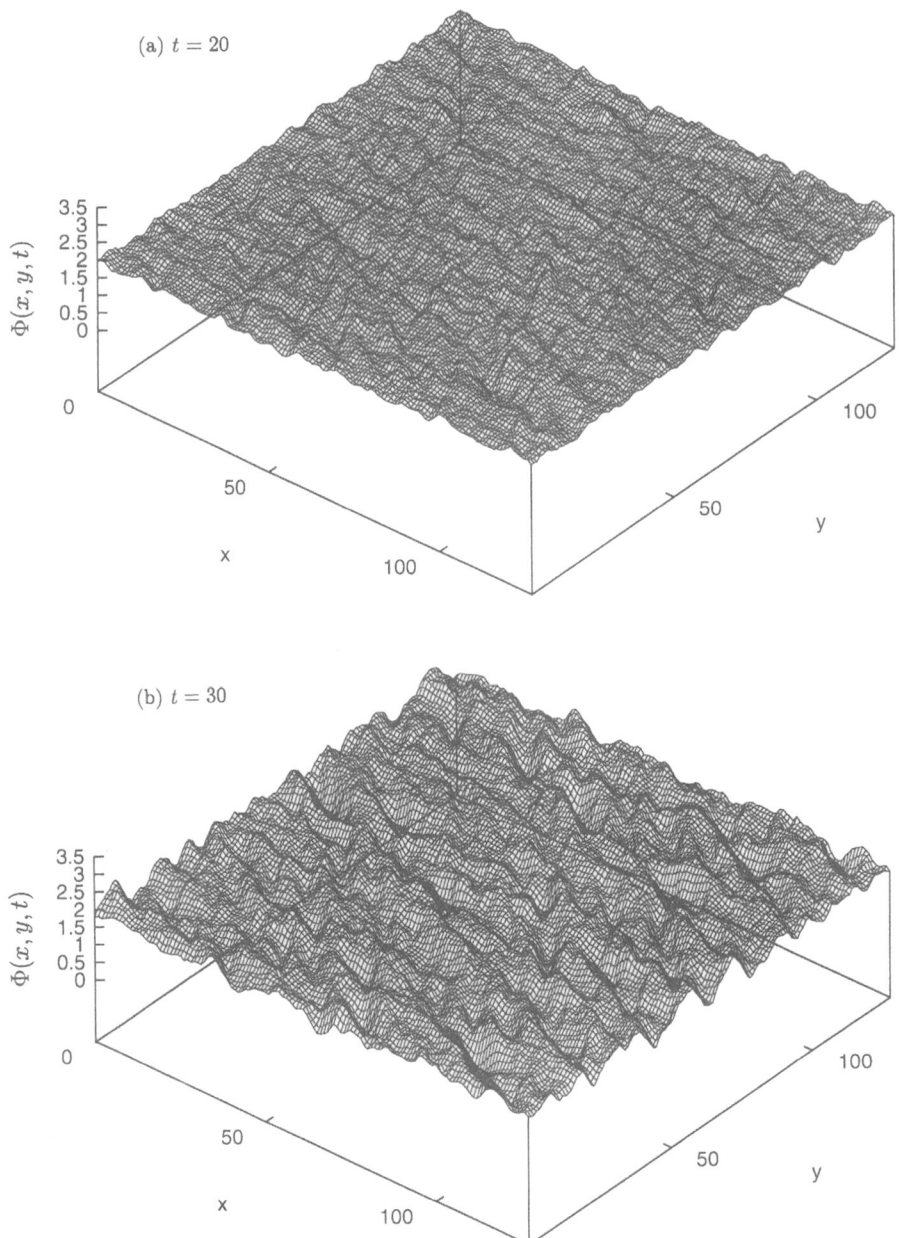

(a) $t = 20$

(b) $t = 30$

$\Phi(x, y, t)$ deviates considerably below 2, it varies smoothly
in space. This is because the small-scale fluctuations created by the random source terms do not grow in solvent-
rich regions.

We also calculate the structure factor,

$$S(q_x, q_y, t) = \langle |\phi_{\mathbf{q}}|^2 \rangle$$

$$= \frac{1}{L} \sum_{\mathbf{r}, \mathbf{r}'} \phi(\mathbf{r}, t) \phi(\mathbf{r}', t) \exp[i\mathbf{q} \cdot (\mathbf{r} - \mathbf{r}')] , \qquad (23)$$

where $L = 128$ is the system length and the summation is
over all the lattice points. It is much enhanced at small q,

but is fluctuating in time because of the small system size in
our calculation, so in Fig. 3 we show the time average of
the structure factor taken over the time interval
$150 < t < 1000$ for $\dot{\gamma}\tau_0 = 0.02$ in (a) and for $\dot{\gamma}\tau_0 = 0.05$ in
(b). We confirm that the structure factor has two peaks in
the $q_x - q_y$ plane in steady states in accord with the scattering experiment [4]. At smaller shear they are located at
$q_x \cong q_y$, while they approach to the q_x-axis as shear is
increased.

In Fig. 4 we show the variance σ defined by (21) in the
region $0 < t < 160$. It increases from the equilibrium value
0.038 and fluctuates around 0.5 in the steady state. There,

Progr Colloid Polym Sci (1997) 106:150–157
© Steinkopff Verlag 1997

Fig. 2 Continued

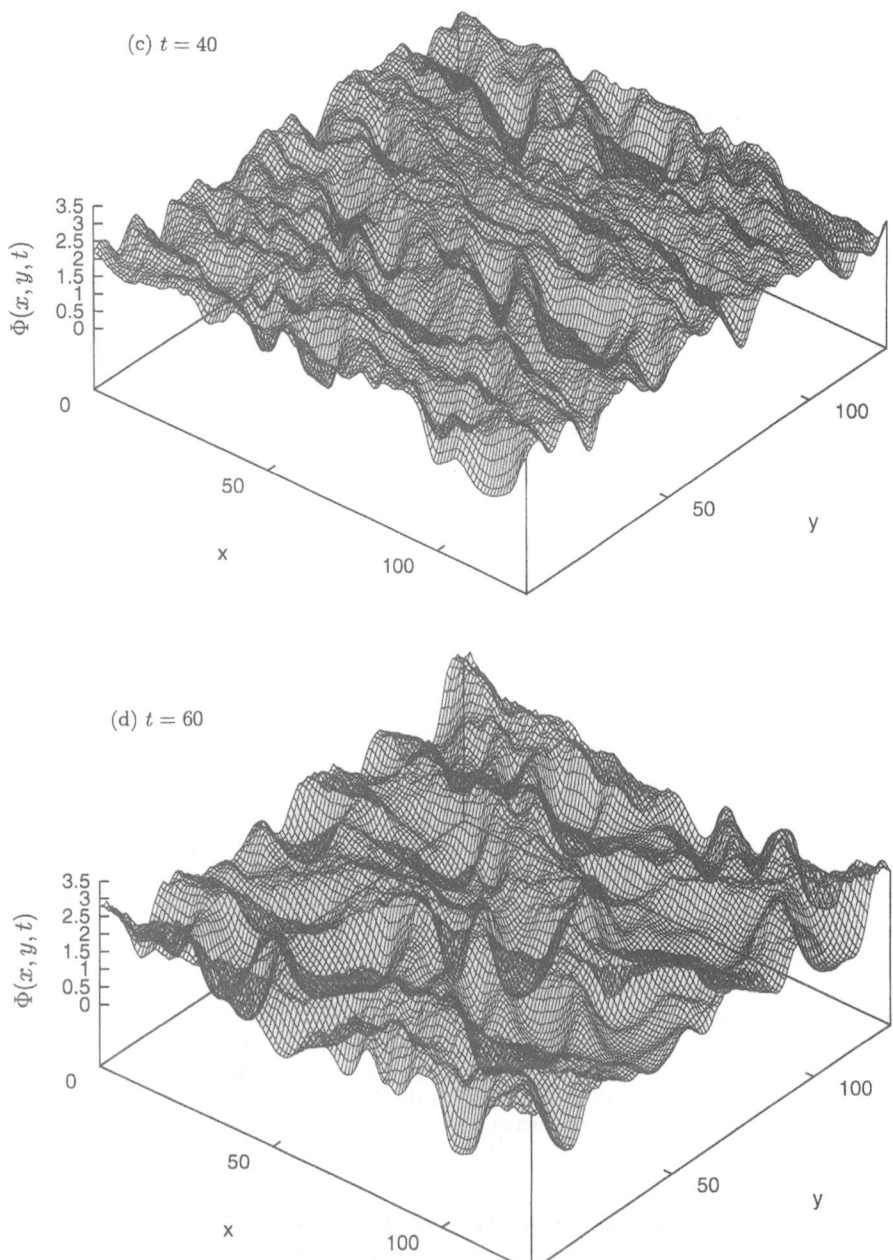

we also show the average shear stress,

$$\sigma_{xy} = \langle \sigma_{pxy} \rangle - \left\langle C(\phi) \frac{\partial \phi}{\partial x} \frac{\partial \phi}{\partial y} \right\rangle , \qquad (24)$$

and the average normal stress difference,

$$N_1 = \langle \sigma_{pxy} - \sigma_{pyy} \rangle + \left\langle C(\phi) \left[\left(\frac{\partial \phi}{\partial y} \right)^2 - \left(\frac{\partial \phi}{\partial x} \right)^2 \right] \right\rangle , \qquad (25)$$

the averages being taken over the lattice points. In our problem we have confirmed that the first terms of (24) and problem we have confirmed that the first terms of (24) and (25) are much larger than the second terms, while the second terms give dominant singular contributions in Newtonian fluids [25]. Figures 5 and 6 display σ_{xy} and N_1 divided by $\eta_0/3\tau_0$ in the longer time region $0 < t < 800$ for $\dot{\gamma}\tau_0 = 0.01$, 0.02, and 0.05. The shear stress first grows linearly in time ($\propto t$) up to the order $\eta_p \dot{\gamma}$ at $t \sim \tau$, but it begins to decrease with growth of the shear-induced fluctuations. The normal stress difference grows at t^2 initially. After the transient stage, they both exhibit considerable fluctuations except for the smallest shear $\dot{\gamma}\tau_0 = 0.01$. At the largest shear $\dot{\gamma}\tau_0 = 0.05$, the network composed of

156

A. Onuki et al.
Shear-induced phase separation

Fig. 3 Contour plots of the time average of the structure factor for $\dot{\gamma}\tau_0 = 0.02$ in (a) and $\dot{\gamma}\tau_0 = 0.05$ in (b) in the q_x–q_y plane. The wave vector is measured in units of $2\pi/128l$. The peak height is 15.7 in (a) and 470 in (b)

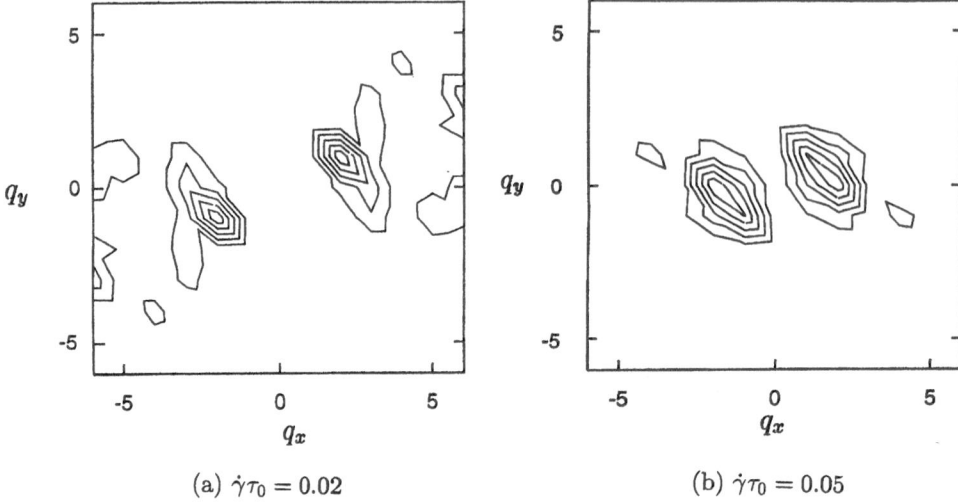

(a) $\dot{\gamma}\tau_0 = 0.02$

(b) $\dot{\gamma}\tau_0 = 0.05$

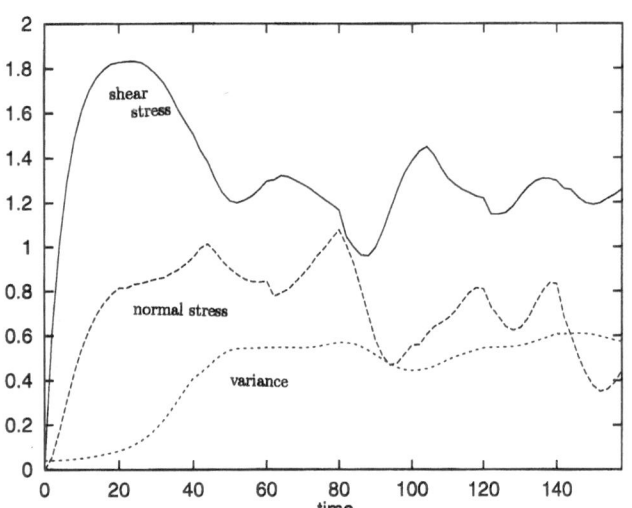

Fig. 4 Time evolution of the variance σ (dotted line), the shear stress σ_{xy} (bold line), and the normal stress difference N_1 (broken line) for $\dot{\gamma}\tau_0 = 0.05$ in the early time region $0 < t < 160$

elongated polymer-rich regions is often extended throughout the system, but is subsequently disconnected. The stress is mostly supported in such a network. This process produces abnormal fluctuations of the stress. Interestingly, in many cases, the normal stress difference takes a maximum (or minimum) when the shear stress takes a minimum (or maximum).

In experiments, the stress components are measured as the force density acting on a surface with a macroscopic linear dimension d. If d is longer than the characteristic size of the network structure of the polymer-rich region, the temporal stress fluctuations will be suppressed. Nevertheless, the shear-induced composition fluctuations will reduce the shear stress on the average and enhance the normal stress difference.

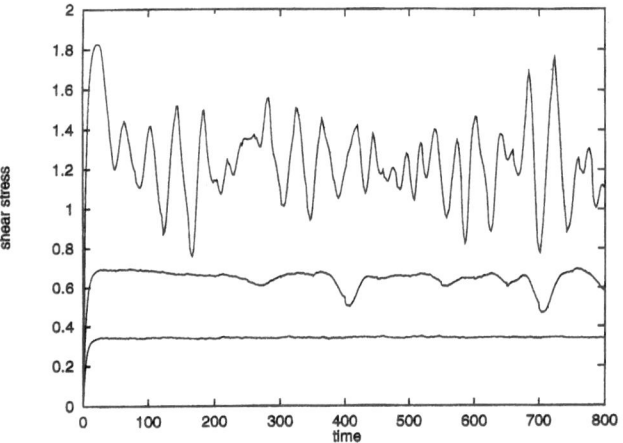

Fig. 5 Shear stress as a function of time for $\dot{\gamma}\tau_0 = 0.01, 0.02$, and 0.05 from below. The fluctuations become larger with increasing shear

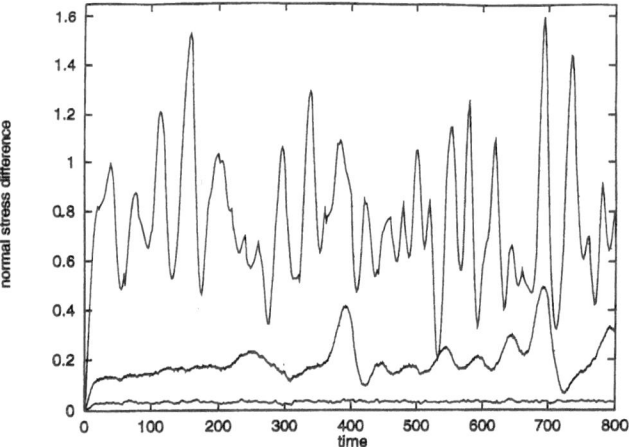

Fig. 6 Normal stress difference as a a function of time for $\dot{\gamma}\tau_0 = 0.01$, 0.02, and 0.05 from below. The fluctuations are even more larger than those of the shear stress

Progr Colloid Polym Sci (1997) 106:150–157
© Steinkopff Verlag 1997

Summary and remarks

In summary, we have numerically examined the effect of shear on polymer solutions above the coexistence curve. Here, macroscopic phase separation cannot be achieved, but the level of the fluctuations increases with increasing shear and can even be comparable to that in spinodal decomposition occurring at lower temperatures. In the nonlinear regime above the coexistence curve, as treated in this paper, polymer solutions can undergo *incomplete phase separation* induced by shear, where anisotropic fluctuations are growing from small to large scales, but large-scale fluctuations are deformed by shear and are eventually dissipated. The random source terms in the dynamic equations are indispensable in this cascade process as sources of the growing fluctuations. Furthermore, in future simulations we should also examine the effects (i) below the coexistence curve and (ii) in the non-Newtonian regime. We expect that the system may phase-separate macroscopically when deeply quenched inside the unstable region under shear. It is also interesting to examine how weak shear can affect nucleation in the metastable region. There seem to be still a number of puzzles in the shear flow effects in phase-separating entangled polymer systems. Analytic theory extending the current theories is also given in [26].

Finally, we mention an early experiment by Lodge [27], who observed abnormal temporal fluctuations of the normal stress difference at a hole of 1 mm diameter from polymer solutions contained in a cone–plate apparatus. He ascribed its origin to the growth of inhomogeneities or gel particles of dimensions about 4 mm. We believe that our results are closely related to his observation. Further experiments on small-scale stress fluctuations are very informative. We may explore the possibility of birefringence and dichroism experiments also.

Acknowledgements One of the authors (A.O.) would like to thank Professor Takeji Hashimoto for showing unpublished microscope pictures of polymer solutions undergoing shear-induced phase separation [28], which resemble the patterns in Fig. 1.

References

1. Ver Strate G, Philippoff W (1974) J Polym Sci Polym Lett 12:267–275
2. Rangel-Nafaile C, Metzner AB, Wissburn KF (1984) Macromolecules 17:1187–1195
3. Krämer, H, Wolf BA (1985) Macromol Chem Rapid Commun 6:21–27
4. Wu XL, Pine DJ, Dixon PK (1991) Phys Rev Lett 68:2408–2411
5. Hashimoto T, Fujioka K (1991) J Phys Soc Jpn 60:356–359; Hashimoto T, Kume T (1992) J Phys Soc Jpn 61:1839–1842; Moses E, Kume T, Hashimoto T (1994) Phys Rev Lett 72:2037–2040
6. Yanase H, Moldenaers P, Mewis J, Abetz V, van Egmond J, Fuller GG (1991) Rheol Acta 30:89–97
7. Helfand E, Fredrickson H (1989) Phys Rev Lett 62:2468–2471; Wittmann HP, Fredrickson GH (1994) J Phys I France 4:1791–1812
8. Grmela M (1988) Phys Lett A 130:81–86
9. Beris AN, Edwards BJ (1994) In: Thermodynamics of Flowing Systems. Oxford University Press, Oxford
10. Onuki A (1989) Phys Rev Lett 62:2472–2475; (1990) J Phys Soc Jpn 59:3423–3426
11. Milner ST (1993) Phys Rev E 48:3874–3691
12. Ji H, Helfand E (1995) Macromolecules 28:3869–3880
13. Doi M, Onuki A (1992) J Phys II France 2:1631–1656
14. Onuki A (1994) J Non-Crystal Solids 172–174:1151–1157
15. Taniguchi T, Onuki (1996) Phys Rev Lett 77:4910–4913
16. de Gennes PG (1985) In: Scaling Concepts in Polymer Physics. 2nd ed, Cornell University Press, Ithaca
17. Larson RG (1986) In: Constitutive Equations for Polymer Melts and Solutions. Butterworths, Boston
18. Adam M, Delsanti M (1984) J Physique 45:1513–1521
19. Brochard F, de Gennes PG (1977) Macromolecules 10:1157–1161
20. Colby RH, Rubinstein M (1990) Macromolecules 23:2753–2757
21. Onuki A (1997) J Phys Soc Jpn 66:1836–1837
22. Beysens D, Gbadamassi M, Moncef-Bouanz B (1983) Phys Rev A 28:2491–2509
23. Onuki A, Kawasaki K (1979) Ann Phys (NY) 121:456–528; Onuki A, Yamazaki K, Kawasaki K (1981) Ann Phys (NY) 131:217–242
24. Onuki A, Yamamoto R, Taniguchi T, J de Physique II 7:295–304
25. Onuki A (1987) Phys Rev A 35:5149–5155
26. Onuki A (1997) J Phys C 9:6119–6157
27. Lodge AS (1961) Polymer 2:195–201
28. Hashimoto T, Kume T, Murase H, unpublished

New results

In a recent review paper on shear flow problems [26], we have chosen more realistic values of the parameters (smaller G and longer τ) in the present Ginzburg–Landau model. There, together with some analytic results, preliminary simulation results are shown for the non-Newtonian case and for the case below the coexistence curve. In the latter case, interfaces become gradually apparent, the solvent regions are highly elongated, and finely divided, two phase states are still realized. Macroscopic phase separation is attained only for deep quenches and at low shear.

Progr Colloid Polym Sci (1997) 106:158–166
© Steinkopff Verlag 1997

F. Tanaka

Phase formation and dynamics of associating polymers

Received: 13 October 1996
Accepted: 4 January 1997

Dr. F. Tanaka (✉)
Department of Polymer Chemistry
Graduate School of Engineering
Kyoto University
Kyoto 606-01, Japan

Abstract This paper reviews possible phase diagrams of associating polymer solutions in which phase separation and molecular association interfere. Paying special attention on the structure and reorganisation of the network junctions, we study competition between phase separation and gelation. The molecular structure of associating micelles, or multiple cross-link junctions, in the networks is analyzed from the sol/gel transition lines. The effect of added surfactants on the formation of reversible gels in hydrophobically modified polymer solutions is also studied under the assumption of the existence of a minimum multiplicity required for stable cross links. To describe dynamics of associating polymers, we introduce a transient network model whose network junctions break and recombine. From the theoretical calculation of dynamic mechanical moduli, average lifetime of a junction, the number of elastically effective chains and of free ends are obtained as functions of the strength of the association constant, polymer concentration and the temperature. We compare the results with experimentally measured moduli of hydrophobically modified test polymers.

Key words Associating polymers – phase separation – thermoreversible gelation – multiple junction – polymer-surfactant interaction

Introduction

Associating polymers are polymers carrying associative groups on the backbone or on the chain side. These groups form aggregates, or micelles, through hydrogen bonds, ionic attraction, hydrophobic interaction, etc. Polymers with associative interactions exhibit a variety of condensed phases, typical examples of which are microscopically ordered phases, gels, and liquids crystals [1–4]. All of these phases have their counterparts formed by covalently connected polymers, but, since association is thermally reversible, associating phases provide a new pathway to modeling statistical clusters, block copolymers, and reversible networks. The time scale for reorganizing transient structures can also be adjusted by the strength of the associative interaction. In this paper, we specifically focus our attention to the network formation by reversible cross linking [5, 6].

Most thermoreversible gels observed so far are cross linked by the formation of network junctions involving polymer segments belonging to several distinct chains (*multiple junctions*). For instance, gelation by micro-crystallization of the chain segments, by ionic aggregation, and by association of multifunctional groups attached on the polymer chains, all belong to this important category. Although experimental observations have been accumulated, theoretical study on phase behavior of thermoreversible gels with multiple-chain junctions remains rather incomplete. In this paper, we study thermodynamic properties and dynamics of reversible gels whose junction multiplicity k is allowed to vary.

Models of network junctions

We consider a model mixture of functional molecules (or primary polymer chains) in a solvent. The molecules are distinguished by the number f of the functional groups ("stickers") they bear, each functional group being capable of taking part in the junctions which may bind together any number k of such groups [5, 7]. Hereafter, we shall call k the *multiplicity* of a junction. We include the value $k = 1$ representing unreacted groups. In what follows, we allow junctions of all multiplicities to coexist, in proportions determined by the thermodynamic equilibrium conditions. Let n_f be the number of the statistical segments on an f-functional molecule and let N_f be the number of f-functional primary molecules in the solution. The weight fraction w_f of the molecules with specified f relative to the total weight is then given by $w_f = f N_f / \sum f N_f$. In thermal equilibrium, the solution has a distribution of clusters with a population distribution fixed by the equilibrium conditions. Following the notation in ref. [7], we define a cluster of type $(\mathbf{j}, \mathbf{l})$ to consist of l_f molecules of functionality $f(f = 1, 2, 3, \dots)$ and j_k junctions of multiplicity $k(k = 1, 2, 3, \dots)$ (see Fig. 1). The bold letters $\mathbf{j} \equiv \{ j_1, j_2, j_3, \dots \}$ and $\mathbf{l} \equiv \{ l_1, l_2, l_3, \dots \}$ denote the sets of indices. An isolated molecule of functionality f, for instance, is indicated by $\mathbf{j}_{f0} \equiv \{ f, 0, 0, \dots \}$, and $\mathbf{l}_{f0} \equiv \{ 0, \dots, 1, 0, \dots \}$. We now introduce three specific models of the multiple junctions:

- *Fixed multiplicity model.* This model allows a single value $k = s(\geq 2)$. We have

$$k = 1 \text{(unassociated)}, \quad k = s \text{(associated)} . \tag{2.1}$$

When $s = 2$, the model reduces to the case of pairwise cross linking.

- *Minimum multiplicity model.* This model allows junctions to take multiplicity greater than s_0. We have

$$
\begin{aligned}
&k = 1 \text{ (unassociated)}, \ k = s_0, s_0 + 1, s_0 + 2, \dots \\
&\text{(associated)} .
\end{aligned}
\tag{2.2}
$$

When $s_0 = 2$, this model reduces to the variable multiplicity model in which junctions of arbitrary multiplicity can coexist at the probability determined by the thermodynamic balance. In the case of micro-crystalline junctions, for instance, it is natural to assume that a minimum number s_0 greater than 2 of the crystalline chains is required for a junction formation. This is because, the surface energy terms will prevent small-k units from being stable, leading to the existence of the critical multiplicity for the nucleation of the crystallites. Similarly, a minimum aggregation number is required for the stability of micelles formed by hydrophobes on water-soluble polymers. As we will see later, surfactants added to the solution cause complex interaction with hydrophobically modified polymers due to the existence of this minimum multiplicity.

- *Saturating junctions.* This model limits the multiplicity below a certain value s_m. We have

$$k = 1 \text{ (unassociated)}, \quad k = 2, 3, \dots s_m \text{ (associated)} . \tag{2.3}$$

Junctions formed by micellization of the hydrophobic polymer segments, for instance, have the maximum multiplicity. The saturation of the junction multiplicity is caused by the dense packing of the chains near the junction zones which prevent access of excessive functional groups.

Fig. 1 A multiplily connected tree made up of polydisperse primary functional molecules. A cluster is specified by the type $(\mathbf{j}; \mathbf{l})$. This example gives $\mathbf{j} = (35, 13, 2, 1, 0, 0, \dots)$ and $\mathbf{l} = (0, 7, 5, 3, 3, 0, 0, \dots)$

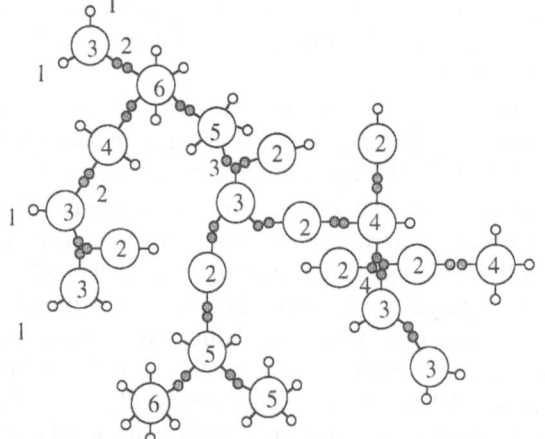

Sol/gel transition

We briefly summarize a lattice theory of network-forming polymer solutions [5, 6]. Let $\phi(\mathbf{j}; \mathbf{l})$ be the volume fraction of the clusters of type $(\mathbf{j}; \mathbf{l})$. The free energy change on passing from the standard reference states (polymers and solvent molecules being separated in hypothetical crystalline states) to the final solution, at equilibrium with respect to cluster formation, is given by the expression

$$
\frac{\beta \Delta F}{\Omega} = v_0 \ln \phi_0 + \sum_{(\mathbf{j};\mathbf{l})} v(\mathbf{j};\mathbf{l}) [\Delta(\mathbf{j};\mathbf{l})
$$
$$
+ \ln \phi(\mathbf{j};\mathbf{l})] + \chi \phi_0 \phi , \tag{3.1}
$$

where $\beta \equiv 1/k_B T$ and ϕ is the volume fraction of the polymer. The subscript zero denotes the solvent, with volume fraction $\phi_0 = 1 - \phi$. The quantity $\Delta(\mathbf{j};\mathbf{l})$ involves the free energy change accompanying the formation of a $(\mathbf{j};\mathbf{l})$-cluster in a hypothetical undiluted amorphous state from the separate primary molecules in their standard states.

By minimizing this free energy with respect to the volume fraction $\phi(\mathbf{j};\mathbf{l})$, the most probable distribution of clusters are found. Using the result of multiple tree statistics [7] for the combinatorial entropy in the free energy $\Delta(\mathbf{j};\mathbf{l})$ of cluster formation, we find the volume fraction of clusters as a function of the temperature and concentration. In the pre-gel regime, the total sum over all volume fractions of clusters must give the volume fraction of polymers in the solution. For example, this normalization relation for the fixed multiplicity model of monodisperse polymers (f and n definite) is given by

$$\lambda(T)\phi/n = \alpha^{1/s'}/f(1-\alpha)^{s/s'}, \tag{3.2}$$

which connects the extent α of association to the (scaled) polymer concentration. Here $s' \equiv s - 1$, and the association constant $\lambda(T)$ is defined by

$$\lambda(T) = \exp(-\beta \Delta f_0), \tag{3.3}$$

where Δf_0 is the standard free-energy change on binding a single functional group into a junction. The extent α of association is defined by the probability for a randomly chosen associative group to be associated. It is the counterpart of the extent of reaction in conventional chemical gels. Let p_k be the probability for a group to be in the junction of multiplicity k. The extent α is then expressed as

$$\alpha \equiv \sum_{k \geq 2} p_k \Big/ \sum_{k \geq 1} p_k. \tag{3.4}$$

Next step is to calculate the weight-average molecular weight of the clusters. From its divergence, we find the sol/gel transition point. It is most generally given by [5, 7]

$$(f_{\mathrm{w}} - 1)(\mu_{\mathrm{w}} - 1) = 1, \tag{3.5}$$

where $f_{\mathrm{w}} \equiv \sum_{f \geq 1} f w_f$ being the weight average functionality of the primary chains, and $\mu_{\mathrm{w}} \equiv \sum_{k \geq 1} k p_k$ the average multiplicity of the junctions.

Specifically for the fixed multiplicity model of monodisperse primary chains discussed above, the gelation condition is given by $\alpha = \alpha^* \equiv 1/f's'$ leading to the critical concentration

$$\lambda(T)\phi^*/n = f's'/f(f's' - 1)^{s/s'}, \tag{3.6}$$

where ϕ^* is the volume fraction of the polymer at gelation, and $f' \equiv f - 1$ and $s' \equiv s - 1$. As the multiplicity is increased, with other parameters being fixed, gelation concentration changes and sol/gel line shifts on the temperature-concentration plane.

Phase equilibrium and thermodynamic stability can also be studied by the chemical potential of the polymer and the solvent. These are derived by differentiating the

free energy. Binodal curves and spinodal curves can then be drawn on the temperature and concentration plane.

Structure of the network junctions

When a functional group on a chain involves ζ sequential repeat units, we can write the standard free-energy change as

$$\Delta f_0 = \zeta(\Delta h_0 - T \Delta s_0). \tag{4.1}$$

Here Δh_0 is the enthalpy of bonding and Δs_0 the entropy of bonding, both measured *per single repeat unit*. Taking the logarithm of the gelation concentration (3.6), we find an important relation

$$\ln \phi^* = \zeta \frac{\Delta h_0}{k_{\mathrm{B}} T} + \ln \left[\frac{f's'n}{f(f's' - 1)^{s/s'}} \right] - \zeta \frac{\Delta s_0}{k_{\mathrm{B}}}. \tag{4.2}$$

We can find s and ζ by comparing this relation with the experimental sol/gel transition concentration. For the micro-crystalline junction formed by homopolymers, each ζ sequence of repeat units along a chain may be regarded as a functional group. A polymer chain is then regarded as carrying $f = n/\zeta$ functional groups. Since we have large n, and hence large f, we can neglect 1 compared to n or f. We are thus led to an equation

$$\ln c^* = \zeta \frac{\Delta h_0}{k_{\mathrm{B}} T} - \frac{1}{s - 1} \ln M + \text{constant}, \tag{4.3}$$

where weight concentration c^* has been substituted for the volume fraction. This equation enables us to find ζ and s independently.

Let us plot $\ln c^*$ against $10^3/T + \ln M$. Then the slope $-B$ of the line at constant T gives $-1/(s-1)$, while the slope $-A$ of the line at constant M gives

$$\zeta = \frac{10^3 k_{\mathrm{B}}}{|\Delta h_0|} A = \frac{10^3 R}{|(\Delta h_0)_{\mathrm{mol}}|} A, \tag{4.4}$$

where $(\Delta h_0)_{\mathrm{mol}}$ is the enthalpy of bonding per mol of the repeat units, and R the gas constant ($10^3 R = 1.9864$ kcal/mol K).

As an example of the analysis, we show in Fig. 2 the results for the gelation of poly(vinyl alcohol) in water [8]. Poly(vinyl alcohol) (PVA) is known to be a typical crystalline polymer, but it also gels in aqueous solution under large supercooling. There are several experimental evidences that the cross links are formed by partial crystallization of the polymer segments in which syndiotactic sequence dominates, while subchains connecting the junctions consist mainly of atactic non-crystalline sequences on PVA chains. The micro-crystals at the junctions are supposed to be stabilized by hydrogen bonds

Poly(vinyl alcohol)/water

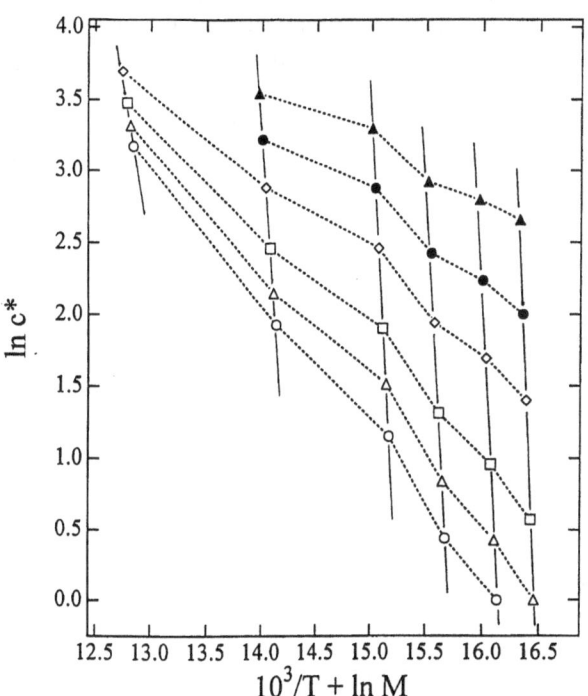

Fig. 2 Modified Eldridge–Ferry plot for poly(vinyl alcohol) gel in water. Gelation concentration at constant molecular weight (solid lines) and at constant temperature (dotted lines) are plotted against a combined variable $10^3/T + \ln M$

between the hydroxy groups. In ref. [9], differential scanning calorimetry (DSC) and viscoelastic measurements were carried out to find the gelation temperature and gel melting temperature for PVA with different molecular weight covering the range 2×10^4–8×10^5 in various concentrations. The gel-melting temperature T_m is estimated from the temperature at which the DSC heating curve shows an endotherm peak. The slope of the solid lines with constant molecular weight gives $-A = 13.43$ almost independently of their molecular weights. Hence, we find $\zeta = 26.7\ \text{kcal/mol}/(\Delta h_0)_{\text{mol}}$. If we use the heat of fusion $(\Delta h_0)_{\text{mol}} = 1.64\ \text{kcal/mol}$ in the bulk crystal, we find $\zeta = 16.3$. On the other hand, the slope of the dotted lines with constant temperature depends on the temperature. At the highest temperature $T = 91\ ^\circ\text{C}$ in the measurement, it is -0.38, while it gives a larger value -0.9 at $T = 71\ ^\circ\text{C}$. The multiplicity is estimated to decrease from 3.6 for high-temperature melting to 2.1 for low-temperature melting, suggesting a very thin junction structure. From thermodynamic stability of the junctions it is only natural that a gel which melts at lower temperature has thinner junctions.

Junction multiplicity and polymer-surfactant interaction

Existence of a limited range in the multiplicity, or average aggregation number, of the network junctions results in the sensitivity of their rheological properties to external perturbation. To see the effect, let us consider how water-soluble thermoreversible gels of hydrophobically modified polymers respond to the added surfactants. Hydrophobic tails of the surfactants may join in the junctions and form mixed micelles. Assume that a minimum multiplicity s_0 is required for a junction to remain thermally stable. When the number of added surfactants is small, most junctions are formed by the hydrophobes carried by the polymers, some of which are stabilized with multiplicity higher than s_0 by the help of the newly joined surfactant tails. As the concentration of the surfactants is increased, however, substantial numbers of hydrophobes in the junctions are replaced by surfactant tails, so that the functionality of a junction (number of network chains coming out of a junction) becomes too small to be a branching point, and the network is eventually broken while the multiplicity of each aggregate exceeds s_0. To describe the situation, let us consider the simplest case where polymers carry a fixed number $f(\geq 2)$ of hydrophobes, and surfactants are modeled by low molecular-weight molecules with a single hydrophobe ($f = 1$, $n_1 = 1$). In the solution, polymers and surfactants interact and mixed micelles, as well as pure surfactant micelles, are formed (see Fig. 3). The critical micelle concentration (cmc) is increased by the existence of the polymers.

General theory developed in sol/gel transition section to find the sol/gel transition point of polydisperse mixtures of associating polymers can directly be applied to this simple but important case. In Fig. 4 the (scaled) gelation concentration $c^* \equiv f\lambda(T)\phi/n_f$ of the telechelic polymer $f = 2$ for the fixed multiplicity model is plotted as a function of the (scaled) concentration $c_s \equiv \lambda(T)\phi_1/n_1$ of the added surfactant [10]. The multiplicity s is changed from curve to curve. For small multiplicities, e.g. $s = 3$, 4, the gelation concentration is a steadily increasing function of the surfactant concentration; gelation is prevented by the surfactants. For larger multiplicities $s \geq 5$, however, there is a minimum gelation concentration as a function of c_s, below which gelation is promoted by the existence of the surfactants, but above which gelation is interfered. The former may be called *surfactant-mediated gelation* and the latter *surfactant-induced melting*. Dotted lines show how cmc for pure surfactant micelles shifts with polymer concentration. As the polymer concentration increases, the number of surfactant molecules absorbed into the network junctions increases, so that formation of pure surfactant micelles is interfered, resulting in the increase of the cmc.

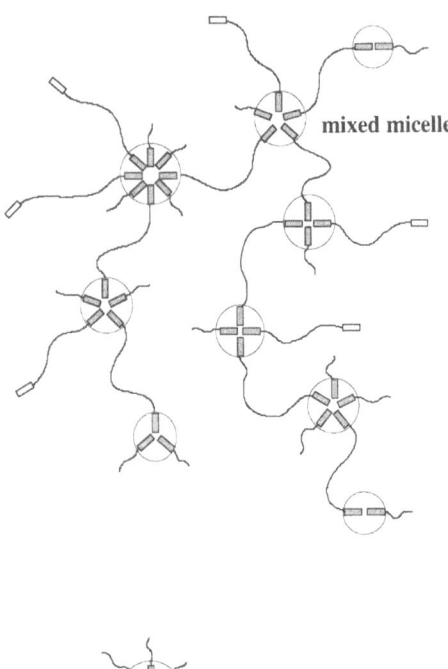

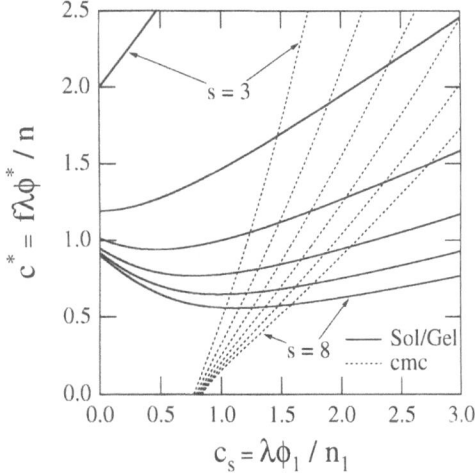

Fig. 3 Polymer-surfactant interaction is sketched. In the network of telechelic polymers ($f = 2$), mixed micelles are formed at the junctions. Isolated micelles made up of pure surfactants also exist in the solution

Fig. 4 The (scaled) gelation concentration (solid lines) of telechelic polymers and cmc of surfactants (dotted lines) plotted against the (scaled) concentration of surfactants for fixed multiplicity model. The multiplicity is varied from curve to curve. For s larger than 4, the gel concentration shows a minimum

lar weight M (or the number n of statistical units) carrying associative functional groups at their both chain ends [11]. We focus our attention specifically on the *unentangled regime* where M is smaller than the entanglement molecular weight M_e, so that each chain obeys Rouse dynamics modified by end association. In the following, we assume the lifetime $\tau_\times$ of a junction is sufficiently long so that it is well separated from the Rouse relaxation spectrum, the longest time of which is given by the Rouse relaxation time $\tau_R \equiv \gamma a^2 n^2 / 3\pi^2 k_B T$, where γ here is the monomer friction coefficient.

Consider a time interval dt *smaller* than $\tau_\times$ but still larger than τ_R. Under a macroscopic deformation $\hat{\lambda}(t)$ given to the network, either end of a chain, being stretched above a critical length, snaps from the junction, and the chain relaxes to a Gaussian conformation, whilst some of the free dangling ends recapture the junctions in their neighborhood (Fig. 5). Since the stress is transmitted only through the chains whose both ends are connected to the network junctions, we call these chains *elastically effective* (or *active*) *chains*.

Fig. 5 Internal reorganization of the transient network induced by a macroscopic deformation. Associative groups on the chains with high-tension disengage from the junctions and form dangling ends, while some dangling chains catch junctions in the neighborhood

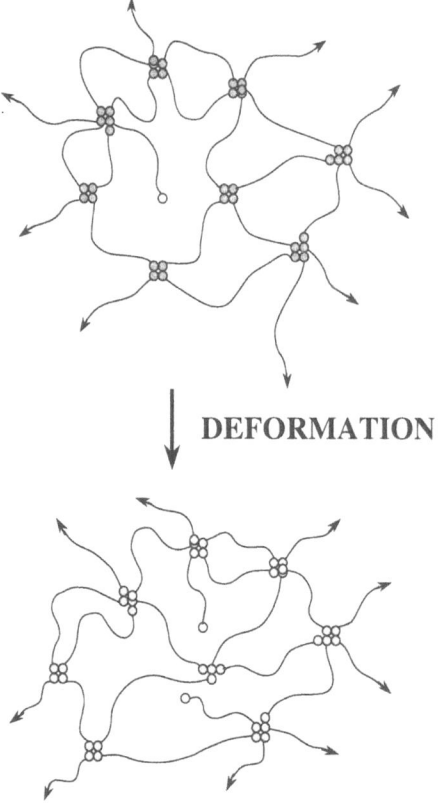

Dynamics of transient gels

To study dynamics of associating polymers, we introduce a model network made up of polymers of uniform molecu-

Let v_0 be the total number of polymer chains in a unit volume, and let $F(\mathbf{r}, t)\,d\mathbf{r}$ be the number of active chains at time t whose end-to-end vectors fall onto a small region $d\mathbf{r}$ around $\mathbf{r}$. The total number of active chains at time t is given by $v(t) \equiv \int F(\mathbf{r}, t)\,d\mathbf{r}$.

By counting the number of active chains created and destroyed during the interval dt, and to integrate it from the initial time $t = 0$ to an arbitrary t, we find the fundamental equation

$$F(\mathbf{r}, t)\,d\mathbf{r} = \Theta(\mathbf{r}, t; \mathbf{r}_0, 0) F(\mathbf{r}_0, 0)\,d\mathbf{r}_0 + p \int_0^t$$
$$\times dt'\Theta(\mathbf{r}, t; \mathbf{r}', t')\,[v_0 - v(t')]\,f_0(\mathbf{r}')\,d\mathbf{r}' \quad (6.1)$$

governing the end-vector distribution of the active chains, where

$$\Theta(\mathbf{r}, t; \mathbf{r}', t') \equiv \exp\left[-\int_{t'}^t \beta(\mathbf{r}_{t'', t'})\,dt''\right] \quad (6.2)$$

is a probability for an active chain created at time t' to remain active until t. The function $\beta(\mathbf{r})$ gives the probability per unit time for an active chain to disengage from the junction and is called *chain breakage rate*. The parameter p in the second term gives the probability per unit time for a dangling end to capture a junction in its neighborhood (called *chain recombination rate*). We have assumed that end-to-end vectors are mapped *affinely* to the macroscopic deformation so that we have

$$\mathbf{r}_{t'', t'} = \hat{\lambda}(t'') \cdot \hat{\lambda}(t')^{-1} \cdot \mathbf{r}', \quad (6.3)$$

etc. The equilibrium distribution $f_0(\mathbf{r})$ of the end-to-end vector is assumed to take the conventional Gaussian form with average square distance $\langle r^2 \rangle_0 = na^2$.

Eq. (6.1) gives an integral equation to find the number of active chains at arbitrary time under a given initial condition. We now transform it into a differential equation by taking differential with respect to the time and find

$$\frac{\partial F(\mathbf{r}, t)}{\partial t} + \nabla \cdot (F(\mathbf{r}, t)\mathbf{v}) = G(\mathbf{r}, t) - \beta(\mathbf{r})F(\mathbf{r}, t), \quad (6.4)$$

where the deformation velocity $\mathbf{v}$ is given by $\mathbf{v} \equiv (d\hat{\lambda}(t)/dt) \cdot \hat{\lambda}(t)^{-1} \cdot \mathbf{r}$ by the affineness assumption. The first term on the r.h.s is the number of active chains created in a unit time and is given by $G(\mathbf{r}, t) \equiv p(v_0 - v(t))f_0(\mathbf{r})$.

From this fundamental equation, we can find the stress tensor $\hat{\Sigma}(t)$ as a function of time under arbitrary deformation.

Dynamic mechanical moduli

In order to derive the dynamic mechanical moduli, let us consider a small oscillatory shear deformation given by the deformation tensor

$$\hat{\lambda}(t) = 1 + \hat{\varepsilon}\sin\omega t, \quad (7.1)$$

where the small tensor $\hat{\varepsilon}$ is given by $\varepsilon_{x,y} = \varepsilon$ with other $\varepsilon_{i,j} = 0$.

The linear approximation allows us to expand the exponential function in the propagator in powers of ε. We find that, to the lowest order of ε, the number of active chains $v(t)$ at arbitrary time t practically remains the same as the initial equilibrium value $v_e = [p\zeta_0/(1 + p\zeta_0)]v_0$, where ζ_0 is a constant defined by $\zeta_0 \equiv \langle 1/\beta(r)\rangle_0$.

Similarly, from the stress–strain relation we find, up to the linear order of ε,

$$\sum_{xy}(t) = \varepsilon k T v_e\,[g_1(\omega)\sin\omega t + g_2(\omega)\cos\omega t]$$
$$+ \text{decaying terms}, \quad (7.2)$$

for the shear stress, where the new functions $g_i(\omega)$ are explicitly given by

$$g_1(\omega) \equiv \frac{1}{\zeta_0\langle r^2\rangle_0}\left\langle\frac{\omega^2 r^2}{\beta(r)\,[\omega^2 + \beta(r)^2]}\left[1 - \frac{2r\beta(r)\beta'(r)}{5[\omega^2 + \beta(r)^2]}\right]\right\rangle_0, \quad (7.3)$$

$$g_2(\omega) \equiv \frac{1}{\zeta_0\langle r^2\rangle_0}\left\langle\frac{\omega r^2}{\omega^2 + \beta(r)^2}\left[1 + \frac{[\omega^2 - \beta(r)^2]r\beta'(r)}{5\beta(r)^2[\omega^2 + \beta(r)^2]}\right]\right\rangle_0, \quad (7.4)$$

These functions $g_i(\omega)$ depend on the temperature indirectly through the chain breakage rate. Keeping only the *secular terms* – i.e., the terms which remain oscillatory without decay –, we find the storage modulus $G'(\omega, T)$ and the loss modulus $G''(\omega, T)$ can be written in the form

$$G'(\omega, T) = v_e(T)kTg_1(\omega, T), \quad (7.5a)$$

$$G''(\omega, T) = v_e(T)kTg_2(\omega, T), \quad (7.5b)$$

In the extreme case where the chain breakage rate $\beta(r)$ is given by a constant value β_0 (called *Green–Tobolsky limit*), our network reduces to a Maxwell fluid with a single relaxation time $\tau_x = \beta_0^{-1}$.

Let us proceed to study the chain breakage rate. Consider the free energy barrier W for an associative group to disengage from a junction W. Since the tension $f = (3kT/na^2)r$ along the effective chain with end-to-end distance r reduces this barrier to $W - f \cdot a$, the chain breakage rate $\beta(\mathbf{r})$ is expected to be proportional to $\exp[-(W - f \cdot a)]$. Hence, $\beta(r)$ takes the form

$$\beta(r) = \beta_0\,e^{\kappa r}, \quad (7.6)$$

where

$$\beta_0(T) \equiv \omega_0\,e^{-W/kT} \quad (7.7)$$

is the reciprocal of the duration time $\tau_\times$ of a bond. Here, ω_0 is the intrinsic frequency of the vibration of the associating group, and $\kappa \equiv 3/na$ a constant depending on the molecular weight of the chain. For this reason the moduli deviate from those of Maxwell fluids with single relaxation time, and depend weakly upon the molecular weight of the polymer.

Taking the Gaussian average in $g_i(\omega, T)$ with the chain breakage rate (7.6), we find that a modulus-frequency curve at any temperature T can be superimposed onto a single curve at the reference temperature T_0, if it is horizontally and vertically shifted properly. Specifically, we have

$$\frac{G(\omega)}{v_e(T_0)kT_0} b_T = g\left(\frac{\omega}{\beta_0(T_0)} a_T \right) \tag{7.8}$$

for both G' and G'', where

$$a_T \equiv \beta_0(T_0)/\beta(T) = \exp\left[-\frac{W}{k}\left(\frac{1}{T_0} - \frac{1}{T} \right) \right] \tag{7.9}$$

is the frequency (horizontal) shift factor, and

$$b_T \equiv v_e(T_0)kT_0/v_e(T)kT \tag{7.10}$$

is the modulus-(vertical) shift factor. The fact that the frequency-shift factor depends exponentially on the reciprocal of the temperature indicates that the linear viscoelasticity is dominated by the activation process of the junction dissociation. In fact, experiments on HEUR estimated the average lifetime $\tau_\times$ of the junction from the peak position of the loss moduli, and found the activation energy as a function of the length of end-chain hydrophobes.

On the other hand, at high frequencies $\omega \gg \beta_0$, the storage modulus (7.5a) gives the equilibrium number v_e of polymer chains in the network. This number, however, must be regarded as the number of elastically effective chains that connect two junctions in the network since only these chains can support stress. Another important feature of the network is the number and structure of elastically inactive chains that are dangling from the network. We call them *dangling ends*. We next study how such topological structures of the network depend on the temperature and concentration.

Structure of the transient networks

To study the network structure, we first specify the type of junctions in more detail. A junction of multiplicity k that is connected to the gel network through i paths is referred to as an (i, k)-junction. Let $\mu_{i,k}$ be the number of junctions specified by the type (i, k) for $k = 1, 2, 3, 4, \ldots$ and for $0 \le i \le 2k$. The total number of junctions with multiplicity

k in unit volume is given by

$$\mu_k = \sum_{i=0}^{2k} \mu_{i,k} . \tag{8.1}$$

We next employ the criterion of Scanlan [12] and Case [13] that only subchains connected at both ends to junctions with at least *three paths* to the gel are elastically effective. We, thus, have $i, i' \ge 3$ for an effective chain. A junction with one path ($i = 1$) to the gel unites a group of subchains dangling from the network matrix whose conformations are not affected by an applied stress. A junction with two paths ($i = 2$) to the gel merely extends the length of an effective subchain. We may call the junctions with $i \ge 3$ *elastically effective junctions*. An effective chain is defined as a chain connecting two effective junctions at its both ends. We, thus, find

$$\mu_e = \sum_{k=2}^{\infty} \sum_{i=3}^{2k} \mu_{i,k} \tag{8.2}$$

for the number of elastically effective junctions in a unit volume, and

$$v_e = \frac{1}{2} \sum_{k=2}^{\infty} \sum_{i=3}^{2k} i\mu_{i,k} \tag{8.3}$$

for the number of elastically effective chains.

A dangling end may consist either of a single subchain or of a group of subchains connected by several branch points. The structure of a dangling end can be described by the number of subchains and branch points it contains. By definition, the number of dangling ends is given by

$$v_{end} = \sum_{k=2}^{\infty} \sum_{i=2}^{2k} (2k - i)\mu_{i,k} . \tag{8.4}$$

The summation is taken over junction for which $i \ge 2$ because a junction with only one path to the gel is just a branch point on an already counted dangling end.

By combinatorial counting we can express $\mu_{i,k}$ as a function of the degree α of association, which is a function of the temperature and concentration through the relation (3.2). Fig. 6 shows the number of elastically effective chains for $f = 2$ with s varied from curve to curve as a function of the extent of reaction (Fig. 6a), and of the reduced concentration (Fig. 6b). The critical behavior obeys the mean-field scaling law

$$v_e/v_0 \simeq (\phi - \phi^*)^t \tag{8.5}$$

with $t = 3$. The cubic power comes of course from the mean-field treatment (tree statistics). According to percolation theory, we should expect a smaller power $t = 1.7$. At the completion of the reaction $\alpha = 1$, (and hence $\lambda(T)\phi/n \to \infty$), the curves asymptotically reach unity. The number of effective chains is proportional to the polymer

Progr Colloid Polym Sci (1997) 106:158–166
© Steinkopff Verlag 1997

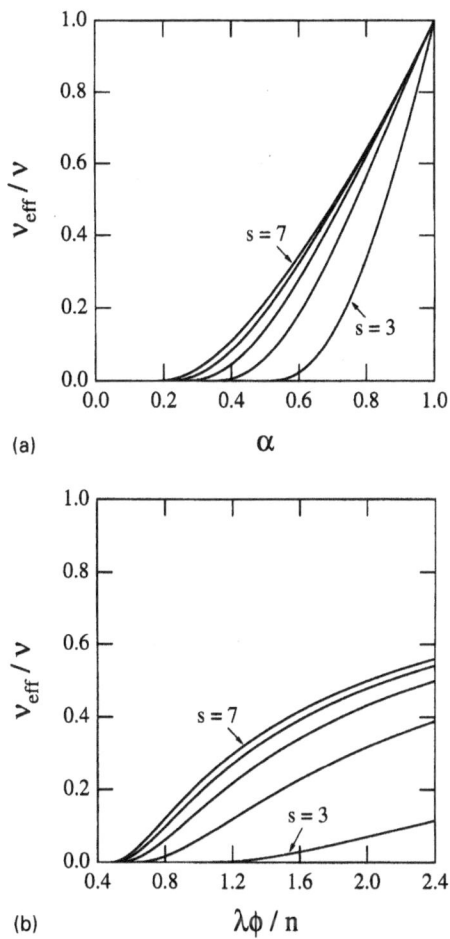

(a)

(b)

Fig. 6 The number of elastically effective chains (relative to the total number of chains) as a function of the extent of association (a), and of the reduced concentration (b). The multiplicity is changed from curve to curve, while the functionality is fixed at $f = 2$. Each curve rises in cubic power of the concentration deviation from the transition point and approaches unity at high limit of the concentration

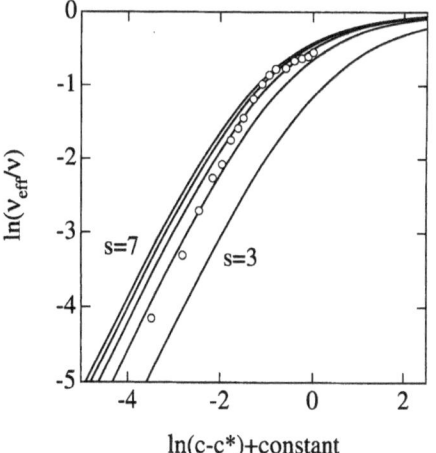

Fig. 7 Comparison of the experimental data on the elastically effective chains in HEUR C16/35 K in water with theoretical calculation. To eliminate temperature pre-factor $\lambda(T)$, the experimental data are horizontally shifted

detailed experimental examination in the critical region is eagerly required. In fitting the data, we have horizontally shifted the experimental data because of the temperature pre-factor $\lambda(T)$ and also of the difference in the unit of the polymer concentration. Although fitting by a single theoretical curve with a fixed multiplicity is impossible due to the existence of polydispersity in the multiplicity, our theory produces correct behavior over a wide range of the concentration.

Conclusions and discussion

In this review, we stressed that junctions in thermoreversible gels have two important structural parameters: multiplicity and sequence length. Regarding dynamics, we have to know the average lifetime of a junction. These are related to the associative force through the binding free energy and the free-energy barrier separating the bonded state from the free one. Controlling these structural and dynamical parameters is the most important factor for functionalize thermoreversible gels. For a chosen associative force and molecular architecture of the primary chains, finding these parameters as functions of the thermodynamic variables is essential. Our attempt presented here is the first step for this purpose.

concentration in this region. These curves can be compared with the experimental data on the high-frequency dynamic modulus for HEUR measured by Annable et al. [14]. Their experimental data for HEUR C16/35K (end-capped with $C_{16}H_{33}$, molecular weight 35000) are compared with our theoretical calculation in Fig. 7 [15]. We have chosen as $c^* = 1.0\%$ for the weight concentration at gelation, so that the scaling power at the critical region gives $t = 1.6$, close to the percolation value. But since this power depends sensitively on how we choose c^*, more

References

1. Russo RS (1987) Reversible Polymeric Gels and Related Systems. Vol 350. American Chemical Society, Washington, DC

2. Kramer O (1988) Biological and Synthetic Polymer Networks, Elsevier Applied Science, London and New York

3. Utracki LA, Weiss RA (1989) Multiphase Polymers: Blends and Ionomers. Vol 395. American Chemical Society, Washington, DC

4. Guenet JM (1992) Thermoreversible Gelation of Polymers and Biopolymers. Academic Press, London
5. Tanaka F, Stockmayer WH (1994) Macromolecules 27:3943
6. Tanaka F (1990) Macromolecules 23:3784; 3790
7. Fukui K, Yamabe T (1967) Bull Chem Soc Japan 40:2052
8. Tanaka F, Nishinari K (1996) Macromolecules 29:3625
9. Nishinari K, Tanaka F (1996) J Chim Phys 93:880
10. Tanaka F, Ishida M, Macromolecules, to be submitted
11. Tanaka F, Edwards SF (1992) Macromolecules 25:1516; Tanaka F, Edwards SF (1992) J Non-Newtonian Fluid Mech 43:247; 273; 289
12. Scanlan J (1960) J Polym Sci 43:501
13. Case LC (1960) J Polym Sci 45:397
14. Annable T, Buscall R, Ettelaie R, Whittlestone D (1993) J Rheol 37:695
15. Tanaka F, Ishida M (1996) Macromolecules 29:7571

Progr Colloid Polym Sci (1997) 106:167–171
© Steinkopff Verlag 1997

Hajime Tanaka

Viscoelastic phase separation in complex fluids

Received: 13 October 1996
Accepted: 7 February 1997

Dr. H. Tanaka (✉)
Institute of Industrial Science
University of Tokyo
Minato-ku, Tokyo 106, Japan

Abstract Phase separation in isotropic condensed matter has so far been believed to be classified into solid and fluid models. When there is a large difference in the characteristic rheological time between the components of a mixture, however, we need a model of phase separation, which we call "viscoelastic model". This model is likely a general model that can describe all types of isotropic phase separation including solid and fluid model as special cases. We point out that this dynamic asymmetry between the components is quite common in complex fluids, one of whose components has large internal degrees of freedom. We also demonstrate that viscoelastic phase separation in such dynamically asymmetric mixtures can be characterized by the order-parameter switching phenomena. The primary order parameter switches from the composition to the deformation tensor, and back to the composition again, reflecting viscoelastic relaxation between a characteristic deformation time of phase separation and the slowest rheological time of the system. This unusual behavior can be explained as follows: the viscoelastic model contains various model of phase separation as special cases and the switching between them is induced by viscoelastic relaxational phenomena. Since the deformation tensor intrinsically has a geometrical nature, the pattern in the elastic regime is essentially different from that of usual phase separation in fluid mixtures; and, thus, there is no self-similarity in the pattern evolution.

Key words Viscoelastic effects – phase separation – critical phenomena – complex fluids – dynamic asymmetry

Introduction

Phase-separation phenomena are commonly observed in various kinds of condensed matter, including metals, semiconductors, simple liquids, and complex fluids such as polymers, surfactants, colloids, and biological materials. The study of these processes of pattern evolution is very important for both engineering applications and basic understanding of nonequilibrium dynamics of pattern formation [1]. Phase separation in each material group of condensed matter is described by specific basic equations describing its dynamic process. From the concept of dynamic universality, types of phase separation are classified into various models by Hohenberg and Halperin [2]. For example, phase separation in solids is known as "solid model (model B)", while phase separation in fluids as "fluid model (model H)" [2]: For the former the local concentration can be changed only by material diffusion, while for the latter by both diffusion and flow. A universal feature of the phase-separation and critical phenomena in each model has been established [1, 2].

Here, we focus our attention on phase separation in complex fluids that are characterized by the large internal degrees of freedom. In all conventional theories of critical phenomena and phase separation, the same dynamics for the two components of a binary mixture, which we call "dynamic symmetry" between the components, has been implicitly assumed [1, 2]. However, this assumption is not always valid especially in complex fluids. Recently, we have found [3, 4] that in mixtures having intrinsic "dynamic asymmetry" between its components (e.g. a polymer solution composed of long chain-like molecules and simple liquid molecules and a mixture composed of components whose glass-transition temperatures are quite different), critical concentration fluctuation is not necessarily only the slow mode of the system and, thus, we have to consider the interplay between critical dynamics and the slow dynamics of material itself. In addition to a solid and a fluid model, we probably need a third general model for phase separation in condensed matter, which we call "viscoelastic model".

Here, we consider the characteristic features of the "viscoelastic model" of phase separation. Further, we demonstrate that these special features of a viscoelastic model lead to a novel phenomenon of "order-parameter switching" during viscoelastic phase separation, although it is always driven by a single thermodynamic driving force.

Relevant theories based on two-fluid model

The coupling between the stress field and the diffusion was first noticed and studied by Brochard and de Gennes [5] for polymer solutions far from the critical point. Relating to this, shear effects on complex fluids have recently attracted much attention because of its unusual nature known as "Reynolds effect": for example, a shear flow that intuitively helps the mixing of components induces phase separation in polymer solutions [6]. There are many examples of such effects in complex fluids such as surfactant systems, block copolymers, and charged colloidal systems. This is caused by the coupling between shear velocity fields and the elastic internal degrees of freedom of complex fluids. To explain this unique feature of polymer solution, there have been considerable theoretical efforts [5, 7–12].

Recently, Doi and Onuki [12] have derived the following kinetic equations including the dynamic coupling between stress and composition, on the basis of the two-fluid model [10], as an extension of the original idea of Brochard and de Gennes [5]. The essential point is that the stress in polymer systems is supported by polymer chains, and, thus, a gradient in the stress causes a net force on the chains and leads to the motion of the chains relative to the solvent. The theory is applicable to both polymer

solutions and polymer blends. We have proposed [3] that viscoelastic phase separation of dynamically asymmetric mixtures can likely be described by their equations. Only, the difference is that the velocity fields are self-induced by phase separation itself in our problem, while they are external shear fields in the problem of shear effects on phase separation. The dynamic equations derived by Doi and Onuki [12] can be written as follows:

$$\frac{\partial \phi}{\partial t} = \nabla \cdot (\phi \mathbf{v}) + \nabla \cdot \frac{\phi(1-\phi)^2}{\zeta} \left[\nabla \cdot \mathbf{\Pi} - \nabla \cdot \mathbf{\sigma}^{(n)} \right], \quad (1)$$

$$\mathbf{v}_p - \mathbf{v} = \frac{(1-\phi)^2}{\zeta} \nabla \cdot \left[\mathbf{\Pi} - \mathbf{\sigma}^{(n)} \right], \quad (2)$$

$$\rho_0 \frac{\partial \mathbf{v}}{\partial t} = -\nabla \cdot \left[\mathbf{\Pi} - \mathbf{\sigma}^{(n)} \right] - \nabla p + \eta_s \nabla^2 \mathbf{v}. \quad (3)$$

Here, $\mathbf{v}_p(\mathbf{r}, t)$ and $\mathbf{v}_s(\mathbf{r}, t)$ are the average velocities of polymer and solvent at point $\mathbf{r}$ and time t, and $\mathbf{v} = \phi \mathbf{v}_p + (1-\phi)\mathbf{v}_s \cdot \phi(\mathbf{r}, t)$ is the component of polymer, and $\mathbf{\Pi}$ is the osmotic stress tensor. Here, the thermodynamic force $\mathbf{F}_\phi$ is given by $\mathbf{F}_\phi = -\nabla \cdot \mathbf{\Pi} = -\phi \nabla(\delta F/\delta \phi)$. ρ_0 is the average density, p is a part of the pressure, η_s is the solvent viscosity, and ζ is the friction constant per unit volume. Here the free energy F is given by the following Flory–Huggins Gennes form [13]:

$$\frac{F}{k_B T} = \int d\mathbf{r} \left[f(\phi) + \frac{C}{2}(\nabla \phi)^2 \right], \quad (4)$$

$$f(\phi) = \frac{1}{N} \phi \ln \phi + (1-\phi)\ln(1-\phi) + \chi \phi(1-\phi), \quad (5)$$

where N is the degrees of polymerization of polymer, and χ is the interaction parameter between polymer and solvent. Here we propose that $\mathbf{\sigma}^{(n)}$ is generally given, as an extension of the relation in ref. [14], by

$$\sigma_{ij}^{(n)} = \int_{-\infty}^{t} dt' \left[G(t-t')\kappa_{ij}^p(t') + K(t-t')(\nabla \cdot \mathbf{v}_p)\delta_{ij} \right] + \cdots,$$

$$(6)$$

where κ_{ij}^p is the deformation velocity tensor given by

$$\kappa_{ij}^p = \frac{\partial v_{pj}}{\partial x_i} + \frac{\partial v_{pi}}{\partial x_j} - \frac{2}{3}(\nabla \cdot \mathbf{v}_p)\delta_{ij}. \quad (7)$$

Here $G(t)$ and $K(t)$ are material functions, which we call the shear and bulk relaxation modulus, respectively. The last term ... represents other higher-order terms and coupling terms. Here it should be noted that $\eta = \int_0^\infty G(t) dt$, where η is the viscosity of polymer solution. The term including $K(t)$ does not exist in the previous theories [10–12], but *we believe that this term plays an*

Progr Colloid Polym Sci (1997) 106: 167–171
© Steinkopff Verlag 1997

Table 1 New classification of phase-separation behavior

Classification	Model name [2]	Examples	Relevant phenomena	Physical origin
Solid model	Model B	Metal alloys	Diffusion	
Fluid model	Model H	Liquid mixtures	Hydrodynamics	
Gel model	None	Chemical gels	Elasticity	Topological & dynamic asymmetry
Viscoelastic model	None	Polymer solutions	Viscoelasticity	Dynamic asymmetry

important role, especially in the two-phase region even for polymer mixtures and polymer solutions. $K(t)$ is likely prerequisite for phase-inversion phenomena characteristic of viscoelastic phase separation. We would like to stress that we cannot assume $K(t) = 0$ in viscoelastic phase separation of complex fluids. The details will be described elsewhere [15]. This theory deals with polymer solutions, but is likely applicable to any dynamically asymmetric mixtures by changing the expressions of $G(t)$, $K(t)$, and $f(\phi)$ to the relevant ones. It should be noted that the rheological relaxation function $G(t)$ and $K(t)$ are functions of local composition $\phi(\mathbf{r})$ and, more importantly, we need some microscopic or phenomenological theories to estimate $G(t)$ and $K(t)$.

Generality of viscoelastic model

Here we point out that the viscoelastic model described by the above basic equations is a quite general model that includes all types of phase separation of isotropic condensed matter as its special cases: (i) If we assume that $\mathbf{v} = 0$ and G and K are independent of the composition in Eqs. (1)–(3), we obtain the basic equation of the "solid model (model B [2])". (ii) If we set $\mathbf{v} = 0$, $G(t) = \mu$ (μ: shear modulus), and $K(t) = K_b$ (K_b: bulk modulus), we obtain the basic equations of the "elastic solid model" [16]. (iii) If we set only $G(t) = \mu$ and $K(t) = K_b$, we obtain the basic equations of the "elastic gel model". Here, we use the fact that the time integration of the velocity becomes the deformation u, and $\sigma_{ij}^{(n)} = \mu[\partial u_{pj}/\partial x_i + \partial u_{pi}/\partial x_j - \frac{2}{3}(\nabla \cdot u_p)\delta_{ij}] + K_b(\nabla \cdot u_p)\delta_{ij}$. (iv) If we assume slow deformation ($\kappa_{ij}^p(t) \sim$ const.) and that G and K are independent of the composition, we obtain the basic equations of the "fluid model (model H [2])".

This generality of the "viscoelastic model" is summarized in Table 1. The viscoelastic model in the classification of isotropic phase separation corresponds to viscoelastic matter in the classification of isotropic condensed matter. Corresponding to the classification of isotropic matter into solids, viscoelastic matter, and fluids, we

can classify isotropic phase separation into three models, namely, solid, viscoelastic, an fluid model.

Concept of order-parameter switching

In the viscoelastic model, the phase-separation mode can be switched between the "fluid mode" and "elastic gel mode'. The dynamic process of viscoelastic phase separation is schematically drawn in Fig. 1. It is characterized by the switching of phase-separation modes between fluid-like and elastic gel-like ones [4]. This switching is likely caused by the change in the coupling between stress fields and velocity fields, which is described by Eq. (6): According to Eq. (6), the two extreme cases, namely, (i) fluid model ($\kappa_{ij}^p \sim$ const.) and (ii) elastic gel model ($G(t)$, $K(t) \sim$ const.), correspond to $\tau_{ts} \gg \tau_d$ and $\tau_{ts} \ll \tau_d$, respectively. Here τ_{ts} is the characteristic rheological time of the slower phase and τ_d is the characteristic time of domain deformation. For $\tau_d \gg \tau_{ts}$ the primary order parameter is the composition in usual classical fluids. For $\tau_d \lesssim \tau_{ts}$, on the other hand, it is the deformation tensor as in elastic gels. The deformation tensor u_{pij} is defined as $u_{pij} = (1/2)(\partial u_{pi}/\partial x_j + \partial u_{pj}/\partial x_i)$. It is well known [17, 18] that the free energy of gel, f, can be expressed only by local deformation tensor as $f(u_{pij})$. Thus, we can say that the order-parameter switching is a result of the competition between two time scales characterizing domain deformation τ_d and the rheological properties of domains τ_{ts}. This is a kind of *viscoelastic relaxation* in pattern evolution.

We estimate $\tau_d \propto R(t)/\Delta\phi(t)$ (R: domain size; $\Delta\phi$: the concentration difference between the two phases) and $\tau_{ts} = g(\phi_p)$, where ϕ_p is the ϕ in the polymer-rich phase and $g(x)$ is an increasing function of x, if we neglect the viscoelastic coupling effect. Although, in reality, this coupling effect does not allow τ_d to become shorter than τ_{ts}, we can understand the order-parameter switching rather straightforwardly on the basis of the above two relations. Note that $\Delta\phi$ and ϕ_p increases only in the initial stage and becomes constant in the late stage, while R is almost constant in the initial stage but increases continuously

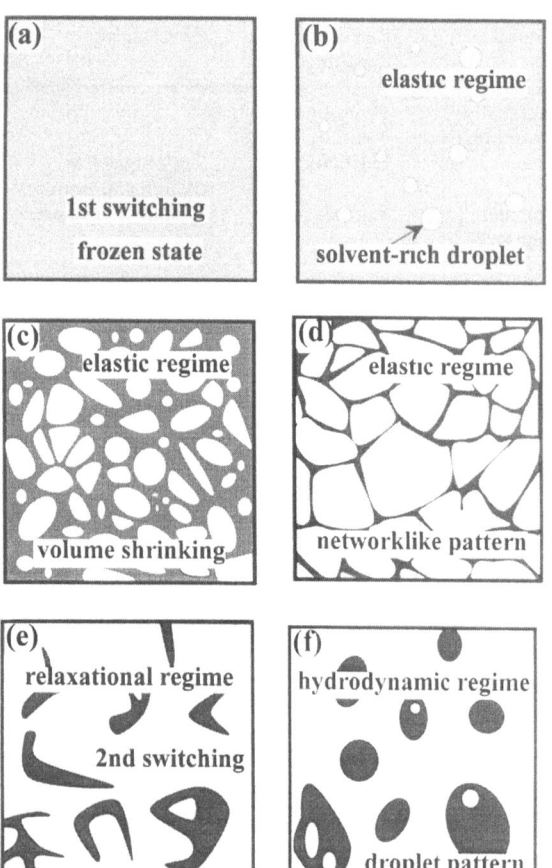

Fig. 1 Schematic figure of pattern-evolution process during viscoelastic phase separation. (a)–(d) correspond to the elastic regime, (f) to the hydrodynamic fluid regime, and (e) to the viscoelastic relaxational regime. The first order-parameter switching occurs around (a), while the second one around (e). The phase-separation process during (a)–(d) is probably essentially the same as that of elastic gel

without saturation in the late stage. The details of the switching mechanism will be described elsewhere. In relation to the order-parameter switching, it should be noted that there is no self-similarity in pattern evolution of viscoelastic phase separation. In viscoelastic phase separation, thus, we cannot use the scaling analysis, which is quite powerful in understanding the late-stage coarsening of the usual phase separation [1].

Difference between elastic effects in solids and viscoelastic effects in fluids

Although the viscoelastic effect looks similar to the elastic effect in solids in a certain time scale, there are essential differences between them. The former is a kinetic effect,

while the latter is a static, energetic effect: The elastic effect appears in the Hamiltonian of the system, while the viscoelastic effect appears in the kinetic equations describing the dynamics of the system. Only when the characteristic viscoelastic time τ_t is much longer than any time scales relevant to phase separation, the system can be treated as an elastic body. A more important difference comes from the difference between solid and fluid. It is worth pointing out the essential difference in pattern evolution between the phase separation influenced by the elastic effect [16] for elastic solid systems and that influenced by the viscoelastic effect for viscoelastic fluid systems. For the former, according to Onuki and Nishimori [16], the softer phase tends to become a continuous phase even for the mixture rich in a harder component to lower the elastic energy of the harder phase, since the deformation of the harder phase costs more energy than the deformation of the softer phase. For the latter, on the other hand, even for the mixture rich in a less viscoelastic component the more viscoelastic phase tends to become the continuous matrix phase during the crossover from the regime dominated by dynamic symmetry to the regime dominated by static symmetry. This is due to the fact that (i) a relaxational nature of the viscoelastic effect causes the continuous change of the apparent phase diagram and leads to the formation of network-like pattern (see Eqs. (1) and (4)), (ii) the connectivity of the phase is preserved till the late stage since the stress is always supported by polymer chains and thus the polymer-rich phase is selectively deformed to keep the stress balance, and (iii) in the two-fluid model the two components can have different velocities. As described below, it should be stressed that the above difference in pattern formation between elastic solid systems and viscoelastic fluid systems likely originates from the difference between a solid (diffusion) model and a fluid (hydrodynamic) model.

However, dynamic asymmetry can be introduced even for solid mixtures through the composition dependence of a diffusion constant. This is only the way to introduce dynamic asymmetry into solids there is no velocity field in solids. It should be noted that in the above argument dynamic asymmetry originates from the asymmetry in the mechanical properties between two components. Dynamic asymmetry is a prequisite to phase-inversion phenomena irrespective of whether a system is solid or fluid.

Another important, but rather apparent, difference between elastic and viscoelastic effect is the final (or very late-stage) morphology of phase separation: An elastic effect affects the final morphology since it is an energetic effect. On the other hand, a viscoelastic effect does not affect the final morphology, which is purely determined by energetic factors such as interfacial energy, since there remain only viscous effects and no elastic effects for $t \to \infty$.

Universal nature of sponge-like pattern

Finally, we discuss the universal nature of the foam-like morphology and its physical origin. It is known that a gel undergoing volume-shrinking phase transition forms a bubble-like structure [17, 19]. The physical origin of the appearance of the honeycomb structure in polymer foams (e.g., polystyrene foam and urethane foam) is also likely similar to ours. We also know that the solvent evaporation of polymer solution induces the honeycomb-like structure, which is used as membrane filters [20, 21]. All these processes have a common feature that the holes of a less viscoelastic phase (gas in foam, water in gels, a solvent in polymer solution, and so on) are nucleated to minimize the elastic energy associated with the formation of a heterogeneous structure in an elastic medium. Then, the more viscoelastic phase decreases its volume. This volume-shrinking process is dominated by the transfer (diffusion or flow) of the less viscoelastic component in the more viscoelastic phase to satisfy the total mass conservation. Thus, the limiting process of evolution is commonly the diffusion or flow process of the fluid component under stress. Among these examples, a polymer foam is unique in the point that the total volume is not conserved since it accompanies the liquid–gas phase transition. In addition to patterns in mixtures of condensed matters, we also point out the similarity of these patterns to the foam-like structure of the universe [22], where the gravitational attractive interaction likely plays a role similar to elasticity in producing the structure. The foam-like structure is probably the universal morphology for the phase separation of a system in which only one component asymmetrically has elasticity stemming from either topological connectivity or attractive interaction. The common feature of these systems originates from the volume phase transition, or more strictly elastic phase separation of dynamically asymmetric mixtures including *fluid* (*liquids and gasses*) at least as a component. As described in the preceding section, this differs from the elastic phase separation of dynamically symmetric solid mixtures (e.g., metal alloys) which does not accompany the volume change of each phase: In this case, a softer phase always forms a continuous phase to minimize the total elastic energy, contrary to our case.

Summary

Viscoelastic phase separation is expected to be universal in any mixture having asymmetry in elementary molecular dynamics between its components. The possible candidates for *dynamic asymmetry* are (1) slow dynamics in complex fluids such as polymer solutions and surfactant solutions, coming from their complex internal degree of freedom (e.g., entanglement effects in polyers) and (2) that near-glass transition. We hope that more examples of viscoelastic phase separation will be found in the family of complex fluids in the near future.

Acknowledgements This work was partly supported by a Grant-in-Aid from the Ministry of Education, Science, and Culture, Japan and also by a grant from Toray Science Foundation.

References

1. Gunton JD, San Miguel, Sahni P (1983) In: Domb C, Lebowitz JH (eds) Phase Transition and Critical Phenomena. vol 8 Academic, London
2. Hohenberg PC, Halperin BI (1976) Rev Mod Phys 49:435
3. Tanaka H (1994) J Chem Phys 100:5253 and references therein; Tanaka H (1995) Int J Thermophys 16:371
4. Tanaka H (1996) Phys Rev Lett 76:787
5. Brochard F, de Gennes PG (1977) Macromolecules 10:1157, Brochard F (1983) J Phys France 44:39
6. see e.g. Ver Strate Philippoff W (1974) Polym Lett 12:267 Rangel-Nafaile C, Metzner AB, Wissbrum KF (1984) Macromolecules 17:1187; Wolf BA (1988) Makromol Chem 189:163. On the recent development, see references in refs. [7–11]
7. Tanaka T, Filmore DJ (1979) J Chem Phys 70:1214
8. Helfand E, Fredrickson GH (1989) Phys Rev Lett 62:2648
9. Onuki A (1989) Phys Rev Lett 62:2427
10. Doi M In: Onuki A, Kawasaki K (eds) Dynamics and Patterns in Complex Fluids. Springer, 1990 p 100
11. Milner ST (1993) Phys Rev E 48:3674
12. Doi M, Onuki A (1992) J Phys II France 2:1631
13. de Gennes PG (1979) Scaling Concepts in Polymer Physics. Cornell Univ Press, New York
14. Doi M, Edwards SF (1986) The theory of Polymer Dynamics (1986). Clarendon Press, Oxford
15. Tanaka H (1997) Prog Theor Phys Suppl 126:333; (1997) Phys Rev E 56:xxx; Tanaka H, Araki T (1997) Phys Rev Lett 78:4966
16. see e.g. Onuki A Nishimori H (1991) Phys Rev B 43:13649
17. Sekimoto K, Suematsu N, Kawasaki K (1989) Phys Rev A 39:4912
18. Onuki A (1993) In: Advances in Polymer Sciences. vol. 109, Springer, Berlin, p 63
19. Matsuo ES, Tanaka T (1988) J Chem Phys 89:1695
20. Song SW, Torkelson M (1994) Macromolecules 27:6390
21. Widawski G, Rawiso M, Francois B (1994) Nature 369:387
22. Geller MJ, Huchra JP (1989) Science 246:897

Progr Colloid Polym Sci (1997) 106:172–174
© Steinkopff Verlag 1997

T. Okuzono

A numerical study of viscoelastic phase separation in polymer solutions

Received: 13 October 1996
Accepted: 15 April 1997

Dr. T. Okuzono (✉)
Department of Physics
Ochanomizu University
Tokyo 112, Japan

Present address: Department of
Computational Science and Engineering,
Nagoya University, Nagoya, 464-01, Japan

Abstract A numerical model is
constructed for the viscoelastic phase
separation in polymer solutions based
upon the two-fluid model using the
method of the smoothed-particle
hydrodynamics. Computer
simulations are carried out with this
model in two dimensions and effects
of the stress relaxation time on
morphology of domains are
examined. It is observed that the more
viscous phase forms network-like
domains when the stress relaxation
time is large although those domains
are not continuous.

Key words Viscoelastic phase
separation – dynamical
asymmetry – two-fluid
model – smoothed-particle
hydrodynamics

Recently, a new type of phase separation called "viscoelastic phase separation" was observed in polymer solutions or dynamically asymmetric fluid mixtures [1–3]. It is an interesting feature of this phenomenon that network-like domains of more viscous phase emerge in a transient regime. It has little been understood what ingredient of physics is crucial to this phenomenon. Various numerical approaches have been made for the phase separation phenomena in binary fluid systems in the last decade [4–6]. Most of these studies have been concerned with classical fluids and have not involved viscoelasticity. A new numerical model was recently proposed by the author [7] based upon the two-fluid model [8,9] using the method of smoothed-particle hydrodynamics (SPH) [10,11]. In this model the Lagrangian picture for fluid is adopted and the viscoelastic effect can easily be incorporated. In this paper we carry out a computer simulation for the viscoelastic phase separation in polymer solutions with this model.

In the SPH method the mass density at **r** of fluid is estimated as a superposition of "smoothing functions" $W(\mathbf{r} - \mathbf{r}_\alpha, h)$, where $\{\mathbf{r}_\alpha\}$ ($\alpha = 1, 2, \ldots$) is a set of positions of fluid particles which is randomly distributed and h defines a size of particle. The smoothing function is typically Gaussian, i.e., $W(\mathbf{r}, h) \propto \exp(-\mathbf{r}^2/h^2)$. In our model there are two types of particles. A and B, corresponding to each fluid in the two-fluid model. The mass densities $\rho_A(\mathbf{r})$ and $\rho_B(\mathbf{r})$ of A- and B-fluid, respectively, are given by

$$\rho_A(\mathbf{r}) = \sum_{j=1}^{N_A} m_A W(\mathbf{r} - \mathbf{r}_j, h), \quad \rho_B(\mathbf{r}) = \sum_{n=1}^{N_B} m_B W(\mathbf{r} - \mathbf{r}_n, h), \quad (1)$$

where m_A and m_B are the mass of a particle of A- and B-fluid, respectively, and N_A and N_B are the total number of A and B particles. The local velocity of A- and B-fluid denoted by $\mathbf{v}_A(\mathbf{r})$ and $\mathbf{v}_B(\mathbf{r})$, respectively, are defined as

$$\mathbf{v}_A(\mathbf{r}) = \sum_j \frac{m_A}{\rho_{A_j}} \mathbf{r}_j W(\mathbf{r} - \mathbf{r}_j, h), \quad \mathbf{v}_B(\mathbf{r}) = \sum_n \frac{m_B}{\rho_{B_n}} \dot{\mathbf{r}}_n W(\mathbf{r} - \mathbf{r}_n, h),$$

$$(2)$$

where $\rho_{A_j} \equiv \rho_A(\mathbf{r}_j)$, $\rho_{B_n} \equiv \rho_B(\mathbf{r}_n)$, and $\mathbf{r}_\alpha \equiv \frac{d}{dt}\mathbf{r}_\alpha$ is the velocity of particle α. From these expressions we can construct a Lagrangian and dissipation function and derive a set of equations of motion of particles as a variational form for the coordinates $\{\mathbf{r}_\alpha, \dot{\mathbf{r}}_\alpha\}$. For details of the model, see [7].

Now, we consider the viscoelastic effect which is not taken into account in our previous study [7]. We write the

Progr Colloid Polym Sci (1997) 106:172–174
© Steinkopff Verlag 1997

equations of motion of A-particle i and B-particle l as

$$m_A \ddot{\mathbf{r}}_i = -\frac{\delta}{\delta \mathbf{r}_i}(F + F_{el}) - \frac{m_A}{\rho_{A_i}}\zeta_i(\mathbf{v}_A - \mathbf{v}_B)_i, \quad (3)$$

$$m_B \ddot{\mathbf{r}}_l = -\frac{\delta}{\delta \mathbf{r}_l}(F + F_{el}) + \frac{m_B}{\rho_{B_l}}\zeta_l(\mathbf{v}_A - \mathbf{v}_B)_l, \quad (4)$$

where $F + F_{el}$ is the total free energy with its elastic part F_{el} and $(\mathbf{v}_A - \mathbf{v}_B)_\alpha$ is the velocity difference between A- and B-fluid at $\mathbf{r}_\alpha$ and ζ_α is the friction coefficient [9] at $\mathbf{r}_\alpha$. The last terms in the above equations describe the friction force between the two fluids, which is a characteristic in the two-fluid model. We assume the variation of the elastic free energy δF_{el} is the sum of each variation of A- and B-fluid, i.e.,

$$\delta F_{el} = \sum_j^{N_A}\frac{m_A}{\rho_{A_j}}\sigma_j^{(A)} : \delta\mathbf{u}_j^{(A)} + \sum_n^{N_B}\frac{m_B}{\rho_{B_n}}\sigma_n^{(B)} : \delta\mathbf{u}_n^{(B)}, \quad (5)$$

where the deformation tensor $\delta\mathbf{u}_j^{(A)} \equiv \frac{1}{2}[(\nabla\delta\mathbf{r}_A)_j + (\nabla\delta\mathbf{r}_A)_j^T]$ is written in terms of the variations of positions $\{\delta\mathbf{r}_j\}$ as $(\nabla\delta\mathbf{r}_A)_i = \sum_j(m_A/\rho_{A_i})(\delta\mathbf{r}_j - \delta\mathbf{r}_i)\nabla_i W_{ij}$ with $\nabla_i W_{ij} \equiv \frac{\partial}{\partial\mathbf{r}_i}W(\mathbf{r}_i - \mathbf{r}_j, h)$. $\sigma_j^{(A)}$ is the stress tensor conjugate to $\delta\mathbf{u}_j^{(A)}$. $\sigma_n^{(B)}$ and $\delta\mathbf{u}_n^{(B)}$ are defined in the same manner. Hence, the elastic force in A-fluid is given by

$$-\frac{\delta F_{el}}{\delta\mathbf{r}_i} = \sum_j m_A^2\left(\frac{\sigma_i^{(A)}}{\rho_{A_i}^2} + \frac{\sigma_j^{(A)}}{\rho_{A_j}^2}\right)\cdot\nabla_i W_{ij}. \quad (6)$$

The stress tensors $\sigma_i^{(A)}$ and $\sigma_i^{(B)}$ are regarded as new variables in this model and appropriate time evolution equations for these variables must be supplemented.

Consider a polymer solution system which consists of polymers (A-fluis) and solvent (B-fluid) as a typical dynamically asymmetric system. The stress tensor $\sigma_i^{(A)}$ is assumed to obey the following Maxwell constitutive model with a single relaxation time τ_A,

$$\frac{d}{dt}\sigma_i^{(A)} + \Omega_i^{(A)}\cdot\sigma_i^{(A)} + (\Omega_i^{(A)}\cdot\sigma_i^{(A)})^T = -\frac{1}{\tau_A}\sigma_i^{(A)} + 2G_e\mathbf{D}_i^{(A)}, \quad (7)$$

where $\Omega_i^{(A)}$ and $\mathbf{D}_i^{(A)}$ are the vorticity tensor and strain rate tensor of A-fluid at $\mathbf{r}_i$ (for the explicit expressions, see [7]), respectively, and G_e is the shear modulus. For the stress tensor $\sigma_n^{(B)}$ of the solvent, its relaxation time τ_B is small ($\tau_B \ll \tau_A$). Thus, $\sigma_n^{(B)}$ is given by the viscous stress, i.e., $\sigma_n^{(B)} = 2\eta\mathbf{D}_n^{(B)}$, where η is the shear viscosity and $\mathbf{D}_n^{(B)}$ is the strain rate tensor of B-fluid at $\mathbf{r}_n$.

Now, we carry out the simulation by using the above model in two dimensions. In order to concentrate our efforts on examining the effect of dynamical asymmetry, we simplify the model assuming that F is the symmetric Ginzburg–Landau-type free energy and τ_A and G_e are

constant. Here we show the results for two systems with different parameters: $\tau_A = 1.0$ and 2.0×10^3 keeping $\tau_A G_e/\eta = 1$. Figs. 1a–d and Figs. 1e–h show sequential snapshots of the systems for smaller and larger τ_A, respectively, after quenched into the spinodal region. In both the

Fig. 1 Snapshots of the domain pattern at $t = 50$ (a, e); 100 (b, f); 200 (c, g); and 400 (d, h). The systems with $\tau_A = 1.0$ and $\tau_A = 2.0 \times 10^3$ are shown in a–d and e–h, respectively. The mass density of A-fluid is shown by gray scales. Bright regions correspond to A-rich domains

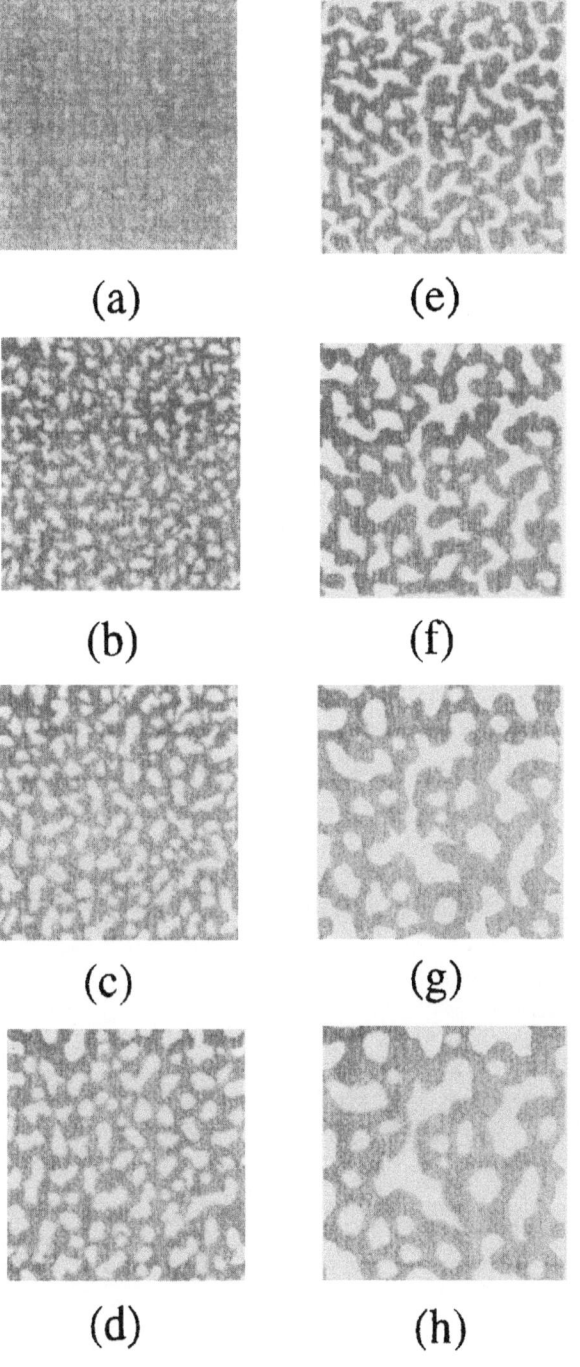

(a) (e)

(b) (f)

(c) (g)

(d) (h)

systems, the total mass fraction of the A-fluid is 0.4 and $N_A = N_B = 10^4$. Bright (dark) regions correspond to A(B)-rich phase in Fig 1. We observe "broken" network-like patterns in the system with larger relaxation time τ_A, although those are not continuous as observed in the real experiments [1-3]. This implies the stress relaxation time is one of important parameters for the viscoelastic phase separations.

This work is supported by a Grant-in-Aid for Encouragement of Young Scientists, from the Ministry of Education, Science and Culture, Japan.

References

1. Tanaka H (1993) Phys Rev Lett 71:3158-3161
2. Tanaka H (1994) J Chem Phys 100:5323-5337
3. Tanaka H (1996) Phys Rev Lett 76:787-790
4. Koga T, Kawasaki K (1993) Physica A 196:389-415 and references therein
5. Shinozaki A, Oono Y (1993) Phys Rev E 48:2622-2654 and references therein
6. Rothman DH, Zaleski S (1994) Rev Mod Phy 66:1417-1479 and references therein
7. Okuzono T, Phys Rev E, submitted
8. De Gennes PG (1979) Scaling Concepts in Polymer Physics. Cornell University Press, Ithaca, New York
9. Doi M, Onuki A (1992) J Phys II France 2:1631-1656 and references therein
10. Monaghan JJ (1992) Annu Rev Astron Astrophys 30:543-574 and references therein
11. Hoover WG, Pierce TG, Hoover CG, Shugart JO, Stein CM, Edwards AL (1994) Computers Math Appl 28:155-174

Progr Colloid Polym Sci (1997) 106:175–179
© Steinkopff Verlag 1997

Phase separations in binary mixtures of a flexible polymer and a liquid crystal

A. Matsuyama
T. Kato

Received: 13 October 1996
Accepted: 7 March 1997

Dr. A. Matsuyama (✉) · T. Kato
Department of Chemistry for Materials
Faculty of Engineering
Mie University
Tsu Mie 514, Japan

Abstract A simple model is presented to describe the phase behavior in binary mixtures of a flexible polymer and a liquid crystal. By using the incompressible approximation to the Onsager theory combining both the excluded volume interactions between the liquid crystals and the attractive interaction due to the alignment of liquid crystals, we obtain the free energy of mixing of the binary mixtures. We discuss the phase separations of binary mixtures of polymers and liquid crystals and that of solutions of rigid rod-like molecules. We also examine the volume phase transition of gels in liquid crystals.

Key words Theory – phase separation – polymers – liquid crystals – gels

Introduction

Recently, polymer-dispersed liquid-crystal materials have played an important role for many practical processes such as the electro-optical displays [1, 2]. Since the miscibility or phase separation of mixtures of polymers and liquid crystals controls the performance of the materials, the phase behavior and the phase separation kinetics have been of experimental and theoretical interest. One of the main problem is to examine the location of various phases, such as isotropic and nematic phases, depending on temperature and concentration.

In the mixture of a polymer and a liquid crystal, the nematic–isotropic phase separation and the isotropic–isotropic one with an upper critical solution temperature has been often observed [3–9]. Typical examples are the mixtures of (p-ethoxybenzylidene)-p-n-butylaniline (EBBA), and polystylene (PS) or polyetylen oxide(PEO) [3] and the mixtures of 4-cyano-4′-n-heptylbiphenyl (7CB) with polymethyl methacrylate (PMMA) or PS [7]. These solutions consist of a low molecular weight liquid crystal and a sufficiently flexible polymer. The main feature of the

phase separation has been described by the lattice theory [8, 9], the Landau-de Gennes expansion [10], and the Maier–Saupe model [11, 12].

In this paper we propose a molecular theory combining both the orientation-dependent attractive interactions [11–15] and the excluded volume interactions [15, 16] between liquid crystals. Our theory predicts that the attractive interaction parameter between liquid crystals controls the peculiar types of phase separations, such as isotropic–isotropic phase separation and nematic–isotropic one. We also examine the phase diagrams for solutions of rigid rod-like molecules [17–19] and the volume phase transition of gels in liquid crystals.

Free energy of mixing of liquid crystals and flexible polymers

We consider a binary mixture of N_r liquid crystals and N_p flexible polymer molecules. On the basis of the Onsager theory, including both the excluded volume interactions and the orientation-dependent attractive interactions between liquid crystals, the free energy of mixing of the binary mixture is given by [20]

$$\beta \Delta F_{mix} = N_t \left[\frac{\phi_p}{n_p} \ln \phi_p + \frac{\phi_r}{n_r} \ln \phi_r + \chi \phi_r \phi_p \right.$$

$$\left. + \tfrac{1}{2}(\chi_a + \tfrac{5}{4})\phi_r^2 S^2 - \frac{\phi_r}{n_r} \ln(I_0(\Gamma_0 \phi_r S)/2) \right], \quad (1)$$

$$\Gamma_0 \equiv (\chi_a + \tfrac{5}{4})n_r , \quad (2)$$

where $\phi_p(\phi_r)$ is the volume fraction of the polymer (liquid crystal). The first two terms in Eq. 1) represent the entropy of mixing and χ shows the isotropic interaction parameter between polymer and liquid crystal. The parameter χ is the well-known Flory–Huggins interaction parameter [21] which controls the miscibility of the two species in the isotropic phase. The last two terms represent the free-energy change due to the alignment of the liquid crystals. The terms $\chi_a n_r$ and $(5/4)n_r$ in Eq. (2) correspond to the attractive [13] and excluded volume [16] interactions between liquid crystals, respectively.

The order parameter S of the liquid crystals is determined by the self-consistent equation [20]

$$S = I_1(\eta S)/I_0(\eta S) , \quad (3)$$

where $\eta \equiv \Gamma_0 \phi_r$ and the function $I_m(\eta S)$ is defined as

$$I_m(\eta S) \equiv \int_{-1}^{1} [\tfrac{3}{2}(x^2 - \tfrac{1}{3})]^m \exp[\tfrac{3}{2} \eta S(x^2 - \tfrac{1}{3})] \, dx . \quad (4)$$

The chemical potentials are given by

$$\beta \Delta \mu_p(\phi, S) = \beta(\mu_p - \mu_p^o) = \beta(\partial \Delta F_{mix}/\partial N_p)_{N_r, T}$$

$$= \ln \phi + \left(1 - \frac{n_p}{n_r}\right)(1 - \phi) + n_p \chi(1 - \phi)^2$$

$$+ \frac{n_p}{2}(\chi_a + \tfrac{5}{4})S^2(1 - \phi)^2 , \quad (5)$$

for the polymer molecules and

$$\beta \Delta \mu_r(\phi, S) = \beta(\mu_r - \mu_r^o) = \beta(\partial \Delta F_{mix}/\partial N_r)_{N_p, T}$$

$$= \ln(1 - \phi) + \left(1 - \frac{n_r}{n_p}\right)\phi + n_r \chi \phi^2$$

$$+ \frac{n_r}{2}(\chi_a + \tfrac{5}{4})S^2(1 - \phi)^2 - \ln(I_0(\eta S)/2) , \quad (6)$$

for the liquid crystals where we define $\phi \equiv \phi_p$.

The condition for the binodal curve is given by the simultaneous equations

$$\Delta \mu_p(\phi_i, 0) = \Delta \mu_p(\phi_i', 0) , \quad (7)$$

$$\Delta \mu_r(\phi_i, 0) = \Delta \mu_r(\phi_i', 0) \quad (8)$$

for the isotropic–isotropic phase separation;

$$\Delta \mu_p(\phi_i, 0) = \Delta \mu_p(\phi_n, S) , \quad (9)$$

$$\Delta \mu_r(\phi_i, 0) = \Delta \mu_r(\phi_n, S) \quad (10)$$

for the nematic–isotropic equilibrium; and

$$\Delta \mu_p(\phi_n, S) = \Delta \mu_p(\phi_n', S') , \quad (11)$$

$$\Delta \mu_r(\phi_n, S) = \Delta \mu_r(\phi_n', S') \quad (12)$$

for the nematic–nematic equilibrium. The order parameter S is given as a function of ϕ and temperature by solving the self-consistent equation (3).

Roles of attractive and repulsive interactions between rod-like molecules

For a numerical calculation, we introduce the temperature parameter τ defined by $\tau \equiv 1/\chi (= k_B T/U_0)$. We then have four parameters characterizing our systems: n_p, the number of segments on a flexible polymer; n_r, the number of segments on a liquid crystal; χ_a, the attractive interaction (Maier–Saupe) parameter between liquid crystals; $\chi = 1/\tau$, the polymer–liquid crystal interaction (Flory–Huggins) parameter whose the origin is the dispersion forces. We here define the nematic interaction parameter $\alpha \equiv \chi_a/\chi$.

From Eq. (3), we can obtain the values of the order parameter $S(\tau, \phi)$ related to a certain temperature τ and concentration ϕ. The nematic phase appears at $\Gamma_0(1 - \phi) = 4.55$ in Eq. (3) [15]. We then obtain the nematic–isotropic transition (NIT) temperature (τ_{NI}) as a function of the polymer concentration ϕ:

$$\tau_{NI} = \frac{\alpha n_r(1 - \phi)}{4.55 - 1.25 n_r(1 - \phi)} . \quad (13)$$

When the temperature τ is below τ_{NI}, the nematic phase is stable and when $\tau > \tau_{NI}$ the isotropic phase is stable. The order parameter S jumps from zero to 0.44 at $\tau = \tau_{NI}$. From Eq. (13), we see that τ_{NI} diverges at the concentration $\phi^* = (1.25 n_r - 4.55)/1.25 n_r$. For a pure liquid crystal $(\phi = 0)$, the NIT temperature τ_{NI}^o is given by

$$\tau_{NI}^o = \frac{\alpha n_r}{4.55 - 1.25 n_r} , \quad (14)$$

and when n_r is larger than 3.64, the τ_{NI}^o diverges to the high-temperature side. For lower values of n_r, the attractive interaction χ_a between rods dominates and so we have the strong dependence of the NIT curve on temperature. For larger values of n_r, however, the repulsive interaction dominates and the attractive interaction plays only an

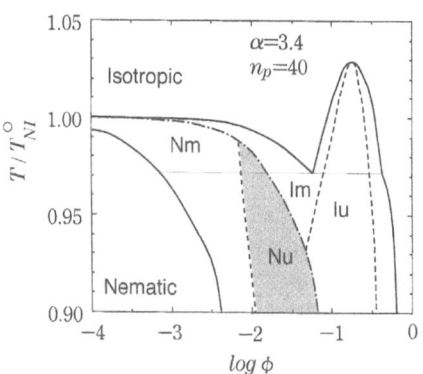

Fig. 1 Phase diagram for $\alpha = 3.4$, $n_r = 1.5$, and $n_p = 40$

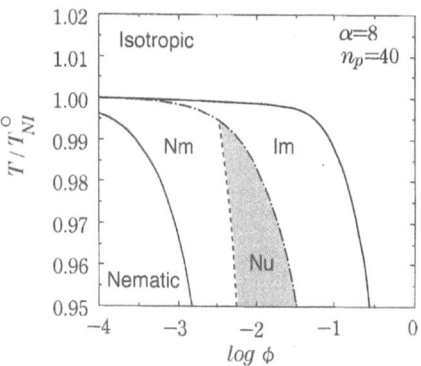

Fig. 2 Phase diagrams for $\alpha = 8$. When the nematic interaction parameter α is increased, the isotropic–isotropic phase separation disappears

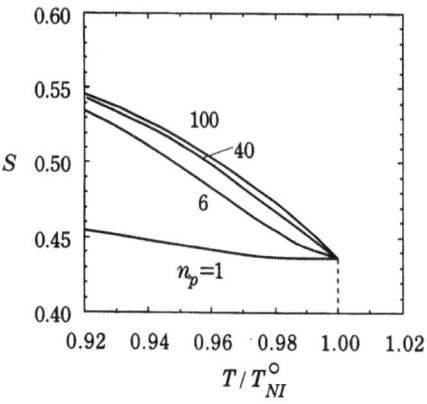

Fig. 3 Order parameter S on the nematic phase boundaries plotted against the reduced temperature

auxiliary role in the NIT. Then we have the weak dependence of the NIT curve on temperature and, consequently, the NIT temperature diverges.

Figure 1 shows the phase diagram for $\alpha = 3.4$, $n_r = 1.5$, and $n_p = 40$. The temperature is normalized by the NIT temperature of the pure liquid crystal. The solid (broken) lines correspond to the binodal (spinodal) lines. The dashed-and-dotted line shows the NIT line. In the region between the binodal and spinodal lines, there are two metastable regions; an isotropic metastable (Im) and a nematic metastable (Nm). In the low temperature of the spinodal line, we have an isotropic unstable (Iu) and a nematic unstable (Nu) regions. At lower temperatures, the biphasic region where a nematic and an isotropic phase coexist appears. The stable nematic phase appears in a dilute region of the polymer concentration. The nematic and isotropic binodal lines intersect at $\phi = 0$ and $T = T_{NI}^\circ$.

We also have the isotropic–isotropic phase separation with an uipper critical solution temperature such as observed for polystylene with 7CB [7]. The critical point (ϕ_c, τ_c) in the isotropic phase is obtained by the condition $\partial \Delta \mu_r / \partial \phi = 0$ and $\partial^2 \Delta \mu_r / \partial \phi^2 = 0$ [21]:

$$\phi_c = \frac{\sqrt{n_r}}{\sqrt{n_r} + \sqrt{n_p}}, \tag{15}$$

$$\tau_c = \frac{2 n_r n_p}{(\sqrt{n_r} + \sqrt{n_p})^2}. \tag{16}$$

We also find the three-phase equilibrium (triple point) where two isotropic phases and a nematic phase can simultaneously coexist. At high temperature, the solution is homogeneous since the entropy of mixing is dominant. As temperature decreases, the isotropic–isotropic phase separation takes place where the interaction parameter χ between solute and solvent molecules becomes dominant. Further decrease in temperature, the attractive interaction χ_a between liquid crystals dominates and so the

nematic–isotropic phase separation occurs at the lower-temperature side of the triple point.

Figure 2 shows the phase diagram for $\alpha = 8$. When the nematic interaction parameter α is increased, the isotropic–isotropic phase separation disappears and we have only the nematic–isotropic phase separation. As the molecular weight of the polymer increases, the nematic phase boundaries are shifted to the lower concentration and the polymers become insoluble in the nematic phase [20]. Figure 3 shows the order parameter on the nematic phase boundaries plotted against the reduced temperature. The order parameter S jumps from zero to 0.44 at $T = T_{NI}^\circ$. For large values of n_p, the order parameter increases with decreasing temperature.

As shown in Figs. 1 and 2, there are two generic types of phase diagrams. The conditions for the existence of these phase diagrams are obtained by comparing the NIT temperature T_{NI}° of pure liquid crystals and the upper critical solution temperature T_c in the isotropic phase.

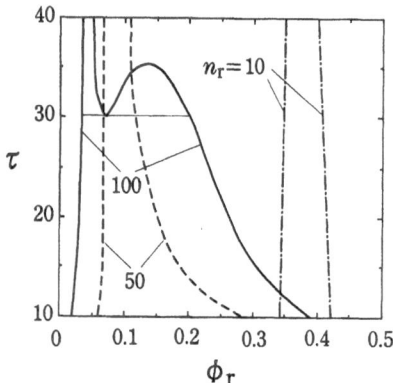

Fig. 4 Phase diagrams in solutions of rigid rod-like molecules with $\alpha = 0$. The number n_r of segments of rods is changed

From Eqs. (14) and (16), the nematic interaction parameter α is given by

$$\alpha = \frac{2n_p(4.55 - 1.25n_r)}{(\sqrt{n_r} + \sqrt{n_p})^2}\left(\frac{T_{NI}^o}{T_c}\right). \tag{17}$$

For $n_p \gg 1$, we have

$$\alpha = \alpha^*\left(\frac{T_{NI}^o}{T_c}\right), \tag{18}$$

where $\alpha^* \equiv 2(4.55 - 1.25n_r)$. When $n_r = 1.5$, we have $\alpha^* = 5.35$. For $\alpha < \alpha^*$ or equivalently $T_{NI}^o < T_c$, we have the isotropic–isotropic phase separation with an upper critical solution point as shown in Fig. 1. For $\alpha > \alpha^*$ or $T_{NI}^o > T_c$, we find the nematic–isotropic phase separation as shown in Fig. 2. Thus, the type of the phase diagram is characterized by the nematic interaction parameter α. We note that when $n_r = 1$ the value of α^* results in that obtained by Holyst and Schick [10].

When $n_r \gg n_p$, where the excluded volume interaction dominates, we can obtain the phase diagram of mixtures of a rigid rod-like molecule and a solvent. Figure 4 shows the binodal curves for $n_r = 10, 50, 100$ with $n_p = 1$ and $\alpha = 0$. The binodal curve for the isotropic solution is on the left and that for the nematic phase on the right. For $n_r = 10$ and 50, the isotropic–nematic phase separation appears. For longer rods, the triple point where two nematic phases and an isotropic phase can simultaneously coexist appears. At the higher temperature side of the triple point, we have the co-occurrence of two nematic phases and the narrow isotropic–nematic biphasic region. At the lower temperature side of the triple point, the broad biphasic region appears. As the strength of the nematic interaction parameter α increases, the width of the broad biphasic region increases and the temperature of the triple point is shifted to the higher temperature side [20]. The form of the

phase diagram becomes similar to that obtained by Flory lattice model [17, 18].

Volume phase transition of gels in liquid crystals

In this section, we extend our theory to networks of flexible chains in a liquid crystal solvent and examine the degree of swelling of the gel as a function of temperature. Although it has been reported that the low molecular weight flexible polymers in a nematic solvent have an anisotropic conformation [22], we here assume that the macroscopic shape of gels (chain) is not changed by the nematic solvent. This assumption will be correct for the gels of the large number of segments between crosslinks. The anisotropic swelling is important for gels under an external field [23] or gels of semiflexible polymer [24].

We consider networks of flexible chains in a nematogenic solvent. Let n_r be the number of segments on the liquid crystal. The equilibrium volume fraction ϕ of the gel can be determined by

$$\mu_r^o(S_b) = \mu_r(\phi, S_i), \tag{19}$$

where μ_r^o is the chemical potential of the liquid crystal outside the gel, S_b the order parameter outside the gel, μ_r the chemical potential of the liquid crytal inside the gel and S_i the order parameter inside the gel. The chemical potential of the liquid crystals outside the gel is given by

$$\beta\mu_r^o(S_b) = \frac{1}{2}\left(\chi_a + \frac{4}{5}\right)S_b^2 - \frac{1}{n_r}\ln\left[I_0(\Gamma_0 S_b)/2\right], \tag{20}$$

where the order parameter S_b is determined by the self-consistent equation

$$S_b = I_1(\Gamma_0 S_b)/I_0(\Gamma_0 S_b). \tag{21}$$

The chemical potential of the liquid crystals inside the gel is given by [25]

$$\beta\mu_r(\phi, S_i) = \frac{\phi^*}{n}\left(Q^{-1/3} - \frac{1}{2}Q^{-1}\right) + \frac{1}{n_r}$$

$$\times (\phi + \ln(1 - \phi)) + \chi\phi^2 + \frac{1}{2}\left(\chi_a + \frac{5}{4}\right)$$

$$\times S_i^2(1 - \phi)^2 - \frac{1}{n_r}\ln\left[I_0(\eta_i S_i)/2\right], \tag{22}$$

where $\eta_i \equiv \Gamma_0(1 - \phi)$, the first term of Eq. (22) represents the elastic energy for the condition of an isotropic swelling [21] and $Q \equiv V/V_0(= \phi^*/\phi)$ is the swelling ratio of the final volume V (volume fraction ϕ) to the initial volume V_0 (volume fraction ϕ^*) of gel. The order parameter S_i inside

Progr Colloid Polym Sci (1997) 106:175–179
© Steinkopff Verlag 1997

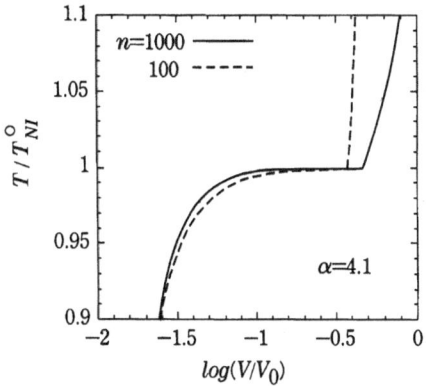

Fig. 5 Swelling ratio V/V_0 of the gel plotted against the reduced temperature T/T_{NI}° for $n_r = 2$, $n = 100$, 1000 and $\alpha = 4.1$

gel is determined by

$$S_i = I_1(\eta_i S_i)/I_0(\eta_i S_i) . \tag{23}$$

Substituting Eqs. (20) and (22) into Eq. (19), we can numerically solve the degree of swelling of gel as a function of temperature.

Figure 5 shows the swelling ratio Q of the gel for $n_r = 2$, $n = 100$, 1000 and $\alpha = 4.1$ plotted against the reduced temperature T/T_{NI}° where T_{NI}° is the nematic–isotropic transition temperature of the pure liquid crystal outside the gel. In contrast to the first-order volume phase transitions of gels in an isotropic solvent [26, 27], we find that the second-order volume phase transition takes place at $T = T_{NI}^\circ$ where the chemical potential (Eq. (20)) of the isotropic and nematic phases are equal: $\mu_r^\circ(S_b > 0) = \mu_r^\circ(S = 0) = 0$. Below T_{NI}°, the nematic ordering of the external solvent takes place and the gel collapses into a globule state. The condensed gel is in an isotropic state ($S_i = 0$) and the volume is almost independent of n. The gel is condensed by the nematic ordering of the external solvent.

Conclusions

On the basis of the molecular theory combining both the excluded volume interactions and the orientation-dependent attractive interactions between liquid crystals, we have theoretically studied the phase transitions of mixtures of a flexible polymer and a liquid crystal. We have found that (1) our theory can qualitatively explain the observed phase diagrams not only in the polymer–liquid crystal systems but also in the solutions of rigid rod-like molecules, (2) gel immersed in a liquid-crystal solvent can undergo the second-order volume transition due to the nematic ordering of the external solvent.

References

1. Drzaic PS (1995) Liquid Crystal Dispersions. World Scientific, Singapore
2. Crawford GP, Zumer S (eds) (1996) Liquid Crystals in Complex Geometries. Taylors Francis, London
3. Kronberg B, Bassignana I, Patterson D (1978) J Phys Chem 82:1714
4. Dubaut A, Casagrande C, Veyssie M, Deloche B (1980) Phys Rev Lett 45:1645
5. Orendi H, Ballauff M (1989) Liq Cryst 6:497
6. Dormoy Y, Gallani JL, Martinoty P (1989) Mol Cryst Liq Crys 170:135
7. Ahn W, Kim CY, Kim H, Kim SC (1992) Macromolecules 25:5002
8. Ballauff M (1986) Cryst Liq Cryst 136:175
9. Ballauff M (1986) Mol Cryst Liq Cryst Lett 4:15
10. Holyst R, Schick M (1992) J Chem Phys 96:721
11. Shen C, Kyu T (1995) J Chem Phys 102:556
12. Chiu HW, Zhou ZL, Kyu T, Cada LG, Chien LC (1996) Macromolecules 29:1051
13. Maier W, Saupe A (1959) Z Naturforsch A 14:882
14. Brochard F, Jouffroy J, Levinson P (1984) J Phys (Paris) 45:1125
15. de Gennes PG (1974) The Physics of Liquid Crystals. Oxford Univ, London
16. Onsager L (1949) Ann NY Acad Sci 51:627
17. Flory PJ (1956) Proc R Soc London Ser A 234:60
18. Flory PJ, Ronca G (1979) Mol Cryst Liq Cryst 54:289
19. Warner M, Flory PJ (1980) J Chem Phys 73:6327
20. Matsuyama A, Kato T (1996) J Chem Phys 105:1654
21. Flory PJ (1953) Principles of Polymer Chemistry. Cornell Univ, Ithaca
22. Kim JY, Muhoray PP (1991) Mol Cryst Liq Cryst 203:93
23. Brochard F (1979) J Phys (Paris) 40:1049
24. Aharnoi S, Edwards SF (1994) Adv Polym Sci 118
25. Detailed calculations will be published elsewhere.
26. Dusek K, Patterson D (1968) J Polym Sci A 2:1209
27. See reviews in Responsive Gels (1993) Adv Polym Sci 109–110

Progr Colloid Polym Sci (1997) 106:180–182
© Steinkopff Verlag 1997

A. Mizutami
M. Kawaguchi
T. Kato

Rheological properties of silica suspensions in the presence of polymer

Received: 13 October 1996
Accepted: 16 April 1997

Dr. A. Mizutami (✉) · M. Kawaguchi ·
T. Kato
Department of Chemistry for Materials
Faculty of Engineering
Mie University
1515 Kamihama
Tsu, Mie 514, Japan

Abstract Storage moduli (G') of hydrophilic and hydrophobic silica suspensions in trans-decalin and in trans-decalin solutions of polystyrene (PS) have been measured by taking account of PS adsorption. PS were adsorbed on the hydrophilic silica. The G' values of the hydrophilic suspensions in the presence of low molecular weight PS are similar to those without PS and they are larger than those in the presence of higher molecular weight PS. On the other

hand, any PS chains cannot adsorb on the hydrophobic silica particles. The G' values of all hydrophobic suspensions are higher than that in trans-decalin due to depletion flocculation and they are two orders of magnitude larger than those for the hydrophilic silica suspensions.

Key words Rheology – suspension – adsorption – polystyrene – depletion flocculation

Introduction

Rheological properties of the fumed-silica suspensions in aqueous media have been investigated by many researchers in the last two decades [1–3]. However, there were few studies for rheology of the silica suspensions in the presence of non-aqueous polymer by taking account of the polymer adsorption [4, 5]. In non-polar media with no hydrogen-bonding ability, the fumed-silica particles form network structure through hydrogen bonding of the surface silanol groups and they behave as gel-like materials [6].

In this study, we have investigated the effects of the presence of polymer on the dynamic moduli of the fumed hydrophilic and hydrophobic silica suspensions in polystyrene (PS) trans-decalin solutions as a function of PS molecular weight.

Experimental

Materials

Four monodisperse PS samples with the molecular weights of 9.68×10^3 (PS-10), 96.4×10^3 (PS-100), 355×10^3 (PS-355), and 706×10^3 (PS-706) were used. The PS-10 and other PS samples were purchased from GL Sciences Co. and Tosoh Co., respectively. Hydrophilic Aerosil 130 and hydrophobic Aerosil R202 were kindly supplied from Japan Aerosil Co. From the manufacture of the Aerosil 130 and the Aerosil R202, the average particle diameter is 16 and 14 nm, the surface area is 130 and 100 m^2/g, and there are 2.5 and 0.3 silanol groups pre nm^2, respectively. The surface on the Aerosil R202 is covered with oligo-dimethylsiloxane (DMS).

Progr Colloid Polym Sci (1997) 106:180–182
© Steinkopff Verlag 1997

Spectra grade quality trans-decalin was used as a solvent for PS and a dispersion medium without further purification.

The 8.6 wt% silica suspensions were mechanically shaken well to obtain a homogeneous mixture in a Yamato BT-23 incubator attached with a shaker for 1 week at 27 ± 0.1 °C. The PS concentration was fixed at 1.72 wt%.

Adsorption of PS

The amounts of PS adsorbed on the silicas were determined by the same method as previously described [6].

Rheological Measurements

Dynamic modulus was measured using an MR-300 Soliquid Meter with a cone-plate geometry at 27 ± 0.5 °C.

Results and discussion

Adsorption of PS

As for the hydrophilic silica suspensions in the PS-100, PS-355, and PS-706 solutions, all the added PS chains were adsorbed on the silica surface, which corresponds to the adsorbed amount of PS, 1.54 mg/m². For the hydrophilic silica suspension in PS-10 solution the adsorbed amount of PS was determined to be 0.808 mg/m². The plateau adsorbed amounts of the PS-10, PS-100, PS-355, and PS-706 in their adsorption isotherms were 0.808, 2.10, 2.45 and 2.45 mg/m², respectively [5]. Though the adsorbed amounts of PS except for the PS-10 were less than the plateau one, the fraction of the silanol groups occupied by PS chains were found to be similar to the plateau fraction of 0.4 [7], indicating that some silanol groups have remained and it allows a direct contact of bare surface of two particles. On the other hand, for the hydrophobic silica suspensions, any PS were not adsorbed on the silica surface. This is due to the presence of DMS chains on the silica surface.

Dynamic moduli of hydrophilic silica suspensions

Figure 1 shows storage moduli (G') of the hydrophilic silica suspensions in various PS solutions and in trans-decalin as a function of frequency. For the silica suspensions in the solutions of PS with lower molecular weights than 100×10^3 loss moduli (G'') are two-orders magnitude

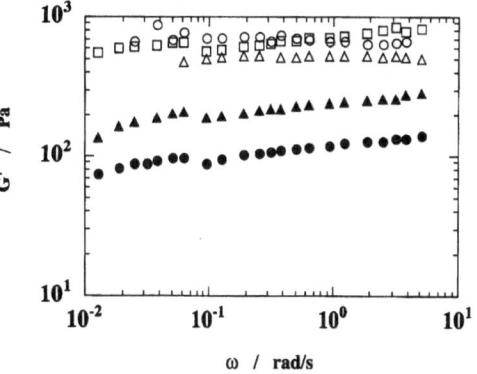

Fig. 1 Frequency dependence of G' for the hydrophilic silica suspensions in trans-decalin and in PS solutions: (○) without PS; (△) PS-10; (□) PS-100; (▲) PS-355; (●) PS-706

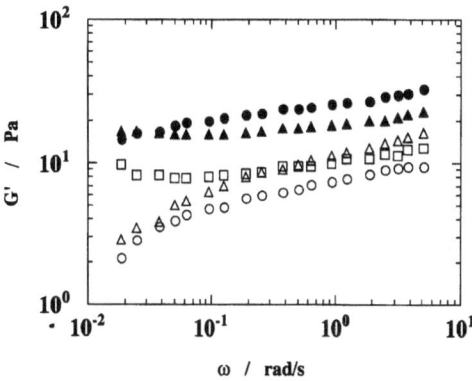

Fig. 2 Frequency dependence of G' for the hydrophobic silica suspensions in trans-decalin and in PS solutions. The symbols are the same as in Fig. 1

less than G', whereas in the presence of higher molecular weight PS the values of G'' are one-order magnitude less than G'. The large G' of the silica suspension in the solvent is strong due to the formation of the network structures through hydrogen-bonding of the surface silanol groups and it is independent of the frequency. Such a frequency dependence of G' looks similar to that of a cross-linked network, such as gel materials. In the presence of the PS-1, the G' value is independent of frequency and it overlays that for the silica suspension in the solvent. On the other hand, for the silica suspensions in the solutions of PS with higher molecular weights than 100×10^3 the respective G' values show a slight frequency dependency. The values of G' in the presence of PS-355 and PS-706 are one order of magnitude less than that in the solvent. Therefore, such rheological properties can be interpreted by taking account of PS adsorption. For the silica suspensions in the lower molecular weight PS than PS-100, a hydrogen-bonding gel structure in the silica suspension is almost not

182

A. Mizutami et al.
Rheological properties of silica suspensions in the presence of polymer

disrupted by PS adsorption. In contrast, the higher molecular weight PS chains than PS-355 is larger than the size of an isolated PS chain and they play an effective role in polymer bridging, leading to partially breaking down the gel-network and to the flocculation of the silica particles.

Dynamic moduli of hydrophobic silica suspensions

Figure 2 shows the G' values of the hydrophobic silica suspensions in the solvent and those in the various PS solutions as a function of frequency. The G'' values are one order of magnitude less than the G' for all silica suspensions and they showed a slight frequency dependence. In trans-decalin the G' value is two orders of magnitude less than that of hydrophilic one. This is caused by the reduction of the silanol groups on the silica surface and the steric stabilization of silica particles by DMS moieties. When PS is present, the G' values of all hydrophobic silica suspensions are higher than that in the solvent due to depletion flocculation [8]. Furthermore, the G' value increases with an increase in the molecular weight of PS.

Conclusions

In trans-decalin the G' value for the hydrophilic silica suspension was two orders of magnitude larger than that of hydrophobic one. The G' values of hydrophilic silica suspensions in the presence of low-molecular-weight PS were almost the same as without PS and in the presence of large molecular weight PS the value of G' becomes lower. On the other hand, depletion flocculation occurred for hydrophobic silica suspensions in the presence of PS.

References

1. Tadros Th F (1980) Adv Colloid Interface Sci 12:141
2. Otsubo Y (1994) J Colloid Interface Sci 53:1
3. Kawaguchi M (1994) Adv Colloid Interface Sci 53:103
4. De Silva GPHL, Luckham PF, Tadros ThF (1990) Colloids Surfaces 50:263
5. Kawaguchi M, Mizutani A, Matsushita Y, Kato T (1996) Langmuir 12:6179
6. Iler RK (1979) The Chemistry of Silica. Wiley-Interscience, Toronto
7. Kawaguchi M, Maeda K, Kato T, Takahashi A (1984) Macromolecules 17:1666
8. Napper DH (1983) Polymeric Stabilization of Colloidal Dispersions. Academic Press, London

Progr Colloid Polym Sci (1997) 106:183–187
© Steinkopff Verlag 1997

A. Takada
N. Nemoto

Dynamics of associating polymers in solution: dynamic light scattering and dynamic viscoelasticity of poly(vinyl alcohol) in aqueous borax solution

Received: 13 October 1996
Accepted: 12 March 1997

Dr. A. Takada (✉) · N. Nemoto
Department of Applied Physics
Kyushu University
Hakozaki, Higashi-ku
Fukuoka 812-81, Japan

Abstract We have studied the dynamics of poly(vinyl alcohol) (PVA) in aqueous borax solution by dynamic light scattering (DLS) and dynamic viscoelastic (DVE) measurements. DLS measurement showed the presence of two dominant modes with decaying rates of Γ_f and Γ_s ($\Gamma_f > \Gamma_s$). Different dynamical behaviors were observed above and below a critical concentration, C_N. The slow mode was manifested to be the diffusive mode for PVA concentration $C < C_N$ and the relaxation mode for $C > C_N$. Dynamical correlation length, ξ_H, estimated from Γ_f exhibited a jump at C_N with increasing C. Detailed analysis revealed the applicability of the dynamic scaling theory to Γ_s for $C < C_N$ and the presence of the dynamic coupling between concentration fluctuation and elastic stress of the transient network for $C > C_N$. From these results, we drew the picture on a growing process of associating PVA chains to a temporally cross-linked network with increasing C and proposed that C_N corresponded to the gel point of the chemically cross-linked gel in the short-time domain. The concept is found to be useful for interpretation of the concentration dependence of the plateau modulus.

Key words Poly(vinyl alcohol) – borax complex – dynamic light scattering – dynamic viscoelasticity – transient network

Introduction

One of the recent topics in studies of dynamic light scattering is finding the slow mode for polymeric systems. First observation of the slow mode was reported on the semidilute solutions of polystyrene in theta solvent by Adam and Delsanti [1]. Since then, the slow mode has been observed in several solutions in theta or poor solvents where polymer chains tend to associate with each other weakly [2–5]. Besides them, in recent years, the slow relaxation mode was also observed in the thread-like micellar system formed by the cationic surfactant in aqueous solutions [6]. If weak interaction between chains should induce the cooperative more of chains corresponding to the slow mode, the intrduction of much stronger interaction is expected to bring such a slow mode, too.

It is well known that addition of a small amount of sodium borate, SB, to an aqueous solution of poly(vinyl alcohol), PVA, enhances the viscosity remarkably [7]. This is due to the formation of a complex between two ols, i.e. hydroxy groups and the borate anion, which plays a role of a transient cross-linker among the PVA chains. Similar behavior has been reported for several other polyhydroxy polymers such as poly(glyceryl methacrylate) [8] and polysaccharides [9–11].

In previous reports [12, 13], we have studied the dynamic properties of the system by dynamic light scattering and found the presence of the slow mode in this system and, furthermore, the presence of a critical concentration point, C_N, at which the properties of the solutions including the slow-mode behavior drastically changed. These should be attributed to formation of the transient network in solution. In this study, in order to manifest the formation of the transient network and mechanism of the growth of the transient network system with increasing polymer concentration, we further studied the molecular-weight dependence of dynamical properties of this system by dynamic light scattering and dynamic viscoelastic measurements.

Experimental

The five PVA samples used in this study are a gift from Kurary Co., Ltd. The weight-average degrees of polymerization, DP_w, of the samples are 350, 600, 2100, 2600 and 3250 with molecular weight distribution of $M_w/M_n = 2.50$, 2.50, 2.14, 2.08 and 2.00, respectively. The degree of saponification is 88.0 mol% for all PVA samples. SB (guaranteed grade, Kishida Chemical Co. Ltd.) was used without further purification. The SB concentration was always adjusted so as to be half that of PVA by weight. In this condition, one boron including all boron species such as boric acid, borate anion and complex in solution always exists per two hydroxy groups of PVA in the solution. The sample code PVA X–Y represents DP_w as X and concentration of PVA, C, as Y in weight percent.

Dynamic light scattering (DLS) measurements were made with a spectrometer (ALV-125) equipped with a multiple τ correlator (ALV-5000FAST). A vertically polarized single-frequency 488 nm line of an argon ion laser (Spectra Physics, Beamlok 2060) was used as a light source. The normalized time correlation function, $A_q(t)$, of the vertical component of the light intensity scattered from the solutions was measured over a range of the scattering angle θ from 15° to 150° at a temperature $T = 25\,°C$.

Dynamic viscoelastic (DVE) measurements were made with a stress-controlled rheometer, CSL100 (Carri-MED, ITS Japan) using a cone-and-plate geometry. The storage and the loss shear moduli, $G'(\omega)$ and $G''(\omega)$, of the samples were measured at a strain of 0.3 over an angular frequency, ω, from 0.07 to 100 rad/s at $T = 25\,°C$.

Results and discussion

The normalized time correlation function, $A_q(t)$, of scattered-light intensity exhibited, at a glance, the presence of two dominant decaying modes for all the solutions studied. The inverse Laplace transformation, ILT, and the multiple exponential function, MEF, methods were applied to obtain a distribution of the decaying rate Γ from $A_q(t)$, which revealed that the decay rate distribution consisted of three major characteristic decay rates ($\Gamma_f > \Gamma_m > \Gamma_s$). The amplitude of the medium mode corresponding to Γ_m was smaller than those of the other modes by an order of magnitude. Analyses of slow and fast modes corresponding to Γ_s and Γ_f divided the solutions into two groups at the characteristic concentration C_N dependent on sample molecular weight. Its determination and meaning will be discussed in more detail later.

Two different scattering vector, $q(= (4\pi n/\lambda_0)\sin(\theta/2))$, dependences of Γ_s were observed for the slow mode, where λ_0 is the wavelength of the incident light in vacuum and n the refractive index of medium. For samples with $C < C_N$, Γ_s had q dependence of $\Gamma_s \sim q^{2-3}$, i.e. Γ_s/q^2 was almost independent of q at lower q and became proportional to q at higher q. This type of q dependence has been observed for solutions of very high molecular weight polymers [14, 15]. Following the dynamic scaling theory of the polymer solution [15], we tried to analyze the observed q dependence of Γ_s/q^2, i.e., we evaluated the cooperative diffusion coefficient, D_c, as the value of Γ_s/q^2 extrapolated to $q = 0$ and superposed the Γ_s/q^2D_c vs. q data on a theoretical curve obtained from Eqs. (1) and (2) derived by Kawasaki [16] by shifting horizontally, in which R_s was treated as an adjustable parameter.

$$\Gamma_s/q^2D_c = F(qR_s)\,, \tag{1}$$

$$F(x) = (\tfrac{3}{4}x^2)\{1 + x^2 + (x^3 - x^{-1})\tan^{-1}x\}\,. \tag{2}$$

All data were well located on the theoretical curve as shown in Fig. 1, which implies that clusters composed of a large number of associating PVA chains are formed in

Fig. 1 Applicability of the dynamic scaling theory to data of the slow mode for solutions with C below C_N. Γ_s/q^2D_s is plotted against qR_s where R_s is determined as an adjustable parameter. ($\bigtriangledown$) PVA350−4.0, −3.8 and −3.6 ($\square$) PVA600−2.7 and −2.5 ($\diamond$) PVA2100−1.6 and −1.4 ($\triangle$) PVA2600−1.4 ($\bigcirc$) PVA3250−1.2 and −1.1

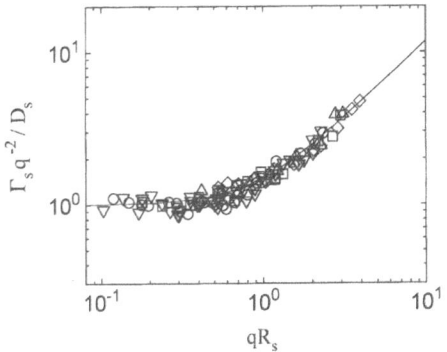

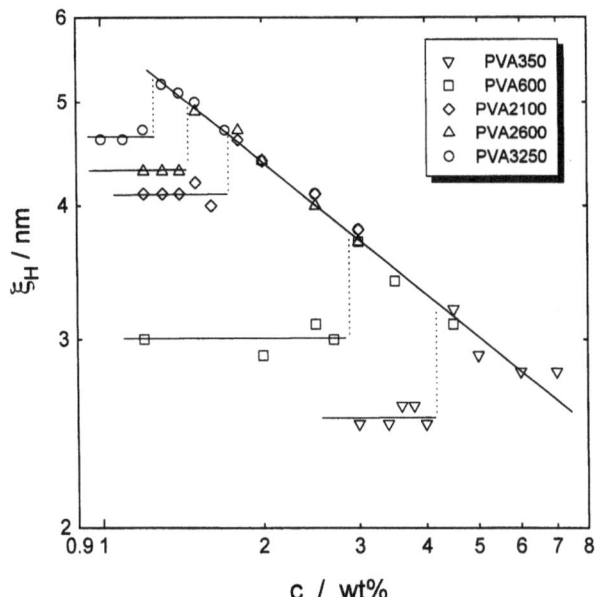

Fig. 2 Plot of the dynamical correlation length ξ_H estimated from D_c using the Einstein–Stokes relationship against PVA concentration C

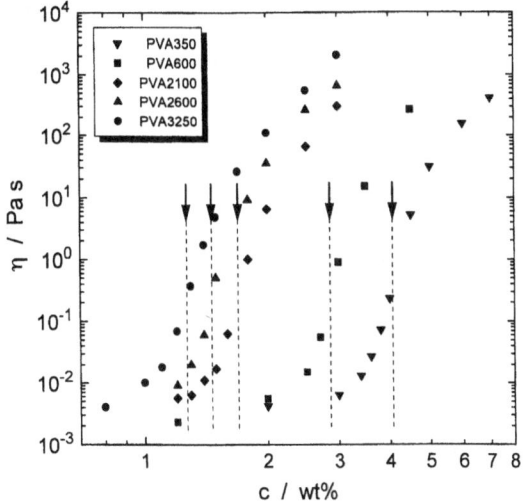

Fig. 3 C dependence of the steady-state viscosity η for five PVA samples with different molecular weight. An arrow indicates the inflection point in the η vs. C curve

the solution with increasing C. At low q, the diffusion of the clusters was observed and, at high q, internal motion in the cluster was observed, being due to consecutive association and dissociation process of di-diol complexes.

For the samples with $C > C_N$, q dependence of Γ_s was entirely different from that of the slow mode described above. Γ_s was independent of q within an error of 20%, which means that the mode is the relaxation mode. For $C > C_N$, all samples were highly viscoelastic, so that the slow mode may be related to dynamical coupling between concentration fluctuation and elasticity of the viscoelastic network.

The fast mode was the diffusive mode for all solutions since Γ_f/q^2 was independent of q. The cooperative diffusion coefficient D_c could be estimated as an average of Γ_f/q^2 and the dynamical correlation length ξ_H was estimated using the Einstein–Stokes relationship, $\xi_H = k_B T/6\pi\eta_s D_c$, to an accuracy of 5%. Here the viscosity of pure water was used as η_s, because C of the polymer solutions studied was low in the range around the overlapping concentration and also ξ_H corresponds to the short-time behavior of concentration fluctuation. A plot of ξ_H against C shown in Fig. 2 reveals that its C and M dependences are different in the two regions which are unambiguously distinguished by a jump of ξ_H.

$$\xi_H \sim C^0 M^{0.25 \pm 0.03} \quad (C < C_N), \tag{3}$$

$$\xi_H \sim C^{-0.42 \pm 0.02} M^0 \quad (C > C_N). \tag{4}$$

The two different groups completely corresponded to the groups in the analysis of the slow mode described above. The values of C_N have been determined from the C dependence of the steady-flow viscosity later.

From these analyses, we propose the picture of a growing process of associating PVA chains to a viscoelastic network with increasing C in solution. At low C, PVA molecules form a small cluster with di-diol complexes as cross-linkers with a finite lifetime. With increasing C, the cluster size increases while the density of the cross-linker in the cluster does not change, as deduced from the independence of ξ_H on C. Consequently, it is supposed that the growth of cluster occurs on the surface of the cluster. This should come from the fact that water is a good solvent of PVA, i.e. PVA chain tends to be apart from other PVA chains, but the transient cross-link prevents the diffusion of PVA chain. When C reaches a certain concentration which depends on molecular weight, the cluster extends to the whole volume of the solution, which induces a drastic change of their properties. In a sense, we can regard this concentration as a kind of gel point. Further increase in C merely increases the number density of the transient cross-links and results in more uniform network.

Figure 3 shows C dependence of the steady-state viscosity η. η increases over five decades sigmoidally for small increase in C for all samples with different molecular weight. The inflection point in η appears to be located in the concentration range in which the DLS behavior drastically changed. Since the C_N value could not be definitely located from the DLS results, we put C_N equal to the concentration at the inflection point. The relationship

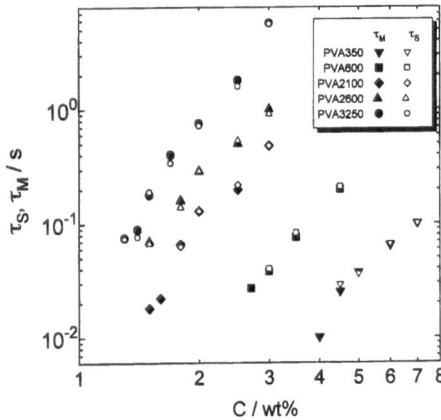

Fig. 4 The characteristic time τ_s of the slow mode above C_N and the mechanical relaxation time τ_M are plotted against PVA concentration C

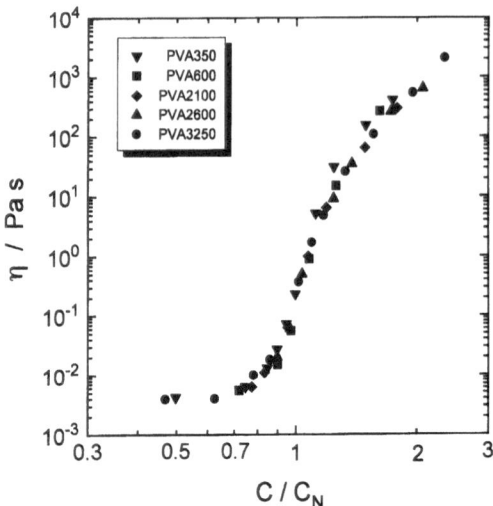

Fig. 5 Plot of the steady-state viscosity η of five PVA samples against C/C_N

between C_N and M was approximately represented as $C_N \sim M^{-0.53}$. We obtained $C^* = 3.7, 2.6, 1.2, 1.0, 0.8$ wt% for DP = 350, 600, 2100, 2600, 3250, respectively, as the reciprocal of the intrinsic viscosity of PVA in water without sodium borate. These values are close to the C_N. According to the picture described above, C_N corresponds to the gel point. Gel point is a critical point in chemically cross-linked gel where the viscosity diverges. For the PVA/Borax case, however, the cross-link point has a finite lifetime. Therefore even after the formation of the network to the whole system, the network can flow by the reassociation process of the di-diol complex and η does not diverge.

Angular frequency ω dependence of the storage and the loss moduli, G' and G'', was in good agreement with the behavior of the Maxwell model with a single relaxation time. At low ω, ω dependences of G' and G'' are characteristic of the steady-flow behavior as given by $G'' = \eta\omega$ and $G' = J_e\eta^2\omega^2$. The mechanical relaxation time τ_M was evaluated as $\tau_M = J_e\eta$, which was subjected to an uncertainty of 30%. Figure 4 shows the concentration dependence of τ_s obtained by DLS as $\tau_s = \Gamma_s^{-1}$ and τ_M. They were in excellent agreement with each other.

As shown in Fig. 3, the five PVA samples look to take almost the same values of $\eta \simeq 0.3$ Pa s at the respective C_N, which suggests that all viscosity data can be superposed on to one master curve by a horizontal shift alone. Figure 5 indicates that such data reduction is quite successful. It is to be noticed that $c[\eta]$ is more appropriate for superposition of η data in the lower C region [17].

It is known that, regardless to the condition of the gelation, dynamical behaviors in the critical region near the gel point can be discussed using the reduced variable, the relative extent of chemical reaction $(p/p_c - 1)$, where

p is the degree of chemical reaction and p_c is the p at the gel point [18]. As described above, we supposed that C_N of this transient network could be regarded, in the time region shorter than the lifetime of the transient cross-link, as the gel point of the chemically cross-linked gels. In order to test this conjecture, we tried to superpose the plateau modulus, G_N, estimated by dynamic viscoelasticity. The G_N vs. C data for five PVA samples could not be superposed onto one master curve by horizontal shifting alone using C/C_N as the reduced variable. Thus, we shifted the

Fig. 6 Superposition of the plateau modulus G_N vs. reduced concentration $C/C_N - 1$ data. Data were shifted along vertical axis using the shift factor a_M

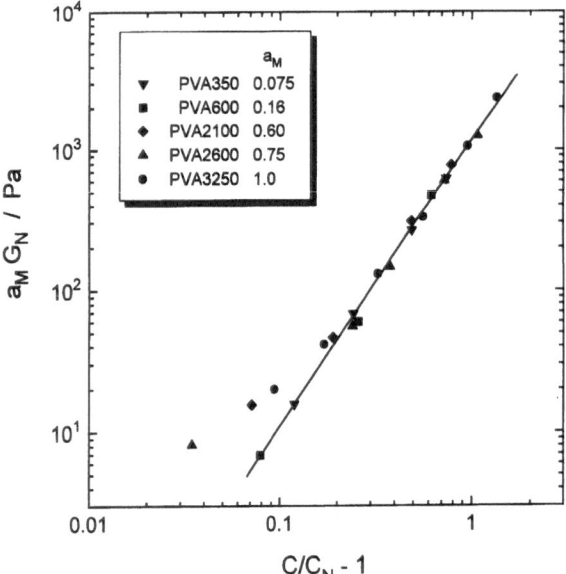

Progr Colloid Polym Sci (1997) 106:183–187
© Steinkopff Verlag 1997

data along the vertical axis and got a master curve. The vertical shift factor a_M is listed in Fig. 6. In permanently cross-linked gel, the modulus is given by the equation $G = vkT$, where v is the number density of chains between two cross-links and is usually related to C and molecular weight of the chain between two successive cross-links, M_c, by the equation $v = C/M_c$. ξ_H is supposed to be proportional to the length between two successive cross-links. The chain between cross-links is assumed to behave like a Gaussian chain, which leads to $\xi_H \sim M_c^{1/2}$. If we apply these relations to the PVA/SB system and invoke the relation $\xi_H \sim M^{-0.42}$ for $C > C_N$, we derive the following relation:

$$G_N M^{0.96} \sim (C/C_N)^{1.84} RT . \qquad (5)$$

This equation predicts that the vertical shift factor is proportional to $M^{0.96}$ and that G_N is located on a straight line against C/C_N in the double logarithmic plot. Actually, the shift factor a_M is proportional to $M^{1.0}$ experimentally. But the master curve was not a straight line. Near the gel point, Eq. (5) can be written in a more appropriate way which reads as

$$G_N M^{0.96} \sim (C/C_N - 1)^{1.84} RT . \qquad (6)$$

When we plot G_N against $C/C_N - 1$ double logarithmically in Fig. 6, the data are well located on a straight line except data with C near C_N, which is in agreement with the prediction of Eq. (6). Furthermore, the slope was 2.0 experimentally, which is also in agreement with the predicted value of 1.84.

References

1. Adam M, Delsanti M (1985) Macromolecules 18:1760
2. Mathies P, Mouttet C, Weisbuch GJ (1980) J Phys 41:519
3. Nemoto N, Makita Y, Tsunashima Y, Kurata M (1984) Macromolecules 17:2629
4. Brown W, Nicolai T, Hvidt S, Stepanek P (1990) Macromolecules 23:357
5. Nicolai T, Brown W, Hvidt S, Heller K (1990) Macromolecules 23:5088
6. Nemoto N, Kuwahara M, Yao M, Osaki K (1995) Langmuir 11:36
7. Deuel H, Neukom H (1949) Makromol Chem 3:13
8. Pezron E, Leibler L, Richard L, Lafuma F, Audebert R (1989) Macromolecules 28:1169
9. Pezron E, Richard L, Lafuma F, Audebert R (1988) Macromolecules 21:1126
10. Sato T, Tsujii Y, Fukuda T, Miyamoto T (1992) Macromolecules 25:3890
11. Nicherson RF (1971) J Polym Sci 15:111
12. Koike A, Nemoto N, Inoue T, Osaki K (1995) Macromolecules 28:2239
13. Nemoto N, Koike A, Osaki K (1996) Macromolecules 29:1445
14. Doi M, Edward SF (1986) The Theory of Polymer Dynamics. Oxford University Press, Oxford
15. Han CC, Akcasu AZ (1981) Macromolecules 14:1080
16. Kawasaki K (1971) In: Green MS (ed) Critical Phenomina. Academic Press, New York, p 342
17. Takada A, Nishimura M, Koike A, Nemoto N Macromolecules (in press)
18. Staufer D, Coniglio A, Adam M (1982) Adv Polym Sci 44:1982

Progr Colloid Polym Sci (1997) 106:188–191
© Steinkopff Verlag 1997

BIOCOLLOIDS

C.J. O'Connor
R.D. Manuel

Calf pregastric lipase catalyzed hydrolysis of short and medium chain-length monoacid triglycerides: temperature, pH and lipid concentration effects

Received: 13 October 1996
Accepted: 1 February 1997

Prof. C.J. O'Connor (✉) · R.D. Manuel
Department of Chemistry
The University of Auckland
PB 92019 Auckland, New Zealand
E-Mail: cj.oconnor@auckland.ac.nz

Abstract The commercial extract from the oro-pharyngeal tissues of calf has been used as the source of pregastric lipase and has been processed to yield a partially purified sample of the pregastric lipase. The activity of this lipase against the short-chain lipid tributyrin has been determined over a range of pH and temperature values. Optimum pH conditions were within the range 5.7–6.4 and the optimum temperature was within 37–48 °C. The lipase was also used to catalyze the hydrolysis of monoacid triglycerides (C4:0–C12:0) at 40 °C, pH 6.0 and maximum activity was obtained against tributyrin (C4:0). The values of the kinetic parameter K_m increase with increasing carbon-chain length of the carboxylic moiety on the triglycerides, thus, indicating a preference for tributyrin under the stated conditions. The results are compared with results previously obtained utilising goat, kid and lamb pregastric lipases.

Key words Calf – pregastric lipase – tributyrin – monoacid triglycerides – Michaelis – Menten parameters

Introduction

In ancient times the nomadic tribes which ranged through eastern Europe and western Asia carried liquids in bags made from animals' stomachs. Milk stored in these containers, warmed by the sun, soured by naturally occurring bacteria and laced with enzymes from the stomach lining, would have been transformed into solid curds and liquid whey. Thus, the first cheeses were made [1].

Later, cheese makers either placed strips of kid, lamb or calf stomach directly into the warm milk or prepared a crude rennet extract by soaking the strips in salt water. World production of rennet now exceeds 25 million litres per year [1]. In the 1960s, however, the Food and Agricultural Organization of the United States predicted that an increased demand for meat would lead to more calves being reared to maturity, so that less rennet would be available. Consequently, over the last 30 years several substitutes for calf rennet have been developed, including both fungal and microbial proteases, whose chief function (as with rennin, the protease in rennet preparations) is to coagulate the milk. As the public's organoleptic demands for a wide variety of cheese have increased, so has the use of fat-degrading enzymes (lipases) which are added to promote the formation of piquant flavors as the cheese ripens. Many traditional Italian cheese, e.g. Parmesan, Romano, Provolone and Mozzarella, benefit from such additions to augment the activity of the naturally occurring lipolytic microbes. One group of additives which is gaining increased interest is the range of enzymes included in the pregastric enzyme extracts from the epiglottal region of calves, lambs and kids, which hydrolyze the fat in milk to promote the formation of these piquant flavors in the cheese as it ripens. Relatively little is known, however, of the systematic chemistry of these enzymes. In this investigation, we describe the ability of a lipase, which has been isolated and partially purified from the extract of calf tongues, to catalyze the hydrolysis of monoacid triglycerides differing in the length of their carboxylic acid carbon-chain length. The hydrolysis of the short-chain lipid, tributyrin, has been further investigated over a range of temperature and pH values and Michaelis-Menten parameters have been evaluated.

Materials and methods

Lipase preparations

The calf pregastric lipolytic extract was prepared from the epiglottal regions (tongue roots) of calves, presented for commercial slaughter in New Zealand, by the N.Z. Rennet Co Ltd, Eltham. The extract was partially purified as described previously for lamb pregastric lipase [2, 3]. The crude calf extract (0.5 g) was dissolved in 50 mM tris-HCl buffer (15 ml, pH 8.0 at 4 °C) and centrifuged at 36 000 rpm for 20 min at 4 °C to remove any insoluble particles. The clear supernatant was applied onto an ion-exchange column of Q-sepharose (1.0 cm × 18 cm) with a flow rate of 1.0 ml/min and washed with a suspension of buffer, 50 mm tris, pH 8.0 at 4 °C, until no more protein was eluted. The column was then washed first with 50 mM acetate-HCl buffer (pH 4.6 at 4 °C) and then with 1 M NaCl/50 mM tris-HCl buffer (pH 8.0 at 4 °C). The protein fraction eluted with this latter wash was collected on ice, dialyzed against degassed milli-Q H_2O at 4 °C and lyophilized. All buffers used for the elution had been degassed. The mass recovered in this lipase fraction was 10%, with specific activity 7.59 μmol/min per mg (1.0 mM tributyrin, pH 6.5, 35 °C) and the purification factor was 3.9. The protein content of the partially purified calf lipase (Bio-Rad protein assay kit using BSA as the standard) was 66%.

Chemicals and reagents

The substrates used for lipase assay were: tributyrin (C4:0), tricaproin (C6:0), tricaprylin (C8:0), tricaprin (C10:0), trilaurin (C12:0) from Sigma. Sodium caseinate (Alanate 185) was supplied by the N.Z. Dairy Board, L-α-lecithin was from Sigma, color key pH calibration buffers were from BDH Chemicals and Q-sepharose was from Pharmacia LKB.

Equipment

pH-stat titrations were carried out in temperature-jacketed vessels with a programmable autotitrator (Mettler DL21). Mathematical data processing was predominantly carried out using IBM personal computers;

3D surface and contour plots were generated using Stanford Graphics for Windows, Version 3.0.

Lipase activity

The technique used for kinetic characterization of lipase activity across a range of pH and temperature values has been described [3–5]. The activity was measured in a sodium caseinate (6.0 mg/ml)/L-α-lecithin (0.5 mg/ml) emulsion base using pH-stat. The emulsion base (40 ml) was added to the lipid which had been weighed into a dry, plastic titrator cup and the mixture was sonicated until it became homogeneous. The enzyme solution (50–100 μl, 5 mg/ml) was added and the release of carboxylic acid was determined by titration with 0.01 M NaOH which had been standardized against potassium hydrogen pthalate.

The activity of the partially pure calf pregastric lipase against tributyrin was measured over a pH range of 5.5–7.5 and at temperatures starting at 30 °C with increments of 5 °C up to 50 °C, or until the initial rate of reaction was non-linear (i.e. inactivation of the enzyme was observed). The activity against tributyrin, tricaproin, tricaprylin, tricaprin and trilaurin was measured at pH 6.0, 40 °C. The emulsion base for these assays contained 6.0 mg/ml sodium caseinate and 1.0 mg/ml L-α-lecithin. For each assay condition, at least seven different concentrations (0 mM–4.5 mM) of lipid were used. The plots of initial rate of hydrolysis (the linear region at the beginning of the volume of NaOH vs. min pH-stat assay where less than 10% of the substrate had been hydrolyzed) were fitted by the curve-fitter function of Sigma Plot™ (1.0), and hence, the Michaelis–Menten substrate affinity constant (K_m) and the apparent maximum rate values, V_{max}, were obtained.

The plot of V_{max} vs. pH and temperature for the tributyrin assay was fitted with a Third-order polynomial against pH and temperature, respectively. The surface net was then generated with unit intervals of 0.1 pH and 1 °C and the contour plot of V_{max} vs. pH and temperature was, subsequently, constructed from the surface net values.

Results and discussion

Figure 1 (upper plot) shows the surface plot of V_{max} for the experimental range of pH and temperature values for calf pregastric lipase catalyzed activity against tributyrin. The point of maximum activity on the smoothed plot is at 42 °C and pH 6.0. These values may be compared with the corresponding values for maximum activity for the lamb (43 °C, pH 6.4) [3], goat (52 °C, pH 6.0) and kid (56 °C, pH 5.5) [4] pregastric lipases, respectively.

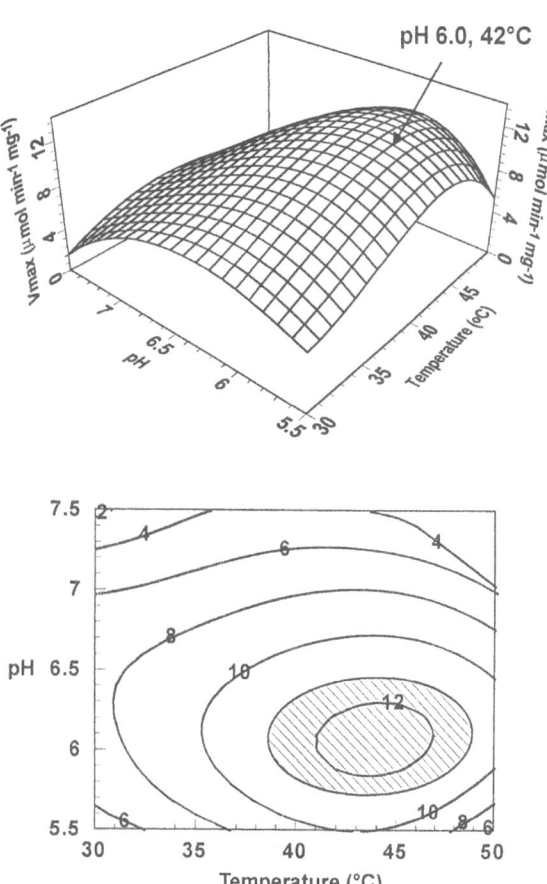

Fig. 1 Upper plot – Kinetic surface plot of apparent maximum rate of hydrolysis (V_{max}) of tributyrin against pH and temperature; Lower plot – Contour plot of lines of constant apparent maximum rate of hydrolysis (V_{max}) against pH and temperature. The shaded portion represents activity within 10% of the optimum value at pH 6.0, 42 °C. Both plots are for hydrolysis catalyzed by pregastric lipase of calf

In all cases a broad region of high activity surrounds these maxima. In the case of the calf enzyme, this is best observed from the contour levels provided in Fig. 1 (lower plot), which shows activity lying within 10% of the maximum over the temperature and pH range of ca. ± 5 °C (37–48 °C) and ± 0.4 pH units (pH 5.7–6.4), respectively. Comparison may again be made with the flatness of the contour plots seen for the other lipases from ruminants: ± 12 °C, ± 0.6 pH units for lamb [3], over the region of 43–60 °C for both goat and kid lipases, and within the range of pH values of 5.6–6.5 for the goat enzyme and 5.5–6.2 for the kid enzyme [4]. The pH maximum for the kid extract lay on the edge of the surface plot, but it was not possible to determine the rate data at a lower pH because the enzyme was precipitated at pH 5.0. Acid precipitation of the enzyme has been noted by others who then used it as a first step in separation and purification of the calf enzyme from calf gullet tissues [6].

At pH 7.5, inactivation within minutes' reaction time was observed at 50 °C. As the pH was lowered, the enzyme exhibited more tolerance to higher temperatures. This extra stability may be affected by the proteins which were co-eluted in our partial purification and we are currently investigating possible protein–enzyme and protein–micelle interactions.

Over the range of pH and temperature values at which the catalysis of tributyrin was investigated, the values of K_m were generally low, frequently < 0.25 mM, with some as low as 0.08 mM, indicating that the enzyme has a very high affinity for this substrate. The values of K_m showed significantly more scatter than V_{max}, in part because of the high substrate affinity (low K_m) and the consequent extreme sharpness of the Michaelis–Menten hyperbolae. At very low substrate concentrations (below K_m values), reproducible rate data were difficult to obtain, partly because of variable loss of substrate onto the walls of the vessel and other surfaces. Conversely, the V_{max} obtained from the substrate saturation region (the extended and very flat plateau region of the Michaelis–Menten plots) were very reliable.

The maximum values of V_{max} was ca. 14.3 µmol/min/mg at pH 6.0, 40 °C compared with 2.1 µmol/min/mg obtained at pH 6.5 and 35 °C for the unprocessed commercial product [7]; K_m values of < 0.1 mM are similar.

The value of V_{max} for the partially purified enzyme may be compared with the corresponding values for partially purified lipases from lamb, goat and kid, i.e. 2.6 µmol/min/mg at 43 °C, pH 6.4 for lamb [3]; 20.3 µmol/min/mg at 52 °C, pH 6.0 for goat [5]; and 10.0 µmol/min/mg at 56 °C, pH 5.5 for kid [4]. In each case the method of partial purification by elution through a Q-Sepharose fast flow ion exchange column was identical.

Bernbäck et al. [8] effected a purification of the pregastric lipase from calf pharyngeal tissues which was based on chromatographies on octyl-Sepharose and lentil-lectin-Sepharose followed by gel filtration. Dodecylsulfate polyacrylamide gel electrophoresis of the product revealed a molecular mass (M_r) of 55 kDa and the enzyme protein was characterized as a glycoprotein with a high content of branched, aliphatic amino acid residues. The M_r on gel filtration was 44–48 kDa. Timmermans et al. [9] have recently cloned the entire coding sequence for bovine pregastric lipase and showed that it exists as a mature 378-amino acid polypeptide with a molecular mass of 42.96 kDa.

De Caro et al. [10] have recently purified lamb pregastric lipase and obtained a protein with an M_r of 50 kDa on SDS-PAGE, and maximal specific activity against a tributyrin emulsion at pH 6.0. They found, as we have also reported for lamb pregastric lipase and for goat and kid pregastric [3–5] lipases, that these pregastric enzymes have a very marked selectivity for short-chain lipids, which is likely to be advantageous in the dairy industry, where it is preferable that short- and medium-chain fatty acids from milk triglycerides be released during cheese ripening.

Figure 2 shows the percent values of the derived values of V_{max} (relative to the value for tributyrin) against the carbon-chain length of the carboxylic acid moieties of the monoacid triglycerides.

The values of V_{max} are seen to decrease as the carbon-chain length increases. For each additional pair of carbon atoms in the carboxylic acid chain, the value of V_{max} decreases by more than a twofold factor. This profile is much steeper than those obtained in similar studies carried out on partially purified lamb [3], goat and kid [4] pregastric lipase. Thus, for the C10:0 substrate only ca. 1% of activity relative to that of C4:0 was determined for the calf enzyme, whereas ca. 10% of the activity remained for the other three enzymes. A study of D'Souza and Oriel [11] on lamb pregastric lipase showed a similar trend but their study used only a single concentration of substrate which lay considerably below the K_m values for the lipids with C8:0 (tricaprylin) and C10:0 (tricaprin), for which they were unable to detect any activity.

As the carbon chain lengthens, the values of K_m increases, e.g. from 0.14 mM for tributyrin (C4:0) to 1.6 mM for tricaprylin (C8:0). Thus, the preferential activity of calf pregastric lipase towards the short-chain lipase, tributyrin, compared with its activity against longer-chain triglycerides, at pH 6.0, 40 °C, is a combination of two factors: a large V_{max} value reflecting a larger turnover number, k_{cat}, ($V_{max}/E_0 = k_{cat}$ where E_0 is the true enzyme concentration) and a small value of K_m, reflecting its preference for saturation of the active site by the physically smaller substrate.

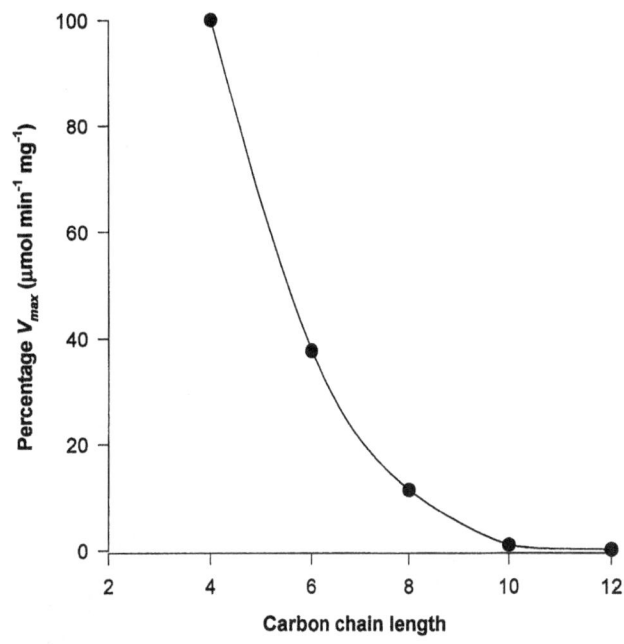

Fig. 2 Calf pregastric lipase catalyzed hydrolysis of monoacid triglycerides at 40 °C, pH 6.0

Acknowledgements Financial assistance from the University of Auckland Research Committee, New Zealand Lottery Science, the New Zealand Agricultural and Marketing Research Development Trust and the New Zealand Rennet Co Ltd and awards from the Maori Education Foundation and Manaki Tauira (to RDM) are gratefully acknowledged.

References

1. Madden D (1995) In: Food Biotechnology: an Introduction, International Life Sciences Institute Press, Washington, p26
2. O'Connor CJ, Lai DT, Barton RH (1996) J Mol Catal B: Enzymatic 1:143–149
3. Barton RH, O'Connor CJ, Turner KW (1996) J Dairy Sci 79:27–32
4. Lai DT, MacKenzie AD, O'Connor CJ, Turner KW (1997) J Dairy Sci (in press)
5. O'Connor CJ, Barton RH, Butler PAG, MacKenzie AD, Manuel RD, Lai DT (1996) Colloids & Surf B: Biointerfaces 7:189–205
6. Sweet BJ, Matthews LC, Richardson T (1984) Arch Biochim Biophys 234:144–150
7. O'Connor CJ, Manuel RD, Turner KW (1992) J Dairy Sci 76:3674–3682
8. Bernbäck S, Hernell O, Bläckberg L (1985) Eur J Biochem 148:238
9. Timmermans MJJ, Teuchy H, Kupers LPM (1994) Gene 147:259–262
10. De Caro J, Ferrato, F, Verger R, De Caro A (1995) Biochim Biophys Acta 1252:321–329
11. D'Souza TM, Oriel P (1992) Appl Biochem Biotech 36:183–198

Progr Colloid Polym Sci (1997) 106:192–197
© Steinkopff Verlag 1997

BIOCOLLOIDS

H. Terada
E. Majima

Important role of loops in the transport activity of the mitochondrial ADP/ATP carrier

Received: 13 October 1996
Accepted: 12 March 1997

Dr. H. Terada (✉) · E. Majima
Faculty of Pharmaceutical Sciences
University of Tokushima
Shomachi-1
Tokushima 770, Japan

Abstract The transport mechanism of the bovine heart mitochondrial ADP/ATP carrier was studied using submitochondrial particles. The modifications of the cysteine residues of the carrier by the SH-reagents eosin-5-maleimide (EMA) and N-ethylmaleimide (NEM), and disulfide bond formation between these cysteine residues catalyzed by copper-o-phenanthroline ($Cu(OP)_2$) under various conditions were studied. In particular, the effects of the transport inhibitors carboxyatractyloside (CATR) and bongkrekic acid (BKA), and fluorescein derivatives were examined. From the results, the topology of the carrier in the membrane, dynamic translocations of the loops of the carrier, and the structure of the primary binding site of the transport substrates ADP and ATP were deduced. The loops are concluded to act as both gates in the transport and binding sites for the substrates. Based on the results, a cooperative swinging-loop model is postulated as the transport mechanism of the ADP/ATP carrier.

Key words ADP/ATP carrier – loop structure – substrate binding site – SH-reagent – mitochondria

Mitochondrial ADP/ATP carrier

Despite extensive studies on the structures of various solute transporters, the molecular mechanism of their transport activities is not well understood. Although a transporter consists of a membrane-spanning region and loops, major interest seems to have been focused on the roles of certain amino acid residues in the membrane-spanning region, while little attention has been paid to the role of its loops [1]. It is generally thought that substrate transport is achieved by conformational change of a transporter, typically postulated by the most simple single-binding center-gated pore model (see Fig. 1 for the case of the ADP/ATP carrier) [2]. However, it is usually difficult to detect a definite conformational change, partly because few reagents that fix a certain conformation of the transporter specifically are known.

The ADP/ATP carrier, which mediates exchange transport of ADP and ATP across the inner mitochondrial membrane, could be very effective for understanding the common mechanism of various transporters, because (1) it shows large conformational changes during the transport process, and (2) specific inhibitors of the transport recognizing different conformations are known. This carier is a 30 kDa protein consisting of six membrane-spanning regions and three large loops M1, M2 and M3 facing the matrix side [2–4]. The carrier is suggested to function as a dimer in the membrane, and it takes two distinct interconvertible conformations, the c- and m-states [2]. The substrate-binding site faces the cytosolic side in the c-state conformation, and the matrix side in the m-state. These two states are fixed by the specific transport inhibitors carboxyatractyloside (CATR) acting from the cytosolic side, and bongkrekic acid (BKA) acting from the matrix

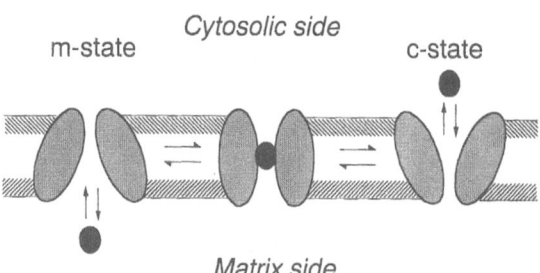

Fig. 1 Single-binding center-gated pore model of the transports of ADP and ATP via the ADP/ATP carrier

side. These two conformational states of the carrier in relation to its transport activity are shown in Fig. 1 schematically.

In the bovine heart mitochondrial ADP/ATP carrier, there are four cysteine residues Cys^{56}, Cys^{128}, Cys^{159} and Cys^{256} which are all at intervals of about 100 amino acid residues except Cys^{128}. This paper deals with the molecular mechanism of substrate transport by the carrier, deduced mainly from the effects of chemical modifications of its cysteine residues by SH-reagents under various conditions. The important role of the loops of the carrier in its transport function is described. Unless otherwise noted, the results were obtained with bovine heart submitochondrial particles, in which the orientation of membrane proteins is inside–out relative to that of mitochondria.

Locations of three loops facing the matrix

The SH-reagents N-ethylmaleimide (NEM) and eosin-5-maleimide (EMA) have often been used for chemical modifications of proteins [5–7]. NEM is a neutral hydrophobic SH-reagent that can permeate the cell membrane, whereas the fluorescein derivative EMA is bulky and impermeable owing to its divalent negative charge (for chemical structures, see Chart). As the cysteine residue embedded in the membrane is thought not to interact with these maleimide SH-reagents [6], NEM and EMA are concluded to recognize cysteine residues in the membrane according to their chemical structures.

We studied the effects of chemical modifications by NEM and EMA of cysteine residues of the ADP/ATP carrier in bovine heart submitochondrial particles. As shown in Fig. 2, NEM labeled Cys^{56} markedly, and Cys^{159} and Cys^{256} moderately, but did not label Cys^{128}. In contrast, EMA labeled Cys^{159} very quickly and markedly, Cys^{56} at a similar rate to that by NEM, Cys^{256} more slowly, and again did not label Cys^{128} at all, although it did not label the carrier in mitochondria. As only Cys^{128} was not labeled at all, by either NEM or EMA, we

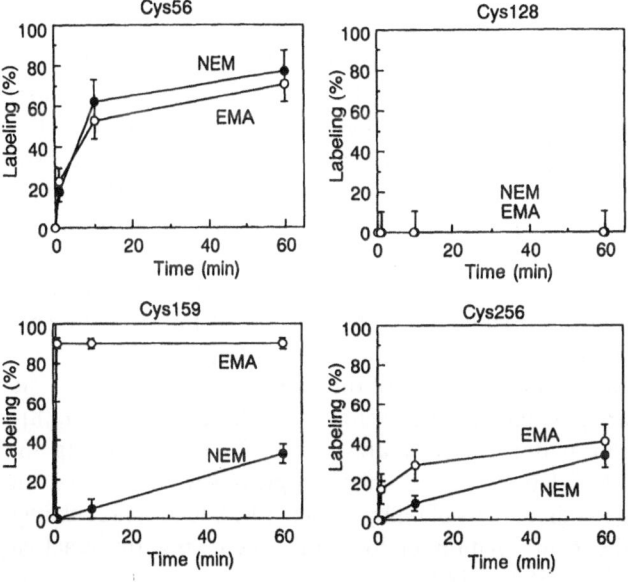

Chart: Chemical structures of fluorescein derivaties, cibacron blue 3GA and NEM. Numericals in parentheses are IC_{50} values (μM) of the effects on EMA-labeling of the carrier (see Fig. 4)

Fig. 2 Time courses of labelings by NEM and EMA of the four cysteine residues of the ADP/ATP carrier in bovine heart submitochondrial particles at 0 °C. Submitochondrial particles (20 mg protein/ml) were treated with EMA (20 nmols/mg protein) or NEM (100 nmols/mg protein) for various periods

presume that this residue is located in the membrane-spanning region, whereas the other cysteine residues are located in matrix-facing loops: Cys^{56}, Cys^{159} and Cys^{256} are in loop M1, M2 and M3, respectively.

Cytosolic side

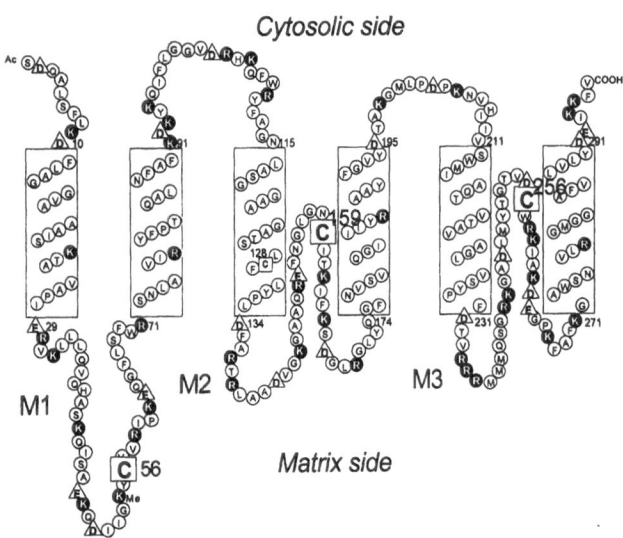

M1 M2 M3

Matrix side

Fig. 3 Topology of the ADP/ATP carrier of bovine heart mitochondria in the membrane. Basic amino acid residues are shown by closed circles, acidic amino acids by open triangles, and cysteine residues by open squares

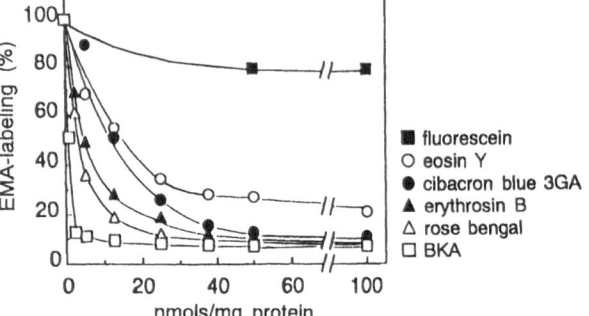

Fig. 4 Inhibitory effects of fluorescein derivatives and cibacron blue 3GA on the labeling by EMA of cysteine residues of the ADP/ATP carrier in bovine heart submitochondrial particles at 0 °C. Submitochondrial particles (2 mg protein/ml) were first incubated with test compound for 10 min at 0 °C, and then with EMA (10 nmols/mg protein) for 30 s at 0°C

As NEM labeled Cys^{56} fastest and its rate of labeling of Cys^{56} was similar to that by EMA, loop M1 should be exposed to the matrix space. Loop M2 is concluded to be located on the matrix side, but it is not exposed to the membrane surface, because Cys^{159} has low reactivity with NEM. The very quick labeling of Cys^{159} by EMA should be due to the electrostatic attraction of divalent anionic EMA to the positively charged amino acid residues in loop M2. Therefore, the labeling of Cys^{159} takes place subsequent to the noncovalent binding of EMA with loop M2. Loop M3 is suggested to be located on the matrix side, and not to be exposed to the membrane surface, because Cys^{256} interacted less than Cys^{56} with both NEM and EMA. Possibly, loop M3 is intruded into the transport path. From these results, the topology of the ADP/ATP carrier and locations of three loops facing the matrix side are postulated to be as shown in Fig. 3, based on the three repeat domain structure model [2, 4]. The finding that EMA labeled Cys^{256} more progressively than NEM suggests that access of bulky anionic EMA to loop M2 changes the conformation of loop M3, increasing the reactivity of Cys^{256} to EMA.

Substrate recognition and binding site

Fluorescein derivatives including EMA have been used as models of adenine nucleotides [8]. Unlike SH-reactive EMA, fluorescein derivatives that do not chemically modify cysteine residues should be effective for study of the interactions of ADP and ATP with their binding site.

Eosin Y, erythrosin B, rose bengal and fluorescein, with structures including divalent negative charges and hydrophobic halogeno groups on the xanthene ring (for chemical structures, see Chart), inhibited the EMA-labeling of Cys^{159} dose-dependently, as shown in Fig. 4, and their effects increased with increase in their hydrophobicities. Cibacron blue 3GA, which is also thought to have similar structural features to adenine nucleotides [8], also inhibited EMA-labeling (Fig. 4). In addition, ADP inhibited EMA-labeling, although its effect was less than that of eosin Y (Fig. 5). As the order of inhibitions by fluorescein derivatives of ADP transport was the same as that of their inhibitions of EMA-labelling [9], their effects were suggested to be due to their interactions with the ADP binding site in loop M2 depending on their hydrophobicities. From the maximum binding of eosin Y to the carrier, we concluded that eosin Y binds to the carrier with 1 : 2 stocihlometry, indicating that it (and ADP and ATP) interacts with a pair of M2 loops in the dimerized carrier [9].

Figure 6 compares the amino acid sequence of M2 loops of 22 sources. In these sequences, more then 40% amino acides are conserved. These amino acid residues may be essential for exhibition of the transport activity. In the bovine heart mitochondrial ADP/ATP carrier, the conserved amino acids are located at both ends of loop M2 connected to the membrane segments (Asp^{134}–Asp^{143} and Gly^{168}–Gln^{174}), and the middle part of the loop (Arg^{151}–Lys^{165}). As the conserved hydorphobic amino acids (Phe^{153}, Leu^{156}, Ile^{160}, Ile^{163} and Phe^{164}) are located in the middle segment, and as the adenine nucleotide binding site should be close to Cys^{159}, the middle peptide segment in loop M2 may constitute the primary binding site of ADP and ATP, and also of BKA and fluorescein derivatives including EMA. Possibly, the

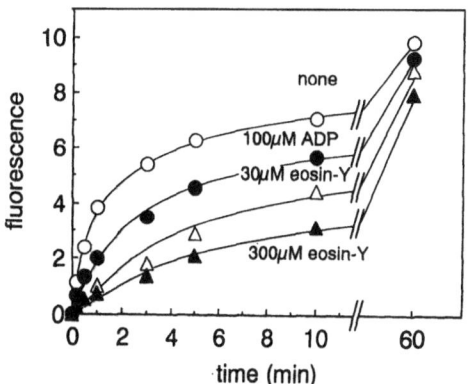

Fig. 5 Time course of EMA-labeling and effects of eosin Y and ADP. EMA-labeling was examined with bovine heart submitochondrial particles at 0 °C. The progress of labeling was determined by measuring the fluorescent intensity of labeled EMA of the carrier band on SDS-PAGE

segment turns at the sequence Gly^{157}–Asn^{158}–Cys^{159}, because by this turn the hydrophobic amino acids are aligned facing together to constitute a hydorphobic domain (Fig. 3). In addition, positively charged amino acid residues in loop M2 could also interact with polyvalent anionic adenine nucleotides. This hydrophobic domain should be stabilized by the salt bridge(s) formed between the positively and negatively charged amino acid residues, and the hydrogen bonds formed between amino acid residues in loop M2. Binding of the negatively charged transport substrates and the transport inhibitors with this domain will alter the conformation of this segment by rearrangement of the salt bridge(s) and hydrogen bonds, leading to change in the conformational state of the car-

rier. Therefore, loop M2 could act as both a primary binding site of the transport substrate and a gate for the transport.

Translocation loops associated with carrier conformation

As shown in Fig. 7, the transport inhibitor BKA greatly inhibited the labelings of Cys^{159} and Cys^{256} by EMA, but it did not have any effect on the labeling of Cys^{56} by EMA or NEM. Interestingly, CATR, when added from the cytosolic side, completely inhibited the labelings of these three cysteine residues by EMA and that of Cys^{56} by NEM, whereas it was ineffective when added from the matrix side. Although the effect of BKA on loops M2 and M3 would be the results of its interactions with these loops, the effect of CATR from the cytosolic side should be due to conformational change of these three loops caused by the binding of CATR to the carrier [10].

The conformational change of the carrier in relation to the state of the loops facing the matrix was examined in more detail by studies on the effect on cysteine residues of copper-o-phenanthroline $(Cu(OP)_2)$, which catalyzes disulfide bond formation between SH-groups [11]. Incubation of the particles with $Cu(OP)_2$ specifically induced formation of a 60 kDa protein band detected by SDS-PAGE with decrease in the 30 kDa carrier band (Fig. 8), but no change in the protein band of the carrier in mitochondria (data not shown). The 60 kDa protein was reversed to the 30 kDa carrier by treatment with 2-mercaptoethanol (data not shown). Therefore, $Cu(OP)_2$ specifically catalyzed the formation of an S–S bond between cysteine residues of the carrier acting from the matrix side. As the ADP transport

Fig. 6 Amino acid sequences of loop M2 of the ADP/ATP carrier from various sources. There are three isoforms of the carrier. Conserved amino acid residues in all the carriers are marked by asterisks (*). Amino acid residues having similar properties are marked by (+). Hydrophobic residues (F,Y,L,I,V and P) are marked by (h). A hydrophobic amino acid located at either position 163 or 164 is considered to be conserved

```
                                      134        144        154        164       174
Bovine 1                              DFARTRLAAD --VGKGAA-QREF TGLGNCITKI FKSDGLRGLY Q
Bovine 3                              DFARTRLAAD --VGKSGS-EREF RGLGDCLVKI TKSDGLRGLY Q
Human 1                               DFARTRLAAD --VGKGAA-QREF HGLGDCIIKI FKSDGLRGLY Q
Human 2                               DFARTRLAAD --VGKAGA-EREF RGLGDCLVKI YKSDGIKGLY Q
Human 3                               DFARTRLAAD --VGKSGT-EREF RGLGDCLVKI TKSDGIRGLY Q
Rat 1                                 DFARTRLAAD --VGKGSS-QREF NGLGDCLTKI FKSDGLKGLY Q
Rat 2                                 DFARTRLAAD --VGKAGA-EREF KGLGDCLVKI YKSDGIKGLY Q
Mouse 1                               DLARTRLAAD --VGKGSS-QREF NGLGDCLTKI FKSDGLKGLY Q
Yeast 1                               DYARTRLAAD ARGSKSTS-QRQF NGLLDVYKKT LKTDGLLGLY R
Yeast 2                               DYARTRLAAD SKSSKKGG-ARQF NGLIDVYKKT LKSDGVAGLY R
Yeast 3                               DFARTRLAAD AKSSKKGG-ARQF NGLTDVYKKT LKSDGIAGLY R
Neurospora crassa                     DYARTRLAAD AKSAKKGG-ERQF NGLVDVYRKT IASDGIAGLY R
Kluyveromyces lactis                  DYARTRLAAD SKSAKKGG-ERQF NGLVDVYKKT LASDGVAGLY R
Chlorella                             DFARTRLAAD --VGSGKS--REF TGLVDCLSKV VKRGGPMALY Q
Chlamydomonas reinhardtii             DYARTRLAND AKSAKKGGGDRQF NGLVDVYRKT IASDGIAGLY R
Oryza sativa                          DYARTRLAND AKAAKGGG-ERQF NGLVDVYRKT LKSDGIAGLY R
Zea mays 1                            DYARTRLAND AKAAKGGG-ERQF NGLVDVYRKT LKSDGIAGLY R
Zea mays 2                            DYARTRLAND AKAAKGGG-DRQF NGLVDVYRKT LKSDGIAGLY R
Potato 1                              DYARTRLAND RKASKKGG-ERQF NGLVDVYKKT LKSDGIAGLY R
Potato 2                              DYARTRLAND AKAAKGGGGRQF DGLVDVYRKT LKSDGVAGLY R
Arabidopsis thaliana 1                DYARTRLAND AKAAKGGGGRQF DGLVDVYRKT LKTDGIAGLY R
Arabidopsis thaliana 2                DYARTRLAND SKSAKKGRGERQF NGLVDVYKKT LKSDGIAGLY R
                                      ******** *   ++     *** ** + * *  +++* +**
                                       h    h                h   h   h  h    h  h
```

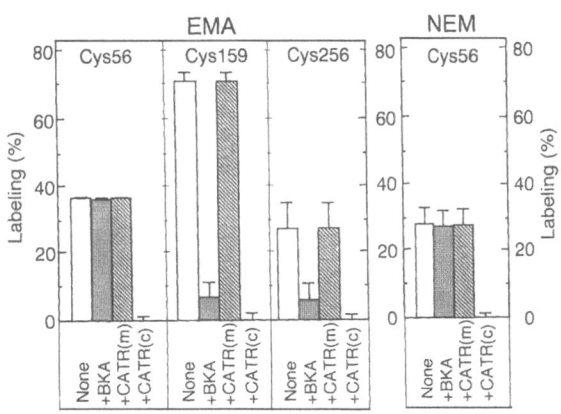

Fig. 7 Effects of BKA and CATR on the labelings by EMA and NEM of cysteine residues of the bovine heart submitochondrial ADP/ATP carrier at 0 °C. Submitochondrial particles (10 mg protein/ml) pretreated with 100 μM BKA or CATR were incubated with 100 μM EMA or 1 mM NEM for 10 min at 0 °C. CATR added from the cytosolic side and from the matrix side are referred to as CATR(c) and CATR(m), respectively

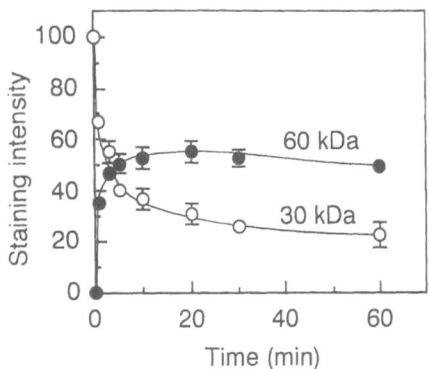

Fig. 8 Cross-linking of the 30 kDa ADP/ATP carrier with decrease in the monomeric carrier catalyzed by Cu(OP)$_2$. The effect of 100 μM Cu(OP)$_2$ on bovine heart submitochondrial particles was examined at 0 °C. 30 kDa, monomeric carrier; 60 kDa, dimeric carrier

was inhibited with progress of cross-linking, cross-link formation of the carrier was directly associated with loss of transport activity of the carrier. Amino acid sequence analysis showed that an S–S bond was formed only between Cys[56] residues in loop M1 [12]. These results showed that the carrier functions as a dimer, in which two carrier molecules are located in the membrane facing each other in the same orientation.

CATR added from the cytosolic side completely inhibited the cross-linking, but CATR and BKA added from the matrix side did not affect disulfide bond formation. These results suggested that the location of loop M1 changes according to the conformational change between the c- and m-state: loop M1 is exposed to the matrix side in the

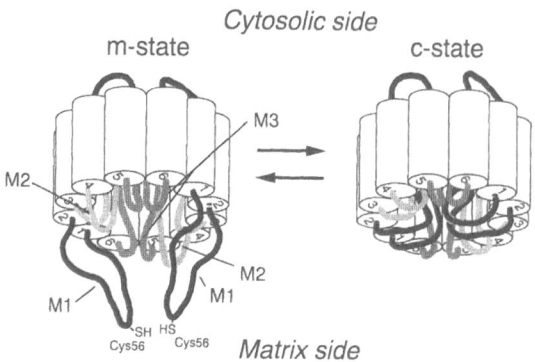

Fig. 9 Schematic representation of the locations of loops of the dimer form of the ADP/ATP carrier in its two conformational states

m-state, but is intruded into the membrane in the c-state. Possibly, that translocation of loop M1 by conversion to the c-state takes place in such a way that this loop lids the transport path, thus inhibiting the transports of ADP and ATP, as shown schematically in Fig. 9. In addition to this marked translocation of loop M1, loop M2 and M3 should change their locations, as suggested by the results in Fig. 2.

Conclusion: a possible transport mechanism

The above results are summarized as follows: (1) the dimer form of the ADP/ATP carrier is a functional unit, (2) three loops, rather than the membrane spanning region, are important for the transport activity of the carrier, (3) pairs of each of these loops act as gates for transport, (4) a pair of the second loop M2 could constitute the primary binding site for transport substrates, and (5) changes in the locations of these loops are responsible for the interconversion of the c- and m-state conformations of the carrier, which is directly associated with its transport activity.

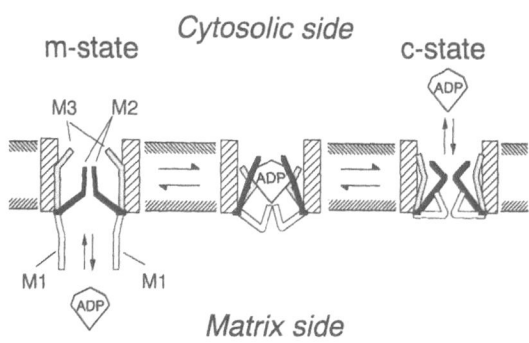

Fig. 10 Cooperative swinging loop model of the transport of ADP via the ADP/ATP carrier

Progr Colloid Polym Sci (1997) 106:192–197
© Steinkopff Verlag 1997

Based on these results, we propose the transport mechanism of the ADP/ATP carrier to be characterized by the cooperative swinging of three loops. As shown in Fig. 10, ADP (and ATP) on the matrix side is attracted electrostatically to loop M2 in the m-state carrier causing opening of the gate of a pair of M2 loops, and binds to its primary site constituted by the hydrophobic and positively charged amino acid residues. This binding changes the location of the loops, and ADP is transported to the cytosolic side, resulting in the c-state of the carrier. Reversal of the conformational states of the loops enables transport of ADP from the cytosolic side to the matrix side.

References

1. Poolman B, Konings WN (1993) Biochim Biophys Acta 1183:5–39
2. Klingenberg M (1989) Arch Biochem Biophys 270:1–4
3. Shinohara Y, Kamida M, Yamazaki N, Terada H (1993) Biochim Biophys Acta 1152:192–196
4. Majima E, Koike H, Hong Y-M, Shinohara Y, Terada H (1993) J Biol Chem 268:22 181–187
5. Houstek J, Pedersen PL (1985) J Biol Chem 260:6288–6295
6. Van Iwaarden PR, Driessen AJM, Konings WN (1992) Biochim Biophys Acta 1113:161–170
7. Majima E, Goto S, Hori H, Shinohara Y, Terada H (1995) Biochim Biophys Acta 1243:336–342
8. Neslund GG, Miara JE, Kang J-J, Dahms AS (1984) Cur Top Cell Regul 24:447–469
9. Majima E, Yamaguchi N, Chuman H, Shinohara Y, Goto S, Terada H (submitted)
10. Majima E, Shinohara Y, Yamaguchi N, Hong Y-M, Terada H (1994) Biochemistry 33:9530–9536
11. Kobashi K (1968) Biochim Biophys Acta 158:239–245
12. Majima E, Ikawa K, Takaeda M, Hashimoto M, Shinohara Y, Terada H (1995) J Biol Chem 270:29548–29554

Progr Colloid Polym Sci (1997) 106:198–203
© Steinkopff Verlag 1997

BIOCOLLOIDS

V.G. Sergeyev
O.A. Pyshkina
M.O. Gallyamov
I.V. Yaminsky
A.B. Zezin
V.A. Kabanov

DNA-surfactant complexes in organic media

Received: 13 October 1996
Accepted: 4 February 1997

Dr. V.G. Sergeyev (✉) · O.A. Pyshkina ·
M.O. Gallyamov I.V. Yaminsky ·
A.B. Zezin · V.A. Kabanov
Department of Polymer Science
Faculty of Chemistry
Moscow State University
Moscow 119899 Russia

Abstract A linear DNA can be dissolved in nonaqueous low-polar organic solvents by forming complexes between DNA and cationic surfactants. The data of high-rate sedimentation, UV-spectrometric and circular dichroism measurements show that the DNA complexes with surfactant in chloroform, heptane and cyclohexane are individual components with a 1:1 stoichiometry. The DNA molecules passing to the organic phase retain a double-stranded helix conformation. By means of scanning tunneling and atomic force microscopy it was found that DNA–surfactant complexes in water and in chloroform exist in compact toroidal conformation. It was demonstrated that such complexes can be transferred through water/organic-phase boundaries.

Key words DNA–surfactant complexes – toroidal conformation – sedimentation – UV and CD spectroscopy – scanning tunneling and atomic force microscopy

Introduction

Recent studies have shown that equimolar complexes formed between various synthetic linear polyelectrolytes with oppositely charged surfactants complexes are soluble in some organic solvents [1–3].

Complexes based on the double-helix DNA and oppositely charged surfactants are formed in aqueous solution due to Coulombic attraction between the polyanion chain units and surfactant ions, and they are stabilized by hydrophobic interaction of nonpolar fragments of the surfactant [4–5]. One can expect that such amphiphilic structure of DNA–surfactant complexes will also promote solubility in low-polarity organic solvents. At present, there is lack of information on the behavior of DNA–surfactant complexes in nonaqueous organic solvents. However, some studies indicate that the complexes based on DNA and cationic dialkyl amphiphiles can be soluble in organic media in the presence of a small amount of water [6]. We have found that carefully dried DNA–surfactant complexes are soluble in nonaqueous low-polarity organic solvents such as chloroform, hexane and cyclohexane [7]. Further studies of the structure of DNA–monocationic lipid-like surfactant (distearyldimethylammonium chloride) complexes in nonaqueous chloroform using UV and CD fluorescence and STM microscope have been undertaken in this work.

Experimental section

DNA from chicken erythrocytes (800–3000 b.p.) (DNAc) and DNA from salmon sperm (300–500 b.p.) (DNA*l*) was purchased from "Soyuzkhimreaktiv" Russia. Cetyltrimethylammonium bromide (CTAB) and distearyldimethylammonium chloride (DDAC) were obtained from Tokio Kasei. CTAB was recrystallized twice from acetone and dried under vacuum; DDAC was used without

Progr Colloid Polym Sci (1997) 106:198–203
© Steinkopff Verlag 1997

additional purification. DNA was purified using standard techniques [8]. Chloroform, hexane and cyclohexane were throughly purified from a trace amount of water, using molecular sieves of 5 Å.

The ultracentrifugation measurements in the scan mode were carried out on a Beckman E ultracentrifige at 48 000 and 60 000 rpm and 20 °C. Scanning was performed at 260 nm (maxima in the DNA absorption spectrum).

The UV spectra were recorded with Specord M40 spectrophotometer. The CD measurements were carried out on a Jasco J-500 dichrograph in a quartz cells. The scanning tunneling microscopy (STM) were performed with a home-made high-stable STM under atmospheric conditions with a Pt/Ir tips

Results and discussions

As has been shown earlier in our publication, interaction between DNA and oppositely charged surfactant molecules can lead to the formation of polymer–colloidal complexes in a water solution [4–5]. In the case of dialkyl cationic surfactant the formation of nonsoluble DNA–SA complexes takes place at an extremely low surfactant concentration and the DNA molecules undergo a marked discrete transition from coiled to the globule state [9].

In order to determine the composition of DNA–DDAC complexes, to aqueous solution of DNA various amounts of DDAC were added. This was followed by an increase in the turbidity of the system due to the formation of insoluble PSC. The complex was separated by ultracentrifugation and the concentration of DNA remaining in the supernatant was measured spectrophotometrically. As can be seen in Fig. 1, in the case of more hydrophobic surfactant DDAC interaction with DNA molecules begins at extremely low surfactant concentration (10^{-6} M).

The concentration of the DNA molecules linearly decreased with the elevation of the concentration of the surfactant, which was indicative that the composition of PSC in the precipitate did not depend practically on the concentration of the reagents. At the molar ratio DNA/Sa close to equimolar, the aqueous phase did not contain any DNA. This result was indicative of formation of polymer–surfactant complex (PSC) of DDAC and DNA with a composition similar to equimolar. In the case of less hydrophobic surfactant (CTAB), the formation of DNA–surfactant complexes starts at a certain critical surfactant concentration (10^{-5} M) Fig. 1b. However, DDAC at the molar ratio DNA/CTAB about equimolar, as well as the aqueous phase did not contain and DNA molecules.

Complexes DNA/Sa with near-equimolar composition as well as complexes of other polyelectrolytes with oppositely charged surfactants are not soluble in water, since

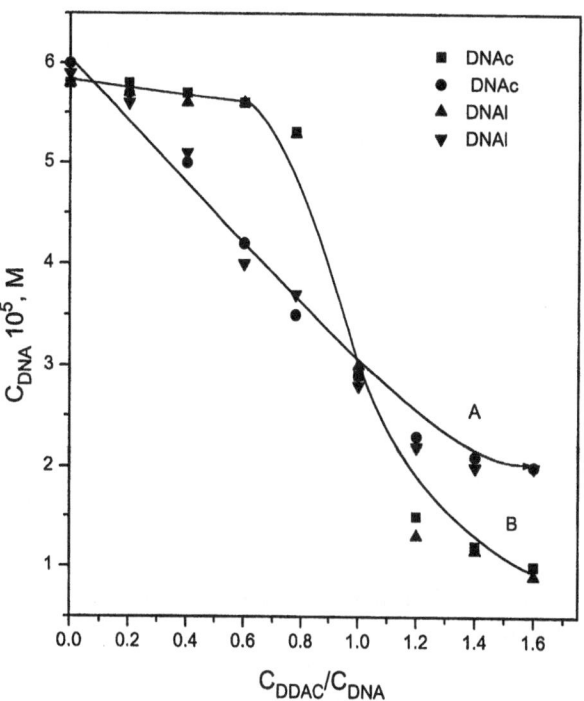

Fig. 1 Dependence of the DNA concentration in the supernatant on the concentration of DDAC-A or CTAB-B added polyanion precipitation from aqueous solution with the surfactant

the ionogenic groups of the polyelectrolytes screened off water by hydrophobic radicals of surfactant molecules. The driving forces for formation of DNA–surfactant complexes are the attractive electrostatic interaction between the phosphate groups of DNA with the counter ions of surfactants. Hydrophobic interactions inside complexes play the major role in complex stabilization.

Polyelectrolyte complexes with oppositely charged surfactants which are soluble in some organic solvents have been synthesized recently [1–3]. One can expect that amphiphilic structure of the DNA–surfactant complexes will also promote solubility in low-polarity organic solvents.

DNA is polyanionic and, so soluble only in aqueous media. We found that carefully dried DNA– surfactant complexes obtained as previously described from an aqueous solution are soluble in organic solvents such as chloroform, heptane and cyclohexane.

Figure 2 represents the UV spectra of DNA–surfactant complexes in chloroform (curves 1 and 2) and the original DNA in water (curve 3). Inspection of the data in Fig 2. reveals that the UV-spectra in chloroform practically coincided with UV-spectrum in aqueous buffer solution. For all spectra the adsorption maximal is observed at 260 nm and the extinction coefficient is about 6500–6700 for both the cases. Similar spectra were observed for DNA–CTAB complexes in chloroform. UV-spectra of DNA–CTAB in

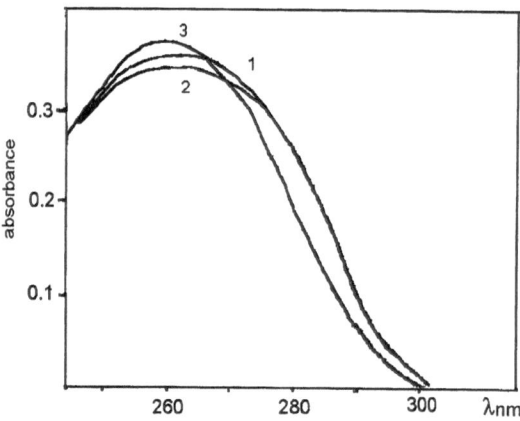

Fig. 2 Absorption spectra of (1) PSC of DNA*l* and DDAC in chloroform, (2) PSC of DNAc and DDAC in chloroform, (3) DNA in aqueous buffer solution

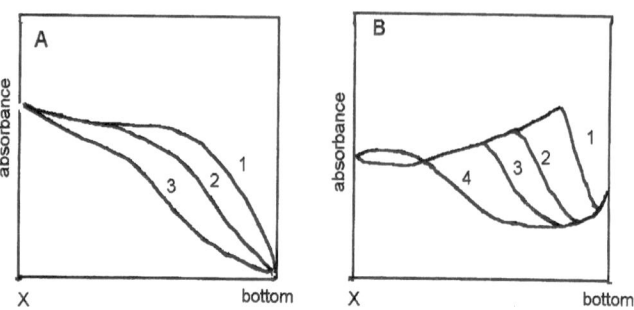

Fig. 3 Typical sedimentation patterns of (1) PSC of DNA*l* and DDAC (2) PSC of DNAc and DDAC in chloroform

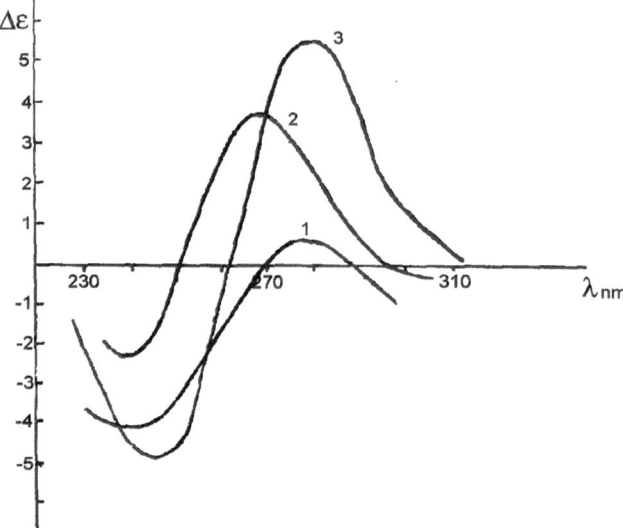

Fig. 4 D-spectra of (1) PSC of DNA*l* and DDAC in chloroform, (2) PSC of DNAc and DDAC in chloroform, (3) DNA in aqueous buffer solution

heptane and cyclohexane are identical to UV-spectra in chloroform.

Figure 3 shows the scan sedimentation patterns of DNA–surfactant complexes in chloroform over the course of 2–30 min against x, the distance from the meniscus in the sedimentation cell. One can see that the pattern has only one step which is moving from right to left in time, i.e. from the bottom to the meniscus of the sedimentation cell. The observed migration may be due to the higher density of chloroform compared to the density of the DNA–surfactant complex. This result indicates that the DNA–surfactant complex is an individual compound with constant composition, since at least one of the components of complexes, i.e. DNA is not soluble in the nonpolar solvents.

One of the most appropriate methods for investigation of different forms of a nucleic acid in solution is CD spectroscopy which is extremely sensitive even to small changes in mutual orientation of the neighboring bases in a polynucleotide. Figure 4 shows the CD spectra of a DNA–DDCAC complex dissolved in chloroform (curves 1 and 2) and the initial DNA in B-form in buffer solution (curve 3). One can see that the CD spectra of DNA–surfactant complexes in chloroform have the CD spectra characteristic of the native DNA in a tight-packed conformation. The CD spectra of DNAc differ from the spectra of native DNA (Fig. 4 curve 2) by the red shift of the positive band maximum (276 nm for free DNA and about 280 nm for DNA in chloroform) and of the crossover point (260 nm for free and 270 nm for DNA in chloroform). The DNAc chloroform spectra is characterized by a decrease in the intensity of the positive band and increase in the intensity of the negative band at 240 nm. CD spectra of DNA*l* display an A-like structure with positive maximum at 270 nm, negative minimum at 240 nm and of crossover point at 255 nm. Similar CD spectra have been discovered

for dense-packed DNA molecules in the heads of phage [10]. Taken together, the UV, CD spectroscopy and ultracentrifugation data suggest that the two-stranded structure of DNA conserved in organic media.

Recent studies have reported that single DNA molecules change their structure in a discrete manner between random coil to compact globule state with the addition of the cationic surfactant cetyltrimethylammonium bromide [4, 5]. Observations by fluorescence microscopy allow only to observe an image of the globule without a more accurate analysis of its structure, due to the resolution limit of the wavelength of the fluorescent light and also from the blooming effect on the side of the high-sensitive TV camera. Thus, we have tried to estimate the actual size and shape of the DNA particles by scanning tunneling and atomic force microscopy (STM/STM). The experiments were carried out with a home-made high-stable STM at ambient conditions [11–12] with Pt/Ir tips prepared by mechanical cutting of a wire 0.4 mm in diameter. A drop of 2 µl of the mixture containing 2 µg/ml of DNA in a 0.5 TBE-buffer and 8×10^{-5} M DDAC was placed on freshly cleaved graphite rinsed with MilliQ-purified water or chloroform and allowed to dry in air for several minutes before imaging in STM. An inert graphite surface was chosen to minimize the substrate effect on the topography of the DNA–surfactant complex.

The DNA condensation induces by the interaction with DDAC results in the formation of toroid-like structures is shown in Fig. 5A. The average diameter of the toroid is about 100 nm, the apparent height is about 10 nm. As can be seen in Fig. 5B, the DNA–DDAC complex in chloroform also has a toroid structure. The diameter of such a toroid structure coincides with that condensed in water, while the apparent height is about 5 nm.

In order to understand the difference between the structures of DNA–surfactant complexes in aqueous and chloroform solutions, we made a quantitative treatment of the toroid-like particles observed. The total volume of the toroid is

$$V_{tor} = (\pi^2/4)(R_{out} - R_{in})^2(R_{out} + R_{in}). \tag{1}$$

The total volume of DNA is

$$V_{dna} = a^2 L, \tag{2}$$

where $a = a_{dna} + a_{surf} = 10A + 30A$, L is total length of DNA ($3.4A$ the number of base pairs). The number of DNA–surfactant complex molecules contained in toroid is

$$N = 0.906 \, (V_{tor}/V_{dna}), \tag{3}$$

where 0.906 is a packing density.

We have estimated that a single toroidal structure in water consists of 5–15 molecules of DNA, while in chloroform the toroid is formed by a single DNA molecule. These data are in good coincidence with the toroid is formed by a single DNA molecule. These data are in good coincidence with the toroid size of DNA, condensed by multivalent cations [13]. It should be pointed out that aggregation of 5–15 DNA molecules in one particle in water is stabilized by the hydrophobic tails of the surfactant molecules. In low-polarity organic media, the liophobic forces play a negligible role and aggregation is not clearly observed.

Our recent independent studies of the hydrodynamical and electro-optical (Kerr effect) properties of DNA–DDCA complexes in chloroform demonstrated that the compact conformation of DNA molecules in the complex maintain in organic media [14].

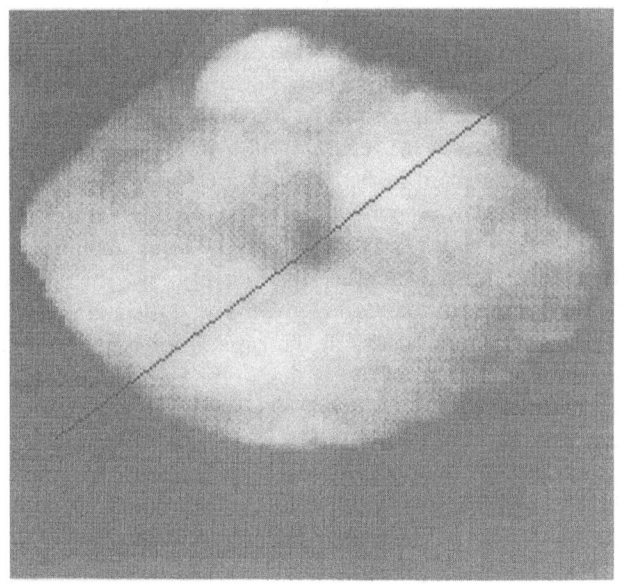

A

B

Fig. 5 STM images of DNA–DDAC complexes absorbed on graphite substrate: (A) – from 0.5 M TBE-buffer (image size 310 × 310 nm, apparent height 13 nm, tunneling current 0.25 nA, bias voltage 100 mV). (B) – form chloroform (image size 130 × 130 nm, apparent height 6 nm, tunneling current 0.25 nA, bias voltage − 150 mV)

The transport of low molecular weight ionic drugs, proteins, polynucleotids, and plazmid DNA through liquid membranes can be achieved by forming ion pairs with detergents [15–17]. In the present work we have explored the possibility of linear transport of DNA through the aqueous/organic interface and *vice versa*.

To this end, a microtube was charged by adding 1 ml of cloroform and then 0.5 ml of an aqueous solution (0.5 M TBE buffer) containing DNA and different amounts of surfactant. The biphasic mixture was shaken for 5 min, then the phases were separated in a microcentrifuge and DNA was detected in the organic and water phases.

We found that complexes based on high molecular weight DNA could not penetrate to the organic phase and they concentrated at the aqueous/organic interphase. However, the complexes based on low molecular weight DNA could be observed in the organic phase. The data obtained are presented in Fig. 6. Curves A and B depict the concentration of DNA in the organic and water phases, respectively, against the initial ratio of DCDA/DNA in water solution. One can seen that a quantitative transfer of the DNA to the organic phase took place and all the DNA can be transferred to the organic phase when the ratio of DCDA/DNA is equal to 1.5.

Zezin and Kabanov [18] reported that dissociation of the polyelectrolyte–surfactant complexes takes place at a certain concentration of added salt. Therefore, it was reasonable to presume the same for DNA–surfactant complexes and expect an extraction of DNA from organic to aqueous phase.

To test this hypothesis, at first we investigated the effect of salt on the DNA–surfactant complexes. DNA–surfactant complexes with stoichiometric composition were placed in NaCl solutions (0.5 ml) of various concentrations and kept for 48 h (it was shown that this period of time was sufficient to attain the equilibrium). The DNA was determined by measuring the absorbance at 260 nm. As can be seen in Fig. 5, at cNaCl > 0.25 M a significant amount of DNA was observed in the solution; importantly, none was detected below this concentration. Inspection of the data in Fig. 5 reveals that the amount of DNA in the solution greatly depends on the concentration of low-molecular salts. Next, we explored whether it is possible to recover DNA from the organic to aqueous phase from which all DNA was transferred to the organic phase by adding DCDACl with a 0.5 ml solution of 0.5 M NaCl. After shaking for 5 min, the phases were separated in a microcentrifuge and DNA was detected in the organic and aqueous phases. As can be seen from Table 1, a significant transport of DNA was observed; importantly, the amount of DNA in the aqueous phase does not depend on the concentration of DNA in the organic phase. It seems that the concentration of DNA in the aqueous phase is mainly controlled by the concentration of the low-molecular salt which causes the DNA–DCDACl complex to dissociate.

In conclusion, we have established the conditions for the formation of hydrophobic DNA/surfactant complexes, soluble in some low-polarity organic solvents with the

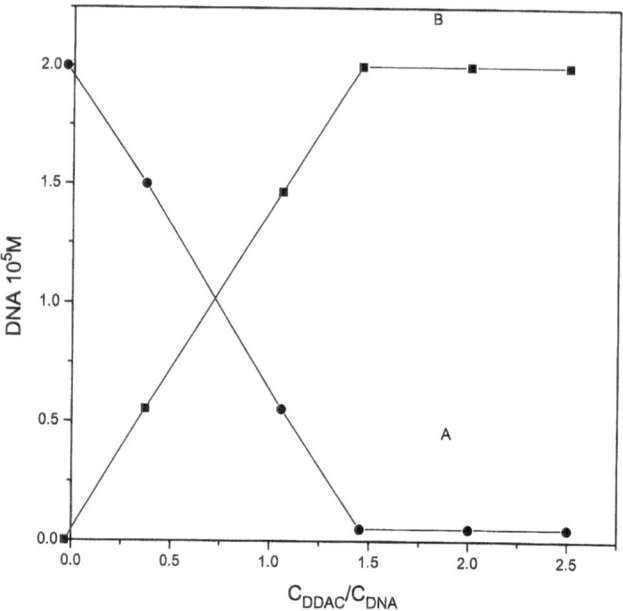

Fig. 6 Effect of increasing amounts of DDAC on the concentration of DNA in the aqueous (A) and organic (B) phase. The amount of DNA used was 20 mkg

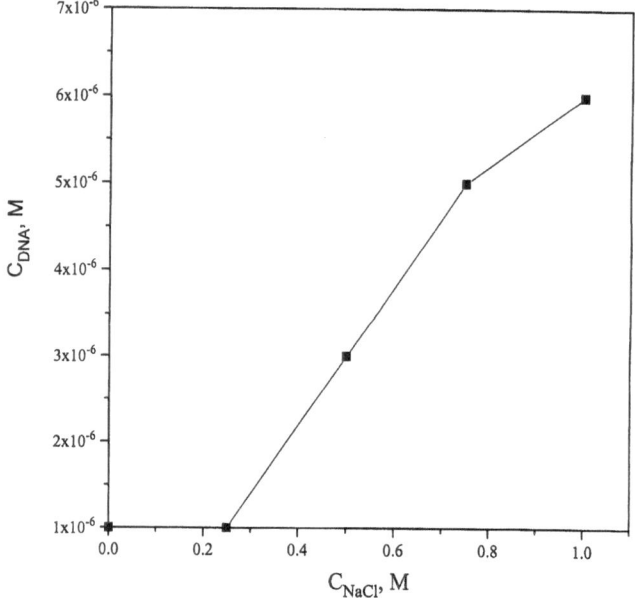

Fig. 7 Effect of the increasing amount of NaCl on the DNA–DDAC complex

formation of molecular solution. The results also demonstrate that surfactant-induced DNA condensation results in close-packed toroid conformation both in water and low-polarity organic media. The observed effect may be caused by the internal mechanical tension in the DNA

Progr Colloid Polym Sci (1997) 106:198–203
© Steinkopff Verlag 1997

Table 1 Effect of 0.5 M NaCl on the recovery of DNA*l* in the aqueous phase from chloroform

Initial concentraton of DNA*l* in chloroform [M]	9×10^{-6}	1.8×10^{-5}	2.5×10^{-5}	3.0×10^{-5}
Equilibrium concentration of DNA*l* in aqueous phase [M]	3.0×10^{-6}	3.0×10^{-6}	3.0×10^{-6}	3.0×10^{-6}

molecules and the reduction of the electrostatic repulsion between the DNA segments leads to a new energetically favourable conformation. Moreover, we have demonstrated the surfactant-induced transport of DNA through aqueous/organic interface. DNA could be recovered in the aqueous phase with the help of a low-molecular salt. Further investigation of the behavior of such complexes with biological membranes are currently in progress.

References

1. Bakeev KN, Yang MS, Zezin AB, Kabanov VA (1993) Dokl Akad Nauk SSSR 332 451
2. Bakeev KN, Shu YM, MacKnight WJ, Zezin AB, Kabanov VA (1994) Macromolecules 27:200
3. Bakeev KN, Yang MS, Zezin AB, Kabanov VA, Lezov AV, Mel'nikov AB, Kolomiets IP, Rjumtsev EI, MacKnight WJ (1996) Macromolecules 29:1320
4. Mel'nikov SM, Sergeyev VG, Yoshikawa K (1995) J Am Chem Soc 117:2401
5. Mel'nikov SM, Sergeyev VG, Yoshikawa K (1995) J Am Chem Soc 117:9951
6. Ijiro K, Okahata Y (1992) J Chem Soc Chem Commun 1339.
7. Pyshkina OA, Sergeyev VG, Zezin AB, Kabanov VA (1996) Dokl Akad Nauk SSSR, 348:496
8. Sambrook J, Fristsch EF, Maniatis T (1989) Molecular Clonning: A Laboratory Manual. 2nd ed., Cold Spring Harbor, New York
9. Mel'nikov SM, Sergeyev VG, Mel'nikova YS, Yoshikawa KJ (in press) Chem Soc Faraday Trans.
10. Karasev AV, Dobrov EN (1988) Int J Biol Macromol 10:227
11. Moiseev Yu N, Panov VI, Savinov SV, Vasil'ev SI, Yaminsky IV (1994) J Vac Sci Technol B 12:1690
12. Moiseev Yu N, Panov VI, Savinov SV, Vasil'ev SI, Yaminsky IV, (1992) Ultramicroscopy 42–44:1596
13. Arscott PG, An-Zhi Li, Bloomfield VA, (1990) Biopolymers 30:619
14. Pyshkina OA, Sergeyev VG, Lezov AV Mel'nikov AB, Rjumtsev EI, Zezin AB, Kabanov VA, (in press) Dokl Akad Nauk SSSR
15. Neubert R (1989) Pharmacol Res 6:743
16. Bromberg LE, Klibanov AM (1994) Proc Natl Acad Sci USA 91:143
17. Reimer DL, Zhang YP, Kong S, Wheeler JJ, Graham RW, Bally MB (1995) Biochemistry
18. Khandyrina Yu V, Rogacheva VB, Zezin AB Kabanov VA (1994) Polym Sci 36:195

Progr Colloid Polym Sci (1997) 106:204–208
© Steinkopff Verlag 1997

BIOCOLLOIDS

K. Yoshikawa
H. Noguchi
Y. Yoshikawa

Folding transition in single long duplex DNA chain

Received: 13 October 1996
Accepted: 21 February 1997

Prof Dr. K. Yoshikawa (✉) ·
H. Noguchi · Y. Yoshikawa
Graduate School of Human Informatics
Nagoya University
Nagoya 464-01, Japan

Abstract Folding transition from elongated coil into compacted globule in a single polymer chain is discussed based on the results of our recent theoretical and experimental studies. As the theoretical approach to this problem, Monte Carlo simulation on the coil–globule transition for a neutral stiff polymer chain has been performed. It has become clear that toroid and rod are the two representative structures as the product of folding transition: toroid is the most stable and rod is a kinetically forzen metastable structure. As for the experimental methodology, single molecular observation with fluorescence microscopy was applied for the coil–globule transition of a single duplex DNA. With this experimental tool, it became evident that individual DNA chains undergo first-order phase transition. In contrast to this, the ensemble of DNA has the characteristics of diffuse or continuous transition. In other words, the coil–globule transition in the ensemble of the chains appears a kind of cooperative transition without any discrete character in spite of the large discrete change in the effective volume of the individual DNA chains. In order to gain further insight on the manner of folding in single chains, we have also performed electron microscopic observation on the morphology of the collapsed DNA chains.

Key words Coil-globule transition – collapsed transition – self-organized nano-structure – single chain dynamics – hierarchical system

Introduction

It is well known that polymer chains exhibit transition between the elongated coiled state and the collapsed-globule state depending on the parameters in the environment, i.e., temperature, salt concentration, dielectric constant, etc. [1, 2]. From theoretical considerations [2, 3], it has been expected that the coil–globule transition can be discrete for stiff polymers, i.e., first-order phase transition. However, previous experimental studies have indicated that the transition is always continuous [4, 5]; in other words, there is no phase transition on conforma-

tional change between the random coil and the collapsed state.

Recently, we have found that individual single chains of double-stranded DNA molecules undergoes a discrete phase transition between elongated coil and collapsed globule, from the direct observation of individual DNA molecules with fluorescence microscopy [6–12]. The essential features of the coil–globule transition discovered in our recent studies are summarized as follows:

1) Individual DNA molecules, with sufficient length of the order of 10^4–10^5 base paris, exhibit discrete coil–globule transition. The change of the packing density of the

Progr Colloid Polym Sci (1997) 106:204–208
© Steinkopff Verlag 1997

DNA segments is more than the order of 10^4. The transition is reversible.

2) The discrete phase transition is induced by various kinds of condensation agents such as neural synthetic polymer (PEG) [6, 7], cationic surfactant (CTAB) [8, 9], alcohol [11], multi-valent inorganic cation [12], and polyamine [10].

3) There lays a coexistence region of coil and globule in the diagram of the phase transition. In other words, the transition is largely discrete individual DNA molecules, whereas it appears continuous in the ensemble average of DNA chains.

In this article, we will show the result of Monte Carlo simulation on the process of collapse from coil state in a single polymer chain. It is demonstrated that a long stiff polymer collapses into compact globule with various shapes such as toroid, rod or somewhat complicated structures. It becomes clear that toroidal shape is the product of collapse with the minimum energy. From the measurement with electron microscopy, it is shown that toroid and rod structures are the shapes with the highest probabilities for the products of collpase from long duplex DNA, corresponding to the result obtained by the Monte Carlo simulation. It is also shown that long DNA chain folds into chromosome-like structure, when a cationic lipid with the head group of spermine, dioctadecylamidoglycylspermine; DOGS, is used as the condensation agent.

Theoretical result

Let us consider a polymer chain which contains N monomer links with persistence length l (Kuhn length: $2l$) and width d. The free energy in a polymer can be written as the summation from the contributions of elastic energy, interacting energy between links (or segments), osmotic term (translational entropy of the counter ion) and Coulomb interaction.

$$F = F_{ela} + F_{int} + F_{tra} + F_{elec}. \qquad (1)$$

In the case of neutral polymer, the third and fourth terms have no contribution. As has been discussed in our previous studies [7, 10, 12], the analysis of Eq. (1), both for neutral polymer and polyelectrolyte, shows that there will be two free-energy minima for stiff polymer chain at the intermediate region of the pair interaction energy between links (see Fig. 1). This implies that individual polymer chains take either coil or globule shape. In the coexisting region, the relative populations of coil and globule are determined by the Boltzmann distribution. Thus, the ensemble average of the size of the (monodisperse) chains exhibits continuous transition, i.e., there appears no

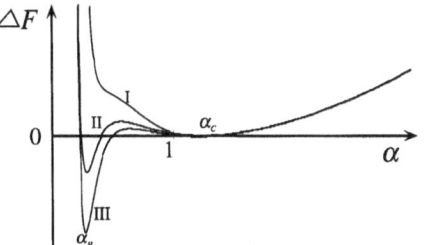

Fig. 1 Schematic representation on the free-energy profile for a single stiff polymer. $\alpha = R_r/R_0$, where R_0 and R_r are the gyration radii of ideal chain and real polymer chain, respectively. I: Coil is stable, II: Coexistence between coil and globule, and III: globule is stable

discreteness in the coil–globule transition on the level of chain ensemble.

We have thus made clear such a hierarchical structure for the ensemble of monodisperse chains. However, the above theoretical considerations on Eq. (1) have been performed based on the framework of the approximation with self-consistent field. In order to make clear the statistics and also kinetics on the transition in an isolated single chain, we have carried out the Monte Carlo simulation on the folding process of a single chain by the Metropolis method. As the detailed numerical procedure of the Monte Carlo simulation has been described in a previous report [13], here we would explain the method only briefly. The polymer chain is modeled by N spherical beads interconnected by $(N - 1)$ sticks with constant length a. A pairwise interaction is evaluated by a well potential $V_b : V_b = \infty (r_{ij} < 0.3a)$, $V_b = -\varepsilon(0.3a < r_{ij} < 0.4a)$, $V_b = 0$ $(0.4a < r_{ij})$, where r_{ij} the distance between the i and j beads. Stiffness potential is introduced as $V_S(\theta) = \kappa\theta^2$, θ is the bending angle between the neighboring links. The essential novel point in our simulation is that we have adapted the length of a link to be $\frac{1}{8}$ of the Kuhn length. (In the usual double-stranded DNA, the Kuhn length is the order of 0.1 μm; ca. 300 base paris.) With this appropriate degree of coarse graining, we have succeeded in obtaining the collapsed structure such as toroid (see Fig. 2). If one adapts the simulation for the chain where one link corresponds to Kuhn length, such a toroidal structure can never be reproduced, because the size of the toroid is similar to that of Kuhn length in this case. On the other hand, if one tries to perform the theoretical simulation with more microscopic level (for example, usual molecular dynamic calculation), it is impossible to follow the whole process of coil–globule transition for such a long chain, even with a super computer with the highest speed available.

It is noted in Fig. 2 that toroidal structure is formed spontaneously from an isolated stiff polymer chain. Beside toroid, it has been confirmed that rod structure and fused structural between toroid and rod are formed [13]. Among

206
K. Yoshikawa et al.
Folding transition in single DNA

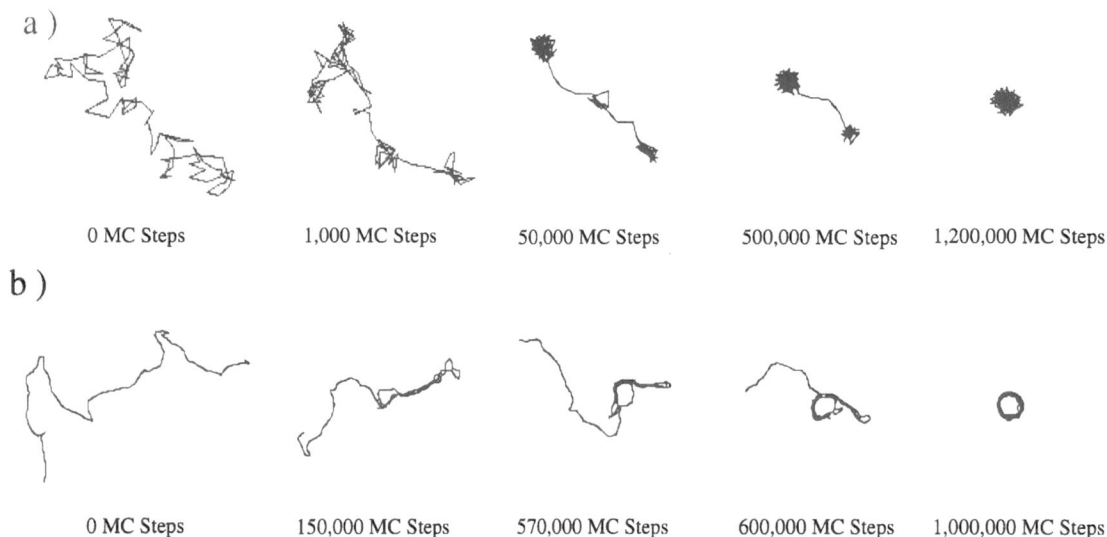

Fig. 2 Snapshot on the collapsing process of single homopolymer with Monte Carlo simulation ($N = 100$, $\varepsilon = 2.5$). (a) Flexible polymer ($\kappa = 0$), and (b) stiff polymer ($\kappa = 2$, corresponding to the Kuhn length with 8.4 times as large as that in a). Where ε is the depth of well potential for pairwise interaction, and κ is the force constant of the angle-dependent potential between the neighboring links, $V_s(\theta) = \kappa\theta^2$ (θ is the bending angle between the links)

these structures, the toroidal structure is found to be most stable. The other structures are metastable or kinetically frozen. In other words, the rod structure once formed may have almost infinite lifetime under thermal agitation at ambient temperatures. In contrast to the manner of collapse in stiff polymer, the spherical globule is formed for flexible polymer. In this case, the arrangement of the segments in a chain becomes rather irregular.

In order to examine the effect of the stiffness on the morphology of collapsed globule from a single chain, we have performed the Monte Carlo simulation on the morphological change in the globule after the abrupt change of the stiffness parameter κ (see Fig. 3). When the stiffness is decreased suddenly at $t = 0$, the toroid transforms gradually into a spherical globule as demonstrated in Fig. 3a. On the other hand, starting from the spherical globule, the toroid is spontaneously generated after the sudden increase of stiffness (Fig. 3b).

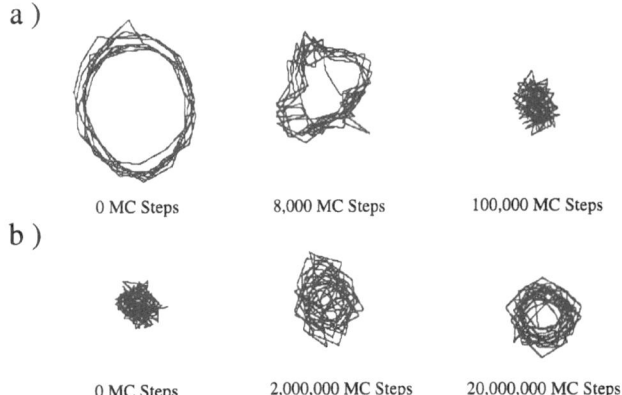

Fig. 3 Morphological change after the sudden change of the chain stiffness with Monte Carlo simulation ($N = 100$, $\varepsilon = 2.5$). (a) Transformation from toroid to spherical globule after the change of the stiffness parameter from $\kappa = 2$ into $\kappa = 0$, (b) transformation from spherical the toroid globule with the change from $\kappa = 0$ into $\kappa = 2$

Experimental result

As for the technique of single molecular observation with fluorescence microscopy, we have already discussed elsewhere [6–12]. Thus, in the present report, we will focus on the experimental trend obtained by the observation with electron microscopy.

Figure 4a shows an electron microscopic picture of the toroidal structure of a T4 DNA chain (166 kilo base pairs)

complexed with spermidine, a tri-amine. Observation on such a toroidal structure has been rather frequently reported in the past [14, 15], i.e., toroid is the typical structure formed with various kinds of condensation agents such as metal cation or neutral polymer (PEG), even for shorter DNA chains. As has been shown in Fig. 2, the formation of toroid seems to be a rather general phenomenon on the collapsed product from stiff polymer chains. Experimental verification on this idea for other

a)

100nm

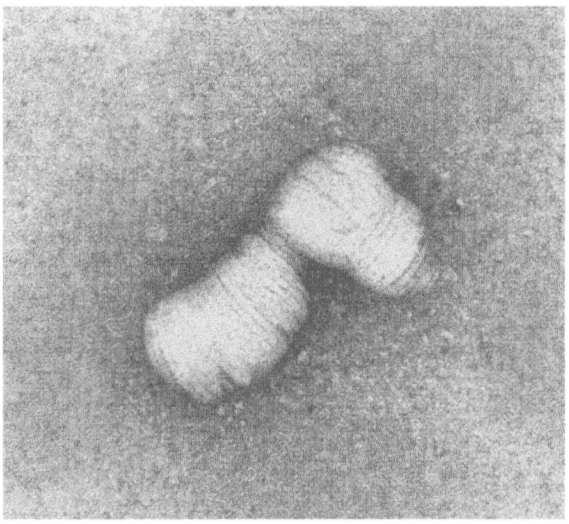

b)

50nm

Fig. 4 Transmission electron micrograph. (a) T4 DNA/spermidine complex prepared from the 1:1000 solution in molar ratio, and (b) pBR322 DNA/DOGS complex prepared from the 1:5 solution

polymers besides DNA will be an exciting future research target. It is also noted in Fig. 4a that a fused structure between rod and small toroid is formed, presumably from a single DNA chain. It has been confirmed in the Monte Carlo simulation that such a complex structure is also encountered in the collapsed product [13].

Fig. 4b shows the complex of pBR322 plasmid DNA with DOGS, dioctadecylamidoglycylspermine. It is expected that the DDNA/DOGS complex is composed of an aggregate of DOGS wrapped by DNA stands, resulting in the formation of a spool or a nucleosome-like structure [16].

Discussion

In the present paper, we have shown from theoretical simulation that toroidal structure is formed spontaneously as the most stable form in the collapsed chain. Although such a result was obtained in a very simple model of polymer chain without any charge or counter ion, the experimental structures obtained for long DNA chain has be reproduced rather well. Thus, it may be promising to extend the study on the collapsing transition of single chains for other natural and synthetic polymers besides DNA, and to try to obtain self-organized nanostructures under the idea, to control the phase transitions. As an actual example of the self-organized nanostructure, we have shown the spool structure from single DNA chain in Fig. 4.

Studies on the coil–globule transition on long DNA chain have biological significance, because the manner of folding of DNA is expected to be deeply concerned with the mechanism of self-regulation is gene expression in living cells [17].

Acknowledgements The present study is partly supported by the Grant-in-Aid from the Ministry of Education, Since, Culture and Sports and by Asahi Glass Foundation.

References

1. Grosberg AYu, Khokhlov AR (1994) Statistical physics of macromolecules. American Institute of Physics Press, New York
2. Lifshitz IM, Grosberg AYu, Khokhlov AR (1978) Rev Mod Phys 50:683
3. Post CB, Zimm BH (1982) Biopolymers 21:2123
4. Geiduschek EP, Gray I (1956) J Am Chem Soc 78:879
5. Swislow G, Sun ST, Nishio I, Tanaka T (1980) Phys Rev Lett 44:796
6. Minagawa K, Matsuzawa Y, Yoshikawa K, Khokhlov AR, Doi M (1994) Biopolymers 34:555
7. Vasilevskaya VV, Khokhlov AR, Matsuzawa Y, Yoshikawa K (1995) J Chem Phys 102:6595

8. Mel'nikov SG, Sergeyev VG, Yoshikawa K (1995) J Am Chem Soc 117:2401
9. Mel'nikov SG, Sergeyev VG, Yoshikawa K (1995) J Am Chem Soc 117:9951
10. Yoshikawa K, Takahashi M, Vasilevskaya VV, Khokhlov AR (1996) Phys Rev Lett 76:3029
11. Ueda M, Yoshikawa K (1996) Phys Rev Lett 77:2133
12. Yoshikawa K, Kodiaki S, Takahashi M, Vasilevskaya VV, Khokhlov AR (1996) Ber Bunsengesellschaft 100:876
13. Noguchi H, Saito S, Yoshikawa K (1996) Chem Phys Lett, in press
14. Gosule LC, Schellman JA (1976) Nature 259:333
15. Plum GE, Arscott PG, Bloomfield VA (1990) Biopolymers 30:631
16. Yoshikawa Y, Emi N, Kanbe T, Yoshikawa K, Saito H (1996) FEBS Lett, in press
17. Alberts B, Bray D, Lewis J, Raff M, Roberts K, Watson JD (1994) Molecular biology of the cell, 3rd ed, Ch 8. Garland Publishing Inc, New York

Progr Colloid Polym Sci (1997) 106:209–214
© Steinkopff Verlag 1997

BIOCOLLOIDS

S.M. Mel'nikov
V.G. Sergeyev
K. Yoshikawa

Cooperation between salt induced globule-coil transition in single duplex DNA complexed with cationic surfactant and sphere-rod transition of surfactant micelles

Received: 13 October 1996
Accepted: 24 March 1997

S. M. Mel'nikov · V.G. Sergeyev
Department of Polymer Science
Faculty of Chemistry
Moscow State University
Moscow 119899, Russia

S.M. Mel'nikov · Prof. K. Yoshikawa (✉)
Division of Informatics for Natural Sciences
Graduate School of Human Informatics
Nagoya University
Furo-cho, Chikusa-ku
Nagoya 464-01, Japan

Abstract The effect of low-molecular salt, sodium bromide (NaBr), on the conformational behavior of T4DNA globules, compacted with cetyl-trimethylammonium bromide (CTAB), was studied in aqueous buffer solution. The conformational dynamics of individual single duplex T4DNA molecules was visualized directly with the use of fluorescence microscopy (FM), whereas viscometry was used for monitoring of the surfactant structure in the aqueous media. We have found that DNA globules, compacted with CTAB, are unfolded into the elongated coil state with the increase of NaBr concentration, exhibiting the character of a discrete first-order phase transition. It is indicated that, accompanied with the increase of NaBr concentration, the unfolding transition is induced simultaneously with the sphere–rod transition of surfactant micelles. These results clearly indicate the cooperative effect between two different types of transitions.

Key words Single duplex DNA – cationic surfactant – fluorescence microscopy – coil-globule transition – sphere-rod transition of micelles

Introduction

During the last several decades, the interaction between polyelectrolyte and oppositely charged surfactant has attracted much attention as an important problem in colloid and polymer chemistry [1–3]. Among the rich variety of the combinations between polyelectrolyte and oppositely charged surfactant, understanding of the mechanism of interaction between DNA and cationic surfactant is of importance from biological and medical perspectives: e.g., cationic surfactants are expected to be effective as a nonviral vehicle for gene transfer [4–13]. For study of the binding of the surfactant ions to DNA polyanions, the application of surfactant ion-selective electrodes has been shown to be useful [14–16]. This method allows to estimate the binding degree of surfactant to polyelectrolyte at a given surfactant concentration, measuring the activity of

free surfactant ions. The results of these studies have shown a continuous change in the binding degree of surfactant ions to DNA. The continuous change has been assessed in terms of the one-dimensional Ising model [17–19].

Contrary to the above theoretical prediction, we have recently obtained clear experimental evidence that the interaction between polyelectrolyte and oppositely charged surfactant can never be described by the classical Ising model, at least for long and stiff polyanion, such as phage T4DNA. Instead of the picture of a continuous transition, it has become evident that individual DNAs exhibit discrete first-order phase transition [20–22]. We have applied a fluorescence microscopy (FM) technique [23–27] to confirm the discrete character of the coil–globule transition in individual DNAs. Although previously there have been reports on the conformational changes of polyelectrolytes induced by interaction with oppositely charged surfactant [28–30], we have presented for the first time the results on the single-macromolecule behavior in its real meaning, avoiding the effect owing to inter-chain interaction.

The effect of salt on the interaction of polyelectrolyte with oppositely charged surfactant is a complicated problem, due to the competition of two opposite effects. Firstly, the interaction between polyions and oppositely charged surfactant species is expected to be weakened, accompanied with the increase of the low-molecular salt concentration [31–34]. This effect is attributed to the screening of the Coulomb interactions between polyion and surfactant. As an opposite effect, increase of the salt concentration may cause decrease of the critical micelle concentration (CMC), inducing the stabilization of the surfactant micelles in the surrounding solution [35]. The decrease in the electrostatic repulsion between the charged headgroups of the surfactant ions will lower the CMC from that in pure water. Thus, the salt effect can work in an opposite manner depending on chemical structures of polyion, surfactant and salt.

In the present study, we will focus our attention on the salt effect on the coil–globule transition of single-DNA globules in the presence of cationic surfactant, cetyltrimethylammonium bromide (CTAB), at high surfactant concentration, equal to 0.137 M, which is almost 2 orders of magnitude higher than the CMC of CTAB [36].

Experimental

Materials

T4DNA, 166 kilobase pairs (contour length 57 μm [37]), was purchased from Nippon Gene. The monodispersity and the actual concentration of the available T4DNA were checked as previously [20]. Cationic surfactant, cetyltrimethylammonium bromide (CTAB), was obtained from Tokyo Kasei. CTAB was recrystallized twice from acetone and dried overnight in vacuum at 25 °C. Fluorescent dye, 4',6-diamidino-2-phenylindole (DAPI), and antioxidant, 2-mercaptoethanol (ME), were purchased from Wako and used without further purification. ME was used as a free radical scavenger to reduce light-induced damage of T4DNA molecules. Sodium bromide from Wako was dried before use.

Methods

Fluorescence microscopic measurements were performed as follows. T4DNA molecules were diluted in $0.5 \times$ TBE buffer solution (45 mM Tris, 45 mM borate, 1 mM EDTA,

pH 8.0) containing 4% (v/v) ME and DAPI. The final concentrations of DNA and DAPI were 0.6 μM in nucleotide units and 0.6 μM, respectively. Under these conditions, the number of DAPI molecules bound per base pair is estimated to be 0.05 [38], and the persistent length is expected to remain nearly the same as in the absence of DAPI [39]. The samples were illuminated with 365 nm UV-light; fluorescence images of DNA molecules were observed using a Zeiss Axiovert 135 TV microscope equipped with a 100 oil-immersed objective lens and recorded on S-VHS videotapes through a high-sensitivity Hamamatsu SIT TV camera. The observations were carried out at 35.0 ± 0.1°C. The apparent length of the long axis, L, which was defined as the longest distance in the outline of the DNA image, was evaluated with an image processor (Argus 10, Hamamatsu Photonics). Due to the blurring effect, L is estimated to be larger than the actual size by about 0.3-0.5 μm [20]. Sample solutions, microscope slides and coverslips were prepared as previously described [20–22]. It has been ascertained that the reported results do not depend on the mixing order of the reagents. The viscosity was measured by Tokimec Visconic ELD viscometer at 35.0 ± 0.1 °C. JS 2.5 calibration liquid (Showa Shell) was used as the standard in the viscosity measurements. For the measurements with both FM and viscometry, the temperature was kept constant using a water jacket, connected to the Neslab RTE-111 thermostating system.

Results and discussion

With the use of FM, we performed direct observation on large single T4DNAs compacted at [CTAB] = 0.137 M with the variation of low-molecular salt concentration, starting from [NaBr] = 0.05 M with the step of 0.05 M. When the NaBr concentration did not exceed 0.30 M, all T4DNAs ehxibited compact globular conformation, showing Brownian motion in aqueous buffer solution, as is exemplified in Fig. 1A. Below the critical concentration, the L values remain essentially constant, similar to those of T4DNA globules compacted with CTAB in the absence of NaBr, where the distribution of L values is narrow and unimodal, as is shown in Fig. 2. At the critical salt concentration of 0.35 M, elongated DNAs appear together with the globular DNAs, i.e., coiled and globular chains coexist in the solution (Fig. 1(B)). The L distribution at this condition is clearly bimodal (Fig. 2). Finally, above [NaBr] = 0.40 M, no compact globular objects were found in the sample; T4DNAs exist only in the unfolded coil conformation (Fig. 1(C)) with a relatively wide unimodal distribution of L (Fig. 2). As is clear from Fig. 2, the unfolded T4DNAs exhibit slightly smaller L values,

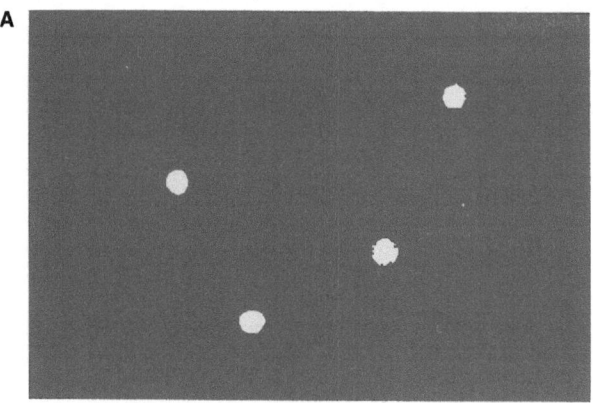

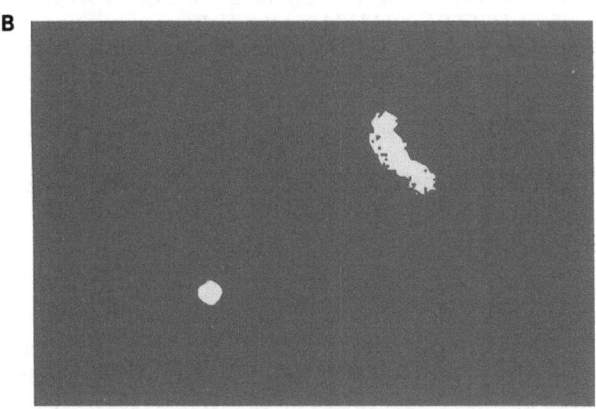

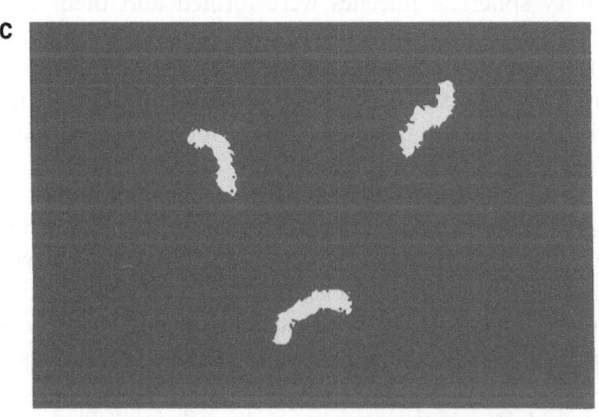

10 μm

Fig. 1 Fluorescence images of single T4DNAs in aqueous buffer solution at [CTAB] = 0.137 M: (A) globular state, at [NaBr] = 0.10 M; (B) coexistence of the coil and globule states, at [NaBr] = 0.35 M; (C) coiled state, at [NaBr] = 0.45 M

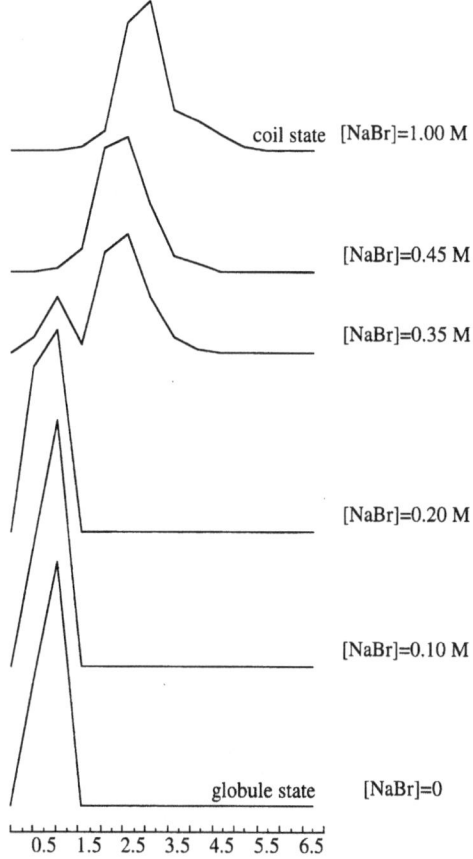

DNA long axis length: **L**, μm

Fig. 2 The long-axis length distribution of T4DNA molecules at [CTAB] = 0.137 M and various NaBr concentrations. The number of the analyzed T4DNA molecules was at least 100 for each NaBr concentration. Each area of histograms is normalized to be equal

about 2.7-3.0 μm, compared to those in the absence of surfactant (about 3.3–3.5 μm in aqueous surfactant-free solution [20]).

Fig. 3 shows the diagram on the change of L in T4DNA depending on NaBr concentration at constant surfactant concentration, equal to 0.137 M, obtained by FM at 35 °C. It indicates the sharp transition from globule to coil with the increase of NaBr concentration. Although the transition in the ensemble of the DNAs looks steep but continuous, individual DNAs exhibit either elongated coiled or compacted globular conformations. In other words, the globule-to-coil transition is a first-order phase transition at the level of the individual DNAs.

Previously, we have reported that individual T4DNA chains undergo a markedly discrete transition between elongated coiled and compacted globular state induced by cationic surfactant. Contrary to the one-dimensional model [17–19], which implied a continuous transition, we proposed a simple model, considering the connection of the total surfactant concentration in the presence of DNA with the relative population of the collapsed globular DNAs. It is expected that no micelles will be formed at

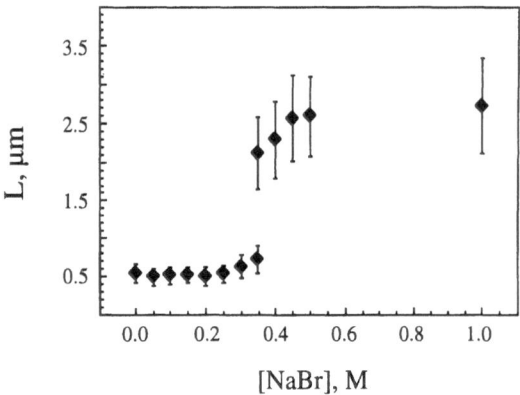

Fig. 3 Dependence of long-axis lengths, L, of T4DNA on NaBr concentration at [CTAB] = 0.137 M. At least 100 DNA molecules were analyzed for each NaBr concentration. Error bars indicate the standard deviation in the distribution

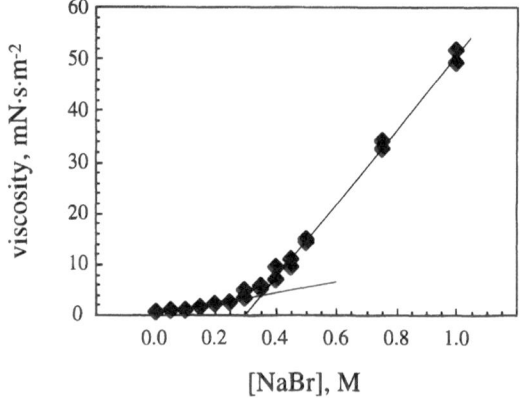

Fig. 4 Dependence of the viscosity of the CTAB-T4DNA-NaBr solutions at [CTAB] = 0.137 M on the NaBr concentration. The concentrations of CTAB, T4DNA and NaBr are analogous to those used for FM observation; $T = 35\,°C$

extremely low surfactant concentrations because the fraction of surfactant inside the coil is below the critical value. When the surfactant concentration reaches a certain critical value, micelles begin to form inside the macromolecule, inducing the collapse of the DNA chain as a result of the cooperative binding of surfactant molecules. The salt-induced unfolding effect may concern with the enhanced screening of the electrostatic charges on the DNA polyanions, resulting in the decrease in attractive Coulombic interactions between DNA and cationic surfactant. It is also to be noted that the gain of the translational entropy [40] due to the release of small ions with the complex formation becomes small in the environment at higher NaBr concentrations.

At such a high CTAB concentration as 0.137 M (5 wt%), the viscosity of sample solution is very significant. Actually, we noted the marked difference in the viscosity of the solutions at various NaBr concentrations during the preparation of samples for FM study. To make clear this aspect, the viscosity of the solutions was measured in similar experimental conditions as those in the FM measurements. Our results are summarized in Fig. 4. At first, the viscosity remains nearly constant, and then increases slightly, being attributed to the formation of ellipsoidal micelles in the surfactant solution [41]. Over [NaBr] = 0.35 M, viscosity tends to increase with a higher rate. The increase of the viscosity at high concentration of NaBr is explained by the sphere–rod transition of surfactant micelles in the surrounding aqueous solution, as will be discussed below. Interestingly, the NaBr concentration of the inflection point in the viscosity of the surfactant solution, 0.35 M, coincides exactly with that of the phase-transition point of T4DNA as is obtained by FM (Figs. 2 and 3).

The formation of CTAB rod-like micelles with molecular weight of 10^6 in the presence of KBr was reported by Debye and Anacker [42]. About a decade ago, the salt-induced sphere–rod transition of alkyltrimethylammonium halides was extensively studied with the use of static light scattering and viscometry [43–46]. The formation of surfactant micelles in low-molecular electrolyte solution has been described as a two-step process: at first, primary spherical micelles were formed and then, with further increase of salt concentration, micelles reorganized to the secondary rod-like shape [47]. It was also concluded that the salt-induced sphere–rod transition of the surfactant micelles is not caused by the change in water structure [48].

The somewhat smaller dimensions of DNA coils in highly concentrated surfactant solution, compared to those in a surfactant-free solution, can be explained with a consideration of two effects. First, by careful observation of unfolded DNAs in the coexistence region, namely at [NaBr] = 0.35 M, the individual DNA macromolecules in the intra-chain segregated state, as is exemplified in Fig. 5, are found. Due to the presence of the coiled unfolded part, these DNA molecules apparently reflect an unfolded state with a relatively slow Brownian motion, similar to the DNAs in the entirely unfolded coiled state. Thus, the smaller L value may be attributed to the existence of the intrachain-segregated DNAs in the solution. Second, the smaller DNA dimensions may be produced by the interaction between the long DNA chains and the rod-like micelles. At the threshold NaBr concentration, the shape transition of micelles from spherical (about 5–6 nm in diameter [49]) to rod-like (persistent length of about 44 nm at 35 °C in 0.5 M NaBr [49]) takes place. As the persistent length of the rod-like micelles is of the same

A

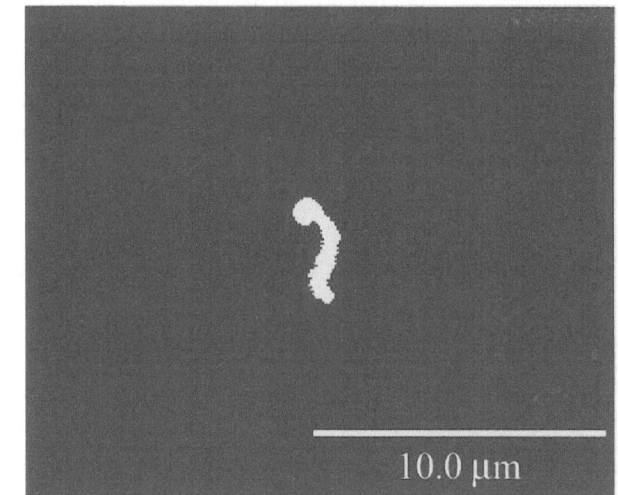

10.0 µm

B

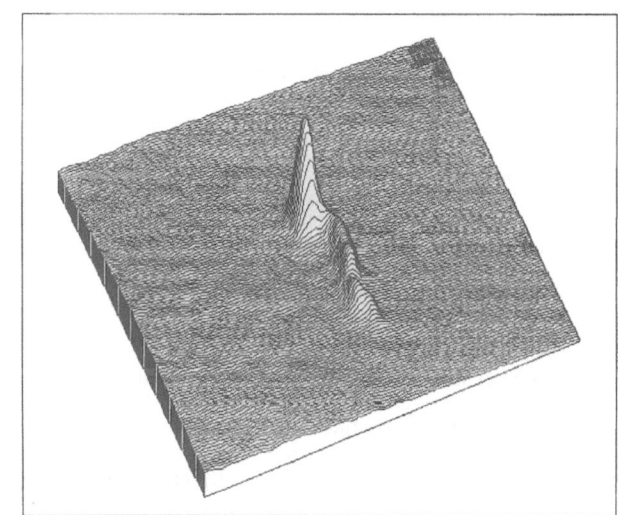

Fig. 5 Fluorescence image of a typical single T4DNA chain, showing intramolecular segregation between unfolded and compacted parts at [CTAB] = 0.137 M and [NaBr] = 0.35 M (A) and the corresponding light-intensity distribution (B)

V_0, is calculated to be 3.3×10^{-20} m³. From the results of the potentiometric titration it has been found that about 2.4×10^5 CTAB molecules are bound to the single T4DNA molecule in the globular state [21]. Thus, the local CTAB concentration inside a T4DNA globule, C_0, is estimated to be 1.2×10^{-2} M. This result indicates that, even at low concentration of surfactant in the solution, i.e., 5×10^{-5} M, CTAB is concentrated to the order of 10 mM, inside the DNA globules. Thus, the DNA polyanion induces the formation of a self-assembled phase of bound surfactant ions, which can be represented as the intramolecular micelles (IM) in DNA-cationic surfactant complex.

The surfactant ions in the IM of T4DNA-CTAB globules are in dynamic equilibrium with those in the outer aqueous solution. Above the threshold concentration of NaBr, the stability of surfactant micelles in the outer solution increases owing to the formation of stable rod-like micelles. In addition, the attractive Coulomb interaction between the cationic surfactant and negatively charged DNA decreases, accompanied with the increase of NaBr concentration. The free energy of "coiled DNA/rod-like micelle" systems becomes, thus, lower than that of "T4DNA-IM complex/rod-like micelle" system. Dynamic equilibrium between IM and rod-like micelles leads to the release of CTA ions from IM and formation of more stable rod-like micelles. This hypothesis seems to be plausible to explain the unfolding of T4DNA-CTAB globules induced by NaBr.

It is also noted that unfolded T4DNA in the rod-like micelle solution with [CTAB] = 0.137 M can be stretched out under a hydrodynamic flow, which is induced by a mechanical stress over the sample (Fig. 6). In this case the

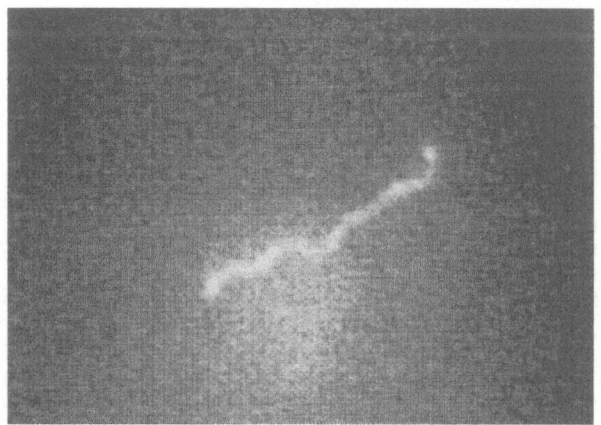

20 µm

Fig. 6 A single T4DNA macromolecule, stretched out by a hydrodynamic flow in the aqueous solution at [NaBr] = 0.50 M and [CTAB] = 0.137 M, observed with FM

order as the persistent length of the double-stranded DNA (about 50 nm [50]), T4DNAs may have slightly diminished dimensions due to the entanglement effect between the DNA chains and the micellar rods.

Let us discuss the mechanism of the decompaction of globular T4DNA in relation to the change on the side of the surfactant micelles. We have previously found [20], from the measurements of the diffusion constant of individual single-DNA globules, that the hydrodynamic radius of T4DNA globule compacted with CTAB, R, is approximately equal to 200 nm. Using this value, the volume of a single-T4DNA globule compacted with CTAB,

214

S.M. Mel'nikov et al.
Cooperation between globule-coil and sphere-rod transition of surfactant micelles

contour length of T4DNA was $17.0 \pm 2.8 \, \mu m$. After an interval of several tens of seconds, the stretched DNA molecules return to their linear dimensions before the stretching induced by hydrodynamic flow. This observation shows that the effect of DNA stretching under the hydrodynamic flow and the slow relaxation rate are attributable to the high viscosity of the medium.

Finally, the effect of low-molecular electrolyte on the surfactant-induced coil–globule transition seems to be important for understanding the mechanism of DNA–surfactant interaction. The detailed study on this effect for various types of surfactants at different concentrations, together with the determination of the IM structure in DNA–cationic surfactant complexes, is in progress in our group [51].

Acknowledgements S.M. and K.Y. acknowledge the financial support by a Grant-in-Aid from the Ministry of Education, Science, Culture and Sports of Japan and by Asahi Glass Foundation.

References

1. Lindman B, Thalberg K (1993) In: Goddard ED, Ananthapadmanabhan KP (eds) Interactions of Surfactants with Polymers and Proteins. CRC Press, Boca Raton, Florida, p 203
2. Goddard ED (1986) Colloids Surfaces 19:301
3. Hayakawa K, Kwak JCT (1991) In: Rubingh DN, Holland PM (eds) Cationic Surfactants. Physical Chemistry. Marcel Dekker, New York, p 189
4. Felgner PL, Gadek TR, Holm M, Roman R, Chan HW, Wenz M, Northrop JP, Ringold GM, Danielsen M (1987) Proc Natl Acad Sci USA 84:7413
5. Felgner PL, Ringold GM (1989) Nature 337:387
6. Lapidot M, Loyter A (1990) Exp Cell Res 189:241
7. Zhou X, Klibanov AL, Huang L (1991) Biochim Biophys Acta 1065:8
8. Behr JP (1993) Acc Chem Res 26:274
9. Gershon H, Ghirlando R, Guttman SB, Minsky A (1993) Biochemistry 32:7143
10. Remy JS, Sirlin C, Vierling P, Behr JP (1994) Bioconjugate Chem 5:647
11. Sternberg B, Sorgi FL, Huang L (1994) FEBS Lett 356:361
12. Zhou X, Huang L (1994) Biochim Biophys Acta 1189:195
13. Reimer DL, Zhang YP, Kong S, Wheeler JJ, Graham RW, Bally MB (1995) Biochemistry 34:12877
14. Hayakawa K, Santerre JP, Kwak JCT (1983) Biophys Chem 17:175
15. Shirahama K, Masaki T, Takashima K (1985) In: Dubin P (ed) Microdomains in Polymer Solutions. Plenum, New York, p 299
16. Shirahama K, Takashima K, Takisawa N (1987) Bull Chem Soc Jpn 60:43
17. Schwarz G (1970) Eur J Biochem 12:442
18. Schmitz KS, Schurr JM (1970) Biopolymers 9:697
19. Satake I, Yang JT (1976) Biopolymers 15:2263
20. Mel'nikov SM, Sergeyev VG, Yoshikawa K (1995) J Am Chem Soc 117:2401
21. Mel'nikov SM, Sergeyev VG, Yoshikawa K (1995) J Am Chem Soc 117:9951
22. Mel'nikov SM, Sergeyev VG, Mel'nikova YuS, Yoshikawa K (1997) J Chem Soc Faraday Trans 93:283
23. Yanagida M, Hiraoka Y, Katsura I (1983) Cold Spring Harbor Symp Quant Biol 47:177
24. Bustamante C (1991) Annu Rev Biophys Biophys Chem 20:415
25. Minagawa K, Matsuzawa Y, Yoshikawa K, Khokhlov AR, Doi M (1994) Biopolymers 34:555
26. Perkins TT, Smith DE, Chu S (1994) Science 264:819
27. Perkins TT, Quake SR, Smith DE, Chu S (1994) Science 264:822
28. Chu D, Thomas JK (1986) J Am Chem Soc 108:6270
29. Binana-Limbele W, Zana R (1987) Macromolecules 20:1331
30. Herslöf Å, Sundelöf LO, Edsman K (1992) J Phys Chem 96:2345
31. Hayakawa K, Kwak JCT (1982) J Phys Chem 86:3866
32. Hayakawa K, Kwak JCT (1983) J Phys Chem 87:506.
33. Hayakawa K, Santerre JP, Kwak JCT (1983) Macromolecules 16:1642.
34. Malovikova A, Hayakawa K, Kwak JCT (1984) J Phys Chem 88:1930
35. Lindman B, Wennerström H (1980) Top Curr Chem 87:1
36. Ananthapadmanabhan KP (1993) In: Goddard ED, Ananthapadmanabhan KP (eds) Interactions of Surfactants with Polymers and Proteins. CRC Press, Boca Raton, FL p 5
37. Yoshikawa K, Matsuzawa Y, Minagawa K, Doi M, Matsumoto M (1992) Biochem Biophys Res Commun 188:1274
38. Matsuzawa Y, Yoshikawa K (1994) Nucleosides Nucleotides 13:1415
39. Matsuzawa Y, Minagawa K, Yoshikawa K, Matsumoto M, Doi M (1991) Nucleic Acids Symp Ser 25:131
40. Vasilevskaya V, Khokhlov AR, Matsuzawa Y, Yoshikawa K (1995) J Chem Phys 102:6595
41. Tartar HV (1955) J Phys Chem 59:1195
42. Debye P, Anacker EW (1951) J Phys Colloid Chem 55:644
43. Ikeda S, Ozeki S, Tsunoda M (1980) J Colloid Interface Sci 73:27
44. Imae T, Ikeda S (1986) J Phys Chem 90:5216
45. Imae T, Ikeda S (1989) In: Mittal KL (ed) Surfactant in Solution. Plenum, New York, p 455
46. Imae T, Abe A, Ikeda S (1988) J Phys Chem 92:1548
47. Ozeki S, Ikeda S (1982) J Colloid Interface Sci 87:424
48. Ikeda S, Hayashi S, Imae T (1981) J Phys Chem 85:106
49. Imae T, Kamiya R, Ikeda S (1985) J Colloid Interface Sci 108:215
50. Grosberg AYu, Khokhlov AR (1994) Statistical Physics of Macromolecules. American Institute of Physics Press, New York
51. Mel'nikov SM, Sergeyev VG, Takahashi H, Hatta I, Yoshikawa K (1997) J Chem Phys, in press

Progr Colloid Polym Sci (1997) 106:215–218
© Steinkopff Verlag 1997

BIOCOLLOIDS

M. Makino
C. Sasaki
K. Nakamura
M. Kamiya
K. Yoshikawa

An overshoot-hump in a π-A curve and the cooperative aggregation among dinitrophenyl moieties in phospholipids at an air/water interface

Received: 13 October 1996
Accepted: 6 March 1997

Dr. M. Makino (✉) · C. Sasaki ·
K. Nakamura · M. Kamiya
Graduate School of Nutritional and
Environmental Sciences
University of Shizuoka
51-2 Yada, Shizuoka 422, Japan

K. Yoshikawa
Graduate School of Human Informatics
464-01 Nagoya, Japan

Abstract We observed the surface pressure (π) – surface area (A) curves of the DNP–DPPE monolayers at an air/water interface. An overshoot-hump and an inflection point were markedly shown in the π–A curve with a successive compression. To elucidate the mechanism of the characteristic changes in the curve, we measured the absorption spectra of the thin film which was deposited on a quartz plate using a Langmuir–Blodgett (LB) technique. It was found that the maximum peak in the spectrum shifted to longer wavelengths at 3 nm in the subsequent overshoot-hump, and that the peak intensity increased threefold after the inflection point. Therefore, it was considered that the overshoot-hump indicated a change of the conformation of the dinitrophenyl moiety in the DNP–DPPE molecule at an air/water interface, and that the inflection point reflected a collapse of the DNP–DPPE monolayer, respectively. To explain the mechanism of the overshoot-hump, we established a kinetic model which included a cooperative aggregation process between DNP–DPPE molecules during compression. The overshoot-hump was successfully reproduced with the help of computer simulation on the basis of this kinetic model.

Key words Overshoot-hump – dinitrophenyl phospholipid – Langmuir–Blodgett technique – cooperative aggregation

Introduction

Dinitrophenyl dipalmitoylphosphatidylethanolamine (DNP–DPPE) is well known for its potential as an antigen in the preparation of immunologically responsive liposomal model [1, 2]. It has been found that this compound was capable of actively sensitizing liposomes toward antibody complement. However, the physicochemical properties of the membrane of DNP–DPPE, especially those of the monolayer at an air/water interface, are not yet clear. As the dinitrophenyl group plays an important role in some immunological reactions, the interaction between the DNP groups in the membrane should be given some attention.

We have studied the viscoelastic behavior of lipid thin films at an air/water interface from surface pressure (π) vs. surface area (A) curves, since an observation of the π–A curve of a monolayer is one of the most convenient methods to elucidate the viscoelastic behavior of the monolayer. Recently, in spite of compression process, we observed an overshoot-hump, a zero surface pressure, and a flat plateau in the π–A curve of a synthetic fatty acid [3]. These characteristic features in the curve were explained by using a kinetic model representing the formation of aggregates of the fatty-acids molecules at an air/water interface.

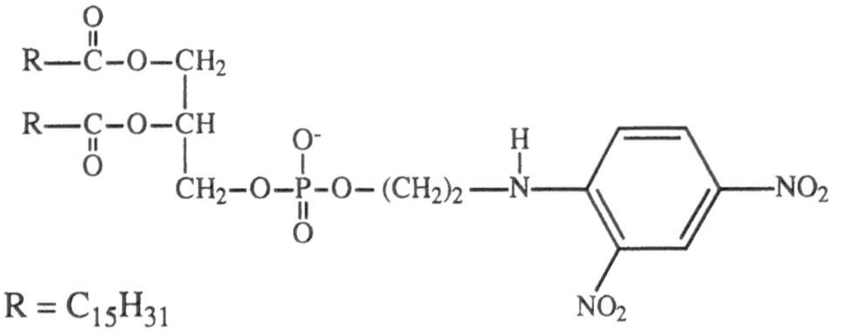

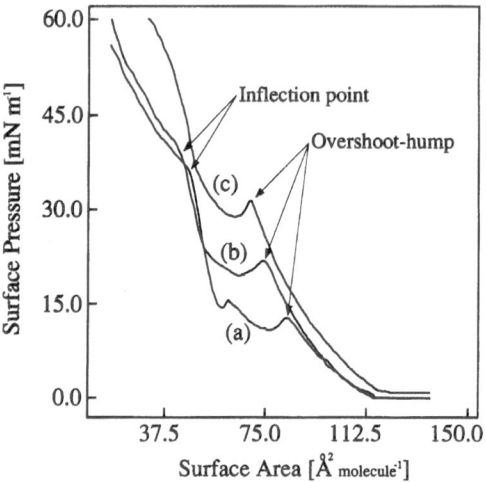

Fig. 1 Chemical structure of the DNP–DPPE molecule

In the present paper, we address the behavior of DNP–DPPE molecules on the air/water interface from a change in shape of the π–A curves and the absorption spectra, especially, the interaction between the dinitrophenyl groups. We try to discuss the change in feature based on the same kinetics model.

Fig. 2 Surface pressure (π) vs. surface area (A) curves of DNP–DPPE monolayer. Subphase temperature is (a) 10 °C, (b) 20 °C, and (c) 30 °C, respectively. The surface area was changed in a constant speed which was 2.71 $\mathrm{\mathring{A}}^2$ molecule^{-1} min^{-1}. The subphase component was potassium bicarbonate (50 μM) and calcium chloride (300 μM)

Experimental section

DNP–DPPE (Fig. 1), a synthetic phospholipid, was obtained from Avanti Polar Lipid Inc. Other reagents were of analytical grade and purchased from Wako Pure Chemical Industries, Ltd. Before each experiment, water was distilled with an all-glass apparatus (SA-2000E, Tokyo Rika Kikai Co. Ltd., Japan), and purified further with a Millipore Milli-Q filtering system maintained at a resistivity above 18.0 MΩ cm. Chloroform was purified by three consecutive distillations. DNP–DPPE (0.455 μM) dissolved in chloroform was used as the spreading solution. π–A curves were obtained using a trough equipped with a pair of moving tapes (Nippon Laser & Electronics Labs., model NL-LB-240S-MWC). The monolayer of DNP–DPPE was deposited on only one layer on a quartz plate, which had been treated with 0.1 M NaOH solution, by using an LB technique. The transfer ratio of the monolayer from the air/water interface to the plate was ca. 100%. The UV-visible absorption spectra of the LB films were obtained using a HITACHI U-3410 spectrometer.

Results and discussion

As shown in Fig. 2, an interesting characteristic, an overshoot-hump, was observed in the π–A curves of the DNP–DPPE monolayer with a constant-speed compression. It is clear that the shape of the overshoot-hump became more marked with an increase in the subphase temperature, and that the surface pressure, at which the overshoot-hump was observed, increased from 12.7 to 31.4 mN/m. Furthermore, it was found that both the

slopes of the curves (a) and (b) changed around 40 mN/m. These inflection points did not show any remarkable dependence on the subphase temperature compared to the overshoot-hump [4].

To elucidate the origin of the characteristic features in the π–A curves such as the overshoot-hump and the inflection point, we observed the absorption spectra of DNP–DPPE thin film by using the LB technique [5,6]. The DNP–DPPE monolayer was deposited onto a quartz plate at different surface pressures (see (a)–(c) in Fig. 3A). It was found that the maximum peak in the absorption spectrum shifted to long wavelengths at 3 nm after the overshoot-hump had been observed. However, the intensity of the peak little increased. On the other hand, we observed that the intensity became threefold after the inflection point.

From these experimental findings, it may be considered that the overshoot-hump in the π–A curve indicates the formation of a condensed layer such as aggregates of the DNP–DPPE molecules, and that the inflection point represents a collapse from a monolayer to some multilayers at the air/water interface. As an overshoot-hump has not been found in the π–A curves of dipalmitoylphosphatidylethanolamine (DPPE), the dinitrophenyl moiety in the phospholipids may play an important role in the mechanism of the aggregation between the DNP–DPPE molecules during the compression process. For example, a stacking of the moieties may be induced at ca. 20 mN/m.

On the basis of the above discussion, we provide a useful kinetic model to describe the formation of the aggregates at an air/water interface [3]. Let [S] and [D] be the

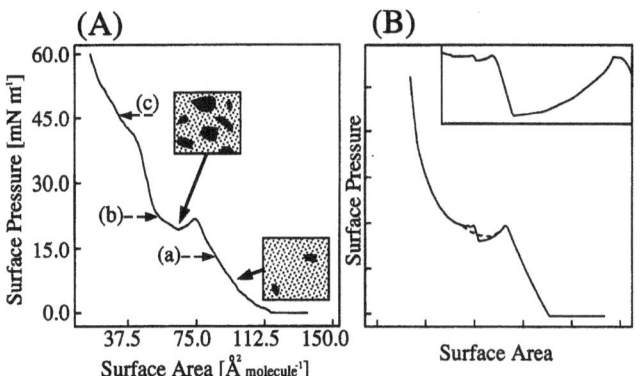

Fig. 3 Comparison between the experimental results and the computer simulation on the π–A curve. (A) is the same as in Fig. 2b. The arrows ((a)–(c)) in Fig. 3A show the surface pressure at which the DNP–DPPE monolayer was deposited on a quartz plate. The schematic representation in Fig. 3A illustrate the formation of DNP–DPPE aggregates at an air/water interface with compression. The enlargement of the oscillation of the surface pressure is shown in the upper right-hand side of Fig. 3B. Parameters used in this simulation are; $k_1 = 1.0 \times 10^{-3}$, $k_{-1} = 1.0 \times 10^{-7}$, $f(\pi) = 0.05\pi$ $(\pi \leq \pi^c)$, $f(\pi) = 6.0$ $(\pi > \pi^c)$

concentration of the DNP–DPPE molecules in the "regular" and "aggregated" states, respectively. The "regular" state corresponds to the usual conformation of lipid molecules at the air/water interface, i.e., the hydrophilic head faces the hydrophilic environment and the hydrophobic tail is expelled from the aqueous surface toward the hydrophobic environment.

$$S + S \underset{k_{-1}}{\overset{k_1}{\rightleftarrows}} D \,, \tag{i}$$

This step corresponds with the aggregation of the DNP–DPPE molecules, and the interaction between the dinitrophenyl moieties in the molecules at the air/water interface. We consider that the aggregation process at 20 °C changes to a much faster process than the dissociation one at ca. 20 mN/m, since a temporal decrease in the surface pressure (an overshoot-hump) was observed at the pressure. In other words, the aggregation process sensitively depends on the surface pressure around 20 mN/m. The rapid increase in the value of k_1 indicates a cooperative behavior on the aggregation process and induces a temporal decrease of the surface pressure inspite of compression. On the basis of this kinetic model, the time dependence of $[S]$ and $[D]$ is given by Eqs. (1) and (2), respectively.

$$\frac{d[S]}{dt} = -k_1 f(\pi)[S]^2 + k_{-1}[D] + N_S \frac{d}{dt}\left(\frac{1}{A(t)}\right), \tag{1}$$

$$\frac{d[D]}{dt} = k_1 f(\pi)[S]^2 - k_{-1}[D] + N_D \frac{d}{dt}\left(\frac{1}{A(t)}\right), \tag{2}$$

where $A(t)$ is the surface area at time t, k_1 the aggregation rate-coefficient, k_{-1} the dissociation rate-coefficient. N_S and N_D the number of molecules in the "regular" and "aggregated" state, respectively. $f(\pi)$ is a discontinuous function of surface pressure (π). In a previous paper, we have introduced $f(\pi)$ as a discontinuous "S"-shape function which includes critical surface pressure (π^c) and it is a simple form to represent the cooperative aggregation between lipid molecules at an air/water interface.

To simplify the analysis, it was considered that the surface pressure was directly proportional to the concentration S and D on the basis of a two-dimensional gas model

$$\pi \propto a[S] + b[D] \,, \tag{3}$$

where a and b are the correction factors representing the contribution of $[S]$ and $[D]$ to the surface pressure, respectively.

In the right-hand side of Fig. 3 is shown the simulation results obtained by using the above kinetic equations. Figure 3B shows the oscillation of surface pressure around the critical surface pressure. If we include the damping factors such as viscous properties of DNP–DPPE monolayer in the simulation, the oscillation would be represented by the dotted line.

Figure 3B is a simulation result corresponding to the actual π–A curve in Fig. 2b. We have considered that the increase of the subphase temperature induces an increase in the critical surface pressure. In other words, the increase changes the shape of the nonlinear "S"-shape function. If we simulate the experimental results (Fig. 2c), it should be used as another appropriate function. In addition to this, it is clear that the rate coefficients (k_1 and k_{-1}) depend on the subphase temperature. In the present work, for convenience of calculation, we used the parameters, k_1 and k_{-1}, as a constant, and estimated the value of k_1 to be much larger than k_{-1} which was arbitrarily determined as 1.0×10^{-7}.

Conclusions

The overshoot-hump in the π–A curve of DNP–DPPE monolayer was found with constant-speed compression. From the absorption spectra of the lipid thin film, the hump reflected the formation of condensed layers at the critical surface pressure. A computer simulation was carried out on the basis of a cooperative aggregation process between the DNP–DPPE molecules at the air/water interface, the characteristic feature in the π–A curve was able to be reproduced.

218

M. Makino et al.
An overshoot-hump in a π-A curve of DNP-DPPE monolayer

References

1. Umemura K, Kinsky SC (1972) Biochemistry 11:4085
2. Hafeman DG, von Tscharner V, McConnel HM (1981) Proc Natl Acad Sci USA 78:4552
3. Makino M, Kamiya M, Ishii T, Yoshikawa K (1994) Langmuir 10(4):1287
4. Laschewsky A, Ringsdorf H, Schmidt G, Schneider J (1987) J Am Chem Soc 109:788
5. Langmuir I (1920) Trans Faraday Soc 15:62
6. Blodgett KB (1935) J Am Chem Soc 57:1007

Progr Colloid Polym Sci (1997) 106:219–222
© Steinkopff Verlag 1997

BIOCOLLOIDS

T. Yoshimura
K. Kameyama
S. Aimoto
T. Takagi
Y. Goto
S. Takahashi

Formation of protein- and peptide-membrane assemblies and membrane fusion

Received: 13 October 1996
Accepted: 26 February 1997

Dr. T. Yoshimura (✉)
Department of Chemistry for Materials
Faculty of Engineering
Mie University
Tsu, Mie 514, Japan

K. Kameyama · S. Aimoto · T. Takagi
Institute for Protein Research
Osaka University
Suita, Japan

Y. Goto
Faculty of Science
Osaka University
Toyonaka, Japan

S. Takahashi
Institute for Chemical Research
Kyoto University
Uji, Japan

Abstract Membrane fusion reactions consist of three steps: close apposition of membranes, mixing of lipid molecules, and formation of new bilayers in a different direction. For elucidation of the molecular mechanisms of these three steps, we have studied membrane fusion induced by amphipathic helical peptides and the protein clathrin using liposome membrane systems. Based on the results, we propose two different mechanisms of membrane close apposition: close apposition of negatively or positively charged membranes occurs through the hydrophobic interaction between proteins or peptides after their electrostatic binding to the membranes, whereas close apposition of neutral membranes occurs through the membrane binding and self-aggregative properties of proteins or peptides that are provided by neutralization of their surface charges. We also propose a possible mechanism for the lipid-mixing reaction to form new bilayers, which is interpreted in terms of destabilization of the bilayer structure within hydrophobic protein- or peptide-membrane assemblies and stabilization of the bilayer structure in their hydrophilic media.

Key words Amphipathic peptide – clathrin – liposome – membrane fusion – fusion mechanism

Introduction

Membrane fusion is an important process in many cellular events, such as fertilization, myoblast fusion, virus infection, secretion, and intracellular vesicle-mediated transport. Proteins are known to be essential for mediating membrane fusion in these events. However, the exact mechanisms for induction of membrane fusion by proteins are still unknown. Membrane fusion reactions consist of three steps: close apposition of membranes, mixing of lipid molecules in the closely apposed membranes, and formation of new bilayers in a different direction. For elucidation of the mechanism, we have studied protein-induced membrane fusion using the protein clathrin and liposome membrane systems.

An amphipathic helix is defined as a helix in which the distribution of amino acid residues forms opposing polar and nonpolar faces. It is an important structural unit included in proteins and peptides and is responsible for interaction with biological membranes to elicit their biological functions such as membrane fusion. Influenza virus hemagglutinin [1], fertilin [1], and meltrin-α [2] contain amphipathic fusion peptides, which are likely to adopt a helical conformation during the fusion reaction. To clarify the role of amphipathic peptides in membrane fusion reactions, we synthesized five types of amphipathic model peptides and examined their helix formation, membrane binding and membrane fusion activities.

In the present communication, we briefly review these studies, make a proposal for the mechanism of membrane close-apposition induced by them, and discuss a possible

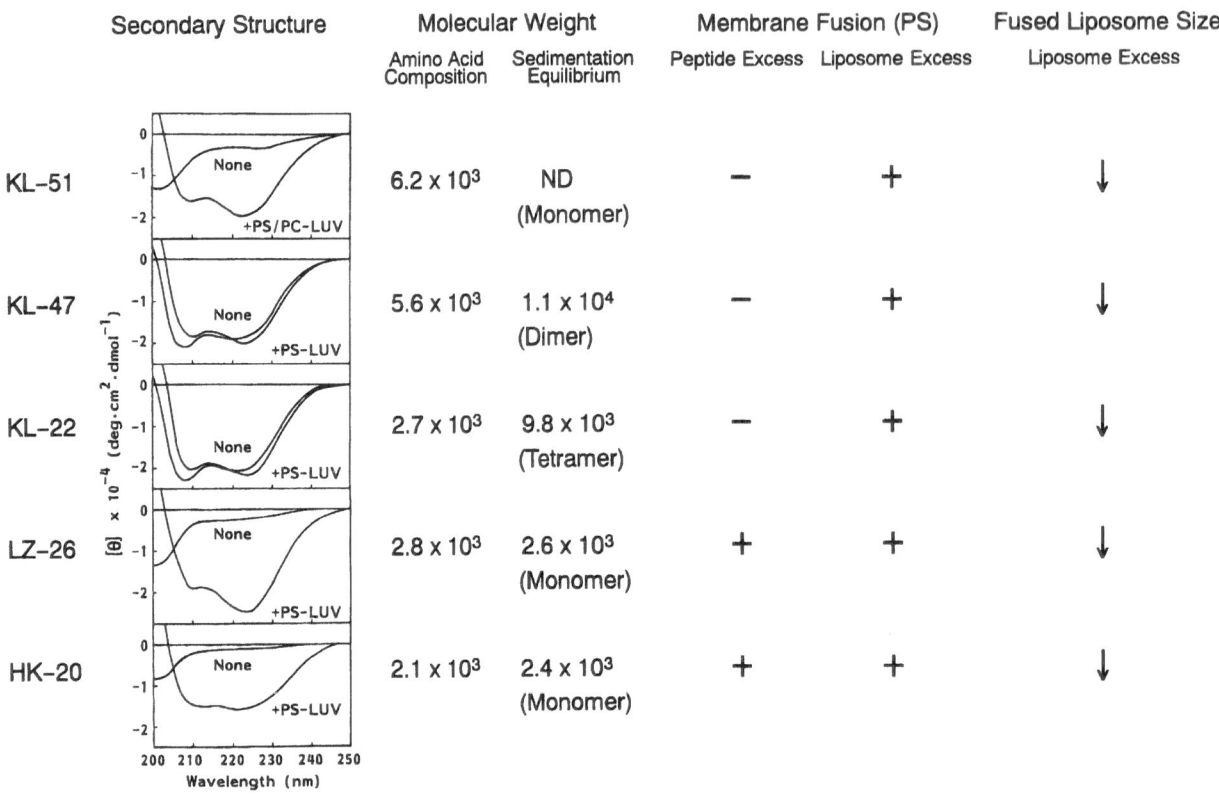

| Secondary Structure | Molecular Weight | | Membrane Fusion (PS) | | Fused Liposome Size |
	Amino Acid Composition	Sedimentation Equilibrium	Peptide Excess	Liposome Excess	Liposome Excess
KL–51	6.2×10^3	ND (Monomer)	−	+	↓
KL–47	5.6×10^3	1.1×10^4 (Dimer)	−	+	↓
KL–22	2.7×10^3	9.8×10^3 (Tetramer)	−	+	↓
LZ–26	2.8×10^3	2.6×10^3 (Monomer)	+	+	↓
HK–20	2.1×10^3	2.4×10^3 (Monomer)	+	+	↓

Fig. 1 Properties of amphipathic peptides and fusion of phosphatidylserine membranes

mechanism of protein-induced membrane fusion occurring through formation of protein- and peptide-membrane assemblies.

Membrane fusion induced by amphipathic helical peptides

The amphipathic peptides used in the present study were model peptides with 51, 47, and 22 amino acid residues (KL-51 [3], KL-47, and KL-22, respectively), all of which consist of tandem repeats of a KKLL sequence, a model peptide of leucine zipper with 26 amino acid residues (LZ-26), and a cationic peptide as an analog of the influenza hemagglutinin fusion peptide with 20 amino acid residues (HK-20) [4].

Liposomes were prepared, and membrane fusion induced by the model peptides, the binding of the peptides to liposome membranes, and the self-aggregation of the peptides were measured at 25 °C as described previously [3]. The size of fused liposomes was evaluated by their diffussion coefficients measured at 25 °C in an Otsuka photon-counting laser light scattering photometer model DLS-700 as described previously [5]. CD spectra were obtained at 25 °C with a Jasco spectropolarimeter model

J-600 essentially as described previously [3]. Sedimentation equilibrium was performed at 20 °C in a Beckman analytical ultracentrifuge model XL-A. Molecular weights and partial specific volumes of the peptides were determined by the H_2O/D_2O sedimentation equilibrium method [6]. All the measurements were carried out in 10 mM Tris-HCl (pH 7.5) with or without 10 mM NaCl.

Results with liposome membranes containing phosphatidylserine (PS) are shown in Fig. 1. In solution, KL-51, LZ-26 and HK-20 were in random and monomeric forms, whereas KL-47 and KL-22 had helical and oligomeric (dimeric and tetrameric, respectively) structures, but they all formed helical structures on association with PS membranes. The monomeric peptides LZ-26 and HK-20 always induced fusion of PS membranes, but intramolecularly interacting KL-51 and oligomeric KL-47 and KL-22 could not trigger fusion of the membranes in the presence of excess peptides and induced fusion only in the presence of excess liposomes, suggesting that the monomeric or dissociated form of amphipathic helical peptides is fusion active [3]. The size of fused PS liposomes decreased in the presence of excess liposomes, suggesting that it is modulated by the density of the fusion-active form on membranes.

Progr Colloid Polym Sci (1997) 106:219–222
© Steinkopff Verlag 1997

Table 1 Properties of amphipathic peptides and fusion of phosphatidylcholine membranes

Peptide	Addition	pH	Net charge at the pH	Helix formation	Self-aggregation	Membrane binding	Membrane fusion	Reference
KL-51	None	Neutral	+	−	−	−	−	[3]
	ATP	Neutral	±	+	+	+	+	
KL-47	None	Neutral	+	+	−	−	−	This work
	ATP	Neutral	±	+	+	−	−	
KL-22	None	Neutral	+	+	−	−	−	This work
	ATP	Neutral	±	+	+	−	−	
LZ-26	None	Neutral	+	−	−	±	±	This work
	ATP	Neutral	±	+	+	+	+	
HK-20	None	Neutral	+	−	−	±	±	This work
(K5)	ATP	Neutral	±	+	+	+	+	
K5	None	> 9	±	+	+	+	+	[7]
E5	None	> 6.5	−	−	−	−	−	[7]
	None	< 6.5	±	+	+	+	+	
GALA	None	> 5	−	−	−	−	−	[8]
	None	< 5	±	+	+	+	+	

Results with liposome membranes containing phosphatidylcholine (PC) alone are shown in Table 1. The amphipathic model peptides had little or no activity for fusion of PC membranes. On the other hand, they adopted helical conformations in the presence of ATP and became self-aggregative above an ATP concentration at which the positive charges of the peptides were neutralized by the negative charges of ATP, and above an ATP concentration at which they became self-aggregative, the peptides, except oligomeric KL-47 and KL-22, interacted with the PC membranes and triggered fusion, suggesting that the hydrophobic helical peptide is fusion active [3].

Similar phenomena on amphipathic helical peptide-induced fusion of PC membranes have been observed on change in pH (Table 1). HK-20 (K5) showed membrane binding and membrane fusion activities above pH 9, the pH at which the lysine residues of the peptide were deprotonated and as a result, the peptide became helical and self-aggregative [7]. An anionic peptide as an analog of the influenza hemagglutinin fusion peptide (E5) [7] and an amphipathic peptide with tandem repeats of an EALA sequence (GALA) [8] become fusion active below pH 6.5 and 5, respectively, the pH at which glutamic acid residues of the peptides are protonated and they acquire helix-forming and self-aggregative properties, also suggesting that the hydrophobic helical peptide is fusion active.

Based on these results, we propose the following mechanism for membrane close apposition induced by amphipathic helical peptides: close apposition of the *negatively (or positively) charged* membranes occurs through the interaction between the exposed nonpolar faces of amphipathic and cationic (or anionic) helical peptides that are formed by their electrostatic binding to the membranes, whereas close apposition of the *neutral* membranes occurs through the membrane-binding and self-aggregative properties of amphipathic and cationic or anionic helical peptides that are provided by neutralization of their positive or negative charges. Membrane fusion is induced through formation of these peptide-membrane assemblies.

Membrane fusion induced by the protein clathrin

Clathrin is a major coat protein of coated pits and coated vesicles formed in receptor-mediated endocytosis and intracellular vesicle-mediated transport and has a unique pinwheel-like structure in solution termed a triskelion composed of three heavy chains and three light chains [9]. Previously, we found that below pH 6, clathrin forms self-aggregates in the absence of liposomes and fusogenic assemblies in their presence [10, 11].

The properties of clathrin for induction of membrane fusion were similar to those of the amphipathic model peptides: (i) the protein-induced fusion of liposome membranes containing acidic phospholipid, but not fusion of liposome membranes composed of neutral phospholipid [10, 12], (ii) the fusion reaction was associated with the exposure of hydrophobic domains of the protein through a conformational change [11–13], (iii) the protein had positive charges at the fusion-inducible pH region because its isoelectric point was about 6 [12], (iv) the size of the fused liposomes decreased in the presence of excess liposomes, also depending on the temperature and ionic

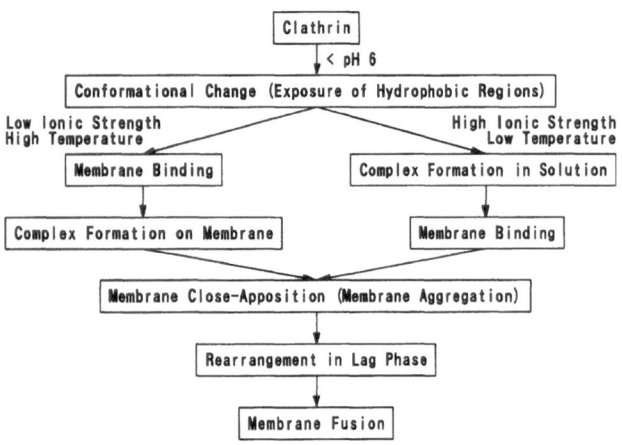

Fig. 2 Sequence of critical events in clathrin-induced membrane fusion

strength of the solution [5]. It is, thus, likely that at the fusion-inducible pH region, clathrin associates electrostatically with negatively charged membranes and apposes the membranes closely through the hydrophobic interaction between the bound protein molecules.

Possible mechanism for membrane fusion induced by proteins and peptides

Moreover, we found that clathrin-induced membrane fusion is initiated through the following sequential events (Fig. 2): (i) conformational change of clathrin, (ii) membrane binding and complex formation (self-association) of the protein, the sequence of these two events depending on the temperature and ionic strength, (iii) membrane close apposition (membrane aggregation), (iv) protein rearrangement in the fusion lag phase, and (v) actual fusion (lipid mixing) [5, 14, 15]. We also found that the proximal portions but not the terminal domains of the clathrin heavy chains have fusogenicity [16]. Thus, the following mechanism can be envisaged for membrane fusion induced by clathrin. At acidic pH, clathrin becomes positively charged and changes its conformation with the exposure of hydrophobic domains. Then, the protein binds electrostatically to negatively charged membranes and as a result, becomes more hydrophobic and assembles with each other on the membrane. Close apposition of the membranes is induced through intermembrane interaction of the hydrophobic clathrin assemblies, and then, lipid molecules in the closely apposed regions of membranes are concentrated into the rearranged clathrin assemblies. Consequently, some defects are formed in the membrane regions located in the assemblies, and once the water molecules pass through the defects, new bilayers are created in the different direction. This mechanism could also be involved in the membrane fusion induced by amphipathic helical peptides. In other words, the lipid-mixing reaction to form new bilayers is interpreted in terms of destabilization of the bilayer structure within hydrophobic protein- or peptide-membrane assemblies and stabilization of the bilayer structure in their hydrophilic media. Further studies from the viewpoint of supramolecular assemblies or systems are important for complete elucidation of the mechanism of membrane fusion induced by proteins and peptides.

References

1. White JM (1992) Science 258:917–924
2. Yagami-Hiromasa T, Sato T, Kurisaki T, Kamijo K, Nabeshima Y, Fujisawa-Sehara A (1995) Nature 377:652–656
3. Yoshimura T, Goto Y, Aimoto S (1992) Biochemistry 31:6119–6126
4. Takahashi S (1990) Biochemistry 29:6257–6264
5. Yoshimura T, Maezawa S, Kameyama K, Takagi T (1994) J Biochem 115:715–723
6. Edelstein SJ, Schachman HK (1973) Meth Enzymol 27:82–98
7. Murata M, Takahashi S, Kagiwada S, Suzuki A, Ohnishi S (1992) Biochemistry 31:1986–1992
8. Parente RA, Nir S, Szoka FC (1988) J Biol Chem 263:4724–4730
9. Yoshimura T, Kameyama K, Maezawa S, Takagi T (1991) Biochemistry 30:4528–4534
10. Hong K, Yoshimura T, Papahadjopoulos D (1985) FEBS Lett 191:17–23
11. Yoshimura T, Maezawa S, Hong K (1987) J Biochem 101:1265–1272
12. Maezawa S, Yoshimura T, Hong K, Düzgünes N, Papahadjopoulos D (1989) Biochemistry 28:1422–1428
13. Yoshimura T (1993) Meth Enzymol 221:72–82
14. Maezawa S, Yoshimura T (1990) Biochem Biophys Res Commun 173:134–140
15. Maezawa S, Yoshimura T (1991) Biochim Biophys Acta 1070:429–436
16. Maezawa S, Yoshimura T (1990) Biochemistry 29:1813–1817

Progr Colloid Polym Sci (1997) 106:223–227
© Steinkopff Verlag 1997

BIOCOLLOIDS

H. Takahashi
M. Imai
Y. Matsushita
I. Hatta

Small-angle neutron scattering study on short-chain phosphatidylcholine micelle in the presence of sorbitol

Received: 13 October 1996
Accepted: 26 February 1997

Dr. H. Takahashi (✉) · I. Hatta
Department of Applied Physics
Nagoya University
Chikusa-ku
Nagoya 464-01, Japan
E-mail: takahasi@hix.nagoya-u.ac.jp

M. Imai · Y. Matsushita
Neutron Scattering Laboratory
Institute of Solid State Physics
Tokyo University
Tokai, Japan

Abstract Sorbitol is a natural cryoprotectant for some organisms. In order to reveal the mechanism of a protection some organisms against low temperatures, we studied the interaction between sorbitol and phosphatidylcholine (PC) which is one of major component of cell membranes. Small-angle neutron-scattering contrast variation method was used to characterize dihexanoyl-PC (dC(6)PC) micelles in the presence of sorbitol (2 M). As a result, the aggregation number for the dC(6)PC micelle and the volume per a single molecule in the micelle were estimated to be ~ 32 and $\sim 640\,\text{Å}^3$, respectively. By comparing these data and those for the dC(6)PC micelle in pure water (Lin et al., J Am Chem Soc 108:3499–3507), it was concluded that sorbitol reduces the contact area between the lipid and the aqueous phase. On the basis of this conclusion, we discuss the role of sorbitol for the ability of some organisms to survive in low temperatures.

Key words Sorbitol – cryoprotectant – phosphatidylcholine – micelle – neutron scattering – contrast-variation method

Introduction

It is well known that certain sugars and sugar alcohols act as neutral cryoprotectants [1]. Insects get a low-temperature tolerance by increasing the concentration of sugars or sugar alcohols in their body fluid. Storey et al. [2] have investigated the levels of some metabolites in gall-fly larvas acclimated to low temperatures. They [2] have found that the concentration of a sugar alcohol, sorbitol, is increased in the insect acclimated to low temperatures. In order to reveal the mechanism of cryoprotectant function of sugars and sugar alcohols, the interaction between sugars or sugar alcohols and cell membranes has been extensively studied. Phospholipid bilayer membranes have been widely used as a model system for cell membranes in such investigations. In particular, phosphatidylcholine (PC) [3, 4] and phosphatidylethanolamine (PE) [5–7] bilayers, and the mixtures of PC and PE [8] have been mainly utilized for

the model systems, because both PC and PE are one of major lipid components of cell membranes.

On the basis of the changes of the phase transition temperatures of these model membranes induced by the presence of sugars or sugar alcohols, Koynova et al. [5] have proposed the following hypothesis: sugars or sugar alcohols act as a kosmotropic reagent stabilizing the water structure, and as a result, the presence of these substances reduces the contact area between the lipid and the aqueous phases.

However, no direct evidence for this hypothesis has been reported. The aim of the present study is to confirm the above hypothesis by investigating dihexanoyl-PC (dC(6)PC) micelles in the presence of sorbitol. PC molecules usually form a bilayer structure in aqueous solutions. However, short-chain PC molecules composed of 6–8 carbons in their hydrocarbon chains form a micelle structure in water [9, 10]. Lin et al. [11] have clarified that dC(6)PC, which has six carbons in each chain, forms globular micelles.

Micelle structures have been extensively studied by contrast-variation small-angle neutron [11–14] and X-ray [15,16] scattering methods. In X-ray scattering experiments, the contrast is usually changed by adding various amount of sugars to the solvents. Therefore, it is difficult to apply contrast-variation X-ray scattering measurements for the study of the interaction with sugars or sugar alcohols. On the other hand, in neutron scattering measurements, the change of the contrast can be achieved by varying the D_2O/H_2O ratio of solvents. Therefore, we used a contrast-variation neutron scattering method for investigating the interaction of the sugar alcohol and dC(6)PC micelle. The mean aggregation number of the micelle and the volume of dC(6)PC molecule in the micelle in the presence of sorbitol are determined by means of the contrast-variation method. From a comparison with the literature values for the dC(6)PC micelle in pure water [11], the effect of sorbitol on the dC(6)PC micelle is discussed.

Materials and methods

Dihexanoylphoshatidylcholine (dC(6)PC) (> 99%) was obtained from Avanti Polar Lipids Inc. (Alabaster, AL, USA). The purity was monitored by thin-layer chromatography (TLC). The sample gave a single spot on a TLC plate and it was, therefore, used without further purification. Sorbitol was obtained from Towa Kasei Kogyo Co. (Tokyo, Japan) with a specified purity better than 99.9%. The purity was checked by the manufacturer by means of liquid chromatography.

Small-angle neutron-scattering measurements were performed with the SANS-U spectrometer at JRR-3M reactor at the Japan Atomic Energy Research Institute (Tokai, Japan). The detail of this spectrometer has been reported by Ito et al. [17]. A polychromatic cold neutron beam from the reactor was monocromated by a mechanical velocity selector. The wavelength of the neutron beam was 7 Å and its resolution was 10%. The samples were contained in a rectangular quartz cell with 1 mm optical path length. The sample cell was sealed by paraffin to prevent the evaporation of water from the solutions. The data correction were performed at $25 \pm 1°C$. The exposure times were 1.5–5.5 h. The scattering data were corrected for an instrumental background, backgrounds from buffers, a detector efficiency, transmissions, exposure times, and the thickness of samples. The determinations of absolute scattering intensities were performed by comparison with the scattering intensity of the standard Lupolen sample with 1 mm thickness which was calibrated against standard vanadium [18].

Results and data analyses

The presence of various salts in solvents affects critical micellar concentrations of dC(6)PC [9]. Therefore, it is expected that sorbitol also affects the critical micellar concentration. To determine the critical micellar concentration of dC(6)PC in the presence of sorbitol, neutron-scattering curves were measured for various dC(6)PC concentrations. Figure 1 shows the scattering curves for dC(6)PC solutions (D_2O) with various concentrations in the presence of sorbitol (2 M). The horizontal axis represents the neutron-scattering vector, $\mathbf{Q} = (4\pi/\lambda)\sin\theta$, where 2θ is the scattering angle and λ the neutron wavelength. The peak at $\sim 0.025 \text{ Å}^{-1}$ of the scattering curves in Fig. 1 is due to the presence of a beam stopper. In this concentration region (35–65 mM), the micelle–micelle interaction can be thought to be negligible, judging from these scattering curves in Fig. 1.

According to a standard theory for a globular micelle system with low polydispersity [11, 14], the neutron scattering cross section per unit volume at $\mathbf{Q} = 0$ ($I(0)$) is given by

$$I(0) = (C - C_{cmc})n(b_m - \rho_s V_m)^2 \,, \tag{1}$$

where, C and C_{cmc} are the concentration of the dC(6)PC and its critical micellar concentration, respectively, n the mean aggregation number, b_m the sum of the coherent neutron-scattering lengths of dC(6)PC, ρ_s the coherent neutron-scattering lengths density of the solvent, and V_m is the volume occupied by a single molecule in the micellar aggregate. From Eq. (1), if the aggregation number n and the volume V_m do not depend on the concentration, the critical micellar concentration can be established from the zero intercept of the plot of $I(0)$ as a function of dC(6)PC concentration with the same contract between the micelle and the solvent. The values of $I(0)$ were estimated by fitting of the experimental data in the small $\mathbf{Q}$ region to the Guinier approximation [19] as

$$I(\mathbf{Q}) = I(0)\exp(-Q^2 R_g^2/3) \,, \tag{2}$$

where R_g is the radius of gyration of the micelle. The Guinier plots of $\ln I(\mathbf{Q})$ vs. $\mathbf{Q}^2$ for different dC(6)PC concentrations in the presence of 2 M sorbitol are shown in Fig. 2. For the fitting by the Guinier approximation, scattering data of $\mathbf{Q} < 0.03 \text{ Å}^{-1}$ were not used because the data in this region affected by the presence of a beam stopper. The R_g values are summarized in Table 1. The R_g values do not depend on the lipid concentration. The values of $I(0)$ estimated from the fitting by the Guinier approximation are plotted in Fig. 3 as a function of the concentration C. The linear relation of $I(0)$ and C indicates that the aggregation number n and the volume V_r do not

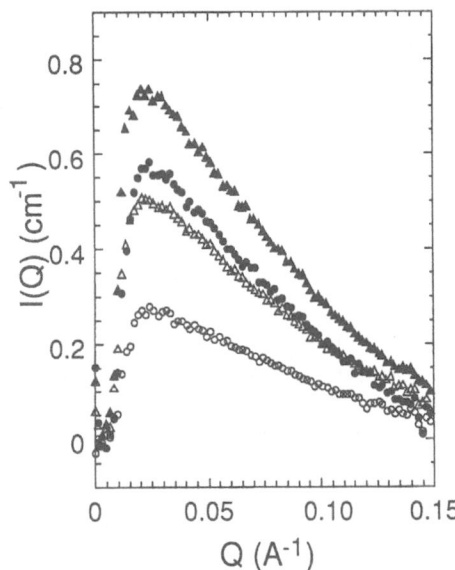

Fig. 1 Scattering curves for dC(6)PC with various concentrations ((○) 35 mM, (△) 50 mM, (●) 55 mM, and (▲) 65 mM) in the presence of 2 M sorbitol

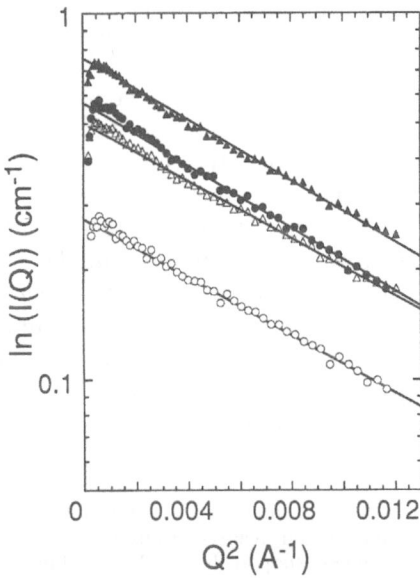

Fig. 2 Guinier plots for dC(6)PC with various concentrations ((○) 35 mM, (△) 50 mM, (●) 55 mM, and (▲) 65 mM) in the presence of 2 M sorbitol

Table 1 The estimated values of R_g from the Guinier plots for dC(6)PC micelles in the presence of 2 M sorbitol at 25°C.

Concentration [mM]	R_g [Å]
35	16.7 ± 0.2
50	16.7 ± 0.2
55	17.7 ± 0.2
65	16.8 ± 0.2

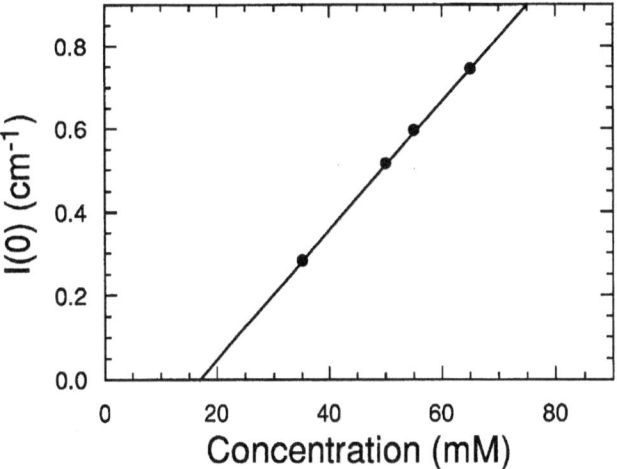

Fig. 3 Scattering intensities at $\mathbf{Q} = 0$, ($I(0)$) plotted vs. the lipid concentration. The straight line results in the critical micellar concentration as 16.7 mM

change with the concentration. From the position of zero intercept, the critical micellar concentration was found to be 16.7 mM. This value is slightly higher than that of dC(6)PC in pure water [9, 11].

In order to determine the aggregation number n and the volume V_m, we performed a contrast-variation experiment at constant dC(6)PC concentration (50 mM). The contrast, $\Delta\rho = (b_m/V_m) - \rho_s$, was changed by various D_2O/H_2O mixtures. Figure 4 presents the Guinier plots for dC(6)PC micelles in the presence of 2 M sorbitol with various contrasts. Equation (1) also indicates that V_m and n can be determined from the position of the zero intensity and the slope of the plot of $I(0)$ as a function of contrasts, respectively. Figure 5 is a plot of the square root of $I(0)$ vs. the volume fraction of H_2O. $I(0)$ vanishes in a solvent containing 16.1 vol% D_2O. This corresponds to $\rho_s = 6.93 \times 10^{-7} \text{Å}^{-2}$. From this value and the slope of the straight line in Fig. 5, we found $n = \sim 32$ and $V_m = \sim 640 \text{ Å}^3$. In this calculation, the value of the micellar concentration estimated from the above measurement was also taken into consideration.

Lin et al. [11] have determined $n = 19$ and $V_m = 670 \text{ Å}^3$ for dC(6)PC micelle in pure water by neutron scattering. They [11] have also reported that the shape of dC(6)PC micelle in pure water is not spherical but prolate elliptical. However, to simplify, we assume that dC(6)PC micelle in both pure water and sorbitol solutions are spherical. From this assumption and the values of n and V_m, the interfacial areas per single lipid molecule can be calculated to be 138 and 113 Å^2 for the micelles in pure water and sorbitol solutions, respectively. This indicates that sorbitol reduces the contact area between the lipid and the aqueous phases.

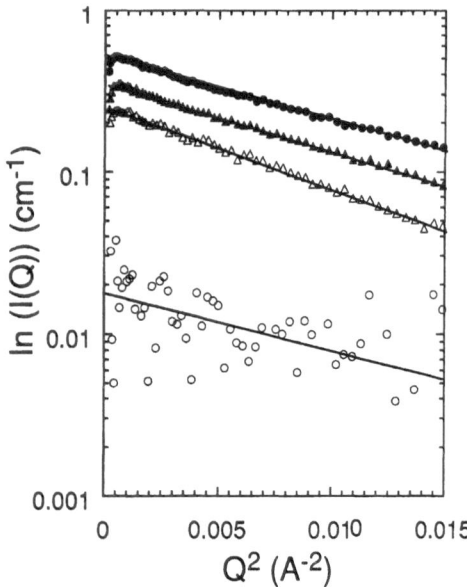

Fig. 4 Guinier plots for dC(6)PC micelles (50 mM) in the presence of 2 M sorbitol. The scattering intensities change with the H_2O/D_2O ratio. (●) 0, (▲) 13, (△) 26, and (○) 100 vol% H_2O

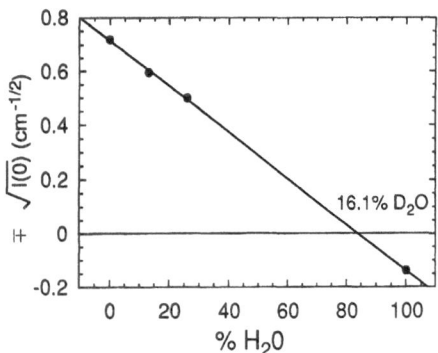

Fig. 5 Result of contrast variation: The zero-intensity intercept occurs at 16.1 vol% D_2O in the solvent

Discussion

The present study provides clear evidence that the surface area at the interface between the water and the phospholipid is decreased by the presence of sorbitol. We could confirm the hypothesis proposed by Koynova et al. [5]. Let us consider the relation of this result and the role of sorbitol in a low-temperature tolerance for some organisms. It has been reported that sugars or sugar alcohols hardly bind to phospholipid monolayers on aqueous phases [20]. This suggests that the sugars and sugar alcohols indirectly affect an assembly of phospholipid molecules. Therefore, it is thought that the effect of sugars and sugar alcohols is due to change of the water structure induced by these substances [5, 7]. Furthermore, the change of the gel-to-liquid crystalline phase transition temperatures of phospholipid membranes, which is induced by sugars or sugar alcohols, hardly depends on the difference of polar headgroups [3–8]. This also suggests the effect of sugars and sugar alcohols is not specific, i.e., the effect might be common for amphiphilic macromolecules or assemblies of amphiphilic molecules. Consequently, it can be expected that sugars or sugar alcohols also reduce the interfacial area of proteins in aqueous phases. It is known that proteins undergo a cold-induced denaturation, i.e., proteins do not adopt a native state at some subzero temperatures (see a review by Privalov and Gill [21]). In the native state of proteins, the hydrophobic amino residues of proteins are localed in an inside region and only hydrophilic amino residues contact with surrounding water molecules. However, because of the break of the secondary and higher-order structures, in the denatured state, the hydrophobic residues of proteins are exposed to water and the interfacial area contracted with water increases in comparison with the native state. Therefore, sugars and sugar alcohols can be considered to stabilize the native state of proteins at some subzero temperatures by reducing the interfacial area. This might be correlated with the low-temperature resistance by producing sugars or sugar alcohols in some organisms.

Acknowledgments We would like to thank Dr. M. Hirai of Gunma University for useful comments on neutron transmissions and Mr. T. Imura in the glass blowing shop at the Graduate School of Engineering of Nagoya University for making sample cells for neutron scattering experiments. This work is supported in part by Grant-in-Aid from Ministry of Education, Science, Sports and Culture, Japan.

References

1. Crowe JH, Crowe LM, Carpenter JF, Wistrom CA (1987) Biochem J 242: 1–10
2. Storey KB, Baust JG Storey JM (1981) J Comp Physiol 144: 183–190
3. Chowdhry BZ, Lipka G, Sturtevant JM (1984) Biophys J 46: 419–422
4. Tsvekov TD, Tsonev LI, Tsvekova NM, Koynova RD, Tenchov BG (1989) Cryobiology 26: 162–169
5. Koynova RD, Tenchov BG, Quinn PJ (1989) Biochim Biophys Acta 980: 377–380
6. Bryszewska M, Epand RM (1988) Biochem Biophys Acta 943: 485–492
7. Sanderson PW, Lis LJ, Quinn PQ, Williams WP (1991) Biochem Biophys Acta 1067: 43–50

Progr Colloid Polym Sci (1997) 106:223–227
© Steinkopff Verlag 1997

8. Tsonev LI, Tihova MG, Brain APR, Yu ZW, Quinn PJ (1994) Liq Cryst 17: 717–728

9. Tausk RIM, Karmiggelt J, Oudshoorn C, Overbeek JThG (1947) Biophys Chem 1: 175–183

10. Burns Jr RA, Roberts MF, Dluhy R, Mendelsohn R (1982) J Am Chem Soc 104: 430–438

11. Lin, TL, Chen SH, Gabriel NE, Roberts MF (1986) J Am Chem Soc 108: 3499–3507

12. Lin TL, Chen SH, Gabriel NE, Roberts MF (1987) J Phys Chem 91: 406–413

13. Magid LG, Trioio R, Johonson Jr JS, Koehler WC (1982) J Phys Chem 86: 164–167

14. Bendedouch D, Chen SH, Koehler WC (1983) J Phys Chem 87: 153–159

15. Kawaguchi T, Hamanaka T, Kito Y, Machida H (1991) J Phys Chem 95: 3837–3846

16. Kawaguchi T, Hamanaka T (1992) J Colloid Interface Sci 151: 41–48

17. Ito Y, Imai M, Takahashi S (1995) Physica B 213/214: 889–891

18. Imai M, Kaji K, Kanaya T, Sakai Y (1995) Rev B 52: 12696–12704

19. Guinier A, Fournet G (1995) In: Small Angle Scattering of X-rays. Wiley, New York

20. Aarnett EA, Harvey N, Johnson EA, Johnston DS, Chapman D (1986) Biochemistry 25: 5239–5242

21. Privalov P, Gill SJ (1988) Adv Protein Chem 39: 191–231

Progr Colloid Polym Sci (1997) 106:228–231
© Steinkopff Verlag 1997

BIOCOLLOIDS

Y.A. Shchipunov
E.V. Shumilina

Molecular model for the lecithin self-organization into polymer-like micelles

Received: 13 October 1996
Accepted: 8 April 1997

Dr. Y.A. Shchipunov (✉) · E.V. Shumilina
c/o Prof. Dr. H. Hofman
Universität Bayreuth
Physikalische Chemie I, NW I
96470 Bayreuth, Germany

Abstract Interactions of lecithin dissolved in alkane with polar solvents and phosphatidyl-ethanolamine have been studied by means of IR and ^{31}P NMR spectroscopy. Comparison of gel-forming compounds with sructurally related ones incapable of inducing jellification of lecithin solutions has provided an insight into the molecular organization of polymer-like micelles. In a developed molecular model, the polar solvents bridge neighboring phospholipid molecules through a ribbon-like hydrogen bonding network consisting of alternate solvent molecules and lecithin phosphate groups.

Key words Lecithin – phosphatidyl-ethanolamine – self-organization – reverse micelles – polymer-like micelles – organogel – hydrogen bonds – molecular model

Introduction

Lecithin organogels are made up of a three-dimensional network of entangled flexible tubular reverse micelles, often called polymer-like micelles. They are prepared by addition of trace amounts of water into a nonaqueous lecithin solution. With dissolving polar solvent, the initial spherical reverse micelles undergo a one-dimensional growth that leads to sequential transformation of the aggregates into worm-like ones [1–3]. Although the water is essential for the formation and stabilization of organogels, its role as well as the molecular mechanism of lecithin self-assembly in non-polar media is still poorly understood.

We have recently ascertained [4, 5] that in addition to water a few polar organic solvents are capable of inducing the jellification of lecithin solutions. Their comparison with nongel-forming structural analogue provides an insight into the mechanism of processes at the molecular level.

Here experimental results indicative of the molecular mechanism are considered. Finally, we suggest a molecular model for the lecithin self-organization into the polymer-like micelles

Experimental

Chromatographically pure egg-yolk lecithin and phos-phatidylethanolamine[1] (PE) was provided by Prof. E. Ya. Kostetsky. Soybean lecithin was from *Lucas Meyer*. Distilled water was prepared in the common manner. Polar organic solvents – *n*-decane, formamide (FA), N,N-dimethylformamide (DMFA), glycerol (GL), ethylene glycol (EG), 1,3-propanediol (1,3-PD), of chemical pure – were dried over sodium sulfate.

Organogels were prepared by the method described in [4, 5] from lecithin–decane solutions containing 1–8% (w/v) phospholipid. Components for each sample were

---

[1] *Abbreviations:* PE, phosphatidylethanolamine; FA, formamide; DMFA, N,N-dimethylformamide; GL, glycerol; EG, ethylene glycol; 1,3-PD, 1,3-propanediol.

dissolved completely and were then allowed to equilibrate for 3–6 h.

Transmission IR spectra were recorded using a Perkin–Elmer model 577 spectrometer, ^{31}P NMR spectra were taken at 32.442 MHz by means of a Bruker WP-80-SY spectrometer. An external deuterated lock was used. The reference was 85% orthophosphoric acid. Oscillatory shear measurements were made with a Bohlin CS-10 rheometer using a double-gap cell with concentric cylinder geometry.

Results and discussion

A list of additives inducing the transformation of initial spherical micelles of lecithin into wormy-like micellar aggregates includes water, GL, EG, FA and formic acid [4, 5]. Factors determining the lecithin self-organization into the polymer-like aggregates can be appreciated from comparison of the gel-forming solvents with their close structural analogue incapable to promote the one-dimensional growth of reverse micelles. For this purpose, let us examine such series as FA-DMFA and GL-EG-1,3-PD. In addition, it is helpful to consider effects of such structural analogue of lecithin as PE on the lecithin organogels.

Formamide-dimethylformamide

The spectroscopic studies showed that in the reverse micelles polar solvents added in trace amounts localize mainly in the vicinity of the phosphate group of lecithin. They are hydrogen bonded with it. The gel-forming ability of polar additives correlates with their hydrogen-bonding capacity. In the case under consideration, FA can participate in as many as four hydrogen bonds owing to the presence of hydrogen-donating and hydrogen-accepting groups, whereas DMFA contains only one hydrogen-

accepting group. This difference may account for a discrepancy in the strength of hydrogen bonding revealed from the widths of the resonance bands at half-signal height and their chemical shifts on ^{31}P NMR spectra (Fig. 1). Comparison of FA and DMFA, along with other series of polar organic substances examined, suggested that *to promote the one-dimensional growth of lecithin micelles the gel-forming species should be capable to* (i) *form multiple hydrogen bonds and* (ii) *create a three-dimensional hydrogen bonding network*.

Phosphate represents a hydrogen donor–acceptor moiety of lecithin. The formation of hydrogen bonding networks is inherent in this group too. For example, the phosphate groups are linked with each other in lecithin crystals via hydrogen bonds generating a ribbon-like network [6]. H-bondings are not found in the reverse lecithin micellar aggregates in nonpolar media because of an increase in the mean separation between molecules after dissolving. They are revealed after addition of polar solvents [4, 5]. We suggest that *additives tend to induce the lecithin self-organization into polymer-like micelles where they are aligned so that two phosphate groups of neighboring phospholipid molecules are hydrogen bonded via a molecule of polar solvent acting as a "bridge" between the former* (Fig. 2). The FA-DMFA series provides an illustrative example.

Glycerol-ethylene glycol-1,3-propanediol

On passing from GL to 1,3-PD, the ability to promote the thickening effect disappears. Since all these organic solvents are liable for multiple hydrogen bonds and for the formation of three-dimensional hydrogen-bonding networks, this brings up the question as to which difference in their properties is essential for the gel-forming ability.

A survey of the physico-chemical properties of solvents showed that *the loss in the ability to induce the organogel*

Fig. 1 ^{31}P line width at half-signal height (A) and chemical shift (B) vs. number (n) of formamide (FA) or dimethylformamide (DMFA) molecules per lecithin molecule. Soybean lecithin concentration in *n*-decane was 70 mg/ml. Temperature 22 °C

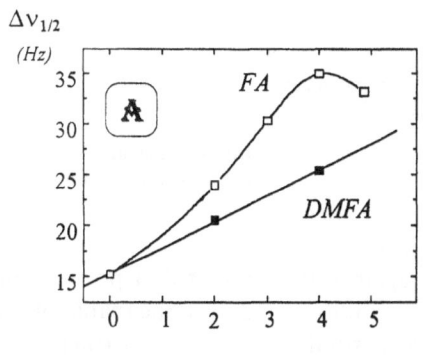

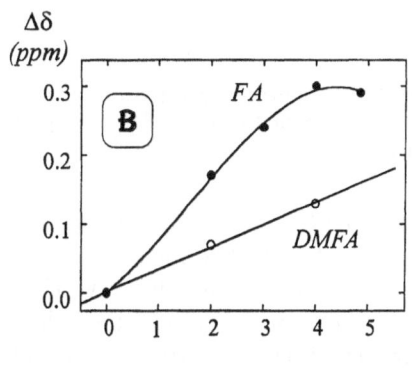

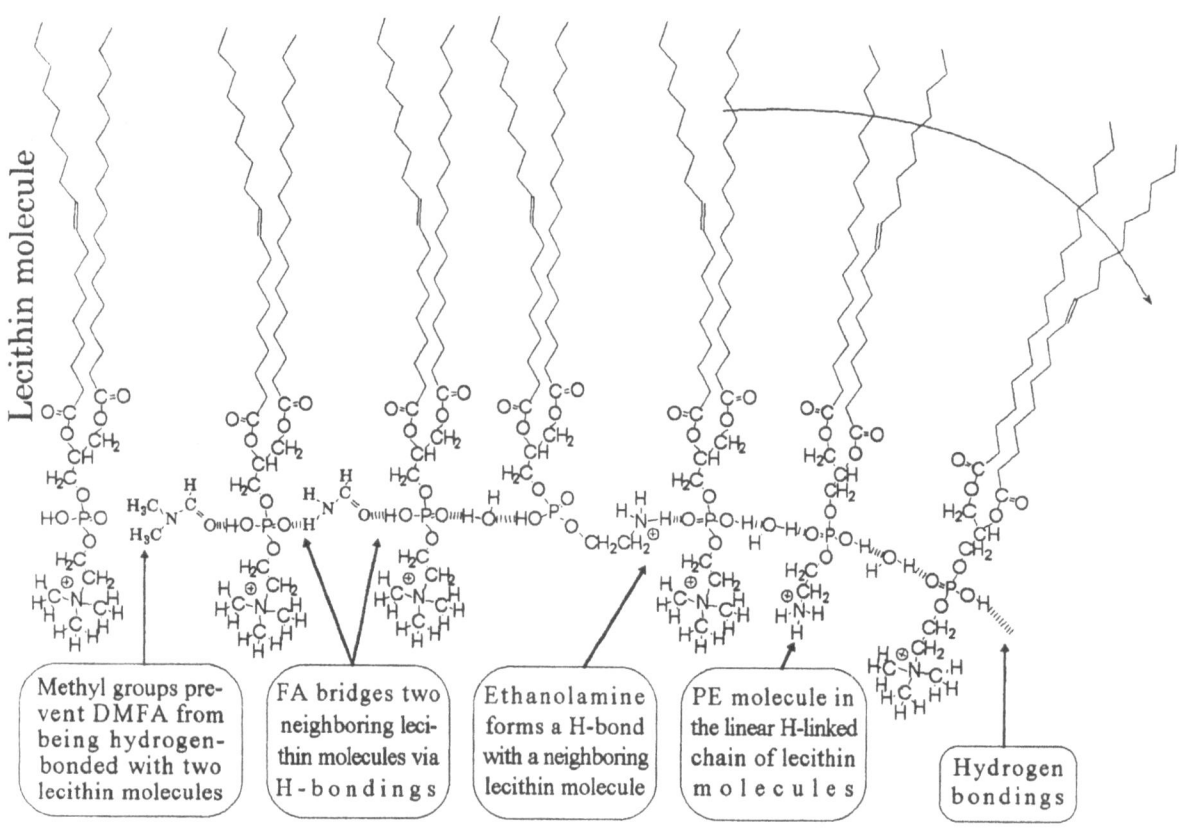

Fig. 2 Schematic representation of the polymer-like micelle formed by lecithin in the presence of DMFA, FA, water and PE. Various types of molecular arrangement and hydrogen bondings are shown

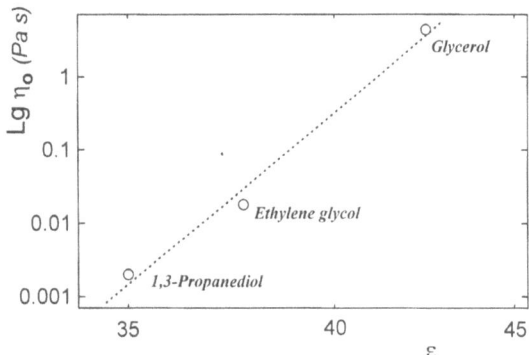

Fig. 3 Plot of the zero-shear viscosity measured at low frequencies vs. relative permittivities (dielectric constants) of the solvents indicated.

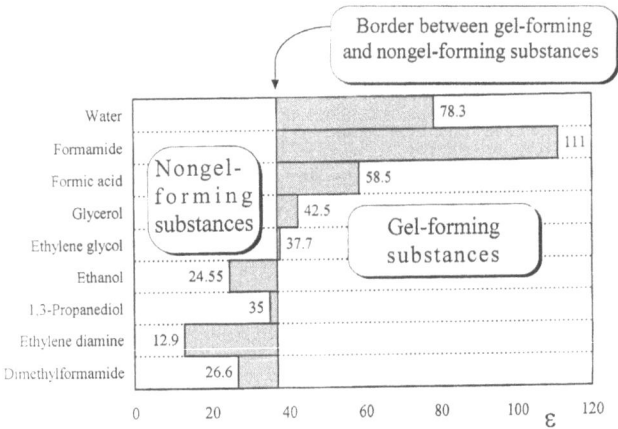

Fig. 4 Dielectric constants (relative permittivity) for the polar organic solvents studied

formation is most well correlated with a decrease in solvent polarity. A correlation is seen in Fig. 3 which represents a plot of zero-shear viscosity of mixtures against the dielectric constants of the three polar additives considered. The dielectric constants of all solvents studied in this work are presented in Fig. 4. A boundary between gel-forming and nongel-forming substances is shown. It

seems that polar additives with dielectric constant under 37 are incapable of inducing jellification of lecithin solutions. Nevertheless, 1,3-PD-like GL forms hydrogen bonds with lecithin phosphate group, though they are weaker. This effect is attributable to the location of small

Progr Colloid Polym Sci (1997) 106:228–231
© Steinkopff Verlag 1997

molecules along the phospholipid molecule. Less polar additives should give preference to a less polar region. The most polar moiety of lecithin molecule is the phosphate residue. With decreasing polarity, the point of solvent location will become displaced from a position close to phosphate. This will cause complications for a solvent to form hydrogen bonds with two neighboring lecithin molecules at once and, as a result, to bridge them.

Phosphatidylethanolamine

This phospholipid is a structural analogue of lecithin. The difference between them lies in the terminal functional group (Fig. 2). The nitrogen atom in choline is surrounded by three methyl groups, whereas these groups are absent in ethanolamine that makes it capable to form hydrogen bonds. This difference was found to be essential for the gel-forming properties. Thus, PE self-assembly into tubular micelles in nonpolar media occurs without the water addition [7]. Adding water, tubular aggregates becomes disintegrated. Furthermore, the jelly-like phase cannot be generated from a PE-lecithin mixture if a PE mole fraction exceeds 0.5.

IR spectroscopic studies demonstrated that hydrogen bonds are formed between an amino and phosphate in PE-lecithin mixtures in the absence of polar solvents. This is supported by the pertinent frequency shift of the P=O stretching band in Fig. 5. Intermolecular hydrogen bondings between ethanolamine and phosphate are typical of PE in self-assembled structures [8]. Water or glycerol replaces ethanolamine in micelles because the amino group as a hydrogen bond donator is weaker than the hydroxyl [9]. Although these solvents provide the bridg-

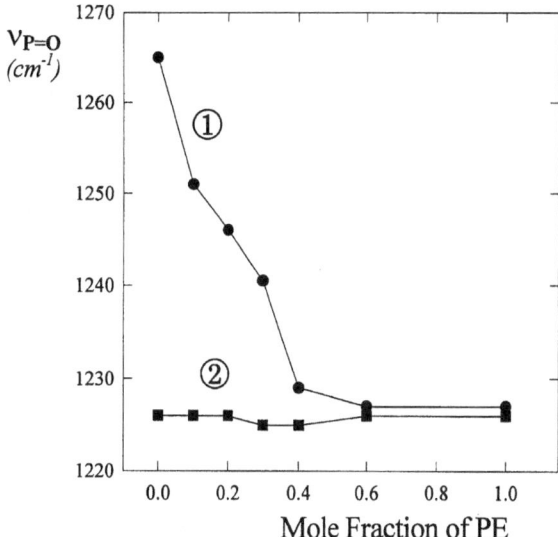

Fig. 5 Frequency shift of the P=O symmetric stretching band in spectra of PE-lecithin mixtures in the absence of added water (1) and in the presence of 4 water molecule per lecithin molecule (2) vs. mole fraction of PE. The net concentration of egg yolk phospholipids in n-decane was 40 mg/ml

ing of phospholipid molecules, an increase of a PE fraction in the PE-lecithin mixture causes a viscosity to drop. The effect is explicable on the basis of molecular geometry. *When a PE molecule of a truncated cone shape is incorporated into a linear chain of lecithin molecules having a cylindrical shape, it will make a "defect"*, tending to curve the chain away from the linear arrangement towards a curved one that should lead to a decrease of the polymer-like micelle length. This is shown schematically in Fig. 2.

Acknowledgements This work was partly supported by Volkswagen Stiftung (grant I/71 486).

References

1. Scartazzini R, Luisi PL (1988) J Phys Chem 92:829–833
2. Schurtenberger P, Magid LJ, Lindner P, Luisi PL (1992) Prog Colloid Polym Sci 89:274–277
3. Schurtenberger P, Cavaco C (1994) Langmuir 10:100–108
4. Shchipunov YuA, Shumilina EV (1995) Materials Sci Eng C 3:43–50
5. Shchipunov YuA, Shumilina EV (1996) Colloid J 58:117–125
6. Pascher I, Lundmark M, Nyholm P-G, Sundell S (1992) Biochim Biophys Acta 1113:339–373
7. Sen A, Yang PW, Mantsch HH, Hui SW (1988) Chem Phys Lipids 47:109–116
8. Boggs JM (1987) Biochim Biophys Acta 906:353–404
9. Jeffrey GA, Saenger W (1991) Hydrogen Bonding in Biological Structures. Springer, Berlin

Progr Colloid Polym Sci (1997) 106:232–236
© Steinkopff Verlag 1997

BIOCOLLOIDS

M. Hirai
T. Takizawa
K. Hayashi
S. Yabuki
M. Imai
Y. Matsushita
Y. Ito

Neutron scattering study of phosphatidylinositol 4,5-bis(phosphate) aqueous dispersion

Received: 13 October 1996
Accepted: 6 March 1997

Dr. M. Hirai (✉) · S. Yabuki
T. Takizawa · K. Hayashi
Department of Physics
Gunma University
4-2 Aramaki
Maebashi 371, Japan
E-mail: hirai@sun.aramaki.gunma-u.ac.jp

M. Imai · Y. Matsushita
Institute for Solid State Physics
University of Tokyo
Tokai 106-1, Japan

Y. Ito
Faculty of Education
Yamanashi University
Koufu 400, Japan

Abstract In this paper we treat the thermotropic phase behavior of phosphatidylinositol 4,5-bis(phosphate) (PIP_2) aggregates in aqueous solution. By using the solvent contrast variation method in neutron small-angle scattering we could determine the absolute value of the PIP_2 molecular volume in the micellar form from the contrast matching point. The shift of the contrast matching point indicates the increase of the PIP_2 molecular volume of $\sim 11\%$ with elevating temperature from $10\,°C$ to $50\,°C$. The decrease of the gyration radius of the PIP_2 micelle observed in the present experiments well agrees with our previous results of X-ray scattering study. The present results suggest that the increase of the molecular volume would be mainly attributable to the change of the hydrophilic polar head region to decrease the critical packing parameter.

Key words Phosphatidylinositol – thermal transition – small-angle scattering – neutron scattering – contrast variation

Introduction

Phosphatidylinositol 4,5-bis(phosphate) (PIP_2) is one of the acidic glycerophospholipids and most abundant in the cell membranes of brains and kidneys, which exhibits high solubility in water. Previously inositol lipids including PIP_2 (PI, Phosphatidylinositol; PIP, Phosphatidylinositol 4-phosphate; IP_3, Phosphatidylinositol 1,4,5-tri(phosphate)) were studied in connection with nerve excitation, and recently have been attracting interests as the source of the second messenger in cellular signal transduction [1–5]. By using small-angle X-ray scattering, calorimetry and NMR spectroscopy we have been studying the structural properties and phase behaviors of the PIP_2/water systems [6–10]. In our previous calorimetric and NMR studies [8], we reported that the endothermal transition of PIP_2 showing a rather broad endothermal peak and a remarkable hysteresis are not attributable to melting of hydrocarbon chains, the so-called gel to liquid-crystal transition. In the recent small-angle X-ray scattering study, we showed that the systematic change of the PIP_2 aggregates in buffer solution at pH 7.0 induced by elevating the temperature would result from a high sensitivity of the interaction between water and PIP_2 polar heads to the variation of temperature [9]. In another paper we also

Abbreviations The following abbreviations are used in the manuscript without definitions.
N-(2-hydroxyethyl)piperazine-*N'*-(2-ethanesulfonic acid) is abbreviated as HEPES. Ethylenediaminetetraacetic acid is abbreviated as EDTA.

reported the effect of mono- and di-valent cations on the structure of PIP_2 aggregates that the PIP_2 aggregates show a drastic structural phase transition from prolate micelles to disordered lamellae with the addition of a small amount of Ca^{2+} ions [10]. These above previous structural and calorimetric studies have elucidated the interaction between PIP_2 and water would be very different from those of other phospholipids. To clarify another aspect of the thermotropic phase behavior of PIP_2 aqueous dispersion system, we have carried out neutron scattering experiments by using a solvent-contrast variation method.

Experimental

Sample preparation

As we described elsewhere [11], PIP_2 used for the present experiments was extracted from bovine brain and the crude extract was purified further by DEAE-cellulose chromatography. After this purification, PIP_2 was obtained as an ammonium salt, which was removed by dialysis against buffer containing EDTA and was lyophilized. Freeze-dried powder of PIP_2 is known to be very hygroscopic and the PIP_2 powder was dissolved in 50 mM HEPES buffer at pH 7.0. To employ a solvent-contrast method in the neutron-scattering experiments [12], we used four different HEPES buffers with different D_2O/H_2O ratios (100% v/v D_2O, 60% v/v D_2O, 41% v/v D_2O, 100% v/v H_2O). The concentrations of PIP_2 of the samples were adjusted to 1.0% w/v.

Neutron scattering experiments

Small-angle neutron-scattering experiments were performed by using a small-angle neutron-scattering spectrometer installed at the C1-2cold-neutron beam line in the research reactor JRR-3M of the Japan Atomic Energy Research Institute (JAERI), Tokai, Japan. The scattering intensity was detected by using a two-dimensional position sensitive detector with an effective area of 55×55 cm^2. The wavelength used was 7.0 Å and the sample-to-detector distance was 200 cm. The samples were contained in quartz cells which were placed in a thermostated cell-holder. The temperature of the samples was varied from 10 °C to 50 °C. The exposure time was 1 h for 100% D_2O solution, 2 h for 60% D_2O solution and 4 h for both 41% D_2O and H_2O solution, respectively.

Scattering data analysis

The scattering curve $I(q)$ was obtained by the following calibration form:

$$I(q) = \frac{I_{sol}(q)}{T_{sol}} - \frac{I_{solv}(q)}{T_{solv}} , \qquad (1)$$

where q is the magnitude of scattering vector defined by $q = (4\pi/\lambda)\sin(\theta/2)$ (θ, the scattering angle; λ, the wavelength), $I_{sol}(q)$, $I_{solv}(q)$, T_{sol} and T_{solv} represent the observed scattering curves and neutron beam transmissions of the solution and solvent, respectively. The beginning of the scattering curve $I(q)$ is known to depend on the Guinier equation in the form

$$I(q) = I(0)\exp(-q^2 R_g^2/3) , \qquad (2)$$

where $I(0)$ designates the zero-angle scattering intensity, and R_g is the radius of gyration [13]. By using the least-squares method for the Guinier plot ($\ln(I(q))$ versus q^2) on the data sets of small q range, we determined the values of both $I(0)$ and R_g.

Results and discussion

Figure 1 shows the full profiles of the observed scattering curves of 1% w/v PIP_2 in four different solvents (100%, 60%, 41% and 0% v/v D_2O buffer at pH 7.0) at different temperatures, where (a) and (b) correspond to 10 °C and 50 °C, respectively. By using Eq. (2) we estimated the zero-angle scattering intensity $I(0)$ and the gyration radius R_g. Figure 2 shows the Guinier plots ($\ln(I(q))$ versus q^2) and the lines obtained by applying least-squares fitting to the scattering curves in the q region 0.03–0.05 Å^{-1} at 10 °C and 50 °C in Fig. 1. The Guinier plots show good linearity in this q region, insuring that we can estimate appropriate values of $I(0)$ and R_g. In the case of a monodispersed aggregate system composed of a single species of solute molecules such as micelles, we can express the $I(0)$ using the average excess scattering density $\bar{\rho}$, the so-called "contrast", of the solute particle as follows:

$$I(0) \propto N(\bar{\rho}V)^2 = N\left[\left(n_a\sum_i b_i + n_w\rho_s V_s\right) - \rho_s V\right]^2 , \qquad (3)$$

where N is the number of the aggregate, n_a the aggregation number, n_w the number of hydrated water within the aggregate, V the aggregate volume, V_s the volume of water molecule, ρ_s the average scattering density of water, and $\sum_i b_i$ the total scattering density of the solute molecule. The ρ_s varies with changing the D_2O/H_2O ratio as

$$\rho_s = (-0.562 + 0.0697x)10^{10}\ cm^{-2} , \qquad (4)$$

234

M. Hirai et al.
Neutron scattering study of phosphatidylinositol aqueous dispersion

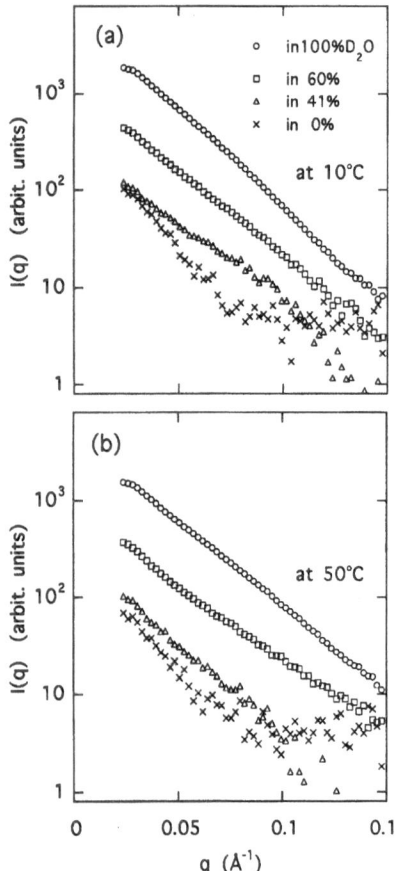

Fig. 1 Neutron-scattering curves of PIP$_2$ in 50 mM HEPES buffer at pH 7.0 (100% v/v D$_2$O, 60% v/v D$_2$O, 41% v/v D$_2$O, 100% v/v H$_2$O). (a) at 10 °C; (b) at 50 °C. The concentration of PIP$_2$ was 1.0% w/v

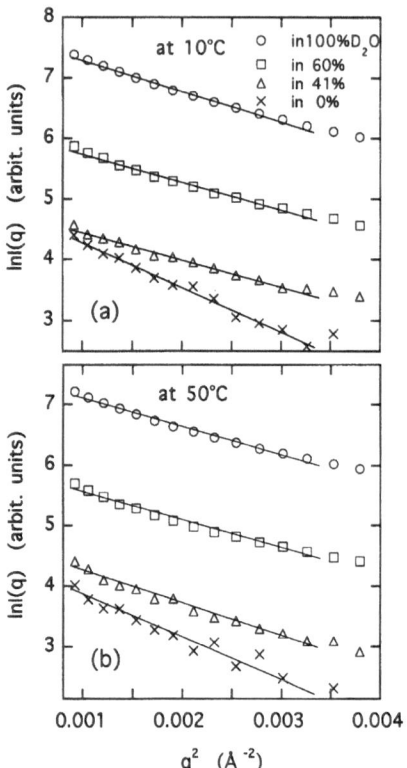

Fig. 2 Guinier plot ($\ln(I(q))$ versus q^2) on the data sets of small q range used for the determination of both values of the zero-angle scattering intensity $I(0)$ and the gyration radius R_g. (a) and (b) correspond the scattering data in Figs. 1a and b, respectively

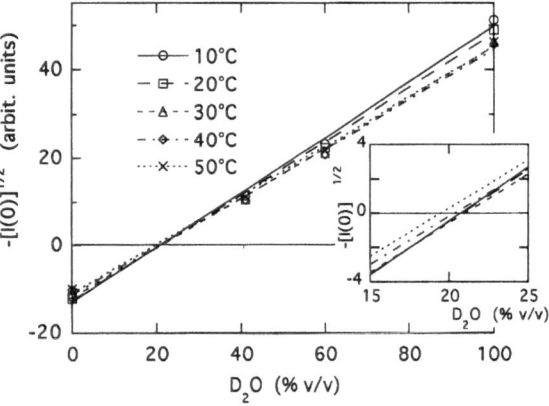

Fig. 3 $[I(0)]^{1/2}$ versus D$_2$O content (% v/v) plots at different temperatures to determine the contrast matching points. The inset displays the intercept points satisfying $[I(0)]^{1/2} = 0$, namely the contrast matching points used for the estimation of PIP$_2$ molecular volume

where x is the volume percentage of D$_2$O in water [12]. The presence of HEPES in water slightly decreases the absolute values of the intercept and the slope in Eq. (4) since the volume fraction of HEPES in the solvents is about 0.01 and HEPES molecules have exchangeable protons with water. Such an effect on the average scattering density of solvent can be neglected in the present analysis, as also discussed below. As the volume V is given as $V = n_a V_m + n_w V_s$ (V_m, the volume of the solute molecule), the $I(0)$ is simply expressed as

$$I(0) \propto N n_a^2 \left(\sum_i b_i - V_m \rho_s \right)^2. \tag{5}$$

Eq. (5) indicates that in the case of the system of aggregates with some hydration inside we can also determine the monomer volume V_m directly by varying the solvent scattering density, namely, by using the solvent contrast variation method, in the same way of the cases of the large

macromolecular systems [12]. Figure 3 shows the plot of the $[I(0)]^{1/2}$ depending on the volume percentage x of D$_2$O. According to the simple linear relation between $[I(0)]^{1/2}$ and x derived from Eq. (5), we can determine the

Progr Colloid Polym Sci (1997) 106:232–236
© Steinkopff Verlag 1997

contrast matching point from the condition of the $[I(0)]^{1/2}$ versus x plots satisfying $[I(0)]^{1/2} = 0$. The contrast matching points at different temperatures are located at around $x \sim 20\%$. The inset of Fig. 3 displays the intercept points of the lines on the above linear relation, where we can recognize the shift of the intercept points, namely the change of the contrast matching points depending on temperature. As the $\sum_i b_i$ of the PIP$_2$ molecule only depends the chemical composition, we can calculate the $\sum_i b_i$ value to be 0.124×10^{-12} cm for the PIP$_2$ molecule, where we assumed the PIP$_2$ molecules take all-protonated forms at pH 7.0 [14]. Then, we can determine the PIP$_2$ monomer volume V_m from the ρ_s values of the contrast matching points by the following relation:

$$V_m = \sum_i b_i / \rho_s . \qquad (6)$$

The temperature dependence of the V_m and R_g values is shown in Fig. 4. With elevating temperature the V_m value once decreases slightly from 1405 ± 57 Å^3 at $10\,°C$ to 1380 ± 48 Å^3 at $30\,°C$ and increases significantly to 1560 ± 54 Å^3 at $50\,°C$. The thermal expansion of water with elevating temperature from $10\,°C$ to $50\,°C$ is expected to reduce the average scattering density ρ_s in about 1%, meaning ca. 1% decrease of both absolute values of the intercept and the slope in Eq. (5). Then, from Eq. (6) the thermal expansion of water in the temperature range $10–50\,°C$ turns out to increase the contrast matching point of PIP$_2$ in about 0.1% D$_2$O, namely to decrease the PIP$_2$ volume V_m. Therefore, the effect of the thermal expansion of water on the change of the contrast matching point can be neglected. By using the empirical formula of the hydrocarbon chain volume [15] and by considering the apparent volumes of the basic chemical elements [16, 17], we can tentatively estimate the volume of the PIP$_2$ molecule to be 1442 Å^3 (1397 Å^3), where the hydrophilic head and hydrophobic tail of the PIP$_2$ molecule are calculated to be 395 Å^3 (376 Å^3) and 1048 Å^3 (1021 Å^3), respectively. Here, in the present temperature range, we assume that the arachidonic acid chain takes a fluid conformation and that the stearic acid chain takes a gel or fluid conformation depending on temperature. The above values and the values in the brackets correspond to those obtained by using molecular volumes at fluid and gel phases [17], respectively. The present experimental V_m value at $10–30\,°C$ is close to the empirical value at gel phase, on the other hand, the V_m value at $50\,°C$ is much larger than that at fluid phase, suggesting that the increase of the V_m value would be mainly attributable to the change of the hydrophilic polar head region. The temperature dependence of the R_g values in Fig. 4 evidently suggests the decrease of the micellar dimension, which agrees well with our pre-

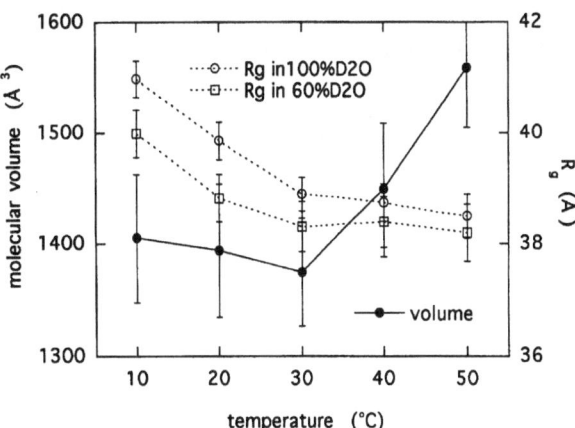

Fig. 4 Gyration radius R_g of PIP$_2$ micelle and molecular volume of PIP$_2$ monomer estimated from Figs. 2 and 3, plotted against temperature. Right ordinate, gyration radius (open marks); left ordinate, molecular volume (full marks). The errors of the obtained values are shown by short vertical bars

vious results. In the former paper, by using both the synchrotron radiation X-ray scattering data and a double-shell modeling method we reported that the PIP$_2$ aggregates in buffer solvent at pH 6.7 takes an elongated ellipsoidal micelle in the temperature range $10–50\,°C$ and that the elevation of temperature from $10\,°C$ to $50\,°C$ induces the shrinkage of the micellar dimension (the decrease of semi-axial ratios) and the change of the intramicellar scattering density distribution primarily occurring in the hydrophilic portion of the micelles [9]. Combined the present results with the previous ones, we assume that the increase of the V_m value primarily occurring in the hydrophilic head region would decrease the critical packing parameter defined by v/sl_c (v, hydrocarbon chain volume; s, optimal surface area per polar head; l_c, critical chain length) [18], inducing the change of the micellar structure from an ellipsoidal shape to a spherical one which appeared as an evident decrease of the gyration radius.

Conclusion

In the present neutron-scattering experiments we have confirmed the decrease of the PIP$_2$ micellar dimension when elevating the temperature from $10\,°C$ to $50\,°C$, which has been already shown in our previous X-ray scattering study [9]. In addition, by using a solvent contrast variation method we have found the change of the contrast matching point of the PIP$_2$ micelles depending on temperature, indicating the increase of the PIP$_2$ molecular volume of $\sim 11\%$. This increase is much larger than the experimental error and cannot be explained by the so-called gel to liquid-crystal transition of hydrocarbon chains with the elevation of temperature. From the present results combined with our previous X-ray scattering ones, we can conclude that the PIP_2 polar head contributes to the increase of the PIP_2 molecular volume, which supports the results of our previous calorimetric and NMR studies [8]. The present results show a unique characteristic of the PIP_2 molecule sensitivity responding to a variation of temperature in the physiological range around 30 °C, may be implying some active role of the PIP_2 molecules as a kind of functional lipids by themselves in cell membranes.

Acknowledgements This work was done under the approval of the Neutron Scattering Program Advisory Committee (Proposal No. 95-5).

References

1. Hokin MR, Hokin LE (1953) J Biol Chem 203:967–977
2. Michell RH (1975) Biochim Biophys Acta 415:81–147
3. Sterb H, Irvine RF, Berridge MJ, Schulz I (1983) Nature 306:67–69
4. Berridge MJ (1987) Ann Rev Biochem 56:159–193
5. Hill TD, Dean NM, Boynton AL (1988) Science 242:1176–1178
6. Takizawa T, Hayashi K, Mitomo H (1988) Thermochim Acta 123:247–253
7. Takizawa T, Mitomo H, Hayashi K (1990) Thermochim Acta 163:133–138
8. Takizawa T, Hayashi K, Mitomo H (1991) Thermochim Acta 183:313–321
9. Hirai M, Takizawa T, Yabuki S, Hirai T, Hayashi K (1995) J Phys Chem 99:17456–17460
10. Hirai M, Takizawa T, Yabuki S, Nakata Hirai T, Hayashi K (1996) J Chem Soc Faraday Trans 92:1493–1498
11. Hayashi F, Sokabe M, Takagi M, Hayashi K, Kishimoto U (1978) Biochim Biophys Acta 510:305–315
12. Stuhrmann HB, Miller A (1978) J Appl Cryst 11:325–345
13. Guinier A (1939) Ann Phys 12:161–237
14. van Paridon PA, de Kruijff B, Ouwerkerk R, Wirtz KWA (1986) Biochim Biophys Acta 877:216–219
15. Tanford C (1972) J Phys Chem 76:3020–3024
16. Zamyatnin AA (1972) Prog Biophys Mol Biol 24:107–123
17. Marsh D, DPhil MA (1990) In: CRC Handbook of Lipid Bilayers. CRC Press, pp 185–194
18. Israelachvili JN, Mitchell DJ, Ninham BW (1976) J Chem Soc Farady Trans II 72:1525–1568

Progr Colloid Polym Sci (1997) 106:237–241
© Steinkopff Verlag 1997

M. Hirai
T. Takizawa
S. Yabuki
Y. Nakata
S. Arai
M. Furusaka

Structural study of protein–ganglioside interaction

Received: 13 October 1996
Accepted: 26 February 1997

Dr. M. Hirai (✉) · T. Takizawa · S. Yabuki
Y. Nakata · S. Arai
Department of Physics
Gunma University
4-2 Aramaki
Maebashi 371, Japan
E-mail: hirai@sun.aramaki.gunma-u.ac.jp

M. Furusaka
National Laboratory for High Energy
Physics
Tsukuba 305, Japan

Abstract In this paper we treat the interaction of acidic-glycolipids with proteins. We used monosialoganglioside, disialoganglioside, and ganglioside mixture. The proteins used were albumins whose surfaces are modified differently with mono- and disaccharides. By using X-ray and neutron-scattering techniques we have found that the interactions of mono- and disialogangliosides with albumins having monosaccharides occur less destructively for ganglioside aggregate structures at low protein concentrations and that the interaction of monosialogangliosides with albumins having disaccharides occurs very destructively, suggesting a strong interaction between them compared with other cases.

Key words Micelle – ganglioside – albumin – molecular recognition – synchrotron radiation – X-ray scattering – neutron scattering

Introduction

Gangliosides, the most complex of the glycosphingolipids, are acidic glycolipids composed of a ceramide linked to oligosaccharide chain containing one or more N-acetyl-neuraminic acid (called sialic acid) residues. Through a numerous variety of the structures of the oligosaccharide chains, gangliosides have been shown to be involved in the self-organization of tissues, immune response and cell differentiation through the molecular recognition [1]. Physicochemical and thermotropic properties of gangliosides have been studied intensively and the presence of large hydrophilic head groups of ganglioside molecules was shown to lead a marked amphiphilic property and a micellar aggregation in aqueous solution [2, 3].

In the previous papers treating ganglioside dispersion systems, we have reported that the ganglioside micelles show the shrinkage of the hydrophilic region and the expansion of the hydrophobic region accompanying the change of the axial ratios of the ellipsoidal micelles at around 20–30 °C, suggesting that the gangliosides themselves can modulate biomembrane surface structures by responding to a local environmental perturbation [4–6]. By using mixing-heat microcalorimetry and X-ray and neutron-scattering methods, we have also shown that the complexation of gangliosides with methylated bovine albumins occurs heterogeneously [7, 8]. In the present paper, by using X-ray and neutron-scattering techniques we will discuss the differences in the complexation processes of three different types of gangliosides samples with bovine serum albumins whose surfaces are chemically modified with mono- and disaccharides.

Experimental

Material

Three different types of gangliosides from bovine brain purchased from Sigma Chemical Company were used, which were monosialoganglioside (II³NeuAcα–

238
M. Hirai et al.
Structural study of protein-ganglioside interaction

GgOse$_4$Cer as G$_{M1}$; 1.56 kDa), disialoganglioside (IV^{3}NeuAcα–, II^{3}NeuAcα–GgOse$_4$Cer abbreviated as G$_{D1a}$; 1.86 kDa) and type-III (containing 20% sialic acid). The major components of the type-III gangliosides are disialo- and monosialogangliosides [9]. We used 0.01 M citrate buffer in H$_2$O adjusted at pH 6.7 and 50 mM hepes buffer at pH 7.0 in H$_2$O for the X-ray scattering experiments and 50 mM hepes buffer at pH 7.0 in 41% v/v D$_2$O/H$_2$O for the neutron-scattering experiments, respectively. Bovine serum albumins whose molecular surfaces are chemically modified with different saccharides were purchased from Sigma Chemical Company and were used as model proteins. Albumin-2-amido-2-deoxy-D-galactose, albumin-N-1-[deoxymaltitol] and albumin-N-1-[deoxycellobiitol] were used. In the following sections we use the abbreviations forms of these albumins as gal-, mal- and cel-albumins, respectively. The samples for the scattering measurements were prepared by mixing equivolume amounts of the ganglioside and albumin stock solutions with different concentrations. The final concentrations of gangliosides were 0.5% w/v for X-ray scattering samples and 1.0% w/v for neutron-scattering samples.

Scattering experiments and data analysis

Small-angle X-ray scattering experiments were carried out by using the small-angle X-ray scattering spectrometer installed at the synchrotron source at the National Laboratory for High Energy Physics (KEK), Tsukuba, Japan. The wavelength and the sample-to-detector distance used were 1.49 Å and 88 cm, respectively. The exposure time of one measurement was 300 s at 25 °C. Neutron-scattering experiments were done by using the time-of-flight wide-angle neutron-scattering spectrometer installed at the pulsed neutron source at KEK [10]. In the neutron-scattering experiments a solvent contrast variation method was employed. The exposure time of one measurement was 6 h. The detail of the scattering data analyses are described elsewhere [4, 5]. The following analyses were done. The distance distribution function $p(r)$ was calculated by the Fourier inversion of the scattering intensity $I(q)$ as

$$p(r) = \frac{2}{\pi} \int_0^\infty rqI(q)\sin(rq)\,\mathrm{d}q , \qquad (1)$$

where q is the magnitude of scattering vector defined by $q = (4\pi/\lambda)\sin(\theta/2)$ (θ, the scattering angle; λ, the wavelength). The maximum dimension $D_{\max}$ of the particle was estimated from the $p(r)$ function satisfying the condition $p(r) = 0$ for $r > D_{\max}$. To estimate a gyration radius R_g, we also used the Glatter's method [11] giving R_g as

$$R_g^2 = \frac{\int_0^{D_{\max}} p(r)r^2\,\mathrm{d}r}{2\int_0^{D_{\max}} p(r)\,\mathrm{d}r} . \qquad (2)$$

Results and discussion

Complexation of G$_{M1}$ and G$_{D1a}$ gangliosides with albumins

Figure 1 shows the X-ray scattering curves $I(q)$ depending on the albumin/ganglioside ratio, where (a) and (b) for the

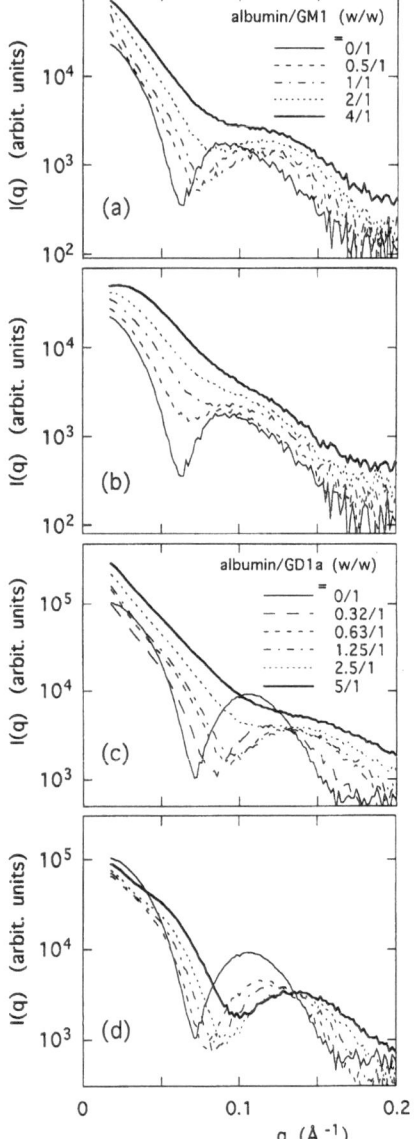

Fig. 1 Protein concentration dependence of X-ray scattering curve $I(q)$ of albumin/ganglioside mixture suspensions at 25 °C. (a) and (b), albumin/G$_{M1}$ ganglioside mixture in 50 mM hepes buffer at pH 7.0; (c) and (d), albumin/G$_{D1a}$ ganglioside mixture in 0.01 M citrate buffer at pH 6.7. The ganglioside concentration was 0.5% w/v and the albumin concentration was varied. Two different types of albumins were added; (a) and (c), albumin-2-amido-2-deoxy-D-galactose; (b) and (d), albumin-N-1-[deoxymaltitol]

complexation of G_{M1} gangliosides with gal- and mal-albumins; (c) and (d) for the complexation of G_{D1a} gangliosides with gal- and mal-albumins, respectively. The scattering curves of the gangliosides without proteins show a saturating tendency of the scattering intensity below $q \sim 0.04 \, \text{Å}^{-1}$ and have an evident minimum at $q \sim 0.06$–$0.07 \, \text{Å}^{-1}$ followed by a rounded peak at $q \sim 0.09$–$0.11 \, \text{Å}^{-1}$. This suggests that both G_{M1} and G_{D1a} ganglioside suspensions are monodispersed and consist of globular particles with an identical structure, namely, micelles. These results agree very well with our previous reports [4–6] that the ganglioside aggregates without proteins from a highly monodispersed micellar solution in shape and in dimension. As we showed in our previous paper [6], under the present experimental condition the effect of the repulsive interparticle interaction between micelles on the scattering curves is rather minor, which can be neglected. By adding the albumins to the ganglioside suspensions the scattering curves vary, losing those initial characteristics. Below the protein/ganglioside ratios of 1/1 for G_{M1} and 1.25/1 for G_{D1a}, the additions of gal-albumins induce change of the scattering curves, shifting the minimum and rounded peak positions, which are almost smeared above the ratios of 2/1 and 2.5/1. The fast increasing tendencies of the scattering intensities below $q \sim 0.04 \, \text{Å}^{-1}$ show that the complexation of gangliosides with albumins induces the formation of polydispersed large aggregates at high protein concentrations, except for the case of the complexation of the G_{M1} gangliosides with mal-albumins. In the case of the addition of mal-albumins, the scattering curve for the G_{D1a} suspension shows a minimum and a rounded peak in spite of the addition of mal-albumins; in contrast, the addition of mal-albumins greatly affects the scattering curve of the G_{M1} suspension from the lowest protein/ganglioside ratio of 0.5/1.

Figure 2 shows the distance distribution functions $p(r)$ obtained by the Fourier inversion using Eq. (1) to the scattering curves in Fig. 1. The $p(r)$ profiles without proteins are characterized both by a shoulder of ~ 25–$40 \, \text{Å}$ and by a peak of ~ 71–$76 \, \text{Å}$, suggesting that the ganglioside micelle has a strong fluctuation of the intraparticle scattering density distribution. As shown previously [4, 5], such $p(r)$ profiles apparently reflect the intramicellar structure composed of ceramides and of oligosaccharide chains with sialic acids since for X-ray these different moieties have large negative and positive excess scattering densities (the so-called contrasts), respectively. The maximum dimension D_{max} of the micelle is $130 \pm 1 \, \text{Å}$ for G_{M1} micelle and $111 \pm 1 \, \text{Å}$ for G_{D1a} micelle. The estimation of gyration radius by using Glatter's method are known to be less sensitive to the presence of an interparticle interaction. Then we estimated the gyration radius R_g by the Glatter's method using Eq. (2) is $51.9 \pm 0.2 \, \text{Å}$ for G_{M1} and $44.4 \pm$

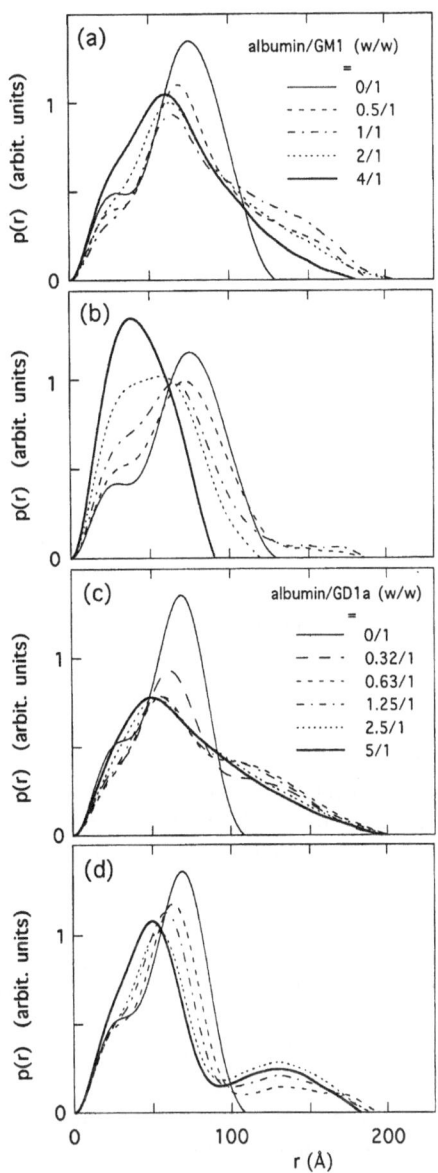

Fig. 2 Distance distribution function $p(r)$ of albumin/ganglioside mixture suspension calculated by the Fourier inversion of the scattering curve $I(q)$ in Fig. 1. (a)–(d) as in Fig. 1

$0.3 \, \text{Å}$ for G_{D1a}. In case of addition of gal-albumins to both the G_{M1} and G_{D1a} suspensions, the $p(r)$ profiles show humps around 140–$150 \, \text{Å}$ below the protein/ganglioside ratios of 1.25/1 for G_{M1} and of 1/1 for G_{D1a}, whose humps become smeared at high albumin concentrations. In case of addition of mal-albumins to the G_{D1a} suspension, the presence of the hump becomes much clear even at the highest albumin concentration. Various types of complexes would be produced; however, the appearance of these humps might be attributable to the presence of complexes with dumbelled structures as a major component [11]. In contrast to the above case, the addition of

240

M. Hirai et al.
Structural study of protein-ganglioside interaction

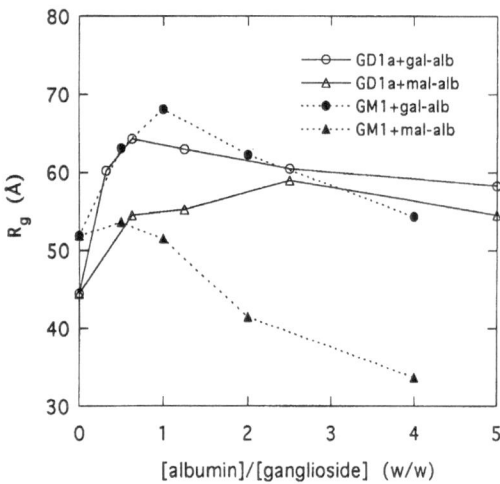

Fig. 3 Variation of gyration radius R_g of albumin/ganglioside mixture suspension depending on protein concentration. Open and full marks correspond to albumin/G_{D1a} and albumin/G_{M1} mixtures, respectively, (○) and (●), albumin-2-amido-2-deoxy-D-galactose; (△) and (▲), albumin-N-1-[deoxymaltitol]

mal-albumins to the G_{M1} suspension significantly changes the $p(r)$ profile, especially at high protein concentration to show a small globular structure. Thus, the interaction between G_{M1} gangliosides and mal-albumins can be assumed to be sufficiently strong to destroy the initial structure of the ganglioside aggregates, compared with those in other cases.

The protein concentration dependency of the R_g values estimated by the Glatter's method using Eq. (2) are shown in Fig. 3, which also reflect the above situation. In case of addition of gal-albumins, the R_g value increases significantly below the protein/ganglioside ratio of $\sim 1/1$ and gradually decreases with increasing protein concentrations. In case of addition of mal-albumins, the R_g value gradually increases for G_{D1a} and, contrarily, decreases monotonously for G_{M1} with increasing protein concentration.

Complexation of type-III gangliosides with albumins measured by neutron scattering

Figure 4 shows the neutron scattering curves $I(q)$ depending on the type-III albumin/ganglioside ratio, where (a) and (b) correspond to the complexation of the gangliosides with gal- and cel-albumins, respectively. In the present neutron scattering experiments, we observed the complexation process at low protein concentration below the protein/ganglioside ratio of 0.5/1 by using 41% v/v D_2O/H_2O solvent, therefore, the scattering curve only reflects the change of the ganglioside aggregate structure since the 41% v/v D_2O/H_2O solvent is the contrast

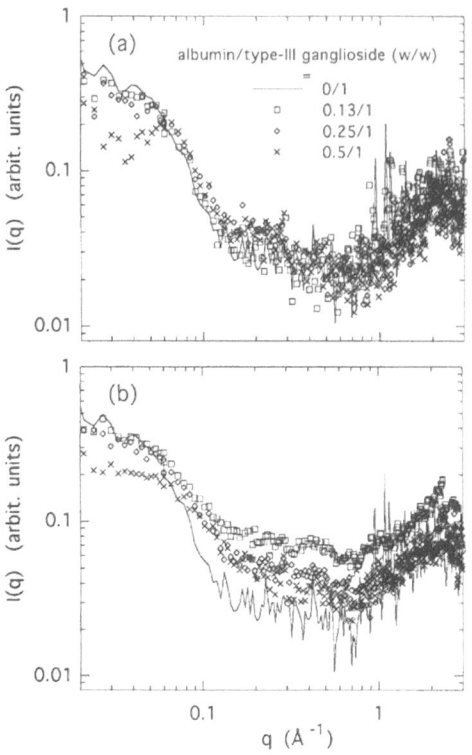

Fig. 4 Neutron-scattering curve $I(q)$ of albumin/type-III ganglioside mixture suspension in 50 mM hepes buffer (41% v/v D_2O/H_2O) at pH 7.0 at 25 °C. 41% v/v D_2O/H_2O is the contrast matching point of albumin. The ganglioside concentration is 1.0% w/v and the albumin concentration is varied from 0 to 0.5% w/v. Two different types of albumins were used; (a), albumin-2-amido-2-deoxy-D-galactose; (b), albumin-N-1-[deoxycellobiitol]

matching point of albumin [7]. The difference of the effect of the additions of the above albumins on the ganglioside aggregate structures is clearly recognized in Fig. 4. Thus, the gel-albumin hardly affects the aggregate structure below the protein/ganglioside ratio of 0.25/1; the cel-albumin gradually changes the aggregate structure with increasing the protein concentration. These results evidently show the difference of the binding affinity of gangliosides with albumins depending on the protein-surface modifications.

Conclusion

As we have shown previously [4–6], the number of sialic acid residues and the interaction between the oligosaccharide chains are essentially important to determine the shape and dimension of the ganglioside aggregate structure through a conformational change of the oligosaccharide chains. In the present results, we have partly clarified that the characteristics of the complexation processes

Progr Colloid Polym Sci (1997) 106:237–241
© Steinkopff Verlag 1997

241

of gangliosides with proteins, i.e., the interactions of mono- and disialogangliosides with proteins modified by monosaccharides occur less destructively for ganglioside aggregate structures at low protein concentrations, and on the other hand, the interaction of monosialogangliosides with proteins modified by disaccharides occurs very destructively. As gangliosides are considered to exist in the outer-biomembranes heterogeneously, we are now carrying out further experiments to elucidate an interaction of proteins with ganglioside/lipid mixtures by using scattering techniques and calorimetry.

Acknowledgements This work was done under the approval of the Photon Factory Programme Advisory Committee (Proposal No. 93G047 & 95G084 & 96G157) and was partly supported by a Grant-in-Aid for Scientific Research from the Ministry of Education, Science and Culture of Japan (No. 08680712).

References

1. Svennerholm L, Asbury AK, Reisfeld RA, Sandhoff K, Suzuki K, Tettamanti G, Toffano G (eds) (1994) Biological Function of Gangliosides. Elsevier, Amsterdam
2. Tettamanti G, Sonnion S, Ghidoni R, Masserini M, Venerando B (1985) In: Corti M, Degiorgio V (eds) Physics of Amphiphiles: Micelles, Vesicles and Microemulsions. North-Holland, Amsterdam, pp 607–636
3. Sonnion S, Cantù L, Corti M, Acquotti D, Venerando B (1994) Chem Phys Lipids 71:21–45
4. Hirai M, Takizawa T, Yabuki S, Nakata Y, Hayashi K (1996) Biophys J 70: 1761–1768
5. Hirai M, Takizawa T, Yabuki S, Hirai T, Hayashi K (1996) J Phys Chem 100: 11675–11680
6. Hirai M, Takizawa T, Yabuki S, Hayashi K (1996) J Chem Soc Faraday Trans 92:4533–4540
7. Hirai M, Takizawa T, Yabuki S, Nakata Y, Mitomo H, Hirai T, Shimizu S, Kobayashi K, Furusaka M, Hayashi K (1995) Physica B 213, 214:751–753
8. Takizawa T, Hirai M, Yabuki S, Nakata Y, Takahashi A, Hayashi K (1995) Thermochim Acta 267:355–364
9. Gammack DB (1963) Biochem J 88: 373–383
10. Furusaka M, Suzuya K, Watanabe N, Osawa M, Fujikawa I, Satoh S (1992) KEK Progress Report 92-2:25–27
11. Glatter O (1982) In: Glatter O and Kratky O (eds) Small Angle X-Ray Scattering. Academic Press, London, pp 119–196

Progr Colloid Polym Sci (1997) 106:242–244
© Steinkopff Verlag 1997

BIOCOLLOIDS

R. Morikawa
Y. Saito
H. Hyuga

Monte Carlo study of shape changes of liposomes at finite temperature

Received: 13 October 1996
Accepted: 6 March 1997

Dr. R. Morikawa (✉)
School of Life Science
Tokyo University of Pharmacy and
Life Science
1432-1 Horinouchi, Hachioji
Tokyo 192-03, Japan

Y. Saito · H. Hyuga
Department of Physics
Faculty of Science and Engineering
Keio University, 3-14-1 Hiyoshi
Yokohama 223, Japan

Abstract Monte Carlo simulation of the three-dimensional liposome is performed under the restriction of an axial symmetry. The self-avoiding effect of the membrane surface is taken into account. At finite temperatures and pressures we obtain various shapes of a liposome such as a sphere, a dumbbell, a discocyte and a triangular-like shape. By annealing the liposome to a very low temperature, the ground state of the system is obtained. On increasing the pressure the shape of the ground state is found to change from the sphere to the dumbbell and then to the discocyte. The sphere–dumbbell transition agrees with the one obtained previously by Jenkins using the variational calculation of the continuum elastic energy.

Key words Membrane – liposome – Monte Carlo – bending rigidity

Liposome is a closed vesicle with the lipid-bilayer membrane, which takes a variety of shapes such as biconcave discocytes, cup-shaped stomatocytes and prolate and oblate ellipsoids, depending on the temperature and the osmotic conditions [1]. Among various attempts made to explain these shapes [2, 3], Helfrich has discussed the bending elastic energy [4] of fluid membranes formed by lipids as

$$F = \frac{\kappa}{2} \int (c_1 + c_2 - c_0)^2 \, \mathrm{d}A + p \int \mathrm{d}V + \lambda \int \mathrm{d}A \,, \tag{1}$$

where κ is the bending rigidity and c_1, c_2 and c_0 are the two principal curvatures and the spontaneous curvature, respectively. The Lagrange multipliers p and λ take account of the constraints of constant volume and area. Jenkins investigated numerically the local minimum configuration of F under the condition of axial symmetry and $c_0 = 0$ [5]. He found four types of the liposome shape, i.e. a sphere, a discocyte, a dumbbell and a shape like a triangle. The dumbbell shape is found more stable than the discocyte. These solutions correspond to the shapes of the liposome with no thermal fluctuation.

In order to obtain various shapes of the liposome under thermal fluctuation, we perform Monte Carlo simulation of liposome at a finite temperature. We construct a tethered–bead model corresponding to Helfrich's continuous model of Eq. (1). In addition, the model is specialized to study axisymmetric deformations of a liposome, as axisymmetric shapes of a liposome are often observed in experiments [6]. With an axial symmetry, only half of the cross section is sufficient to describe the conformation of liposome. The model is then constructed by joints $i = 1-N$ and segments $i = 1-N-1$. The segment i connects the neighboring joints i and $i + 1$ (Fig. 1). The two end joints $i = 1$, N move on the z-axis of symmetry and the intermediate joints $i \, (\neq 1, N)$ move on the positive r half-plane. In order to incorporate the self-avoiding effect of the membrane, joint motion which leads to the segment crossing is not accepted in the simulation. The length of the segment s_i is restricted to vary in the range $a < s_i < 2a$ where a is the unit length. In the simulation, when $s_i \leq a$ and the distance between the second neighboring joints $i - 1$ and $i + 1$ is shorter than $2a$, the joint i is removed and the new segment is formed between the joint $i - 1$ and

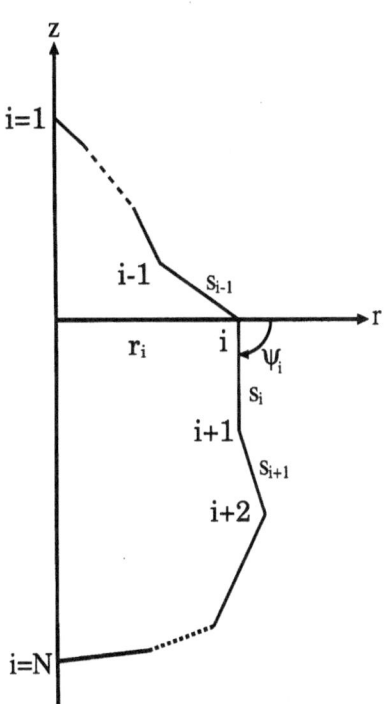

Fig. 1 Schema of the axisymmetric tethered-bead model

$i + 1$. On the other hand, when $s_i \geq 2a$, a new joint is created in the middle point between the joint i and $i + 1$. Therefore, the number of joints, N, changes during the course of the simulation.

The energy of the system E is written as

$$E = \frac{\kappa}{2} \sum_{i=1}^{N} (c_{1i} + c_{2i} - c_0)^2 \Delta A_i + p \sum_{i=1}^{N-1} \Delta V_i$$
$$+ \gamma T \left(\sum_{i=1}^{N} \Delta A_i - A_0 \right)^2 . \tag{2}$$

The first term represents the curvature elastic energy of the membrane with the bending rigidity κ. c_{1i}, c_{2i} are two principal curvatures at a joint i written as

$$c_{1i} = \frac{2 \sin(\psi_i - \psi_{i-1})}{\sqrt{s_{i-1}^2 + s_i^2 + 2s_{i-1}s_i \cos(\psi_i - \psi_{i-1})}},$$
$$\tag{3}$$

$$c_{2i} = \frac{\sin \psi_i}{r_i}, \quad (i \neq 1, N),$$

$$c_{11} = c_{21} = \frac{2 \sin \psi_1}{s_1}, \qquad c_{1N} = c_{2N} = \frac{2 \sin \psi_{N-1}}{s_{N-1}}, \tag{4}$$

and c_0 is the spontaneous curvature. r_i is the distance between the joint i and the z-axis of symmetry and ψ_i is the angle variable between the segment i and the r-axis, as shown in Fig. 1.

ΔA_i is the area element around the joint i written as

$$\Delta A_i = \frac{\pi}{4} \{ (r_{i-1} + 3r_i)s_{i-1} + (3r_i + r_{i+1})s_i \} \quad (i \neq 1, N) , \tag{5}$$

$$\Delta A_1 = \frac{\pi}{4} s_1 r_2, \qquad \Delta A_N = \frac{\pi}{4} s_{N-1} r_{N-1} . \tag{6}$$

In the second term of Eq. (2), p is an osmotic pressure difference between the outer and inner media. ΔV_i is the volume element defined as a volume of the truncated cone ringed by segment i,

$$\Delta V_i = \frac{\pi}{3} (r_i^2 + r_i r_{i+1} + r_{i+1}^2)(z_i - z_{i+1}) . \tag{7}$$

The third term takes account of the constraint of constant area $A_0 = 4\pi R_0^2$ of the liposome. γT is Lagrange multiplier including temperature of the system. By taking large γ, the area of liposome changes little for any temperature T in the simulation.

In a Monte Carlo trial, a joint i is chosen randomly with a probability $\Delta A_i / \sum_{i=1}^{N} \Delta A_i$, and is tried to be shifted randomly on (r, z)-half-plane. These trials are accepted or rejected by the standard Metropolis algorithm in which the fluctuation with the energy change δE occurs with a Boltzmann weight $\exp(-\delta E / T)$ at a temperature T. The maximum value of shift is so chosen that about 50% of the shift trials are accepted. Simulations are run up to 4×10^6 Monte Carlo steps (MCS) at various osmotic pressure p. Parameters chosen are $R_0/a = 20.0$, $c_0 = 0$, $\kappa = 1/\pi$ and $\gamma = 1.0$. The energy of a spherical liposome is $E_0 = 8$ with these parameter values.

To realize the equilibrium state at each temperature, we anneal the liposome from $T = 0.05$ to $T = 0.001$ at various p's. The annealing proceeds as follows: First, we prepare a spherical liposome and perform simulation at $T = 0.05$ up to 4×10^6 MCS. Then simulation is continued at a lower temperature $T = 0.02$ for the same MCS. Likewise, it is iterated by lowering temperatures as $T = 0.01$, 0.005, 0.002 down to $T = 0.001$. Figure 2 shows a reduced volume of the liposome $\langle V \rangle / V_0$ against a scaled osmotic pressure $\bar{p} \equiv pR_0^3 / 2\kappa$ at various temperatures T. Here $\langle V \rangle$ is a statistical average of the volume. A curve in Fig. 2 depicts the ground state value obtained by Jenkins by the variational method. It is unity for $0 \leq \bar{p} \leq 5.69$ and decreases sharply for larger $\bar{p}$. Although the linear stability analysis of Eq. (1) predicts an instability of spherical shape at $\bar{p} = 6$ [7,8], Jenkins predicted the first-order phase transition at a lower osmotic pressure. The result of our simulation approaches to the curve of Jenkins when the temperature T is lowered.

The shape of the liposome changes on increasing the osmotic pressure as shown in Figs. 3. At $T = 0.001$ and for $0 \leq \bar{p} < 5$, the shape of the liposome is spherical (Fig. 3a)

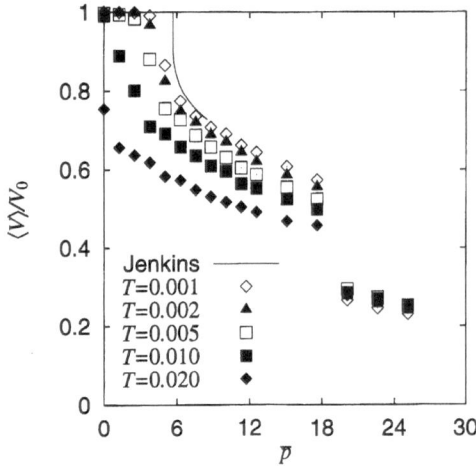

Fig. 2 Reduced volume V/V_0 versus reduced $\bar{p}$. A curve represents the variational calculation of the continuum elastic energy by Jenkins

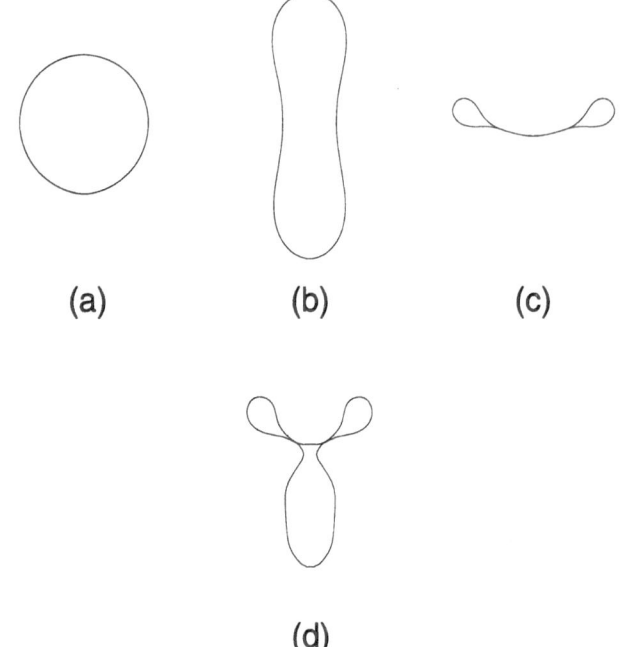

Fig. 3 Shapes of a vesicle under various pressures $\bar{p}$ and temperatures T: (a) A spherical shape ($\bar{p} = 0$, $T = 0.001$); (b) a dumbbell ($\bar{p} = 10.05$, $T = 0.001$); (c) a discocyte ($\bar{p} = 25.13$, $T = 0.001$); and (d) a shape like a triangle ($\bar{p} = 20.11$, $T = 0.05$)

and fluctuates little. At $\bar{p} = 5$–6, the fluctuation of the shape becomes large and then the spherical shape changes into the dumbbell shape (Fig. 3b) for $6 < \bar{p} < 18$. This shape behavior was also found by Jenkins. At a still larger pressure $\bar{p} > 18$, the second shape transition is observed to the discocyte (Fig. 3c). This behavior has not been discovered previously. In addition, a shape like a triangle (Fig. 3d) which is observed as the asymmetric family bifurcation by Jenkins [5] is also realized, but only as an excited state at $T = 0.05$ and $\bar{p} \simeq 20$.

In summary, at the lowest temperature, we successfully recover the ground-state morphology of the membrane

such as a sphere, a dumbbell and a discocyte which is obtained previously by the variational method [2, 5]. At a high temperature $T = 0.05$ and high pressures $\bar{p} \simeq 20$, a triangular shape is observed, which looks similar to the shape observed experimentally [6].

References

1. Leibler S (1989) In: Statistical Mechanics of Membranes and Surfaces. World Scientific, Singapore, pp 45–103
2. Deuling HJ, Helfrich W (1976) J Phys (Paris) 37:1335–1345
3. Seifert U, Berndl K, Lipowsky R (1991) Phys Rev A 44:1182–1202
4. Helfrich W (1973) Z Naturf C 28:693–703
5. Jenkins JT (1977) J Math Biol 4:149–169
6. Hotani H (1984) J Mol Biol 178:113–120
7. Zhong-can OY, Helfrich W (1987) Phys Rev Lett 59:2486–2488
8. Zhong-can OY, Helfrich W (1989) Phys Rev A 39:5280–5288

Progr Colloid Polym Sci (1997) 106:245–248
© Steinkopff Verlag 1997

BIOCOLLOIDS

A. Goto
H. Hakamata
Y. Kuwahara
R. Goto
P. Walde
P.L. Luisi
T. Imae

Functional nano-structure of aggregates self-organized on the liquid/solid interface–enzymatic polymerization of ADP

Received: 13 October 1996
Accepted: 31 March 1997

Dr. A. Goto (✉)
School of Informatics
University of Shizuoka
52-1 Yada, Shizuoka-shi
Shizuoka-ken, 422, Japan

H. Hakamata · Y. Kuwahara · R. Goto
Graduated School
University of Shizuoka

P. Walde · P.L. Luisi
Polymer Institut in ETH

T. Imae
Fac. Sci. in Nagoya Univ.

Abstract The enzymatic polymerization of ADP was carried out in sodium bis(2-ethylhexyl)sulfosuccinate (AOT)-reversed micellar solutions. The poly (adenylic acid), poly(A), being formed in the water pools precipitated out of the AOT solution together with the enzyme, whose activity was maintained for a long time. The process of the precipitation was studied in comparison with the polymerization in cationic surfactant reversed micelles and the precipitate aggregates were observed by atomic force microscopy (AFM).

Key words Poly(adenylic acid) (poly(A)) – sodium bis(2-ethylhexyl) sulfosuccinate reversed micelle – liquid/solid interface – polynucleotide phosphorylase (PNPase) – atomic force microscopy (AFM)

Introduction

Reversed micellar enzymology has been studied by many researchers [1, 2]. The conversion of apolar compounds by enzymes entrapped in aqueous cores of reversed micelles has drawn the most attention. Reversed micellar enzymology requires not only a large-scale operation, but also enzyme and product recoveries for its practical application. There is still a lack of understanding of many aspects of enzyme behavior in reversed micelles.

We have studied the enzymatic polymerization of ADP in a sodium bis(2-ethylhexyl) sulfosuccinate, AOT, reversed micellar system [3]. This enzymatic polymerization by polynucleotide phosphorylase(PNPase) and Mg^{2+} ion was greatly different from the general aspect of the enzy-matic reactions in reversed micellar solutions [3]. This enzymatic reaction in an AOT reversed micellar solution showed that the poly(A), poly(adenylic acid), being formed in the water pools precipitated out together with the enzyme, whose activity was maintained for a long time. The process of the precipitation in AOT-reversed micelles was compared with the polymerization in cationic surfactant-reversed micelles and the precipitated aggregates (PNPase, poly(A) and AOT in various molar fractions) were observed by atomic force microscopy (AFM).

Experimental

AOT and PNPase from *Micrococcus luteus*, commercial products from Sigma were used. Hexadecyltrimethyl ammonium chloride (HTAC, Merck) was recrystallized in a mixed solvent of MeOH and acetone.

The PNPase-catalyzed synthesis of poly(A) in AOT-reversed micelles was carried out according to the method described in the previous paper [3]. The reaction progress was followed by measuring an unreacted ADP [3]. After a 50 μl of the reaction mixture was added to 1 ml of 1 N HClO₄, the solution being centrifuged to precipitate poly(A), the absorbance of unreacted ADP was estimated by measuring the absorbance of a mixture of the supernatant and MeOH (1:1) at 260 nm. The synthesis of poly(A) in 0.2 M HTAC/iso-octane:octanol (88:12) reversed-micellar solution was carried out in a similar manner to that in the AOT and the reaction progress was measured by the procedure described for AOT with the exception that a 50 μl of the HTAC reaction solution was dissolved in 0.5 ml of 1% n-octyl-β-D-glucopyranoside solution to prevent an interaction between ADP and HTAC.

HPLC measurement was carried out to detect the produced poly(A) on a Shimazu LC-10A with a nucleogen-DEAE 4000-7 anion exchange column in the same manner as described in the previous paper [3]. After 3 ml of the reaction solution was separated into the supernatant and the precipitate or residue, the solvents being evaporated, 1 ml of ethanol was added to deposit the poly(A). The precipitate was dissolved in 0.5 ml of 50 mM Tris buffer (pH 7) and chloroform was added to remove AOT. After the centrifugation of the solution, the above procedure was repeated. The precipitated poly(A) was dissolved in 100 μl of 10 mM Tris-1 mM EDTA and the aliquot (25 μl) was injected into the HPLC column. The maximum base number for poly(A) which could be analyzed on the column was 30 bases, but poly(A) above that could not be estimated.

The samples for electrophoresis were first prepared according to the procedure of the HPLC sample preparation described above. After the solution was treated with a 1000 μl of ISOGEN solution containing thiocyanate and phenol to remove the PNPase, and with chloroform to remove AOT, 2 ml of 2-propanol was further added to the aqueous phase with the addition of a small amount of glycogen as a coprecipitation agent. The centrifugation and the addition of ethanol were repeated to refine the poly(A). The size of the poly(A) was estimated under 1.2% agarose denaturating gel electrophoresis [4] by using SYBR Green II at 254 nm. RNA Molecular Weight Maker I (7.4, 5.3, 2.8, 1.9, 1.6, 1.0, 0.4, and 0.3 kb) was used. The AFM image of the precipitated aggregate was observed on a Nanoscope III from Digital Instruments using a tapping mode. A mica sheet was set at the bottom of the reaction vial containing 0.2 M AOT/iso-octane reaction solution for 12 h at 37 °C. After its removal, it was soaked in

iso-octane for 30 min and then in a mixture of ethanol:water (4:1) for 30 min to remove the excess of AOT attached thickly on the mica. The mica was then dried in vacuo at room temperature for 2 h. In the case of PNPase, a mica sheet was soaked in an aqueous PNPase solution for 1 min, and the mica was dried under reduced pressure at room temperature for 2 h.

Results and discussion

It is well known that the physical properties of reversed micellar solutions depend significantly upon the molar ratio of water to surfactant, W_0 [5] and the catalytic activity of solubilized enzymes depends also to a greater or lesser extent upon the degree of W_0 [6]. The enzymatic polymerization of ADP was carried out at W_0 of 10, 20, 30 and 40 of 0.3 M AOT/iso-octane reversed micelles at 25 °C. From the reaction mixture, samples were withdrawn at given intervals and the concentration of unreacted ADP was determined, as shown in Fig. 1a, where $[OD]_t/[OD]_0$ (a ratio of OD at any time (t) to OD at the initial time (0)) was plotted against time. From this, it was assumed that the polymerization proceeded the most quickly at $W_0 = 20$. At the end of the polymerization at $W_0 = 10$, 20, and 30, the poly(A) was overwhelmingly found in the precipitate by HPLC, but every supernatant involves scarcely poly(A). The activity of the precipitated PNPase was examined by replacement with a fresh AOT micellar solution containing ADP. Figure 1b shows that since the precipitated PNPase had an activity at $W_0 = 10$, 20 and 30, respectively, but scarcely at $W_0 = 40$, the precipitation of PNPase occurred at $W_0 = 10$, 20 and 30. The polymerization in AOT-reversed micelles seems to need an appropriate size of the water pools and an appropriate ratio of free water to bound water, and the larger the W_0 and the closer it approaches emulsion, the scarcer the tendency to precipitate.

The enzymatic polymerization occurred also in cationic surfactant-reversed micelles of HTAC, as shown in Fig. 2, but it was found by HPLC and electrophoresis that no poly(A) precipitated out of the HTAC micellar solution. Therefore, negative charges of sulfonate group of AOT might play an important role in the precipitation. Gel electrophoresis of poly(A) revealed a broad band on the whole, as shown in Fig. 3. In the case of AOT reversed micelles, the size of poly(A) in the precipitate was in the range 0.6–4.0 kb, and that in the supernatant was in the range 0.4–2.5 kb. In the case of HTAC reversed micelles, no poly(A) was observed in the residue on the bottom of a glass test tube, but the size of poly(A) in the supernatant was in the range 1.0–7.4 kb. The size of poly(A) synthesized in HTAC tended to be larger than that in AOT because the

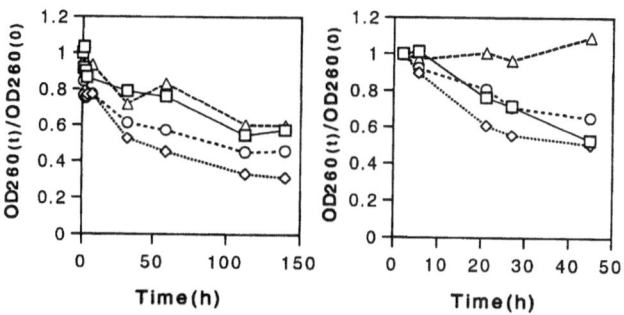

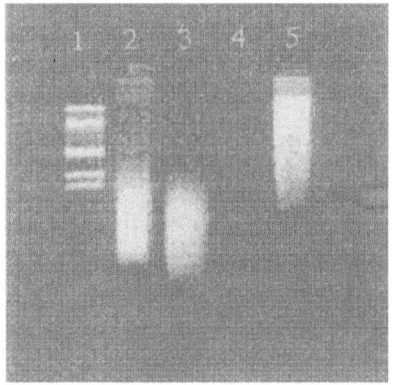

Fig. 1 (a) Effect of W_0 on PNPase catalyzed synthesis of poly(A) in 0.3 M AOT/iso-octane-reversed micelles at 25 °C: $[PNPase]_w = 0.76$ mg/ml, $[ADP]_w = 27.2$ mM, $[MgCl_2]_w = 10$ mM, $W_0 = 20$ (50 mM Tris-HCl, pH 9.5). (b) Activities of the PNPase in the precipitate at various W_0 at 25 °C. After the above reaction proceeded for 140 h, the supernatant was replaced by a fresh AOT solution containing ADP $\{[ADP]_w = 14$ mM, $[MgCl_2]_w = 10$ mM, pH 9.5, $W_0 = 20$ (50 mM Tris-HCl)$\}$. The arrow shows the replacement of a fresh AOT solution; ($\square$) $W_0 = 10$, ($\Diamond$) $W_0 = 20$, ($\bigcirc$) $W_0 = 30$, ($\triangle$) $W_0 = 40$

Fig. 3 Gel electrophoretic pattern of poly(A) formed in AOT and HTAC micellar solutions: PNPase-catalyzed synthesis of poly(A) was carried out in 0.2 M AOT or HTAC reversed micelles of 5 ml at 25 °C: $[PNPase]_w = 1$ mg/ml, $[ADP]_w = 20$ mM, $[MgCl_2]_w = 10$ mM, $W_0 = 20$ (50 mM Tris-HCl, pH 9.5). After the polymerizations proceeded for 24 h (the yields were 70% for AOT and 60% for HTAC), the respective reaction solutions were separated into the supernatant and the precipitate, and they were treated according to the experiment. The purified poly(A) was dissolved in 50 μl of 10 mM Tris buffer (pH 7), was diluted 10 times, and the constant volume was injected. (precipitate: 1 μl for AOT and 10 μl for HTAC; supernatant; 10 μl for AOT and 0.5 μl for HTAC) 1: RNA Marker, 2: AOT (precipitate), 3: AOT (supernatant), 4: HTAC (precipitate), 5: HTAC (supernatant)

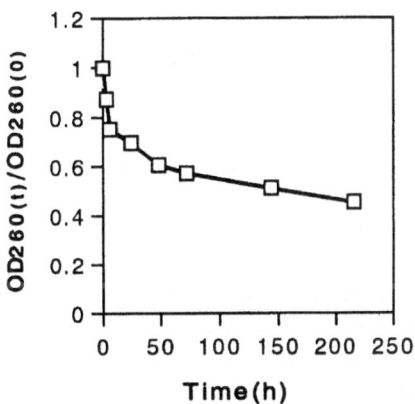

Fig. 2 PNPase-catalyzed synthesis of poly(A) in 0.2 M HTAC/iso-octane/octanol-reversed micelles at 25 °C: $[PNPase]_w = 0.83$ mg/ml, $[ADP]_w = 11$ mM, $[MgCl_2]_w = 10$ mM, $W_0 = 20$ (50 mM Tris-HCl, pH 9.5)

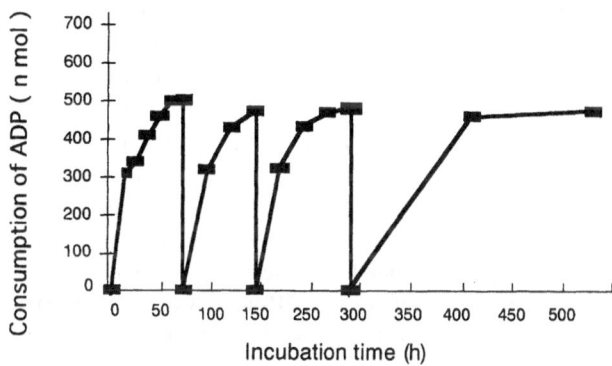

Fig. 4 Activities of PNPase on the interface of glass/0.2 M AOT-reversed micellar solution. After the reaction (the total volume of 0.2 M AOT solution, 3 ml, $[PNPase]_w = 1$ mg/ml, $[ADP]_w = 10$ mM, $[MgCl_2]_w = 10$ mM, $W_0 = 20$ [50 mM Tris-HCl, pH 9.5]) proceeded for a definite time, the supernatant was removed and the precipitate was rinsed three times with AOT solutions containing buffer ($W_0 = 20$) without ADP, and a fresh AOT solution containing ADP ($[ADP]_w = 10$ mM, $[MgCl_2]_w = 10$ mM, $W_0 = 20$ [50 mM Tris-HCl, pH 9.5]) was added to the precipitate. The arrows show an addition of a fresh AOT solution. The ADP was determined from the absorbance at 259 nm ($\varepsilon_{max} = 15.4 \times 10^3$)

polymerization proceeds on the interface of the water pool due to the electric attraction between the positive HTAC monolayer surrounding the water pool, and negative poly(A).

The addition of a fresh AOT-reversed micellar solution containing ADP to the precipitate resulted in a new start to the reaction, showing that the PNPase precipitated with the poly(A), as shown in Fig. 1b. Interestingly, each repeated addition of a fresh reversed micellar solution to the precipitates resulted in a new polymerization of ADP. The activity of the PNPase on the interface of the glass surface/0.2 M AOT solution was then examined, as shown in Fig. 4. The ca. 500 nmol of ADP was consumed per 72 μg PNPase, in which 720 nmol of the ADP was contained as a substrate in 1 ml of 0.2 M AOT reaction solution. This

revealed that the yield was maintained to be constantly ca. 70% for 3 weeks, and poly(A) was successively accumulated on the glass surface. Therefore, it is reasonable to consider that the precipitate corresponds to functional aggregates.

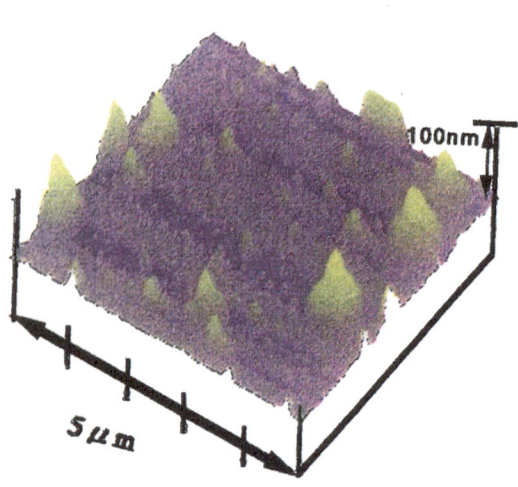

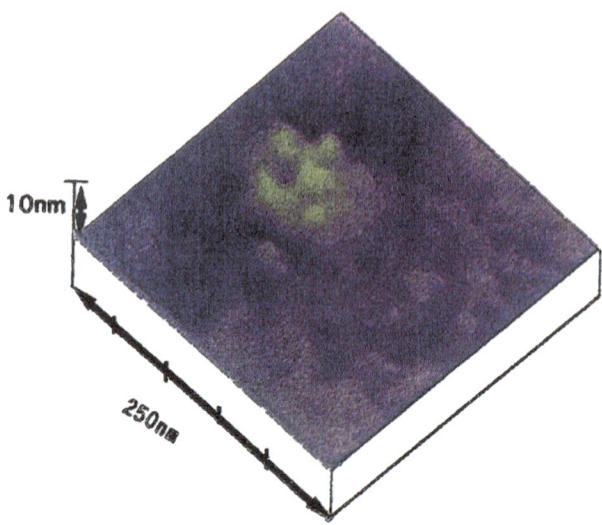

Fig. 5 AFM image of PNPase-catalyzed synthesis of poly(A) on the solid surface. A mica sheet was soaked in 0.2 M AOT/iso-octane solution ([PNPase]$_w$ = 1 mg/ml, [ADP]$_w$ = 10 mM, [MgCl$_2$]$_w$ = 10 mM, W$_0$ = 20 [50 mM Tris-HCl, pH 9.5]) at 37 °C for 12 h

Fig. 6 AFM image of PNPases. [PNPase] = 0.1 mg/ml (50 mM Tris-HCl, pH 9.5)

The nonostructure of the functional precipitate was studied by AFM. Figure 5 shows the AFM image of the precipitated product. Large particles with sizes 200–500 nm and heights of 100 nm were observed, but the control picture scarcely showed such particles (data not shown). Such large particles must be covered thickly with AOT layers because they can be observed by rinsing of the AOT. The PNPase dissolved in the buffer was observed by AFM. As shown in Fig. 6, the AFM image of PNPase molecules in the absence of ADP shows several masses, of which a unit corresponds to be ca. 20 nm. The unit can be assumed to be a trimer of PNPase because it is known that PNPases forms a trimer [7]. Since the 85 Å [8] PNPase molecules in the absence of ADP were found by electron microscopy, the size in the AFM image seems to be larger.

One of the factors is considered; the PNPase sample for AFM seems to be more hydrated compared with that for electron microscopy. The above result shows that the PNPases are much smaller than the precipitated articles in Fig. 5, and the PNPases and the poly(A) which form the particles could not be discriminated by AFM.

It may be concluded that the particles in the precipitate observed by AFM are nano-structured functional aggregates which are self-organized on the interface between the solid phase of glass and the liquid phase of AOT-reversed micellar solution. Our results show not only a novel synthesis of biopolymer on the interface of solid/oil liquid, but also a possibility of enzyme and product recovery which can lead to a large-scale application procedure.

References

1. Luisi PL, Gimini M, Pileni MP, Robinson H (1988) Biochim Biophys Acta 7:209–246
2. Hilhorst R (1989) In: Pileni MP (ed) Structure and Reactivity in Reverse Micelles Elsevier p 323–341
3. Walde P, Goto A, Monnard P-A, Wessicken MJ, Luisi PL (1994), J Am Chem Soc 116:7541–7547
4. Lehrach H, Diamon D, Wozney JM, Boedteker H (1977) Biochemistry 16:4743–4751
5. Zulauf M, Eicke H-F (1979) J Phys Chem 83:480–486. Hauser H, Hearing G, Pande A, Luisi PL (1989) J Phys Chem 93:7869–7876
6. Martinek K, Levashov AV, Khmelnitsky YuL, Klyachko NL, Berezin LV (1986) Eur J Biochem 155:453–468
7. Barbehenn EK, Craine JE, Chrambach A, Klee CB (1982) J Biol Chem 257:1007–1016
8. Valentine RC, Thang MN, Grunberg-Manago M (1969) J Mol Biol 39:389–391

Progr Colloid Polym Sci (1997) 106:249–251
© Steinkopff Verlag 1997

BIOCOLLOIDS

K. Nakano
K. Tani
K. Sada
M. Miyata

Molecular architecture of steroidal acids and their derivatives: selective acquisition of polymorphic inclusion crystal of cholic acid

Received: 13 October 1996
Accepted: 12 March 1997

K. Nakano (✉)
Nagoya Municipal Industrial Research
Institute
Rokuban, Atsuta-ku
Nagoya 456, Japan

K. Tani · K. Sada · M. Miyata
Material and Life Science
Faculty of Engineering
Osaka University
Yamadaoka, Suita
Osaka 565, Japan

Abstract Selective acquisition of polymorphic inclusion crystals of cholic acid with acrylonitrile and methacrylonitrile is described. Two polymorphic inclusion crystals of **1** with methacrylonitrile have different hydrogen-bonded networks from those with acrylonitrile.

Key words Solvent – recrystallization – polymorph – inclusion crystal – cholic acid

Real polymorphic inclusion crystals, which consist of not only identical host-to-guest ratios but also different hydrogen-bonded networks, were found by two groups. The one deals with the crystals of deuterated urea (ND_2COND_2) with sebaconitrile [1], and the other deals with those of cholic acid(**1**) (Fig. 1) with acrylonitrile [2]. It is empirically known that selective acquisition of any desirable crystal from multiple polymorphic crystals on the basis of recrystallization accompanies poor reproducibility [3]. In fact, we happened to obtain the latter crystals by addition of a third component. But now, we can surely and selectively acquire the crystals in the presence of different kinds of butanols under definite temperatures and concentrations. This method led us to the finding of another polymorphic crystals of **1** with methacrylonitrile having different hydrogen-bonded networks from those of **1** with acrylonitrile.

Here, we report the first selective preparation of the polymorphic inclusion crystals of **1** as well as their crystal structures.

The inclusion crystals of **1** are usually produced by direct recrystallization from liquid guests (Method A) [4] and/or by using some solvents (Method B) [5]. In the case of acrylonitrile [2], Methods A and B gave the crystals with a crossing structure [6] and a bilayered structure [7], respectively. Method B enables us to control the recrystallization conditions as compared to Method A. Four different items were applied to Method B as mentioned below, leading to selective acquisition of the polymorphic crystals.

First, we checked solvent effects with a definite temperature in the presence of an equimolar amount of a guest and a solvent. For example, the inclusion crystals of **1** with acrylonitrile were prepared by using four butanol isomers as the solvents. **1** was dissolved in the hot butanols and

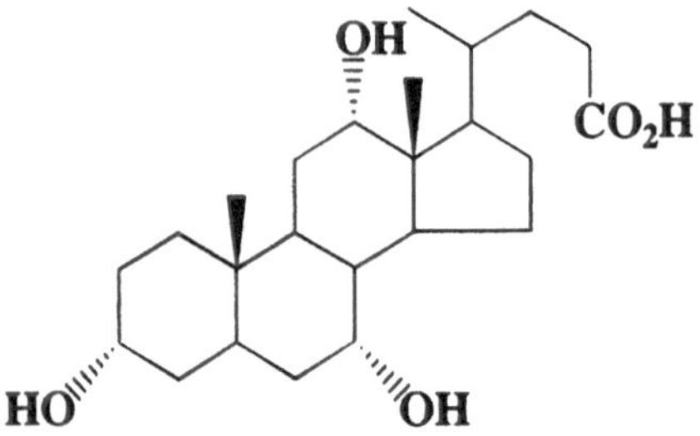

Fig. 1 Cholic Acid(**1**)

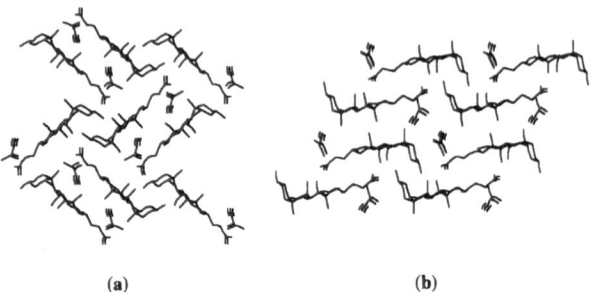

(a)   (b)

Fig. 2 Crystal structures of the inclusion compounds of **1** with methacrylonitrile (a) 1 : 1 crossing structure (b) 1 : 1 bilayered structure

cooled to 25°C. Neat acrylonitrile, which was kept at 25°C, was added into the solution. The resulting solution was maintained at 25°C. The precipitated crystals had an equimolar ratio of host-to-guest by means of ^{1}H NMR spectroscopy and thermal analysis. They were classified into two groups by means of X-ray powder diffraction. As shown in Table 1, the crossing structure appeared in the case of 2-butanol and 2-methyl-2-propanol, while the bilayered structure did in the case of 1-butanol and 2-methyl-1-propanol.

Secondly, we examined an effect of seed crystals. For example, when the solution involving 2-butanol or 2-methyl-2-propanol was seeded with the bilayered crystals, only the crossing crystals were obtained. On the other hand, when the solution involving 1-butanol and 2-methyl-1-propanol was seeded with the crossing crystals, a mixture of the crystals was obtained. These results indi-

cate that the solvents rather than the seeding crystals are more effective for determining the assembly modes.

Thirdly, we studied temperature effects. As shown in Table 1, the assembly modes of **1** with acrylonitrile depended on the temperature of recrystallization in the presence of the same solvent. In the cases of 1-butanol and 2-methyl-1-propanol, the borderline for separation of the two assembly modes lies in about 30°C. On the other hand, in the cases of 2-butanol and 2-methyl-2-propanol, the only crossing crystals were obtained at any temperatures.

As for concentration effects, the bilayered crystals were obtained in the presence of an excess amount of the liquid guests compared to the amount of the solvents.

Next, we investigated the inclusion compound of **1** with different nitriles. The resulting crystals were exam-

Table 1 Assembly modes[a] of inclusion crystals of **1** with Nitriles in the Method B

Guest	Temperature of recrystallization/°C	Solvent[b]			
		1-Butanol	2-Butanol	2-Methyl-1-propanol	2-Methyl-2-propanol
Acrylonitrile	5	Bilayered	Crossing + GF[d]	Bilayered	c)
	15	Bilayered	Crossing	Bilayered	c)
	25	Bilayered	Crossing	Bilayered	Crossing
	35	Crossing	Crossing	Crossing	Crossing
	50	Crossing	Crossing	Crossing	Crossing
Methacrylonitrile	5	Bilayered	Bilayered	Bilayered	c)
	15	Bilayered	Bilayered	Bilayered	c)
	25	Bilayered	Bilayered	Bilayered	Crossing
	35	Bilayered	Bilayered	Bilayered	Crossing
	50	Bilayered	Bilayered	Bilayered	Crossing
Propionitrile	5	Crossing	Crossing + GF	Crossing	c)
	25	GF	GF	GF	GF
Isobutyronitrile	5	Bilayered	Bilayered	Bilayered	d)
	25	Bilayered	Bilayered	GF	GF

a) Assembly modes were determined by X-ray powder diffraction as soon as the crystals were obtained
b) 100 mg of **1** was dissolved in 0.4 ml of 1-butanol, 0.5 ml of 2-butanol, 0.4 ml of 2-Methyl-1-propanol and 1 ml of 2-Methyl-2-propanol, respectively. The same amount of nitriles were added to the saturated butanol solution
c) 2-Methyl-2-propanol could not be used as solvent because of the melting point
d) GF means homocrystals of **1** without incorporation of the guest

Progr Colloid Polym Sci (1997) 106:249–251
© Steinkopff Verlag 1997

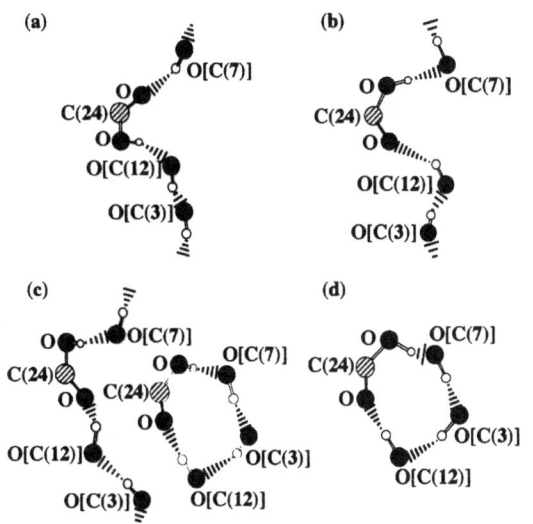

Fig. 3 Hydrogen-bonding networks of the 1:1 crossing structure of **1** with (a) methacrylonitrile and (b) acrylonitrile, 1:1 bilayered structure of **1** with (c) methacrylonitrile and (d) acrylonitrile

ined by X-ray powder diffraction method. A partial list is shown in Table 1. Methacrylonitrile gave a reproducible result with respect to the polymorphic crystals, which depend on the kind of solvents. X-ray crystallographic study confirmed that each crystal consists of a 1:1 crossing (Fig. 2a) and 1:1 bilayered structure (Fig. 2b) [8]. The former belongs to the same space group (orthorhombic,

$P2_12_12_1$) as the crossing crystal of **1** with acrylonitrile. The assembly modes are very similar, but the donor–acceptor relation of helical hydrogen-bonding networks were reverse (Fig. 3a and b). On the other hand, the latter belongs to the different space group (triclinic, P1) from the bilayered crystal of **1** with acrylonitrile (monoclinic, $P2_1$). The hydrogen-bonded networks were also different. That is, the latter crystal has both cyclic and helical networks (Fig. 3c), although most of the bilayered crystals of **1** have only cyclic one (Fig. 3d) [7].

Saturated nitriles also yielded the inclusion compounds. The inclusion crystal of **1** with propionitrile had only the crossing structure under the conditions mentioned in Table 1, while it had the bilayered structure in the presence of seed crystals or different concentrations. On the contrary, only the bilayered structure was formed in the case of isobutyronitrile.

In conclusion, this study demonstrates that the polymorphic inclusion compounds can selectively be obtained by changing the recrystallization methods. An identical unsaturated nitriles were accommodated into the different inclusion spaces. This face would enable us to evaluate an effect of molecular-level spaces for the reaction, such as inclusion polymerization.

This work was supported by Grants-in Aid for Scientific Research from the Ministry of Education, Science, Sports and Culture, Japan and by the Research Foundation for the Electrotechnology of Chubu.

References

1. Hollingsworth DM, Santarsiero DB, Harris MDK (1994) Angew Chem Int Ed Engl 33:649–652
2. Nakano K, Sada K, Miyata M (1996) J Chem Soc Chem Commun 989–990
3. Dunitz DJ, Bernstein J (1995) Acc Chem Res 28:193–200
4. Miyata M, Shibakami M, Goonewardena W, Takemoto K (1987) Chem Lett 605–608
5. Nakano K, Sada K, Miyata M (1994) Chem Lett 137–140
6. (a) Miki M, Kasai N, Shibakami M, Chirachanchai S, Takemoto K, Miyata M (1990) Acta Crystallogr C46:2442–2445; (b) Johnson LP, Schaefer PJ (1972) Acta Crystallogr B28:3083–3088; (c) Jones LE, Nassimbeni RL (1990) Acta Crystallogr B46:399–405

7. (a) Miki K, Masui A, Kasai N, Miyata M, Shibakami M, Takemoto K (1988) J Amer Chem Soc 110:6594–6596; (b) Miki K, Kasai N, Shibakami M, Takemoto K, Miyata M (1991) J Chem Soc Chem Commun 1757–1759; (c) Nakano K, Sada K, Miyata M (1995) J Chem Soc Chem Commun 953–954; (d) Shibakami M, Sekiya A (1994) J Chem Soc Chem Commun 429–430; (e) Shibakami M, Tamura M, Sekiya A (1995) J Amer Chem Soc 4499–4505; (f) Caira RM, Nassimbeni RL, Scott LJ (1994) J Chem Soc Perkin Trans 2 623–628; (g) Scott LJ (1995) J Chem Soc Perkin Trans 2 495–502

8. X-ray crystal structure analyses (a) **1**-methacrylonitrile (1:1) (−65°C), $C_{24}H_{40}O_5 + C_4H_5N$, $M = 475.67$, orthorhombic, $P2_12_12_1$, $a = 16.708(2)$, $b = 17.848(1)$, $c = 8.648(1)$ Å, $V = 2578.77(4)$ Å^3, $Z = 4$, $D_C = 1.205$ g/cm^3, $R = 0.057$, $R_W = 0.058$, 2256 unique reflections with $[|F_0| > 3\sigma(|F_0|)]$; (b) **1**-methacrylonitrile (1:1) (−65°C), $C_{24}H_{40}O_5 + C_4H_5N$, $M = 475.67$, triclinic, P1, $a = 12.533(6)$, $b = 14.16(1)$, $c = 8.277(2)$ Å, $\alpha = 90.94(4)$, $\beta = 94.86(3)$, $\gamma = 107.23(4)°$, $V = 1396(1)$ Å^3, $Z = 2$, $D_C = 1.092$ g/cm^3, $R = 0.047$, $R_W = 0.049$, 5013 unique reflections with $[|F_0| > 3\sigma(|F_0|)]$.

Progr Colloid Polym Sci (1997) 106:252–256
© Steinkopff Verlag 1997

M. Yonese
S.H. Xu
S. Kugimiya
S. Sato
I. Miyata

Light scattering studies of soluble complexes between hyaluronate and bovine serum albumin

Received: 13 October 1996
Accepted: 23 May 1997

Dr. M. Yonese (✉) · S.H. Xu
S. Kugimiya · S. Sato · I. Miyata
Faculty of Pharmaceutical Sciences
Nagoya City University
Tanabe-dori, Mizuho-ku
Nagoya 467, Japan

Abstract Soluble complexes between sodium hyaluronate (NaHA) and bovine serum albumin (BSA) were studied by dynamic light scattering DLS and an electrophoretic light scattering ELS. Various amounts of BSA were added to NaHA solution ($1.2 \, \mu\text{mol dm}^{-3} = 0.1 \, \text{w/v\%}$) and their electrical mobility U were measured under $J = 0.001 \, \text{mol dm}^{-3}$. Their spectra of U were found to be one sharp peak in the experimental concentration region (the maximum concentration: $C_{BSA} = 128 \, \mu\text{mol dm}^{-3}$). The value of U attained to a constant negative value in the region of $C_{BSA} > 120 \, \mu\text{mol dm}^{-3}$ in which a saturated soluble complex was found to be formed. From the results of viscosity and diffusion coefficients of the mixed solutions, with increasing C_{BSA} the shape of complexes was found to change from a worm-like to random-coil structure. The ratio of BSA to NaHA of the saturated complex was 120 and 18 repeating units bind to one BSA molecule, i.e., 18 carboxylic groups participate in the binding.

Key words Soluble complex – acid polysaccharide – bovine serum albumin – electrophoretic mobility – dynamic light scattering

Introduction

Acid polysaccharides have carboxylic and/or sulfuric groups. The kinds of the charged groups and the combination affect their characteristics, such as counterion bindings [1] and the complex formation with a protein [2]. Recently, the interaction and macrostructure of protein-polyelectrolyte complexes have been reported using a dynamic light scattering DLS and a electrophoretic light scattering ELS [3,4]. In this paper, the soluble complexes between sodium hyaluronate (NaHA) and bovine serum albumin (BSA) were studied by ELS and DLS. NaHA is a linear polysaccharide whose repeating disaccharide units are composed of N-acetyl-D-glucosamine and D-glucuronic acid, i.e., NaHA have one carboxylic group per repeating unit composed of two sugar rings.

Experimental

Materials

NaHA offered from Seikagaku Kogyo Co. Ltd. was used without any more purification. The weight averaged molar mass was $8.50 \times 10^5 \, \text{g mol}^{-1}$. Purified BSA was obtained by removing fatty acids from BSA Fraction V (Seikagaku kougyo) according to R.F. Chen [5]. The weight average molar mass of BSA was determined to be $7.0 \times 10^4 \, \text{g mol}^{-1}$ using a light scattering (DLS700Ar, Otsuka Electronics).

Measurement of electrophoretic mobility

Electrophoretic mobilities of NaHA, BSA and the soluble NaHA–BSA complexes were measured under various

Progr Colloid Polym Sci (1997) 106:252–256
© Steinkopff Verlag 1997

ionic strengths adjusted by NaCl ($J = 0.001–0.1$ mol dm^{-3}) at 25 °C using an electrophoretic light scattering photometer LEZA-600 (Otsuka Electronics Co.). The true mobility U was obtained by correcting the electro-osmotic effects. The changes of U due to pH were measured using the cell for pH titration equipped with an autotitrator.

Measurement of diffusion coefficient and hydrodynamic diameter

Diffusion coefficients of the HA–BSA complexes were measured also at 25 °C in the range of angle $\theta = 35–90°$ using a dynamic light scattering DLS (DLS 700Ar: Otsuka Electronics Co., $\lambda = 188$ nm).

Measurement of viscosity

Viscosity of the mixed solutions composed of NaHA and BSA were measured at 25 °C using Ostward-type viscometer in which the falling time of water was 92 s.

Results

Electrical mobility of hyaluronate and bovine serum albumin

Electrical mobility U of NaHA was measured in the region of pH = 1.5–10 at 25 °C under various ionic strengths adjusted by NaCl ($J = 0.001–0.05$ mol dm^{-3}). As one of the typical examples, the spectrum of mobility at pH = 7.16 and $J = 0.001$ mol dm^{-3} is shown in Fig. 1b and was found to be one sharp peak. The peak value was taken as the mobility. The values of U obtained under various J are shown in Fig. 1a as a function of pH. As NaHA has carboxylic groups, the values of U decreased with increasing pH and in the region of pH > 5 became almost constant. With increasing ionic strengths, the absolute values of U decreased due to the shielding effects around carboxylic groups.

Figure 1a shows also the mobilities of sodium alginate (NaAlg) and chondroitin 4-sulfate (NaChS-A) under $J = 0.001$ mol dm^{-3}. Their absolute values of them were found to be more than that of NaHA in alkaline pH region. This is considered to result from the difference of their charge densities. NaAlg has one carboxylic group per one sugar ring, which is two times as much as that of NaHA. NaChS-A has a carboxylic and a sulfuric groups per repeating unit. Then, the densities of charged groups of NaAlg and NaChS-A are two times as much as that of NaHA in the alkaline region. However, the difference of U was found to

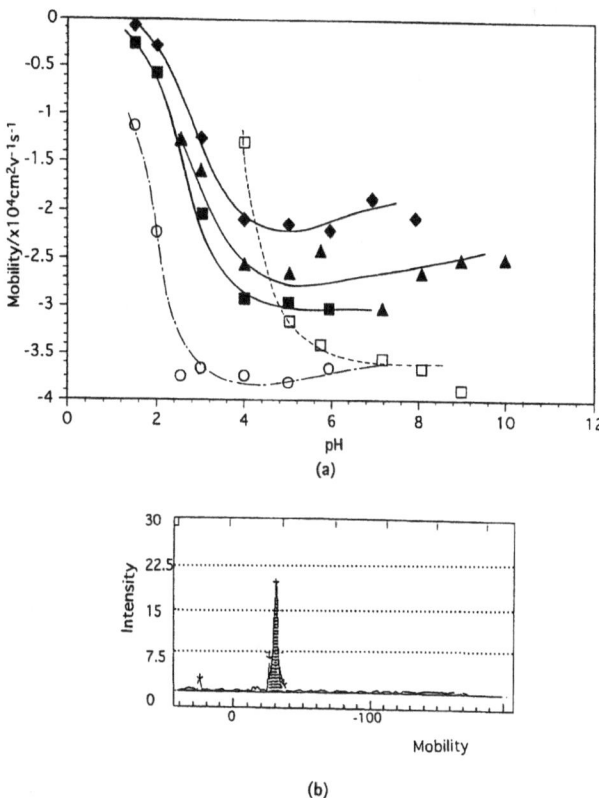

Fig. 1 Electrical mobility of NaHA, NaAlg and NaChS-A under various ionic strengths. (a) Effect of pH on electrical mobility: NaHA under $J = 0.001$ mol dm^{-3} (■), $J = 0.01$ mol dm^{-3} (▲), $J = 0.1$ mol dm^{-3} (◆). NaAlg under $J = 0.001$ mol dm^{-3} (○). NaChS under $J = 0.001$ mol dm^{-3} (□). (b) Spectrum of NaHA mobility. $C_{HA} = 1.2$ μmol dm^{-3}, pH = 7.16, $J = 0.001$ mol dm^{-3}

be not so much, i.e., in the alkaline region the values of NaHa was -3.1, and those of NaAlg and NaChS-A were -3.8 cm^{-2}V^{-1}s^{-1}.

Electrical mobility U of BSA was measured in the region of pH = 3–11 at 25 °C under various ionic strengths adjusted by NaCl ($J = 0.001–0.05$ mol dm^{-3}). Figure 2b shows the spectrum of U at pH = 5.0 and $J = 0.001$ mol dm^{-3} as a typical result. The values of U decreased with increasing pH as shown in Fig. 2a and with increasing J the absolute values of U decreased. The isoelectric point was found to be pH = 5.2 irrespective of J.

Electrical mobility of NaHA-BSA complex

Various amounts of BSA ($C_{BSA} = 0–128$ μmol dm^{-3}) were added to the NaHA solution of which concentration was 0.1 w/v% ($C_{HA} = 1.2$ μmol dm^{-3}) and that of the carboxylic groups was 2.5 mmol dm^{-3}. Phase separation did not occur even at the maximum C_{BSA}. The electrical mobilities of the mixed solutions were measured under

$J = 0.001 \text{ mol dm}^{-3}$ and their spectra of U were found to be one sharp peak in the experimental concentration region as shown in Fig. 3 for the cases of $C_{BSA} = 42$ and $C_{BSA} = 128 \text{ μmol dm}^{-3}$. In the case of $C_{BSA} = 128 \text{ μmol dm}^{-3}$, small shoulders appeared at the lower region of the

absolute value of U, which shows the presence in small amounts of the complexes containing more BSA than that of the peak. The peak of BSA near at $U = 0$ can be neglected. Then, almost all BSA molecules added to the NaHA solution should bind to NaHA molecules and form the soluble complexes.

Figure 4 shows the results of U of the complexes as a function of the concentration of BSA C_{BSA}. In the region of $C_{BSA} < 60 \text{ μmol dm}^{-3}$, the values were almost constant. In the region of $C_{BSA} > 60 \text{ μmol dm}^{-3}$, they increased with increasing C_{BSA} and attained the constant value at $C_{BSA} = 122 \text{ μmol dm}^{-3}$ at which a saturated complex is considered to be formed.

The estimated numbers of BSA molecules bound to a NaHA molecule n_{BSA}/n_{HA} are shown in an upper abscissa of Fig. 4. The complexes in the region of $n_{BSA}/n_{HA} = 0$–50 were found to show almost a constant negative mobility $(-3.3 \text{ cm}^2 \text{ V}^{-1} \text{ s}^{-1})$. The value of n_{BSA}/n_{HA} of the

Fig. 2 Electrical mobility of BSA under various ionic strengths. (a) Effect of pH on electrical mobility $J/\text{mol dm}^{-3}$: 0.001 (■), 0.01 (●), 0.03 (▲), 0.05 (◆). (b) Spectrum of BSA mobility $C_{BSA} = 7.14 \text{ μmol dm}^{-3}$, pH = 5.0, $J = 0.001 \text{ mol dm}^{-3}$

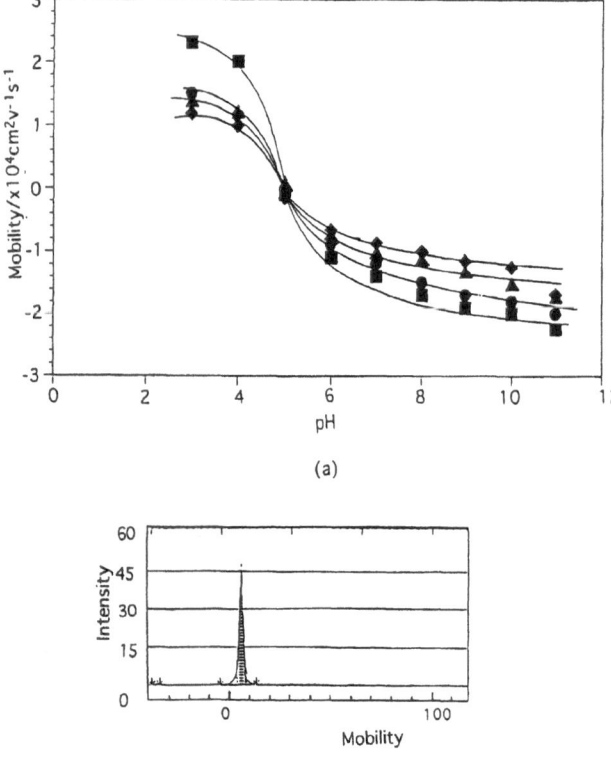

(a)

(b)

Fig. 4 Electrical mobility of soluble complexes as a function of pH. $C_{HA} = 1.2 \text{ μmol dm}^{-3}$, $J = 0.001 \text{ mol dm}^{-3}$

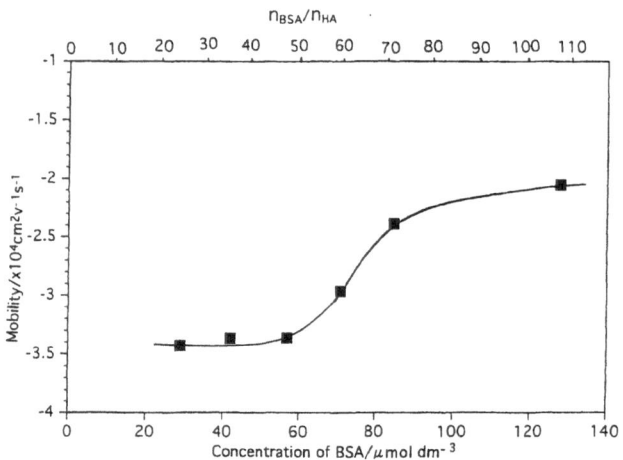

Fig. 3 Spectra of mobility of soluble complexes (a) $C_{HA} = 1.2 \text{ μmol dm}^{-3}$, $C_{BSA} = 42 \text{ μmol dm}^{-3}$, $n_{BSA}/n_{HA} = 35$; (b) $C_{HA} = 1.2 \text{ μmol dm}^{-3}$, $C_{BSA} = 128 \text{ μmol dm}^{-3}$, $n_{BSA}/n_{HA} = 107$

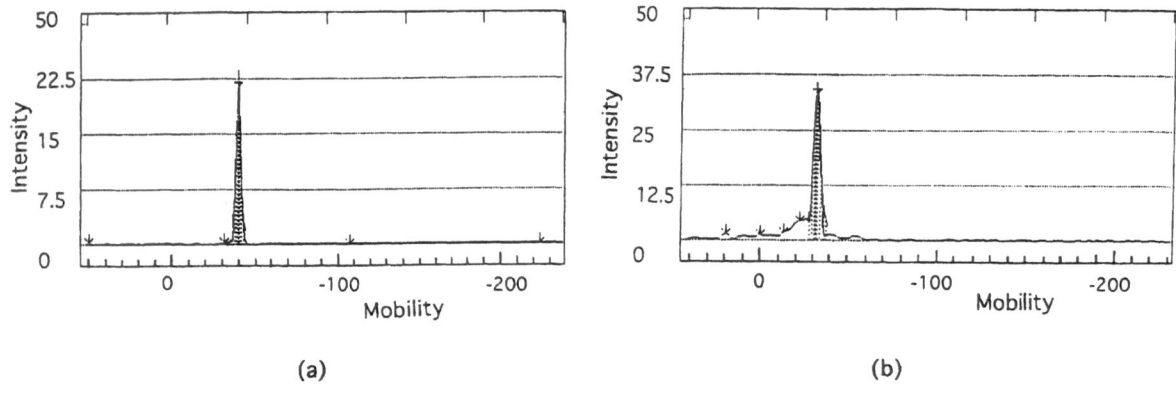

(a)

(b)

Progr Colloid Polym Sci (1997) 106: 252–256
© Steinkopff Verlag 1997

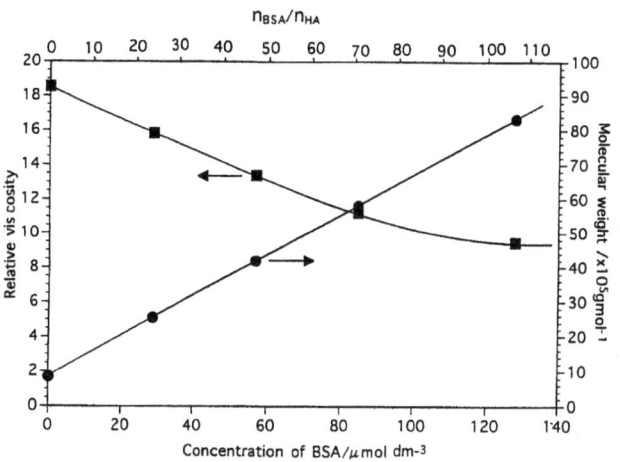

Fig. 5 Relative viscosity of mixed solutions composed of BSA and NaHA, and molar masses of soluble complexes. $C_{HA} = 1.2\ \mu\mathrm{mol\,dm}^{-3}$, $J = 0.001\ \mathrm{mol\,dm}^{-3}$. Molar masses were estimated by approximating all added BSA bind to NaHA

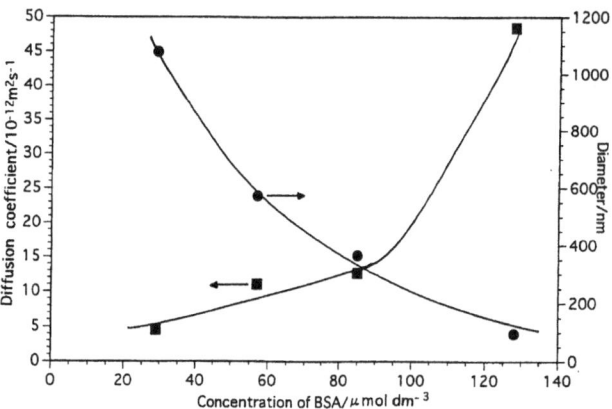

Fig. 6 Diffusion coefficients and diameters of soluble complexes. $C_{HA} = 1.2\ \mu\mathrm{mol\,dm}^{-3}$, $J = 0.001\ \mathrm{mol\,dm}^{-3}$

saturated complex was 120 and the value of U was $-2.2\ \mathrm{cm^2\,V^{-1}\,s^{-1}}$.

Viscosity of complex solution

The relative viscosity η_{rel} of the mixed solutions composed of NaHA and BSA was measured at 25 °C under $J = 0.001\ \mathrm{mol\,dm}^{-3}$. The results are shown in Fig. 5 as a function of C_{BSA} added to the NaHA solution $(1.2\ \mu\mathrm{mol\,dm}^{-3})$ which was the same solution as that for the electrical mobility. The values of η_{rel} were found to decrease with increasing C_{BSA}. As mentioned above, with increasing C_{BSA} the values of n_{BSA}/n_{HA} of the complexes and the molar masses increased. The molar masses were obtained by approximating all added BSA molecules bind to NaHA molecules. As shown in Fig. 5, the molar masses of the complexes increased linearly with increasing C_{BSA} from 8.5×10^5 to $9.25 \times 10^6\ \mathrm{g\,mol}^{-1}$. In spite of the increase of the molar mass of the saturated soluble complex as much as almost ten times of NaHA, the viscosity was found to decrease. This is considered to result from the change of the shape of the complexes, i.e., from a rod to a random-coil type.

Diffusion coefficients and diameters of complexes

Diffusion coefficients D of the complexes were measured in the same mixed solutions as those used in measuring U and η_{rel}. The values of D measured at $\theta = 90°$ are shown

in Fig. 6 as a function of C_{BSA} and were found to increase with increasing C_{BSA} as expected from the results of viscosity shown in Fig. 5. Assuming that the complexes are a sphere-like structure, their apparent diameters were estimated using Einstein–Stokes's equation and are shown in Fig. 6 together with the results of D.

Discussion

Mobility of acid polysaccharides and charge densities

As shown in Fig. 1b, in the region of pH > 5 the mobility of NaHA, NaAlg and NaChS-A became constant, respectively, as expected from perfect dissociation of their carboxylic groups in the pH region. The distance between neighboring charged groups b can be estimated from the activity coefficients of counterions or osmotic coefficients using Manning's limiting theory [1, 6]. The counterion activity coefficients at an infinite dilution $\gamma°$ of NaHA, NaAlg and NaChS-A were 0.75, 0.46 and 0.54 and their values of b estimated from $\gamma°$ were 1.2, 0.50 and 0.64 nm, respectively. The value of b is the inverse of a line-charge density. The mobility of NaAlg and NaChS-A was almost equal as expected from the values of b. However, the differences of U between NaHA and NaAlg were not significant as expected from that of b. This is considered to result from the condensation of Na ions on charged groups [1, 6].

Model of complexes

NaHA molecule is a rod- or a worm-like structure and BSA molecule is a prolate one. From the results of the viscosity and the diameters, with increasing n_{BSA}/n_{HA} the shape of the complexes should change from a worm-like to random-coil structure. The value of n_{BSA}/n_{HA} of the saturated complex was 120. Then, 18 repeating units bind to one BSA molecule, i.e., 18 carboxylic groups participate in the binding.

References

1. Yonese M, Tsuge H, Kishimoto H (1987) J Phys Chem 91:1971–1977
2. Yonese M, Yano M, Kishimoto H (1991) Bull Chem Soc Jpn 64:1814–1820
3. Xia J, Dubin PL, Kim Y, Muhoberac BB, Klimkowski VJ (1992) J Phys Chem 97:4528–4534
4. Yamaguchi K, Hachiya K, Moriyama Y, Takeda K (1996) J Colloid Interface Sci 179:249–254
5. Chen RF (1967) J Biol Chem 242:173–180
6. Manning GS (1969) J Chem Phys 51:924–933

Progr Colloid Polym Sci (1997) 106:257–261
© Steinkopff Verlag 1997

BIOCOLLOIDS

S. Sugie
T. Nagata
M. Inukai
S. Kugimiya
S. Sato
I. Miyata
M. Yonese

Effects of gelation of alginate around chymotrypsin on reactivities

Received: 13 October 1996
Accepted: 26 May 1997

S. Sugie · T. Nagata · M. Inukai
S. Kugimiya · S. Sato · I. Miyata
Dr. M. Yonese (✉)
Faculty of Pharmaceutical Sciences
Nagoya City University
Tanabe-dori, Mizuho-ku
Nagoya 467, Japan

Abstract To elucidate adding effects of alginate (Alg) and its gelation on the reactivities of chymotrypsin (Chy), three kinds of Chy-immobilized capsules were prepared by an interfacial condensation, i.e., (a) a capsule containing only Chy in the core (Chy capsule), (b) a capsule containing Chy in NaAlg (Chy capsule(NaAlg)) and (c) a capsule containing Chy in CaAlg gel (Chy capsule(CaAlg)). The coated film was poly(ethylenediamine-terephthaloyl) (PEDT) and NaAlg in the core was gelled in the state adhered tightly to the PEDT film. These Chy-immobilizing capsules were found to be useful for determining the effects of the gelation around Chy. 4-nitrophenyl acetate was used as a substrate. From the rates of efflux of the product (p-nitrophenol), apparent maximum velocities v'_{max} and the apparent Michaelis constants K'_M were obtained. The values of v'_{max} showed the maximum at pH = 7.8 for all three capsules and were found not to show any pH shift. The magnitude of v'_{max} was in the following order: Chy capsule (CaAlg) > Chy capsule (NaAlg) ≅ Chy capsule. The adding effects of Alg and its gelation on the reactivities of Chy are discussed.

Key words Reactivities of chymotrypsin – gelation of alginate – immobilized enzyme – effects of gelation – capsule

Introduction

When polyelectrolytes are gelled, the circumstance in the network changes from that in the solution. For example, water in the network is arrested more tightly than in the solutions [1] and the counterion bindings are enhanced by the formation of network, i.e., the absolute values of the effective charge densities decrease [2]. Therefore, when the environment around an enzyme changes from a sol to a gel state, the reactivities of the enzyme should be affected by the formation of the network. A multimembrane [3, 4] and a capsule can keep an enzyme and/or a polyelectrolyte in the inner phase and are useful to elucidate the network effects on the reactivities of the enzyme using a gelling polyelectrolyte.

In this report, using capsules containing α-chymotrypsin (Chy) and alginate (Alg) prepared by an interfacial condensation, effects of adding Alg and its gelation on the reactivities of Chy were studied. Alg is a major structural polysaccharide and is gelled by the addition of divalent metal ions without changing the temperature [5–7].

Experimental

Materials

Purified sodium alginate (NaAlg) was prepared from commercial origin (Kibun Food Chemifa Co., 150 M) by dialyzing against distilled water for 3 days and by filtrating to remove insoluble substances, as described in the previous papers [5, 6]. The purified NaAlg was stored in a refrigerator after freeze drying. The weight-average molar mass was determined to be $1.85 \times 10^5 \, \text{g mol}^{-1}$ using a light scattering (DLS700Ar, Otsuka Electronics). α-chymotrypsin (Chy) was of commercial origin (Sigma Co. Ltd., special grade, EC 3.4.21.1) and used without any purification.

Preparation of CaAlg beads and capsules containing Chy and NaAlg

Alginate gel (CaAlg) beads were prepared by dropping aqueous solutions of NaAlg (2.0 w/v%) from a capillary (outer diameter: 0.10 cm, inner diameter: 0.08 cm) into CaCl$_2$ solutions ($0.1 \, \text{mol dm}^{-3}$) and were immersed in them for at least 3 days.

Capsules were prepared by coating poly(ethylenediamine-terephthaloyl) (PEDT) on the drops containing NaAlg and Chy using an interfacial condensation. The method of the preparation is shown schematically in Fig. 1. The mixed aqueous solution composed of NaAlg, Chy, ethylenediamine and NaOH was dropped drop by drop from the same capillary as mentioned above into a mixed solvent composed of chloroform and cyclohexane (3 : 5), containing terephthaloyl chloride and trimesoyl chloride as a cross-linking agent. The mixed solvent was held at 4 °C during the condensation for 1 h. As the capsules were apt to adhere to each other in the mixed solvent during the polymerization of the film, the vessel for preparing the capsules was made of Teflon, partitioned into small rooms as shown in Fig. 1. After condensation, the mixed solvent was exchanged into water gradually using the mixed solutions composed of ethanol and water.

Three kinds of Chy-immobilized capsules were prepared, i.e., (a) a capsule containing only Chy in the core (Chy capsule), (b) a capsule containing Chy and NaAlg (Chy capsule(NaAlg)) and (c) a capsule containing Chy in the

CaAlg gel (Chy capsule(CaAlg)). For the gelation of NaAlg in the capsule, Chy capsule(NaAlg) was dipped in CaCl$_2$ solution ($0.1 \, \text{mol dm}^{-3}$). The NaAlg solution was gelled by Ca^{2+} ion diffusing through the PEDT film of the capsule. Furthermore, capsules containing only NaAlg (capsule (NaAlg)) and CaAlg (capsule (CaAlg)) were prepared in the same manner to elucidate effects of the network formation of NaAlg on the diffusion of solutes.

Solute releases from CaAlg beads and capsules

CaAlg bead, capsule(NaAlg) and capsule(CaAlg) were equilibrated with the sodium benzoate (NaBA) solution ($0.03 \, \text{mol dm}^{-3}$). Release of NaBA from them was measured at 25 °C by the absorbance ($\lambda = 230 \, \text{nm}$) under stirring (400 rpm) in an absorbance cell (cell length: 1 cm).

Product release from Chy-immobilized capsules

4-nitrophenyl acetate was used as a substrate and effluxes of the product (p-nitrophenol) from three kinds of Chy-immobilized capsules mentioned above were measured at 25 °C by the absorbance in the region of pH = 7.0–8.6 using Tris buffers.

Results and discussion

Characteristics of poly(ethylenediamine-terephthaloyl) capsules containing alginate

The mean diameter of CaAlg beads prepared in CaCl$_2$ solution ($0.1 \, \text{mol dm}^{-3}$) was 0.26 cm and that of the capsule (NaAlg) was 0.29 cm. The capsule (NaAlg) was dipped in CaCl$_2$ solution and the NaAlg solution in the inner

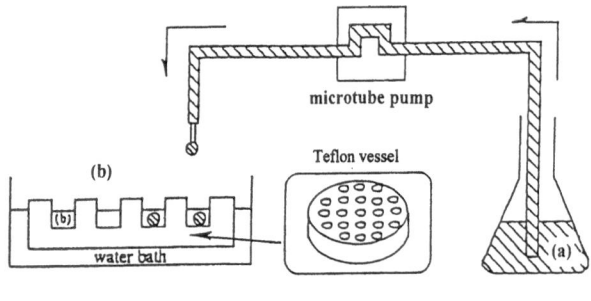

(b)
chloroform-cyclohexane (3:5)
terephthaloyl chloride (3 mmol dm^{-3})
trimesoyl chloride (0.15 mmol dm^{-3})

(a)
NaAlg (2w/v%) and/or Chy solution (2.5×10^{-5} mol dm^{-3})
ethylenediamine (0.5 mol dm^{-3})
NaOH (0.1 mol dm^{-3})

Fig. 1 Preparation method of capsules containing Chy and NaAlg (Chy capsule(NaAlg))

Progr Colloid Polym Sci (1997) 106: 257–261
© Steinkopff Verlag 1997

phases was gelled. NaAlg was found to be gelled in the state adhered tightly to the PEDT film. After the gelation, the mean diameter of the capsule (CaAlg) was 0.29 cm which was the same as that of the original capsule (NaAlg). These results show the affinity between the PEDT film and CaAlg gel was so excellent that the film prevented from the shrinkage of CaAlg gel during the gelation. Optical micro-graph of the capsule (CaAlg) and the coated PEDT film removed partly from CaAlg gel are shown in Fig. 2a. Figure 2b shows outer and inner surface structure of the coated PEDT film observed by a scanning electron micro-scope. The film was found to be asymmetric, i.e., the outer surface was porous and the inner surface was denser structure.

Fig. 2 Optical micrograph of capsule (CaAlg) and scanning electron micrograph of PEDT-coated film. (a) Optical micrograph of capsule (CaAlg) and PEDT-coated film removed partly from CaAlg gel in the core, (b) scanning electron micrograph of PEDT film, (b) – 1 the outer surface, (b) – 2 the inner surface

a

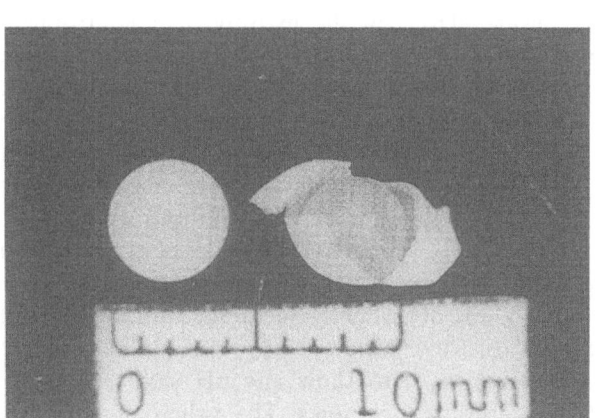

1   b   2

NaBA release from CaAlg beads, capsule (NaAlg) and capsule (CaAlg)

Time course of NaBA effluxes from the CaAlg bead, the capsule (NaAlg) and the capsule (CaAlg) are shown in Fig. 3. The ordinate is the relative values of the release amount Q/Q_∞, in which Q_∞ is the limiting release amount at $t = \infty$. Their release rates were in the following order; CaAlg bead > the capsules (NaAlg) > the capsule (CaAlg), and that of the capsule (CaAlg) was found to be much smaller than that of the capsule (NaAlg). As the diameters of the capsule (NaAlg) and the capsule (CaAlg) were equal as mentioned above, this difference results from the following reasons; i.e., as the inside of the capsule (NaAlg) is a solution, the concentration of NaBA in it is considered to be almost constant due to the perturbation of NaAlg, but as the inside of the capsule (CaAlg) is a gel state, the concentration of NaBA is controlled by the diffusion.

Reactivities of Chy-immobilized capsules

Chy was confirmed not to leak from the capsules. Efflux rates of the product (p-nitrophenol) from three kinds of Chy-immobilized capsules, i.e., (a) Chy capsule, (b) Chy capsule (NaAlg) and (c) Chy capsule (CaAlg), were measured in various initial concentration of the substrate (4-nitrophenyl acetate) $(C_s^0 = 5 \times 10^{-5} – 5 \times 10^{-4} \text{ mol dm}^{-3})$ in the region of pH = 7.0–8.6. The efflux rates of the product J_p obtained from the linear slopes of time courses of the concentration in bulk solutions were analyzed by $C_s^0 – C_s^0/J_p$ plots and apparent maximum velocities v'_{max} and apparent Michaelis constants K'_M were obtained. Figure 4 shows v'_{max} as a function of pH. The results showed the maximum at pH = 7.8 for all three capsules. The addition of NaAlg and its gelation were found not to

affect the optimum pH of Chy. The magnitudes of v'_{max} at the optimum pH were in the following order: Chy capsule(CaAlg) > Chy capsule(NaAlg) $\cong$ Chy capsule. The values of K'_M at the optimum pH were 1.2×10^{-3}, 0.68×10^{-3} and $0.53 \times 10^{-3} \text{ mol dm}^{-3}$ for the Chy capsule(CaAlg), the Chy capsule(NaAlg) and the Chy capsule, respectively, and were in the same order as that of v'_{max}.

Kinetics of hydrolysis of 4-nitrophenyl acetate in Chy solution

To elucidate the effects of adding NaAlg to Chy solutions on the hydrolysis of p-nitrophenol, the kinetics in the solution were studied in the region of pH = 7.0–8.6. The initial velocities for various substrates concentrations C_s^0 were analyzed by $C_s^0 – C_s^0/v$ plots in which v is a velocity of reaction, and the maximum velocities v_{max} and Michaelis–Menten constants K_M were obtained. The results of Chy solutions were found to agree with those of ref. [8]. The values of v_{max} are shown in Fig. 5 as a function of pH. The values of v_{max} at the optimum pH increased slightly due to the addition of NaAlg. The optimum pH of Chy was pH = 7.8 and by adding NaAlg it was found not to show any pH shift as well as the Chy-immobilized capsules. The values of K_M at the optimum pH were 6.3×10^{-5} for the Chy solution and $3.8 \times 10^{-5} \text{ mol dm}^{-3}$ for the NaAlg added solution.

Effects of Alg on optimum pH of Chy in solutions and capsules

The optimum pH of Chy in solutions did not show any pH shift due to the addition of NaAlg as shown in Fig. 5.

Fig. 3 Time courses of NaBa efflux from CaAlg gel, capsule (NaAlg) and capsule (CaAlg). CaAlg gel bead: (———), capsule (NaAlg): (– – –), capsule (CaAlg) (· · · ·)

Fig. 4 Apparent maximum velocities v'_{max}–pH profiles of Chy-immobilized capsules: Chy capsule ($\square$), Chy capsule(NaAlg) ($\bigcirc$), Chy capsule(CaAlg) ($\triangle$)

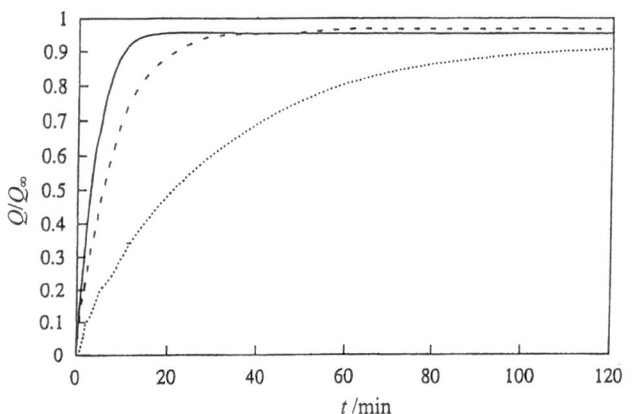

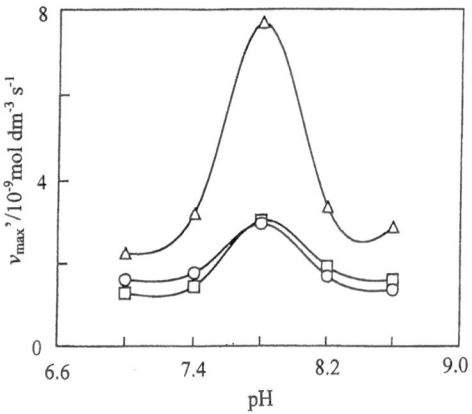

Progr Colloid Polym Sci (1997) 106:257–261
© Steinkopff Verlag 1997   261

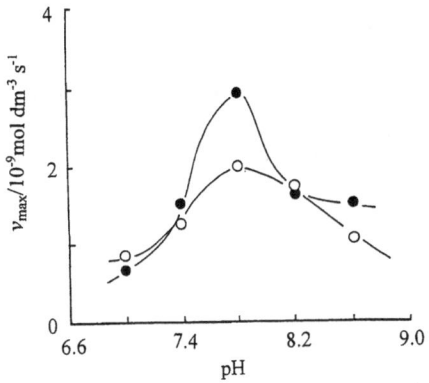

Fig. 5 Maximum velocities v_{max}–pH profiles of Chy in solution: Chy only (○), Chy in NaAlg (●)

the active sites was estimated to be almost -0.03 V which was almost equal to the ζ potential of NaAlg. Then, the active sites of urease and glucoamylase were considered to be surrounded by carboxylic groups of Alg. On the other hand, the active sites of Chy are considered to be apart from the carboxylic groups of Alg, contrary to urease and glucoamylase.

Furthermore, in the Chy-immobilized capsules, their optimum pH did not show any pH shift due to the addition of NaAlg and the gelation, as shown in Fig. 4, and were equal to that in the solution. However, the optimum pH of urease and glucoamylase shifted to alkaline pH by $\Delta pH = 0.62$ due to the addition of NaAlg [3, 4], differing from the results of Chy. The pH shifts are considered to result from the decreases of pH in the microenvironment around the active sites of urease and glucoamylase. From the results of ΔpH, the local electrostatic potential around

Effect of gelation of Alg on reactivities of Chy

The magnitudes of K'_M and v'_{max} of Chy-immobilized capsules were in the following order: Chy capsule(CaAlg) > Chy capsule(NaAlg) $\cong$ Chy capsule. The movements of the substrate and the product in the Chy capsule(CaAlg) are controlled by a diffusion in CaAlg gel phase and slower than those in the Chy capsule(NaAlg) and the Chy capsule. Due to this effect, the value of K'_M of the Chy capsule(CaAlg) is considered to be much more than those of the Chy capsule(NaAlg) and the Chy capsule. Furthermore, this effect should decrease v'_{max}. However, the gelation of NaAlg in the capsule enhanced the values of v'_{max}. This is considered to result from the interaction between Chy and CaAlg, and the effects of the increase of an arrested water in the network [1]. The precise mechanism should be studied furthermore.

References

1. Ohta H, Ando I, Fujishige S, Kubota K (1991) J Mol Struct 245:391–397
2. Yonese M, Baba K, Kishimoto H (1988) Bull Chem Soc Jpn 61:1077–1083
3. Yonese M, Murabayashi H, Kishimoto H (1990) J Memb Sci 54:145–162
4. Yonese M, Baba K, Murabayashi H (1991) In: DeRossi D, Kajiwara K, Osada Y, Yamauchi I (eds) Polymer Gels. Plenum Press, New York, pp 319–338.
5. Smidsrod O, Haug A, Gwhittington S (1972) Acta Chem Scand 26:2563
6. Yonese M, Baba K, Kishimoto H (1992) NIKKASHI, 1992, 198–202
7. Yonese M, Baba K, Kishimoto H (1988) Bull Chem Soc Jpn 61:1857–1863
8. Nohara D, Wakamatsu M, Goto M, Sakai T (1989) Chem Pharm Bull 37:1685–1690

Progr Colloid Polym Sci (1997) 106:262–265
© Steinkopff Verlag 1997

C.D. Dushkin
K. Kurihara

Nanorheology of thin liquid crystal film studied by shear force resonance method

Received: 3 December 1996
Accepted: 24 February 1997

C.D. Dushkin · Dr. K. Kurihara (✉)
Department of Applied Physics
School of Engineering
University of Nagoya
Furoo-cho, Chikusa-ku
Nagoya 464-01, Japan

Present address:
C.D. Dushkin
Mitsubishi Chemical Corp.
Yokohama Research Center
100 Kamoshida-cho, Aoba-ku
Yokohama 227, Japan

K. Kurihara
Institute for Chemical Reaction Science
Tohoku University, Katahira, Aoba-ku
Sendai 980-77, Japan

Abstract The structure and mechanics of very thin liquid crystal films depend on the intermolecular interactions in confined dimensions. The rheology of such films has been investigated by a shear force apparatus constructed as an attachment to the surface forces apparatus. The novelty of this method is that the rheological parameters are extracted from the amplitude and the phase of the output signal as a function of the resonance frequency. The apparent viscosity of the liquid crystal film is calculated from the damping coefficient by using a simple theoretical model. The viscosity of nanometer thin films of 4-cyano-4- hexylbiphenyl was found to be larger than the bulk value due to increased interactions in the molecular layers adjacent to the solid surfaces transferring shear. Further decreasing of the film thickness (increasing of the normal load) results in an increase of the resonance frequency due to the transition from viscous to contact friction. The latter observation opens the door to gaining tribological information at the nanodimension.

Key words Nanorheology – shear force apparatus – surface forces – mechanical resonance – thin liquid crystal films – cyano-alkylbiphenyls

Liquid crystal films are widely used for displaying information and images [1] and as lubricants [2]. Their molecules can align parallel or perpendicular to a solid interface depending on the substrate modification [1–4]. Once aligned in molecular layers, they exhibit oscillatory structural force of periodicity equal to the molecular layer thickness [3]. This force probably promotes the terrace formation in a liquid crystal drop spreading on a substrate [4]. The rheology of very thin liquid films is studied usually by the surface force apparatus (SFA) [5] equipped with shear force attachment [6–8] to create lateral disturbance in the film: aperiodical [6] or periodical one [7, 8]. Usually, the oscillating frequency is much smaller than the resonance frequency [7, 9, 10] which allows ones to utilize only the phase shift of the signal at the expense of an electrical correction to the mechanical model of the system.

To avoid this complication we developed a new method for determination of the rheological parameters from the full-frequency dependence of the amplitude and the phase of the output signal at resonance conditions [11]. The shear-force unit shown in Fig. 1 was constructed following the principle developed by Klein [8, 12]. By applying a sinusoidal input signal the upper lens oscillates with an angular frequency ω. At separation between the lenses in air the output signal passes through a maximum at certain frequency $\Omega_0 = 208.2 \text{ s}^{-1}$ (33.1 Hz), as shown in Fig. 2a. Simultaneously, the phase changes in the vicinity of Ω_0 (Fig. 2b). If the two lenses are brought in contact the resonance peak jumps to larger resonance frequency $\Omega = 360.5 \text{ s}^{-1}$ (57.4 Hz) at almost fivefold increase of the amplitude (not shown in Fig. 2).

The presence of a liquid crystal film between the mica sheets dramatically changes the character of oscillations.

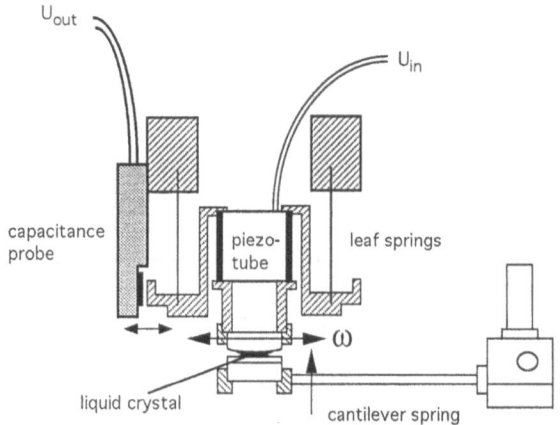

U_{out}

U_{in}

capacitance
probe

piezo-
tube

leaf springs

ω

liquid crystal

cantilever spring

Fig. 1 Unit for shear force measurements attached to the surface force apparatus. The sectioned piezo tube (Morgan Matroc) bends depending on the amplitude and polarity of the applied input voltage U_{in}. The symmetrical leaf springs transfer this bending into lateral oscillations of the liquid crystal film which is sandwiched between two mica sheets glued to the crossed quartz lenses: the upper lens attached to the piezo tube and the lower lens mounted on the cantilever spring. The magnitude of these oscillations is detected as the output voltage U_{out} of the capacitance probe (Japan ADE). In this configuration our shear force unit is a modification of the one proposed first by Klein et al. [8]

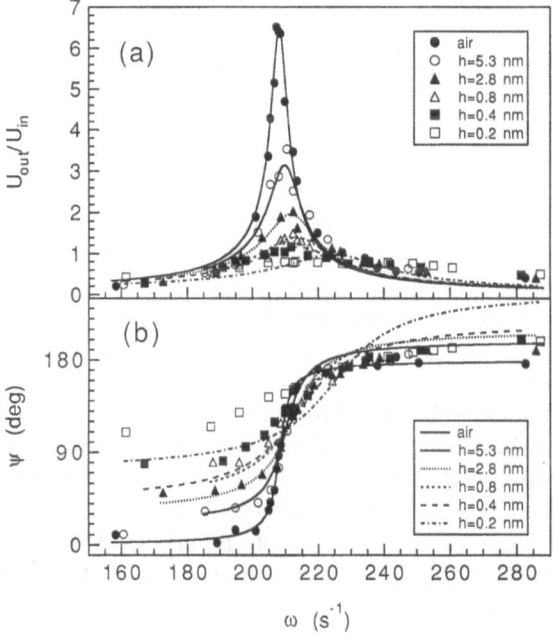

Fig. 2 Experimental data and theoretical fit illustrating the effect of the viscous friction in thin liquid crystal film of 4-cyano-4-hexyl-biphenyl (6CB): (a) Amplitude. The lines are theoretical fits by Eq. (1); (b) Phase delay of the output signal. The lines are drawn by Eq. (2) using the same parameters and the following values for the phase θ (in deg): 0, 20, 30, 40, 40, 70. Liquid crystal volume 10 µl, input voltage $U_{in} = 200$ mV, cantilever spring constant $K = 56.7$ N/m, temperature 24°C assuring nematic state for 6CB

With decreasing the film thickness h less than about 50 nm the magnitude of oscillations decreases due to the viscous friction in the film. At certain finite thickness the resonance peak stops to decrease and starts to move toward larger frequency due to increasing contribution of the cantilever spring. In the new resonance state of frequency Ω the peak starts to increase again never reaching the one for free-of-liquid contact in air. At h close to zero the viscous friction is replaced by friction in contact. This picture was qualitatively the same for various liquid crystal compounds, 4-cyano-4-n-alkylbiphenyls (BDH Co.) with $n = 5$, 6, 8, studied in either nematic or smectic state. Any contribution of the intermolecular interactions and orientation of the liquid crystals can be revealed by the rheological parameters of the film extracted from the experimental data taking into consideration of a theoretical model.

The oscillating unit itself represents a system of elastic elements (springs) and dissipative elements (contacts) which determine its response to the driving force exerted by the piezo. Therefore, it can be considered as a mechanical circuit composed of masses, springs and dashpots accounting for the energy losses [12]. Assuming the piezo tube as a pendulum hanged by the leaf springs and adding the contribution of the cantilever spring in contact one obtains for the amplitude

$$\frac{U_{out}}{U_{in}}$$

$$= \frac{C}{m_0 + m} \frac{1}{\sqrt{\left(\frac{k_0 + k}{m_0 + m} - \omega^2\right)^2 + \left(\frac{b_0 + b + b_{LC}}{m_0 + m}\right)^2 \omega^2}}$$

(1)

and for the phase

$$\psi = \theta + \arctan\left(-\frac{(b_0 + b + b_{LC})\omega/(m_0 + m)}{(k_0 + k)/(m_0 + m) - \omega^2}\right). \quad (2)$$

Here C is an apparatus constant proportional to the piezo elastic constant. m_0 and m are (effective) masses of the upper unit (comprising the piezo tube, lens and counterweights) and of the lower lens, respectively ($m \ll m_0$). k_0 is the effective spring constant of the leaf springs mounted in the upper unit. k is an effective constant accounting for the elastic contribution of the cantilever spring under surface shear. Since the contact between the upper and lower mica surfaces is not rigid, k could increase with decreasing h and to become equal to the transversal constant of the cantilever spring. It is, however, always larger than the constant of vertical bending of the cantilever spring K referred for the normal force measurements [3, 5]. b_0 and b are damping coefficients accounting for the energy losses in the contacts of the upper unit and of the cantilever spring, respectively. b_{LC} is the constant of viscous friction in the liquid-crystal film which can be related to the film viscosity η by means of [9]

$$\eta = \frac{h b_{LC}}{A_c},$$ (3)

where $A_c = \pi R_c^2$ is the area of contact and R_C is the contact radius. θ is a phase angle constant.

Equation (1) predicts the maximum of the amplitude observed experimentally at the resonance frequency $\Omega = \sqrt{(k_0 + k)/(m^0 + m)}$ in contact and $\Omega_0 = \sqrt{(k_0/m_0)}$ at separation ($k = 0$, $m = 0$). Comparing the last two expressions one can attribute the increase of the resonance frequency in contact to the contribution of the cantilever spring constant k.

Here, we explore only the transition from bulk to surface rheology which corresponds to the decrease of the resonance peak in Fig. 2a. The model parameters calculated from the signal amplitude at separation in air are $C = 113$ N/m, $m_0 = 18.1$ g and $b_0 = 84.3$ dyn s/cm. The calculated constant $k_0 = 786$ N/m (7.86×10^5 dyn/cm) coincides well with its experimental value of 712 N/m. The parameters of the cantilever spring obtained from fit of data in air contact are $k = 1525$ N/m and $b = 30.9$ dyn s/cm.

Using these constants fixed we have further processed the amplitude of the liquid crystal film by two parameters: k and b_{LC}. k increases from zero at separation to 134 N/m for the smallest film thickness which correspondingly increases the resonance frequency to $\Omega = 225.5$ s^{-1} seen also from Fig. 2a. Since this increase is much less than the respective value in air, the contact between the mica sheets in the presence of liquid crystal is much looser than that at adhesion in air.

Concerning the phase angle ψ (Fig. 2b) the agreement between theoretical and experimental values seems very good at infinite separation and films of thickness $h \sim 5$–10 nm. The discrepancy at smaller h can be explained by the increased contribution from the lower support, not accounted for in a consistent way in our simple model, to which the phase seems more sensitive than the amplitude.

With decreasing of the film thickness b_{LC} increases sufficiently implying on the increased viscous friction (Fig. 3). The behavior of the film viscosity η, however, is less straightforward. To understand it we compare η with the bulk value 0.25 dyn s/cm^2 stemming as the average from literature data for 5CB [13], 7CB [14] and 8CB [14, 15]. Being first larger than then bulk value, η decreases at very small thickness h were one has only few

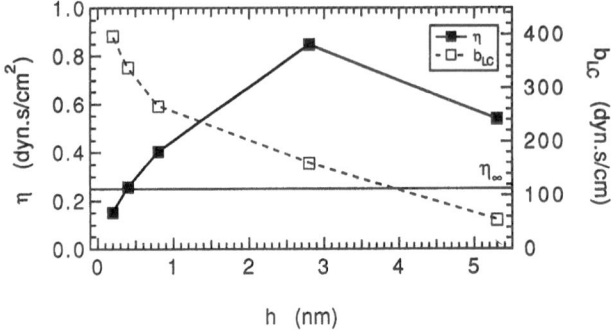

Fig. 3 Apparent viscosity η of a thin liquid crystal film of 6CB versus the film thickness h. The viscosity is calculated by Eq. (3) at contact radius $R_C = 0.0041$ cm. The horizontal straight line marks the most probable literature value of η_∞ for a bulk sample. For comparison is plotted also the coefficient of viscous friction b_{LC} obtained from the data fit in Fig. 2a

layers of liquid crystal molecules aligned parallel with their long axis to the mica surface [2]. Any increase of the thin film viscosity compared to its bulk value is in accord with the usual expectation already proven for polymer melts [10]. For the nematic 6CB it can be related to the increase of the repulsive force at very small separation [2, 16]. At further decreasing of the thickness, however, the viscosity seems to decrease although the friction coefficient b_{LC} is increasing (Fig. 3). Here, it is questionable to apply the phenomenological law of Eq. (3) for films as thin as one–two layers of liquid crystal molecules. Probably, at so small separations the liquid crystal layer is not uniform in dynamic conditions thus locally allowing a direct interaction between the mica surfaces or asperities created in the course of friction. In addition, the viscosity at $h \to 0$ can be influenced by the elastic contribution of the supporting mica sheets [12]. η depends also on the area of contact A_C which is not well defined in our experiments.

The shear resonance behavior of the other homological members (5CB and 8CB) follows the general trends described already for 6CB. We need, however, more data to conclude for the existence of any effects of the liquid crystal state (nematic or smectic) on the rheological parameters.

Competent technical assistance of Mr. Hirano and Mr. Yamauti (University of Nagoya) aiding the construction of the shear attachment is appreciated. C. Dushkin is greatly thankful to L'Oreal Company for the grant to conduct this research as well as to Prof. K. Nagayama (University of Tokyo) for the encouragement.

Progr Colloid Polym Sci (1997) 106:262–265
© Steinkopff Verlag 1997

References

1. Janning JL (1972) Appl Phys Lett 21: 173–174
2. Tichy JA (1990) Tribol Trans 33: 363–370
3. Horn RG, Israelachvili JN, Perez E (1981) J Phys 42:39–52
4. Valignat MP, Vilette S, Li R, Barberi R, Bartolino R, Dubois-Violette E, Cazabat AM (1996) Phys Rev Lett 77:1994–1997
5. Israelachvili JN (173) J Colloid Interface Sci 44:259–272
6. Israelachvili JN, McGuiggan PM, Homola AM (1988) Science 240:189–190
7. Van Alsten J, Granick S (1988) Phys Rev Lett 61:2570–2573
8. Klein J, Perahia D, Warburg S (1991) Nature 352:143–145
9. Peachey J, Van Alsten J, Granick S (1991) Rev Sci Instrum 62:463–473
10. Van Alsten J, Granick S (1990) Macromolecules 23:4856–4862
11. Dushkin C, Kurihara K (1996) Proc 49th Symp Colloid Surface Sci, Tokyo, p. 416
12. Dushkin CD, Kurihara K (1996) Rev Sci Instrum, submitted
13. Kiry F, Martinoty P (1977) J Phys 38: 153–157
14. Tozaki K, Kimura M, Itou S (1994) Jpn J Appl Phys 33:6633–6636
15. Panizza P, Archambault P, Roux D (1995) J Phys II France 5:303–311
16. Dushkin CD, Kurihara K, to be published

Progr Colloid Polym Sci (1997) 106:266–269
© Steinkopff Verlag 1997

M. Mizukami
M. Kurihara

Direct measurements of interaction forces between amine-modified glass surfaces using atomic forces microscope

Received: 3 December 1996
Accepted: 18 March 1997

M. Mizukami · Dr. K. Kurihara (✉)
Department of Applied Physics
School of Engineering
Nagoya University
Chikusa-ku
Nagoya 464-01, Japan

Abstract Interaction forces between amine-modified glass surfaces in aqueous solution were measured using atomic force microscope (AFM). The surface charge of the amine-modified glass was regulated by varying the pH. It has positive and negative sign at pHs' below and above the isoelectric point (pI), respectively, where the electrical double-layer repulsion was observed. The pH value where the apparent surface charge disappeared, pI, shifts within the range 6–9, depending on the surface coverage by the amine group. At the isoelectric point, the long-range attraction being several or ten times greater than the conventional van der Waals attraction appeared at the expense of the double-layer repulsion. The amine-modified surfaces, exhibited relatively low hydrophobicity, the contact angle of water was $32 \pm 7°$. The origin of long-range attraction was discussed based on the data taken at various salt concentrations and temperatures.

Key words Surface force – amine-modified surface – long-range attraction – atomic force microscope – isoelectric point

Introduction

Chemical modification of solid surfaces such as glass and silica using modifying reagents (i.e. silane-coupling reagents [1–3]) are popular in material sciences and technology, and modified surfaces are used for reinforcement, catalysts, and binding proteins [3, 4], etc. Characterization of these surfaces at the molecular level, especially in solvents, is important in order to control their properties and application processes more precisely. In this study, we employed the atomic force microscope to examine the amine-modified glass surfaces, by using the colloidal probe method [5], which uses a cantilever-attached colloidal particle as one surface, and monitored the surface forces, to obtain information on surface modification.

Importance of the present study lies also in its relevance to the long-ranged (extending over 100 nm, much longer than the conventional van der Waals force) attractions observed previously between hydrophobic surfaces using surface force apparatus (SFA) [6–10], atomic force microscope (AFM) [11, 12] and by another apparatus (MASIF) [13]. In spite of increasing and active researches in both experiments and theories [13–16], the origin of the attraction is still under controversy. A generally held view was that stable, highly hydrophobic surfaces (contact angle of water $> 90°$) were necessary to bring about such a long-ranged attraction. However, we have demonstrated recently that nucleic-acid-base monolayers, of which contact angle of water is $\sim 70°$, exhibit a long-range attraction comparable to those between hydrophobic surfaces [17]. The hydrophobic surface is one form of the electrically neutral surfaces. Thus, it should be necessary to examine interactions between electrically neutral surfaces with different hydrophobicities and chemical compositions in order to define the conditions where the long-range attractions appear. The amine-modified glass surfaces suit well for this purpose, because their isoelectric points (pIs') are in the range of neutral pHs, which is easy to work, on the other hand, pI value of glass or silica are highly acidic (less than 2).

Experimental section

Colloidal-glass spheres (radius 3–10 µm, Polyscience) and glass plate (Matsunami, microcover glass) were washed in a mixture of sulfuric acid and hydrogen peroxide (4:1, v/v), followed by rinsing with Millipore water ($\sim 18\,M\Omega\,cm$) thoroughly. A colloidal-glass sphere was then attached to the end of the AFM cantilever tip with epoxy resin (Shell, Epikote1004) [5]. A pair of a colloidal-glass sphere and a glass plate was treated with water-vapor plasma (20 W, 13.56 MHz rf source in 0.6 Torr of argon and water, 50 ml/min flow rate) for 3 min [3], immediately prior to silanization. This treatment ensured that glass surfaces contained a high density of silanol groups. Subsequently, the glass sphere and plate pair was placed in a closed vessel which contained a silanization reagent (3-aminopropyl-triethoxysilane) diluted with toluene. A silanization reaction was performed with a vaporized reagent in order to prevent deposition of silane polymers [3]. The vessel was warmed (around 70 °C) in order to facilitate vaporization of the silanization reagent and its reaction with sample surfaces. After the reaction, samples were rinsed with ethanol and toluene, in order to remove physically adsorbed reagents, and further rinsed with 1 mM NaOH, and 1 mM HNO_3 aqueous solution to remove the reagents bound by ionic or hydrogen bonding interactions [2]. Samples prepared were mounted in a closed AFM fluid cell, which was homebuilt and equipped with a liquid inlet and outlet. A force–distance relationship was measured with atomic force microscope (AFM, Seiko II, SPI3700-SPA300). Forces obtained were normalized by the radius (R) of a sphere using Derjaguin approximation [18]

$$F/R = 2\pi G_f. \tag{1}$$

Here, G_f is the interaction free energy between two flat surfaces. The radius of the sphere was measured using optical microscope. The rectangular-shaped Si_3N_4 cantilever spring (Olympus, spring constant $\sim 0.75\,N/m$) was used for the force measurement. The spring constants of the cantilever were determined by measuring the shift of the resonant frequency before and after adding the load (glass sphere) at the end of the cantilever [19]. The dynamic force mode unit (DFM, Seiko II) was used. The AFM imaging of the modified surface was conducted by AFM using Si_3N_4 triangle cantilever (Olympus, spring constant $\sim 0.09\,N/m$). The silanization condition (reaction time of 1 h, and the concentration of the silanization reagent in toluene of 1:7, v/v) was chosen to ensure the

surface flatness and the uniform modification with the amine groups, which was monitored by AFM imaging and the reproducibility of the double-layer repulsion at different sample positions. The surfaces prepared in this manner showed the height deviation of $\pm 0.3\,nm$ (for an area of $500 \times 500\,nm^2$). This deviation was negligible compared with the surface separation distances discussed in this work.

3-Aminopropyltriethoxysilane purchased from Shinnetsu Chemical Co. was used as delivered, and toluene from Nacalai Tesque was dried with molecular sieve (Nacalai Tesque, 4 A 1/16) prior to use.

Results and discussion

Interaction between water-vapor-treated glass surfaces, a colloidal sphere and a plate, were measured as a reference in water at various pHs' and salt (NaBr) concentrations. The repulsion appeared at all pH values studied (pH > 2). The decay length of the repulsion was equal to the Debye length ascertaining that interaction is the electrical double-layer repulsion due to the dissociated, negatively charged, silanol groups on the glass surfaces, as expected from the highly acidic isoelectric point of glass (close to 1). No attraction was observed over the pH range 2–11, and the salt (NaBr) concentration of 0–100 mM.

A colloidal glass sphere and a plate were then modified with the amine groups, and subjected to the forces measurement. Figure 1 showed a typical pH dependence of the interaction in 1 mM aqueous NaBr. At pH 3.2, the interaction was repulsive and decreased with increasing pH. The decay length of the repulsion indicated again the repulsion to be the double-layer interaction due to the protonated, positively charged, amine groups on the surfaces. The amine groups were then deprotonated at higher pHs, resulting in decrease in the repulsion. These interaction changes were quite different from those between unmodified glass surfaces. Clearly, the amine groups were introduced on glass surfaces. At a pH range around 6 the interaction changed to attraction, followed by the reappearance of repulsion upon further increase of pH to 8.9. The attraction was much longer ranged than the conventional van der Waals force (see below), and appeared when two surfaces were electrically neutral. This was confirmed independently from force profiles using mica (for sphere), and Si_3N_4 tip (for plate) as a counter plate. The pH range where attraction appeared changed from 6 to 9, depending on the surface preparation. The aminosilane reagent reacts with the silanol groups of the glass surface. It is easy to expect, in the case of a solid-surface reaction, that certain percentages of the silanol groups remained unreacted. The actual surfaces bore both of positively chargeable sites

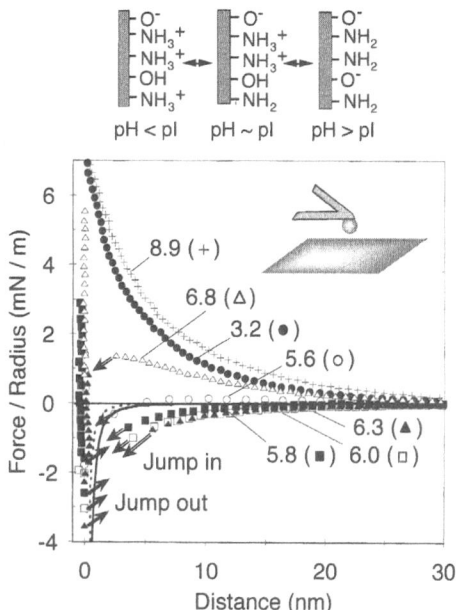

Fig. 1 Profiles of interaction forces between amine-modified surfaces at various pHs in 1 mM NaBr aqueous solution at the room temperature. pH was adjusted by adding HNO_3 or NaOH. The isoelectric point varies depending on the preparation, and has been obtained to be 6 for this sample. The van der Waals attraction calculated using nonretarded Hamaker constants of 1.0×10^{-20} (solid line) and 0.4×10^{-20} J (dashed line), respectively are also plotted

(amino groups, $pK_a \sim 9$) and negatively chargeable sites (silanol groups, $pK_a \sim 1$). The difference in the number density between the amine groups and remaining silanol groups should lead the shift in the isoelectric point of the amine-modified surfaces. It is interesting that the measurement of forces can demonstrate such a high sensitivity to a difference in the surface modification.

Around the isoelectric point the electrical double-layer repulsion was reduced, finally vanished, and the additional attraction appeared. The observed attraction was unexpectedly strong, being several to ten times stronger than the conventional van der Waals force estimated by the equation,

$$F/R = - A/6D^2, \tag{2}$$

where A was the Hamaker constant, 0.4×10^{-20} J (between octane in water) or 1.0×10^{-20} J (between glass surfaces in water). Then what can be the origin of this long-range attraction?

Long-range attractions extending separation distances longer than 100 nm have been reported previously for hydrophobic hydrocarbon [6–13] (contact angle of water $\geq 90°$) and nucleic-acid-base monolayers [17] (contact angle $\sim 70°$), although their origins have been still under active discussion. Our present system is considerably dif-

ferent from these previous systems, because the amine-modified surface exhibits low hydrophobicity, the contact angle of water being $32 \pm 7°$. The long-range attraction between amine-modified surfaces can be described by the exponential function

$$F/R = - B \exp(-D/D_0), \tag{3}$$

similarly to long-range hydrophobic attraction. Here, B is the preexponential intensity factor, and D_0 is the decay length. Several mechanisms have been proposed to explain the long-range hydrophobic attraction. Typical ones are: (i) electrostatic correlation of opposed surfaces [9, 14, 15], (ii) water structuring [16], and (iii) cavitation [13]. Different mechanisms afford different dependence of the force on the salt concentration and the temperature. A possible origin of the observed long-range attraction can be discussed qualitatively based on the salt concentration dependence. Mechanism (iii) of cavitation [13] requires high hydrophobicity, thus should be excluded because it does not fit with electrically neutral surfaces with *low* hydrophobicity. Mechanism (i) have several versions considering different correlating units such as ion fluctuating domains [14] and dipole domains [15], though these mechanisms provide an identical formula,

$$F \propto k_B T \exp(-2\kappa D), \tag{4}$$

where k_B and $1/\kappa$ are Boltzman's constant and Debye length, respectively. The decay length of the attraction decreased with increasing salt concentration in the analogous manner as Eq. (4), however, its salt dependence is not as steep as Debye length (Fig. 2). This makes mechanism (i) less likely. However, the situation may not be so simple, because the theoretical model of dipole domain [15], for example, has not counted the contribution of more than zero-frequency components which are not sensitive to the presence of salt; and the domain size possibly determines the interaction distance range at low salt concentrations. It should be mentioned that the decay lengths taken for samples exhibiting different isoelectric points (pIs') showed identical salt concentration dependence, indicating the difference in the chemical composition of surface species (amine and silanol groups), including different ions adsorbing them, does not influence the basic characteristics of the attraction. Increasing temperature to 40 °C suppressed the attraction, this behavior does not agree with the relation Eq. (4) predicts. Mechanism (ii), which expects decrease of the attraction both by adding salt and elevating the temperature, does not show any such discrepancy with the results. Nevertheless, it is not obvious whether water structuring can extend very long distances. Further experiments and theoretical development are needed to confirm the origin.

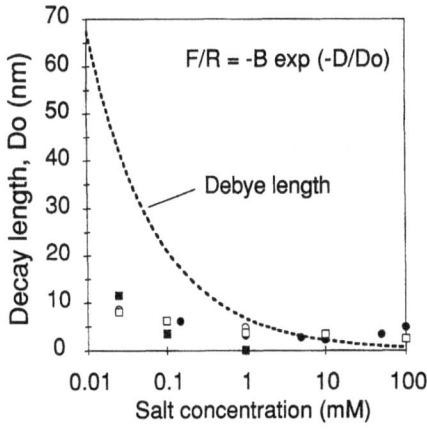

Fig. 2 Salt concentration dependence of the decay length (D_0) of the long-range attraction between amine-modified surfaces. Data were taken for samples exhibiting attraction at different pHs' (pIs'): pH 9 (l), pH 5.6 (m and n), pH 6 (o). Debye length is also included as a reference (dotted line)

Conclusions

This study has demonstrated that the colloidal-probe method employing an atomic force microscope is useful for characterizing chemically modified surfaces at the molecular level. Our major accomplishment is observation of a long-range attraction, several to ten times stronger than the conventional van der Waals attraction, between electrically neutral surfaces with *low* hydrophobicity (amine-modified surfaces). Previously, it has been believed that high hydrophobicity is required to bring about the long-ranged attraction. Obviously, attraction between electrically neutral surfaces in water is much more complicated than we have thought. More than one origin are needed to explain these observed long-range attraction. More experiments and theories are necessary to understand all of long-range attractions which cannot be explained with the conventional van der Waals force.

References

1. Tripp CP, Hair ML (1993) J Phys Chem 97:5693–5698
2. Vrancken KC, Possemiers K, Voort PVD, Vansent EF (1995) Colloids Surf A 98:235–241
3. Okusa H, Kurihara K, Toyoki K (1994) Langmuir 10:3577–3581
4. Kurihara K, Mizukami M, Suzuki S, Oosawa K (1996) Colloids Surf A 109:375–384
5. Ducker AW, Senden TJ (1992) Langmuir 8:1831–1836
6. Christenson HK, Claesson PM (1988) Science 239:390–392
7. Claesson PM, Christenson HK (1988) J Phys Chem 92:1650–1655
8. Kurihara K, Kato S, Kunitake T (1990) Chem Lett 1555–1558
9. Tsao Y-H, Evans DF, Wenneström H (1993) Langmuir 9:779–785
10. Kurihara K, Kunitake T (1992) J Am Chem Soc 114:10927–10933
11. Tsao Y-H, Evans DF, Wenneström H (1993) Science 262:547–550
12. Ravinovich YI, Yoon R-H (1994) Langmuir 10:1903–1909
13. Parker JL, Claesson PM, Attard P (1994) J Chem Phys 98:8468–8480
14. Podgornick R (1989) J Chem Phys 91:5840–5849
15. Attard P (1989) J Phys Chem 93:6441–6444
16. Eriksson JC, Ljunggren S, Claesson PM (1989) J Chem Soc Faraday Trans 85:163–176
17. Kurihara K, Abe T, Nakashima N (1996) Langmuir 12:4053–4056
18. Israelachvili JN (1985) Intermolecular and Surface Forces. Academic Press, London
19. Cleveland JP, Manne S, Bocek D, Hansma PK (1993) Rev Sci Instrum 64:403–405

Progr Colloid Polym Sci (1997) 106:270–273
© Steinkopff Verlag 1997

OTHER RELATED TOPICS

J. Yamanaka
T. Koga
N. Ise
T. Hashimoto

Order–disorder transition in aqueous dispersions of ionic colloidal silica particles

Received: 13 October 1996
Accepted: 6 March 1997

Dr. J. Yamanaka (✉) · T. Koga
T. Hashimoto
Hashimoto Polymer Phasing Project
ERATO, JST
15 Morimoto-cho
Shimogamo, Sakyo-ku
Kyoto 606, Japan

N. Ise
Central Laboratory
Rengo Co., Ltd., 186-1-4 Ohhiraki
Fukushima, Osaka 553, Japan

T. Hashimoto
Department of Polymer Chemistry
Graduate School of Engineering
Kyoto University, Kyoto 606-01, Japan

Abstract Order (crystal structure)–disorder (liquid structure) transition of aqueous dispersion of colloidal silica particles (average particle diameter $= 0.12 \times 10^{-6}$ and 0.11×10^{-6} m) was examined at a silica volume fraction of 3×10^{-2}, and various (analytical) surface charge densities of the particle, σ_a, and salt concentrations, C_s. The value of σ_a was changed by varying concentration of the coexisting sodium hydroxide. Phase diagram for the order–disorder transition was determined at $\sigma_a < 1.4 \times 10^{-6}$ C cm^{-2}. It was found that there existed lower and upper limits of σ_a for the ordering, at given C_s's. The grain size was large near the order–disorder transition points in the $\sigma_a - C_s$ diagram. The structure of the ordered state was studied using an ultra-small-angle X-ray scattering method. Several orders of Bragg reflection having a 6-fold symmetry was observed for large grains. For small grains, the scattering profile was powder-like. In both cases, the ordered structure had a body-centered-cubic lattice symmetry. The observed closest interparticle distances were smaller than the average interparticle distance, suggesting the non-space-filling nature of the ordered structure and existence of grain-boundary regions.

Key words Order-disorder transition – crystallization – ionic colloid – silica – dispersion – ultra-small-angle X-ray scattering

Introduction

Order (crystal structure)–disorder (liquid structure) transition in ionic colloidal dispersions has been intensively studied [1]. Since the driving force of the ordering is electrostatic, the order–disorder transition point for the colloidal system is largely determined by the surface charge density of the particle and the salt concentration of the dispersion, C_s, in addition to the volume fraction of the particles, ϕ. A number of experimental studies have so far been made to determine the transition point as a function of C_s and ϕ. However, little attention has been paid on the influence of the surface charge density. This seems to be due to experimental difficulties in preparing particles with various charge densities at a constant size.

On the other hand, it has been reported that the analytical charge density, σ_a, of colloidal silica can be controlled by addition of hydroxides of alkali metal in the dispersion [2]. The surface of silica particle is covered by weakly acidic silanol groups. By adding the hydroxide, the degree of dissociation of the silanol groups, and thus σ_a value, increases. Here, we study the influence of surface charge density on the order–disorder transition of ionic colloidal system by using aqueous dispersions of silica particles, whose σ_a value is changed by varying concentration of the coexisting sodium hydroxide. We find that there exist lower and upper limits of σ_a for the ordering, at given C_s's.

Progr Colloid Polym Sci (1997) 106:270–273
© Steinkopff Verlag 1997

Previous studies have shown that the ordered structure took cubic (bcc or fcc) lattices with a lattice constant of 0.1–1 μm [1]. Recently, an ultra-small-angle X-ray scattering (USAXS) method, by which an electron density fluctuation of several μm is detectable, has been developed and has made it possible to perform detailed studies on the ordered structure [3,4]. In the present paper, we report an USAXS study at several surface charge densities and salt concentrations.

Experimental

The colloidal silica particles, Seahoster KE-P10W and Cataloid SI-80P, were purchased from Nippon Shokubai Co., Ltd. (Osaka, Japan) and Catalysts & Chemicals Co., Ltd. (Tokyo, Japan), respectively, in the form of aqueous dispersions. The diameters of the particles were $(0.12 \pm 0.01) \times 10^{-6}$ and $(0.11 \pm 0.01) \times 10^{-6}$ m, which were estimated by fitting the USAXS profiles under high-salt conditions to the form factors for isolated spheres. They were purified by dialysis and ion-exchange methods, as described earlier [5]. The samples thus purified had hydronium ions as counterions (hereafter designated by H-type). The effective charge densities of the H-type silica particles, $\sigma_{e,H}$, were determined by conductivity measurements as described earlier [6]. The values of $\sigma_{e,H}$ for KE-P10W and SI-80P were 0.08×10^{-6} and 0.23×10^{-6} C cm^{-2}, respectively. The analytical charge densities, $\sigma_{a,H}$, were 0.1×10^{-6} and 0.5×10^{-6} C cm^{-2}, which were estimated from $\sigma_{e,H}$ values by using a relation between the analytical and effective charge densities for H-type latices [6]. Preparation of samples for measurements was performed as described in ref. [5]. The volume fraction of silica was 3.0×10^{-2}.

The scattering experiments were carried out at 25 ± 0.5 °C under argon atmosphere. A Bonse–Hart USAXS apparatus with a newly designed channel-cut germanium single crystals was constructed, whose details are fully described elsewhere [7,8].

Results and discussion

Phase diagram

Figure 1 is the phase diagram, which was constructed by observing the iridescence due to formation of the ordered structure. The phase boundary between the order and disorder states is shown at various σ_a's in the $\sigma_a - C_s$ coordinates, by rectangles. The surface charge densities of the particle are changed by varying the concentration of the coexisting sodium hydroxide, from 0 to 2.5×10^{-4} M.

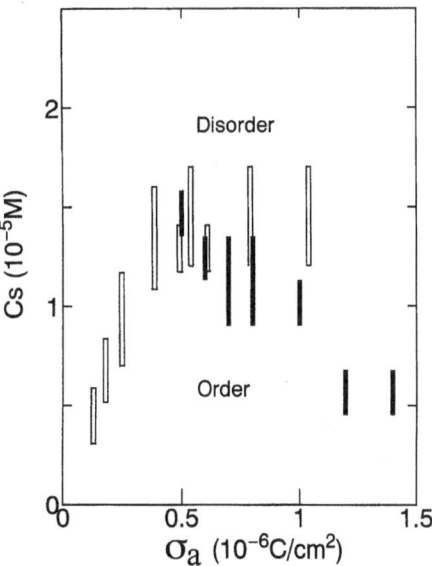

Fig. 1 Phase diagram for order (crystal)–disorder (liquid) phase transition in colloidal silica dispersion at room temperature. C_s, salt concentration; σ_a, the analytical surface charge density. Open symbols, KE-P10W; Filled symbols, SI-80P. Volume fraction of silica was 3×10^{-2}

By performing conductometric titrations with hydrochloric acid, it was confirmed that concentrations of excess Na and OH ions in the bulk were negligibly small under the present condition. The values of σ_a at various [NaOH]'s were calculated from $\sigma_{a,H}$ values and concentration of the coexisting sodium hydroxide, taking the influence of ion-exchange between hydronium and Na ions into accounts. The value of C_s was estimated from the sum of concentrations of coexisting sodium chloride and hydronium ions in the water used (2×10^{-6} M). It was also confirmed that dissolution of silica particles to monosilicates was negligible in the present condition.

At $\sigma_a = 0.1 \times 10^{-6}$ C cm^{-2}, the ordering did not take place even at $C_s = 2 \times 10^{-6}$ M. On the other hand, at $\sigma_a \approx 0.2 \times 10^{-6}$ C cm^{-2}, the ordered structure was formed in a few minutes. The order–disorder transition point shifted towards higher C_s with increasing σ_a, which was presumably due to augmented electrostatic interparticle interactions. With further increase in σ_a, a maximum or plateau region appeared, after which C_s value at the transition point become smaller. This might be explained in terms of an increase in the electrostatic screening effect, since the ionic strength of the system, which is usually taken as the sum of C_s and counterion concentration, increases with σ_a. The present finding implies that there exist lower and upper limits of σ_a for the order–disorder transition, at given C_s's.

The ordered structures consisted of grains. The grain size, at the time when the grains impinged to one another

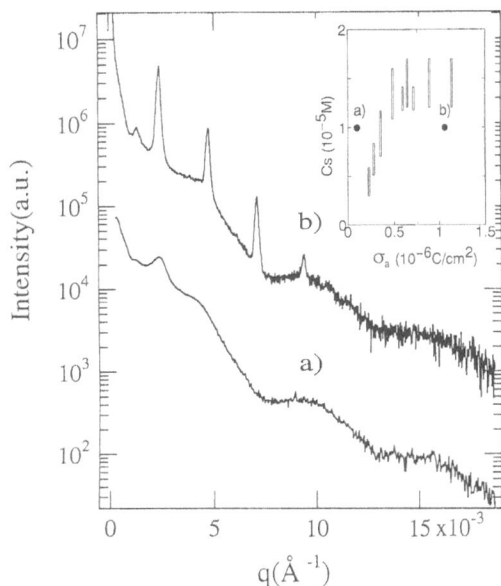

Fig. 2 Ultra-small-angle X-ray scattering (USAXS) profiles for aqueous dispersions of KE-P10W silica at (a) $\sigma_a = 0.1 \times 10^{-6}$ C cm^{-2} and (b) 1.1×10^{-6} C cm^{-2}. $C_s = 10^{-5}$ M. Inset shows the phase diagram for KE-P10W

to fill most of the space, increased when the order–disorder transition points were approached. This trend on the grain size is consistent with that expected from currently accepted theory on nucleation and growth for atomic and molecular crystals [9].

USAXS measurement [5]

The USAXS measurements were performed for KE-P10W dispersions at several C_s's and σ_a's. Figure 2 shows the scattering intensity as a function of scattering vector (q) at $C_s = 10^{-5}$ M and at two σ_a's, which corresponded to the order and disorder regions. In the disorder region ($\sigma_a = 0.1 \times 10^{-6}$ C cm^{-2}), a broad peak and a shoulder caused by interparticle interference from a liquid-type spatial arrangement of particles were observed in addition to the broad peaks from the form factor of isolated particles. We note that these interference peak and shoulder disap-

pear at sufficiently high salt conditions ($C_s > 10^{-3}$ M), suggesting that the liquid-type structure observed here results from an electrostatic interparticle interaction. The characteristic length of the disorder structure, d ($d = 2\pi/q_m$) estimated from the q value at the first-order peak (q_m), was 0.250×10^{-6} m. On the other hand, sharp peaks were observed in the order region ($\sigma_a = 0.96 \times 10^{-6}$ C cm^{-2}, grain size = 1–2 mm) at $q_{m,n} = 2.4n \times 10^{-3}$ Å^{-1}, n being an integer from 1 to 4. These were attributed to the first to the fourth orders of Bragg diffraction.

The measurements were further performed by rotating the sample with respect to the capillary axis. Similar scattering profiles were observed at each multiple angle of 60°, which indicated that the ordered structure had a 6-fold symmetry. Presence of a 6-fold symmetric structure in the colloidal silica crystals has been reported by Konishi et al. [3]. They found out that the diffraction of the 6-fold symmetry was attributable to the (1 1 0) plane of a body-centered-cubic (bcc) structure, which was maintained parallel to the capillary wall, in other words, with the [1 $\bar{1}$ 1] direction being parallel to the capillary axis. The scattering profile presently obtained can be ascribed to the same structure. The lattice spacing of the (1 1 0) planes, d_{110}, and a lattice constant of bcc, a, were 0.266×10^{-6} and 0.376×10^{-6} m, respectively. The value of the closest interparticle distance, $2d_{exp}$ ($2d_{exp} = (\frac{3}{2})^{1/2} d_{110}$), was calculated to be 0.326×10^{-6} m. The measurements were further carried out under conditions where small grains were formed. A bcc powder-like scattering pattern was observed, which suggest that the grains are randomly oriented. At $C_s = 2 \times 10^{-6}$ M and at $\sigma_a = 0.64 \times 10^{-6}$ C cm^{-2} (grain size < 0.1 mm), e.g., the values of d_{110}, a and $2d_{exp}$ estimated by assuming a bcc lattice, were 0.248×10^{-6}, 0.351×10^{-6} and 0.304×10^{-6} m, respectively.

As shown above, the USAXS profiles observed under various conditions were consistent with the finding on the order–disorder transition mentioned in the previous section. It should be noted that the closest interparticle distances, $2d_{exp}$, in the ordered structures were smaller than the average interparticle distance, $2d_0$ (0.331×10^{-6} m, in the present case), which was estimated from silica concentration by assuming uniform space filling of the particles with bcc symmetry ($2d_0 = 3^{1/2}(\pi/3\phi)^{1/3} R$, R being an average radius of the particle). This confirms the non-space-filling nature of the ordered structure [3, 4].

Progr Colloid Polym Sci (1997) 106:270–273
© Steinkopff Verlag 1997

References

1. For review articles, see Sood AK (1991) Solid State Phys 45:1–73; Dosho S et al. (1993) Langmuir 9:394–411; Schmitz KS (1993) Macroions in Solution and Colloidal Suspension. VCH Inc., New York; Arora AK, Tata BVR (eds) (1996) Ordering and Phase Transition in Charged Colloids. VCH Inc., New York

2. For a review article, see Iler RK (1979) The Chemistry of Silica. Wiley, New York, Chs 3 and 6

3. Konishi T, Ise N, Matsuoka H, Yamaoka H, Sogami I, Yoshiyama I (1995) Phys Rev B 51:3941–3944

4. Konishi T, Ise N (1995) J Am Chem Soc 117:8422–8424

5. Yamanaka J, Koga T, Ise N, Hashimoto T (1996) Phys Rev E 53:R4314–4317

6. Yamanaka J, Matsuoka H, Kitano H, Ise N (1990) J Colloid Interface Sci 134:92–106; Yamanaka J et al., unpublished results

7. Koga T, Hart M, Hashimoto T (1996) J Appl Crystallogr 29:318–324

8. Hart M, Koga T, Takano Y (1995) J Appl Crystallogr 28:568–570

9. For a review article, Gunton JD, San Miguel M, Sahni PS (1973) In: Domb C, Lebowitz JL (eds) Phase Transition and Critical Phenomena, Vol 8. Academic Press, New York, pp 267–466

Progr Colloid Polym Sci (1997) 106:274–276
© Steinkopff Verlag 1997

T. Taniguchi

Shape deformation dynamics induced by phase separation on two component membranes

Received: 13 October 1996
Accepted: 8 April 1997

Dr. T. Taniguchi (✉)
Department of Physics
Kyoto University
Kyoto 606-01, Japan

Abstract We numerically investigate a shape deformation and an intramembrane phase separation dynamics of a two-component closed membrane using a purely dissipative dynamical model. It is found that deformed shapes at the late stage after off critical quenches closely resemble the echinocytsis of erythrocytes. Two effects (i) the local coupling of the curvature and the composition of the membrane and (ii) the line tension at domain boundaries strongly influence the formation of projections on the membrane.

Key words Phase separation – membrane – amphiphile – echinocytosis

A wide variety of shape transformations of fluid membranes has been extensively studied theoretically in the past two decades using a bending elasticity model proposed by Canham and Helfrich [1]. The model has succeeded in explaining equilibrium shapes of the erythrocyte. However, much attention has recently been paid to shape deformations induced by internal degrees of freedom of membranes. For example, the bending elasticity model cannot explain the deformation from the biconcave shape of the erythrocyte to the *crenated* one (*echinocytosis*) [2, 3]. It is pointed out [3] that a local asymmetry in the composition between two halves of the bilayer plays an important role in the crenated shape. It has been observed [4] that a lateral phase separation occurs on an artificial two-component membrane where domains prefer local curvatures depending on the composition. In order to study the shape deformation accompanied by the intramembrane phase separation, we consider a two-component membrane as the simplest case of real biomembranes composed of several kinds of amphiphiles.

The mixed vesicle is represented by a two-dimensional (2D) closed surface $\mathscr{S}$ labeled by $\{u\} = \{u^1, u^2\}$. A position vector at a time t of a material point specified by $\{u\}$ is expressed as $\mathbf{r} = \mathbf{r}(u, t)$. The free-energy functional $F = F_1 + F_2 + F_3$ of the mixed vesicle is given by [5–9]

$$F_1 = \frac{\kappa}{2} \int H^2 \sqrt{g}\, \mathrm{d}^2 u, \tag{1}$$

$$F_2 = \int \{\tfrac{1}{2}\xi^2 (\nabla\phi)^2 + f(\phi)\}\sqrt{g}\, \mathrm{d}^2 u, \tag{2}$$

$$F_3 = \Lambda \int \phi H \sqrt{g}\, \mathrm{d}^2 u. \tag{3}$$

F_1 is the bending elastic energy where κ is the bending elasticity modulus, $H/2$ the mean curvature, and $\sqrt{g}\, \mathrm{d}^2 u$ the area element. F_2 denotes the Ginzburg-Landau free-energy functional with $f = -\phi^2/2 + \phi^4/4$ and a correlation length ξ. Here, the order parameter ϕ is given by the relative concentration per unit area $\phi = \phi_A - \phi_B$ in a mixed layer. F_3 provides a coupling energy which yields a local spontaneous curvature, Λ being a coupling constant. Such a local spontaneous curvature comes from the local asymmetry in the composition [3–5] between two halves of the bilayer. Assuming that the motion of the membrane is purely dissipative and its local area is incompressible, the equation of motion is written as [9]

$$\frac{\partial \mathbf{r}(u, t)}{\partial t} = -L_r \frac{\delta}{\delta \mathbf{r}(u, t)} [F + \int \gamma(u, t)\sqrt{g}\, \mathrm{d}^2 u], \tag{4}$$

Fig. 1 Orthogonal projections of vesicle shapes ($\kappa = 2$ and $\Lambda = 0.5$) and surface patterns of ϕ at a late stage $t = 300$ for three cases (a)–(c), (a) a critical quench $\langle\phi\rangle = 0$, off critical quenches (b) $\langle\phi\rangle = 0.3$ (c) $\langle\phi\rangle = -0.3$. In these figures, filled circles represent cites where $\phi < \langle\phi\rangle$

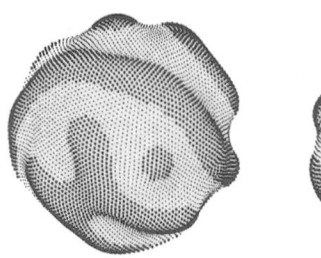
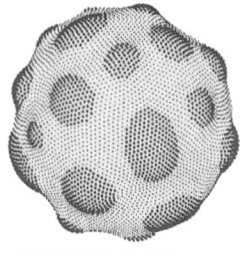
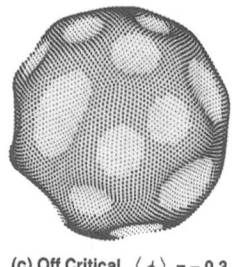

(a) Critical $\langle\phi\rangle = 0$

(b) Off Critical $\langle\phi\rangle = 0.3$

(c) Off Critical $\langle\phi\rangle = -0.3$

where L_r denotes a kinetic coefficient. The incompressibility of the local area throughout deformations is guaranteed by a local Lagrange multiplier $\gamma(u)$ [10]. The change of $\phi(u)$ per a unit time at a material point $\{u\}$ is expressed by the difference between the amount of amphiphiles diffusing into the material point and out of it. Hence, $\phi(u, t)$ satisfies the equation of lateral continuity [9]

$$\frac{\partial\phi(u, t)}{\partial t} = L_\phi\Delta_{\mathrm{LB}}[-\xi^2\Delta_{\mathrm{LB}}\phi + f'(\phi) + \Lambda H] \tag{5}$$

where L_ϕ is a kinetic coefficient. This equation may be read as an extension of a time-dependent Ginzburg-Landau (TDGL) equation which describes the phase separation in Euclidean spaces to that in curved spaces. The operator Δ_{LB} is the Laplace–Beltrami operator which is used as the Laplacian in curved spaces. The equation for γ can be obtained from the incompressibility condition $\partial\sqrt{g}/\partial t = 0$ [11].

Using the coupled equations (4) and (5), we perform simulations of shape deformations induced by intramembrane phase separations after quenching from a high-temperature disordered state in ϕ to a low-temperature unstable region. In the initial disordered state, the shape of vesicles is a sphere and ϕ at each lattice point is a Gaussian random number with a surface average $\langle\phi\rangle$ and $\langle(\phi - \langle\phi\rangle)^2\rangle = 0.01$. The vesicle surface is discretized by triangular lattices and is made of N-lattice points ($N = 9002$). Parameters used here are $L_r = 0.5$, $L_\phi = \xi = 1$, $\Delta t = 0.01$, and the initial radius $r_0 = 22.27$. We use the Lagrange multiplier γ in order to satisfy the area incompressibility, but a numerical error accumulates during the simulation which violates the fixed local area constraint. In order to suppress the error below a harmless level, we use a penalty functional for local areas [12]. The total area $S(t)$ is nicely conserved ($|S(t)/S(0) - 1| < 0.0026$) and a maximum deviation of a local area from its initial area in suppressed within 6% of its initial area throughout simulations. Figure 1 shows vesicle shapes ($\Lambda = 0.5$) at a late stage ($t = 300$) for different lateral averages in ϕ, (a) a critical quench $\langle\phi\rangle = 0$ and off-critical ones (b)

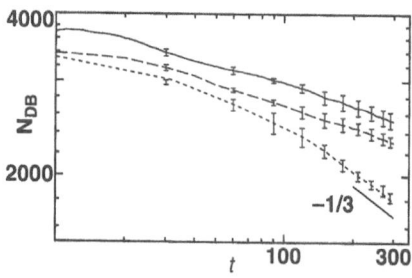

Fig. 2 Log–log plots of N_{DB} versus time t for off-critical quenches ($\langle\phi\rangle = 0.3$), (i) $\Lambda = 0.8$ (solid line), (ii) $\Lambda = 0.5$ (dashed line) and (iii) $\Lambda = 0$ and without deformations (dotted line), where N_{DB} is the number of lattice points located at domain boundaries. The last case (iii) is the spinodal decomposition on a rigid perfect sphere. The guide line is $N_{\mathrm{DB}} \sim t^{-1/3}$

$\langle\phi\rangle = 0.3$ (c) $\phi = -0.3$. The vesicle shape in Fig. 1b is qualitatively very similar to the echinocytosis, and that in Fig. 1c also corresponds to an invagination of erythrocytes. States at $t = 300$ for (a)–(c) are not in equilibrium yet, but their coarsenings are gradually slowing down. The total length of domain boundaries is coarsened as $t^{-1/3}$ for the conserved system on the fixed perfect sphere ((iii) in Fig. 2) and also for the 2D conserved system in Euclidean spaces. Whereas in the present system the coarsening is significantly slower than that in the system with no deformations. As seen from Fig. 2, the larger Λ, the slower the coarsening [9]. Shape deformations induced by phase separations come from two effects (i) the local coupling of the curvature and the composition of the membrane in F_3 and (ii) the line tension at domain boundaries stemming from $\xi^2(\nabla\phi)^2$ in F_2. Fig. 3a shows a shape deformation induced only by (i) the line tension of domain boundaries ($\Lambda = 0$, $\kappa = 0.5$ and $\langle\phi\rangle = 0.3$) at $t = 500$. $N_{\mathrm{BD}}(t)$ is also plotted for the case $\Lambda = 0$, $\kappa = 0.5$ and $\langle\phi\rangle = 0.3$ in Fig. 3b, which shows that $N_{\mathrm{BD}} \sim t^{-1/3}$ holds in the late stage. These results indicate that the slowing down of the coarsening in the cases $\Lambda \neq 0$ and $\kappa = 2$ (see Fig. 2) is mainly due to the local coupling of the curvature and the composition in F_3.

Fig. 3 Shape deformation dynamics for an off-critical quench $\langle\phi\rangle = 0.3$, $\kappa = 0.5$, $\Lambda = 0$. (a) Orthogonal projections of vesicle shapes and surface patterns of ϕ at a time $t = 500$. (b) Log–log plots of N_{DB} versus time t. The guide line is $N_{DB} \sim t^{-1/3}$. In this case, the shape deformation is induced only by the line tension coming from domain boundaries

(a) $\Lambda = 0$ $<\phi> = 0.3$

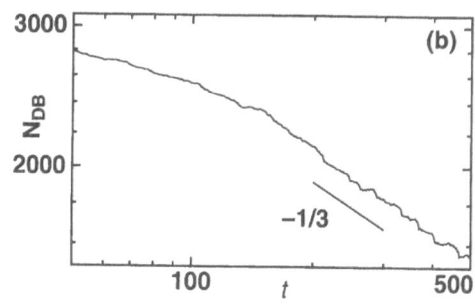

This work is supported by the Grant-in-Aid for Encouragement of Young Scientists, from Ministry of Education, Science and Culture, Japan. The computation has been done using the facilities of the Supercomputer Center, Institute for Solid State Physics, University of Tokyo.

References

1. Canham PB (1970) J Theor Biol 26: 61–81; Helfrich W (1973) Z Naturforsch 28c:693–703
2. Deuticke B (1968) Biochim Biophys Acta 163:494–500
3. Sheetz MP, Singer SJ (1974) Proc Nat Acad Sci USA 74:4475–4461; Sheetz MP, Painter RG, Singer SJ (1976) J Cell Biol 70:193–203
4. Gebhardt C, Gruler H, Sackmann E (1977) Z Naturforsch 32c:581–596
5. Leibler S (1986) J Phys (France) 47:507–516; Leibler S, Andelman D (1987) 48:2013–2018
6. Andelman D, Kawakatsu T, Kawasaki K (1992), Europhys Lett 19:57–62
7. Kawakatsu T, Andelman D, Kawasaki K, Taniguchi T (1993) J Phys II (France) 3:971–997
8. Taniguchi T, Kawasaki K, Andelman D, Kawakatsu T (1994) J Phys II (France) 4:1333–1362; (1993) Cond Matt Materials Comm 1:75–79
9. Taniguchi T (1996) Phys Rev Lett 76:4444–4447
10. Jenkins JT (1976) J Math Biol 4: 149–169
11. Foltin G (1994) Phys Rev E 49: 5243–5248
12. Edberg R, Evans DJ, Morris GP (1986) J Chem Phys 84:6933–6939

Progr Colloid Polym Sci (1997) 106:277–280
© Steinkopff Verlag 1997

S. Mori
Y. Kajinaga

Polymerized membranes in poor solvent

Received: 13 October 1996
Accepted: 31 March 1997

Dr. S. Mori (✉) · Y. Kajinaga
Polymerized Membranes in Poor Solvent
Department of Physics
Graduate School of Science
University of Tokyo
Hongo 7-3-1, Bunkyo-ku
Tokyo 113, Japan

Abstract We discuss the mechanism of the phase transitions of polymerized membranes with attractive interactions or in poor solvent. Depending on the local flexibility of the membrane, the membranes with attractive interactions are known to show a discontinuous or a continuous flat–compact phase transitions. We study a model for D-dimensional polymerized membrane with

attractive long-range interactions ($r^{-\gamma}$) and a square-lattice model with bending rigidity (u) and repulsive or attractive interactions ($\pm\omega$). Based on the results, we give a theoretical interpretation to the flat–compact phase transition of polymerized membranes.

Key words Polymerized membrane – poor solvent – flat–compact transition – critical state – folding

Introduction

Recently there has been considerable interest in the phase transition of polymerized (tethered) membranes with attractive interactions [1–4]. In a pioneer work [1], Abraham and Nelson found by molecular dynamics simulations that the introduction of attractive interactions between monomers leads to a collapsed membrane with fractal dimension 3 at a sufficiently low temperature. Subsequently, Abraham and Kardar [2] showed that for open membranes with attractive interactions, as the temperature decreases, there exists a well-defined sequence of folding transitions and then the membrane ends up in the collapsed phase. They also presented a Landau theory of the transition. Grest and Petsche [4] extensively carried out molecular dynamics simulations of closed membranes. They considered flexible membranes; the nodes of the membrane are connected by a linear chain of n monomers. For short monomer chains, $n = 4$, there occurs a first-order transition from the high-temperature flat phase to

the low-temperature collapsed phase, but no intermediate crumpled phase. For longer chains (more flexible), $n = 8$, the transition is either continuous or weakly first order. With the assumption of the continuous transition, the fractal dimension of the membrane at the transition is estimated as $d_f = 2.4$. Experimentally, graphite-oxide membranes are studied and the collapsed phase and the flat phase have been observed [5, 6]. There is also an evidence in favor of the sequential folding transition [6].

In this report, we review two theoretical models for the phase transitions of polymerized membranes with attractive interactions [7, 8] and try to understand the mechanism why the behavior of the transition depends on the local flexibility of the membrane.

Polymerized membrane with long-range attractive interactions

We study the phase transition of polymerized membrane with attractive interactions [7]. We denote the bending rigidity of the membrane by κ and the elastic Lamé coefficients by μ and λ. The interaction between different positions, which is rotationally invariant, is represented by V. We take

$$V(r^2) = u/(r^2)^{\gamma/2} , \qquad (2.1)$$

where r is the distance, and u and γ are some constants. The Hamiltonian of the tethered membrane with long-range interaction is given by

$$\mathcal{H}[\mathbf{X}(\sigma)] = \int d^D\sigma \left[\frac{d}{2} \kappa \Delta \mathbf{X} \cdot \Delta \mathbf{X} + \frac{1}{4} \mu d \{ \partial_\alpha \mathbf{X} \cdot \partial_\beta \mathbf{X} - \delta_{\alpha\beta} \}^2 \right.$$

$$\left. + \frac{1}{8} \lambda d \{ \partial_\alpha \mathbf{X} \cdot \partial_\alpha \mathbf{X} - D \}^2 \right]$$

$$+ d \int d^D\sigma \, d^D\sigma' V((\mathbf{X}(\sigma) - \mathbf{X}(\sigma'))^2) , \qquad (2.2)$$

where $\mathbf{X}(\sigma)$ represents the position vector in the embedding space with d dimension, while σ denotes the internal coordinate on the D-dimensional membrane. The first term represents the bending elasticity, the second and the third terms the stretching elasticity and the last term the long-range interactions. The model should also have a three-body interaction, which prevent the system from collapsing to a point. Otherwise, a flat phase or a crumpled phase becomes always unstable by the attractive interaction ($u < 0$). For a while, we study the above model Hamiltonian Eq. (2.2) and discuss the stability of the flat phase.

Using the large-d limit and defining the flatness order parameter ζ for the flat phase $\langle \mathbf{X}(\sigma) \rangle = \zeta \sigma^\alpha \mathbf{e}_\alpha$, where $\langle \ \rangle$ means the thermal average, one can derive a set of saddle point equations [9]. We only present the most important one about the full inverse propagator $K(k) \equiv \langle \mathbf{X}(-k) \cdot \mathbf{X}(k) \rangle^{-1}$,

$$K(k) \equiv \kappa k^4 + \chi k^2$$

$$+ 4\gamma u \int d^D\sigma [1 - \cos(\mathbf{k} \cdot \boldsymbol{\sigma})] \left(\frac{1}{A(\sigma)} \right)^{1+\gamma/2} . \qquad (2.3)$$

Here χ is a constant solution of other saddle point equations. These saddle-point equations have been analyzed in the repulsive force case ($u > 0$) [9, 10]. We also recall that, in the phantom case, i.e. $V(r^2) = 0$, the flat phase ($\zeta > 0$) exists only for $D > 2$ and $\kappa > \kappa_C$, where κ_C is a finite constant [9]. Here, phantom means that the self-avoiding interactions are neglected. By a formal expansion of Eq. (2.3) in powers of k, we obtain

$$K(k) = \tau_{\text{eff}} k^2 + \kappa_{\text{eff}} k^4 \quad \text{for small } k , \qquad (2.4)$$

where the effective rigidity κ_{eff} are given by

$$\kappa_{\text{eff}} = \kappa + \frac{\gamma u}{4(D^2 + 2D)} \int d^D\sigma \, \sigma^4 [A(\sigma)]^{-1-\gamma/2} . \qquad (2.5)$$

We find the attractive force $u < 0$ reduce the rigidity κ. By increasing the strength of the attractive interaction $|u|$, the flat phase becomes unstable and the crumpling transition occurs at $\kappa_{\text{eff}} = \kappa_C$.

Studying the above saddle-point equations, we obtain the phase diagram and the behavior of the squared-order parameter $B(u) \equiv \zeta^2$. These are presented in Figs. 1 and 2. In the region $\gamma > D/(1 - D/4)$ and $D > 2$, the order parameter vanishes continuously as we increase the strength of $|u|$ (Fig. 1b). On the other hand, in the domain $\gamma < D/(1 - D/4)$ and $\gamma > D + 2$, $B(u)$ shows discontinuous behavior (Fig. 1a).

From these analyses, it is clear that if the attractive interaction is finite ranged, then the membrane without self-avoiding interaction has the possibility to exhibit a continuous crumpling transition from the flat phase to the compact phase and there may exist a critical crumpled state at the transition point. However, in order to discuss the nature of the transition we must compare the corresponding free energies between the flat phase and the compact phase. It is a difficult task and we only make some reasoning based on the numerical studies of ref. [4] hereafter.

Fig. 1 Behavior of the order parameter in the (B, u) plane. (a) Discontinuous case and (b) Continuous case. B is the square of the order parameter ζ and $|u|$ is the strength of attractive interaction

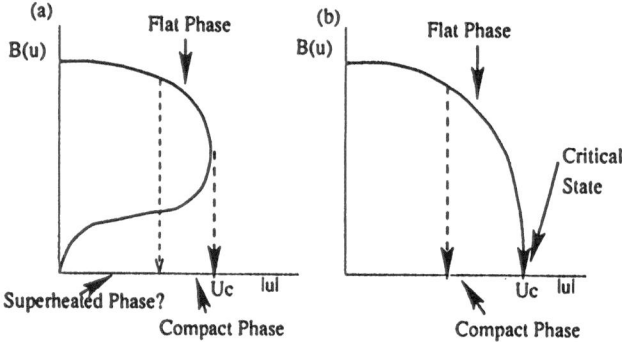

Fig. 2 Phase diagram in the (γ, D) plane: (1) compact state; (2) crumpled state; (3) flat state + compact state; (4) flat state + critical state + compact state

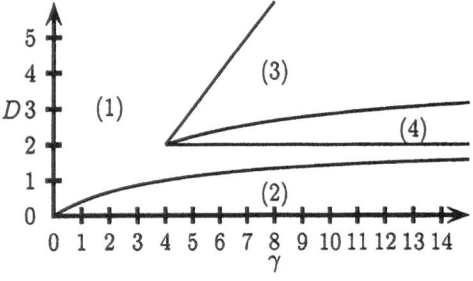

Our analysis deals mainly with the instability of the flat phase of the phantom membrane caused by the attractive interaction. On the other hand, numerical simulations treat the self-avoiding tethered membranes with short-range attractive interactions. The self-avoidance induces the entropic rigidity [1] and because of its effect, the membrane is in the flat phase. When the membrane is in the flat phase, it is widely believed that the self-avoidance is irrelevant and can be ignored. In other words, the bare rigidity κ in the Hamiltonian [Eq. (2.2)] include the effect of the self-avoidance and this contribution does not depend so much on the temperature. Our study showed that the phantom polymerized membrane with short-range attractive interaction can show a continuous flat–compact transition, if the free energy for the compact phase does not become lower than that of the flat phase before the order parameter becomes zero. We cannot compare the free energy between the flat phase and the compact phase and we do not know whether the transition is continuous or not in our model. However the result of ref. [4] means that the transition in the self-avoiding membrane is continuous at $n = 8$. Phantom membrane is more flexible than the self-avoiding membrane, we can conclude that the phantom membrane shows a continuous transition, and the continuous transition of the very flexible self-avoiding polymerized membrane can be interpreted as a crumpling transition induced by the complete cancellation between the entropic bending rigidity and the negative bending rigidity from attractive interactions. When the membrane is less flexible, the flat phase become unstable before the cancellation becomes complete.

Square lattice with attractive interactions

Next we study a simple model which describes the sequential folding transition of the polymerized membrane with attractive interaction [8]. In order to describe the transition, the folding degrees of freedom of the membrane are important. We consider a model of foldings of a two-dimensional square lattice with $L \times L$ size [11]. We consider all possible foldings of the lattice and each folding maintains the correct distances between the neighboring sites. Two configurations are identical if the positions of all corresponding sites (or vertex) coincide and this definition of the identical configuration does not distinguish between the different manners of folding which lead to the same final state. Fig. 3 shows the section of the lattice in a folded state along its x-axis.

The membrane has of course two such sections, i.e., x- and y-directions. The section is represented by a line of zero thickness which folds in $N = 1, 2, \ldots, L$ segments of successive length $x_1, x_2, \ldots, x_N$. Correspondingly, the

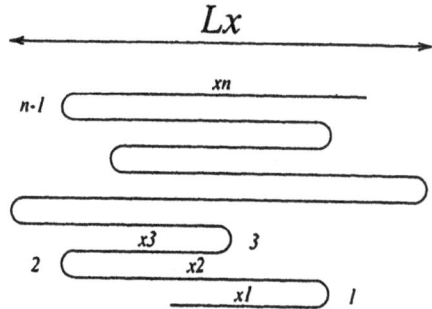

Fig. 3 Typical configuration of a section of the lattice. The folds have a vanishing length

other section, which folds in M segments, is represented by $y_1, y_2, \ldots, y_M$. These segments' lengths x_i and y_i are multiples of lattice constant $a = 1$. The total lengths are fixed:

$$\sum_{i=1}^{N} x_i = L, \qquad \sum_{j=1}^{M} y_j = L. \tag{3.1}$$

A configuration of the system is thus determined by a set of natural numbers (N, M) and a set of positive numbers $x_1, x_2, \ldots, x_N; y_1, y_2, \ldots, y_M$.

The potential energy will be taken to consist of the sum of two terms. First, as a bending energy, we assign energy κ per unit length of a fold. We denote an interaction between the elementary squares of the lattice as U_I and the potential energy $U_{N,M}$ for the system is written as

$$U_{N,M}(x_1, x_2, \ldots, x_N; y_1, y_2, \ldots, y_M)$$
$$= (N - 1)\kappa L + (M - 1)\kappa L + U_I \tag{3.2}$$
$$= U'_{N,M} + U_I. \tag{3.3}$$

As an interaction between different parts of the square lattice, we consider a potential which is proportional to contact area of the lattice. The contact area is determined by the area difference between the initial stretched form and the configuration considered. The potential is written as

$$U_{I,CA}(x_1, \ldots, x_N; y_1, \ldots, y_M) = u(L^2 - L_x L_y). \tag{3.4}$$

Here, L_x and L_y are the widths in each direction of the lattice (Fig. 3).

For convenience we will use the reduced bending rigidity and potential strength,

$$K = \kappa/k_B T, \quad w \equiv u/k_B T. \tag{3.5}$$

The partition function is given as

$$Z(u, \omega, L) = \sum_{N,M=1}^{\infty} \int_a^\infty dx_1 \cdots dx_N \int_a^\infty dy_1 \cdots dy_M$$
$$\times \delta\left(\sum_{i=1}^{N} x_i - L\right) \delta\left(\sum_{j=1}^{M} y_j - L\right)$$
$$\times \exp(- U_{N,M}/k_B T). \tag{3.6}$$

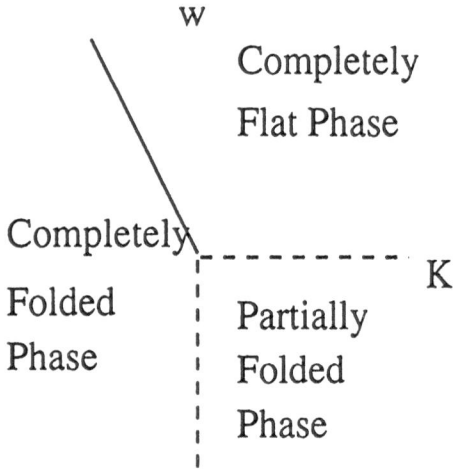

Fig. 4 Phase diagram in the (K, w) plane for the square lattice with CA type interaction. Three first order transition lines $w = -2K$ ($K < 0$), $K = 0$ ($w < 0$) and $w = 0$ ($K > 0$) separate the three phases, completely flat phase, completely folded phase and partially folded state

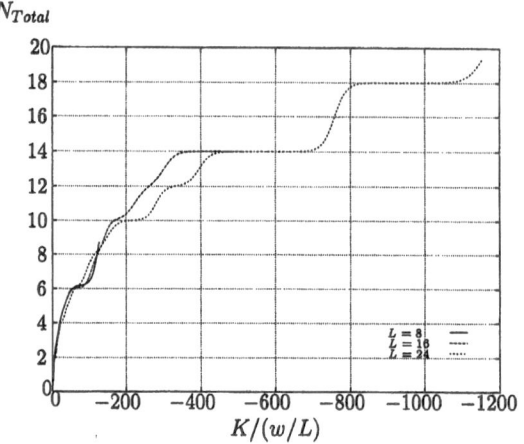

Fig. 5 Mean bending number N_{Total} versus strength of attractive interaction w, which is scaled with w/L for $L = 8, 16, 24$ at $K = 0.5$. Mean bending number are quantized and plateaus emerge

The evaluation of the configuration sum is very difficult, we treat the interaction term U_I at mean field level. Phase diagram is summarized in Fig. 4. Each phase is characterized by the mean bending number $N_{Total} \equiv \langle N \rangle + \langle M \rangle$ and the mean width $\langle L_x \rangle$ ($= \langle L_y \rangle$). In the domain ($K > 0, w > 0$), the membrane is in the completely flat phase ($N_{Total} = 0$, $\langle L_x \rangle = L$) and in the domain ($K < 0, w < 0$), the membrane is completely folded ($N_{Total} = 2L, \langle L_x \rangle = a$). The completely flat phase ($K > 0$) becomes unstable by attractive interaction ($w < 0$). In the region, the membrane is partially folded and $N_{Total} \sim L^{1/3}$. The mean area behaves $\langle L_x L_y \rangle \sim L^{4/3}$. When the system size is finite ($L < \infty$), the system shows the sequential folding transition. In Fig. 5, we show the results of numerical studies. We have fixed the bending rigidity at $K = 0.5$. One sees the behavior of the mean bending number N_{Total} as we change the strength of attractive interaction w, which is scaled with w/L, for system size $L = 8, 16, 24$. In each system size, the curve runs up in a rather discrete manner like a staircase and its value at each plateau takes integer. The length of each plateau becomes longer as w becomes large. This means the sequential folding transition of the membrane.

In the domain ($K \langle 0, w \rangle 0$), the folded state also becomes unstable. There occurs a first-order phase transition from the completely folded phase to the completely flat phase at the line ($2u + \omega = 0$).

Concluding remarks

In this report, we have discussed the phase transition of polymerized membrane with attractive interactions. In order to understand the flexibility-dependent behavior of the transitions, we have studied two models of polymerized membrane with attractive interactions. Especially, we have investigated the instabilities of the flat (phantom) polymerized membrane by attractive interactions. We conclude that the cancellation between the "entropic" bending rigidity from self-avoidance and the negative bending rigidity from attractive interactions plays the essential role in understanding the phase transitions. That is, if the complete cancellation does occur, the continuous crumpling transition occurs and the membrane passes through a critical crumpled state. On the other hand, if the flat phase becomes unstable before the cancellation becomes complete, the membrane shows discontinuous transition.

References

1. Abraham FF, Nelson DR (1990) J Phys (Paris) 51:2653–2672
2. Abraham FF, Kardar M (1991) Science 252:419–422
3. Liu D, Plischke M (1992) Phys Rev A45:7139–7144
4. Grest GS, Petsche IB (1994) Phys Rev E50:1737–1740
5. Hwa T, Kokufuta E, Tanaka T (1991) Phys Rev A:R2235–2238
6. Spector MS, Naranjo E, Chiruvolu S, Zasadzinski J (1994) Phys Rev Lett 21:2867–2870
7. Mori S, Wadati M (1995) Phys Lett A201:61–65
8. Mori S, Kajinaga Y (1996) Phys Rev E53:124–133
9. Le Doussal P (1992) J Phys A Math Gen 25:L469–476
10. Guitter E, Palmeri J (1992) Phys Rev A45:734–744
11. Di Francesco P, Guitter E (1994) Phys Rev E50:4418–4426

Progr Colloid Polym Sci (1997) 106:281–286
© Steinkopff Verlag 1997

C. Sun
M. Ueno

Formation and property of HCO-10 vesicles New method to prepare HCO-10 vesicles with high entrapment efficiency

Received: 13 October 1996
Accepted: 23 May 1997

C. Sun · Dr. M. Ueno (✉)
Faculty of Pharmaceutical Sciences
Toyama Medical and Pharmaceutical
University
2630 Sugitani
Toyama 930-01, Japan

Abstract When the vesicle suspension of HCO-10 was heated to 40–45 °C and then cooled to room temperature, larger vesicles were produced, which not only encapsulated an increased fraction of the solutes but also was narrow in size distribution. The improvement of their size uniformity was dependent on the rate of cooling. During the heating and cooling of HCO-10 suspensions, the transformation process of the vesicles was investigated by observing the freeze fracture electron microscopic appearances and the temperature dependence of the calcein release from those vesicles. It was found that HCO-10 vesicles were transformed into the small vesicles during heating and the aggregation and fusion of those small vesicles formed a larger vesicle during cooling. This method is characterized by the physicochemical features of the dehydration and rehydration of HCO-10 molecules and does not require extensive use of organic solvents, detergents or dialysis systems.

Key words HCO-10 vesicles – heating–cooling cycle – high entrapment efficiency – narrow size distribution – calcein release

Introduction

NSVs are defined as vesicular nonionic surfactant bilayers that enclose a space of aqueous solution. Recently, various surfactants have been used for the preparation of NSVs. These surfactant molecules include polyoxyethylene alkylethers [1], polyglycol alkylethers [2] and glucosyl dialkylethers [3], etc. In many pharmaceutical studies on NSVs, it was demonstrated that using NSVs allows a wide study of the influence of chemical composition on the physicochemical features and the biological fate of vesicles. The ability of NSVs to entrap various solutes and to interact with cells by endocytosis or fusion has led to their application as a vesicle for intracellular delivery [4]. The mean size and size distribution of vesicles were important factors affecting physicochemical stability, encapsulation efficiency, tissue distribution, *in vivo* circulation lifetimes and transfer of lipid onto cells [5–10]. A number of different methods to prepare vesicles have been developed. Although several of those methods, which are being used usually in the preparations of liposomes, are available for the preparation of NSVs, the entrapment efficiency was relatively low [2], and there remain some difficulties in the increment of efficient encapsulation and the uniformity of size distribution of those vesicles.

In the study of physicochemical properties of HCO-10 vesicles, the phase transition of the membrane caused by the dehydration of HCO-10 molecules was suggested [11–14]. In this research, we found that the fusion of the vesicles with dehydrated bilayers made HCO-10 vesicles larger in sizes, higher in entrapment efficiency and relatively uniform in size distribution.

Materials

Poly(oxyethylene) hydrogenated castor oil ether (HCO-10) (Fig. 1) was purchased from Nikko Chemicals, Japan, and used without further purification. All other reagents were of analytical grade.

Methods

The vesicle suspension of HCO-10 was prepared by dispersing HCO-10 in 20 mM Tris buffer (pH 7.4), in some cases, a same buffer solution containing calcein; HCO-10 was placed in a grass test tube and an aqueous solution was added. This dispersion was stirred by a vortex mixer for 10 min. The resulting vesicle suspension, if the size control of vesicles was needed, was frozen in liquid nitrogen and thawed in hot water 5 times, respectively, and then extruded 5 times through two stacked Nuclepore filters (0.6 or 0.2 μm) [11]. In the case of the heating and cooling of HCO-10 suspensions, those vesicle suspensions of HCO-10 were incubated at 40–45 °C for 1 h and subsequently cooled to room temperature. Reverse-phase-evaporation vesicles (REVs) of HCO-10 were prepared according to the method of Szoka et al. [15] using diethyl ether and methanol as solvents. HCO-10 vesicles were also prepared by removal of SDS from mixed micelles of HCO-10/SDS by dialysis.

The calcein release from the vesicles was measured as follows: the vesicle suspensions of HCO-10 were prepared in 100 mM calcein/20 mM Tris buffer (pH 7.4, 388 mOsm), in which the fluorescence intensity of calcein is self-quenched. The vesicles were separated from untrapped calcein by gel-permeation chromatograph using a Sephadex G-75 gel (0.5 × 10 cm column) equilibrated with an isotonic buffer, glucose/20 mM Tris buffer (pH 7.4). The osmolarity of buffers was monitored with an Osmometer (Semi-micro Osmometer, Knauer). The separated vesicles, suspended in a cold isotonic buffer, were rapidly diluted (1 : 200) into the well-stirred isotonic buffer equilibrated to the experimental temperature. The fluorescence emission intensity of a sample was recorded continuously subsequent to this dilution (zero time) on a RF-5000 fluorescence spectrophotometer, Shimadzu (excitation $\lambda = 490$ nm, emission $\lambda = 520$ nm) equipped with temperature control accessories and a magnetic stirrer [11].

Freeze fracture electron micrographs were taken according to the method described previously [16]. The mean diameter of the vesicles was measured by quasielastic light scattering (QLS) using Otsuka Electronics LPA 3000/3100. The calcein-entrapment efficiency of HCO-10 vesicles were determined as described [11].

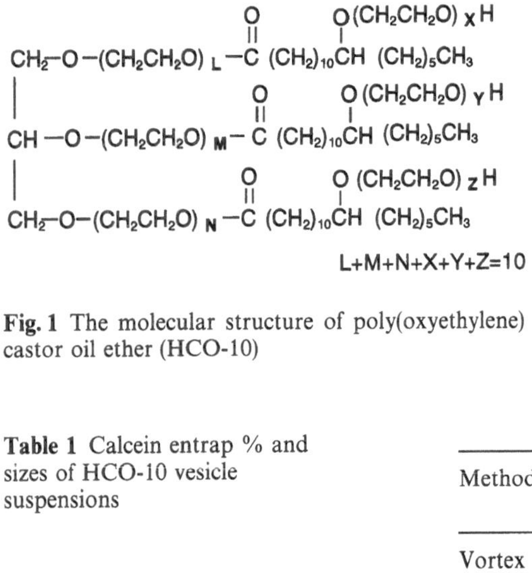

Results

The entrapment efficiencies and sizes of HCO-10 vesicles are shown in Table 1. When a dispersion of HCO-10 prepared either by vortex or by extrusion was heated and then cooled, both the entrapment efficiency and size of HCO-10 vesicles became increased and the size distributions were narrow (Fig. 2). We call the above procedure a heating–cooling cycle. It is apparent that the method of

Fig. 1 The molecular structure of poly(oxyethylene) hydrogenated castor oil ether (HCO-10)

Table 1 Calcein entrap % and sizes of HCO-10 vesicle suspensions

Methods of prep.	Composition [w/w%]	Mean diameter [nm]	Calcein entrap [%]
Vortex	5 HCO-10	240	6.20
	10 HCO-10	368	11.6
Sonication*	5 HCO-10	n.d.	n.d.
REV	5 HCO-10	595	5.66
Detergent removal**	1 HCO-10	170	1.60
	5 HCO-10	76	n.d.
Extrusion	5 HCO-10	187	6.86
Heating–cooling cycle	5 HCO-10	440	15.4

*) The suspension was separated after sonicated
**) The detergent was SDS (Sodium dodecyl sulfate)
n.d.) Not detectable

Progr Colloid Polym Sci (1997) 106:281–286
© Steinkopff Verlag 1997

Fig. 2 The sizes and autocorrelation functions in QLS measurement of HCO-10 vesicles. The vesicles were prepared by extrusion (A)–(C) and prepared by vortex (D) (E) (F); (A) (D) before the heating–cooling cycle, (B) (E) after step-by-step cooling and (C) (F) after direct cooling from 40 °C to room temperature

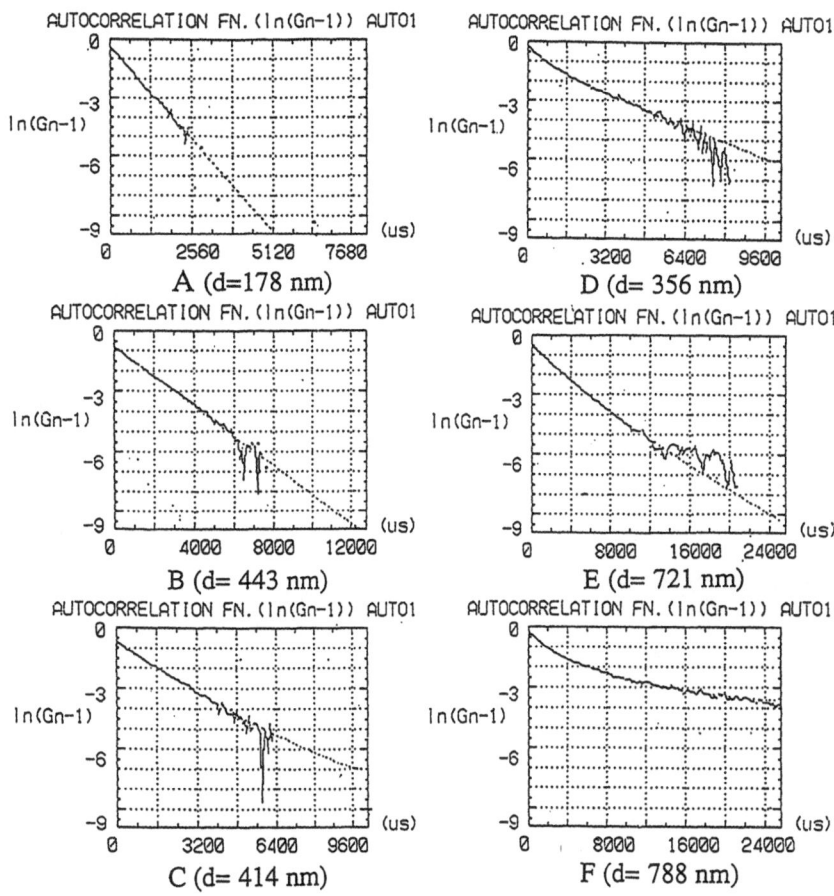

heating–cooling was distinguished from other preparations for HCO-10 vesicles.

In general, the entrapment efficiencies for the preparations of HCO-10 vesicles in Table 1 were lower than those described in the literatures for the corresponding liposome preparations. When HCO-10 vesicles prepared by vortex were sonicated, they were destructed and the suspensions separated.

HCO-10 vesicles can be seen as particles like liposomes with an aqueous compartment. The sizes and organization of HCO-10 vesicles prepared by vortex or the heating–cooling cycle are shown in Table 1 and Fig. 3. At room temperature, the vesicles appear to consist of a few concentric bilayers with a large internal aqueous space in Figs. 3A and 3C. While, during the heating of HCO-10 suspensions small vesicles are observed, some appear as small unilamellar vesicles (SUVs), as shown in Fig. 3B.

The temperature dependence of the sizes of HCO-10 vesicles is shown in Fig. 4. While the suspension was being heated, the vesicle size decreased steeply from about 220 to 50 nm around 37 °C. In order to investigate further the morphology of the vesicles, a freeze-fracture electron microscope was used because ultrarapid cryofixation was able to preserve the morphology of HCO-10 vesicles at

40 °C. When those small vesicles of HCO-10 were cooled to room temperature, larger vesicles were formed, as shown in Fig. 4. Apparently, there was a transformation of the inhomogeneous large vesicles into small ones, followed by the formation of the homogeneous larger vesicles during the heating–cooling of a HCO-10 suspension.

Discussion

Each of those preparative methods described in Table 1 produced close, discrete structures as judged from the freeze fracture electron microscopic appearances or the latency of calcein fluorescence. HCO-10 vesicles prepared by the heating–cooling cycle had a higher entrapment efficiency than those prepared by the other methods. The differences in entrapment efficiency between HCO-10 vesicles presumably reflect the size or type of HCO-10 vesicles formed by each method, e.g., the vesicles prepared by the heating–cooling cycle (5% HCO-10, 440 nm, 15.4%) and the vesicles prepared by vortex (5% HCO-10, 240 nm, 6.20%). The increase in the entrapment efficiency is accompanied by an increase in the size of the HCO-10

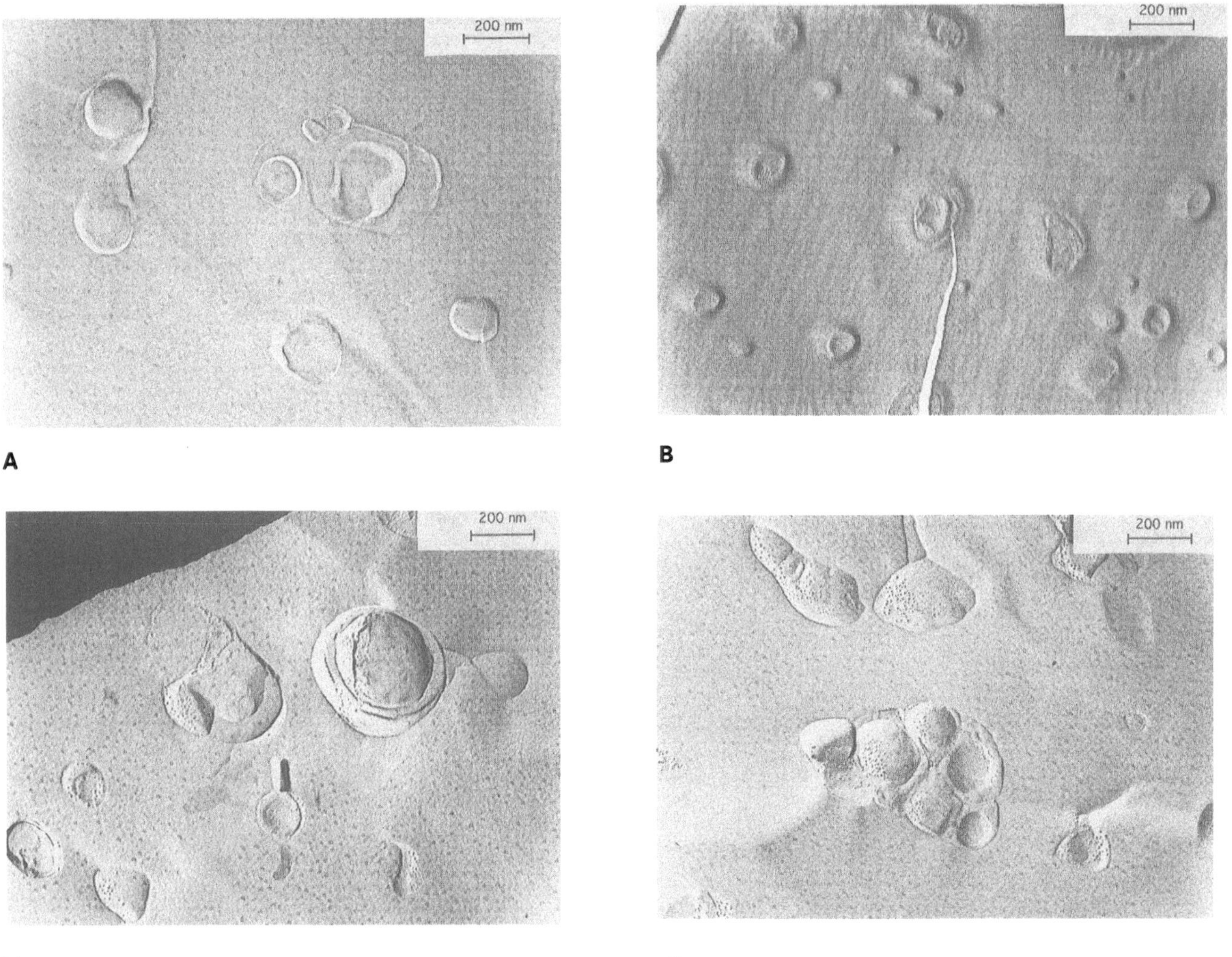

200 nm

A

200 nm

B

200 nm

C

200 nm

D

Fig. 3 The freeze-fracture electron microscopic appearances of HCO-10 vesicle suspensions. (A) HCO-10 vesicles prepared by vortex, (B) the vesicles being incubated at 37 °C or more, (C) the vesicles after step-by-step cooling and (D) the vesicles after direct cooling from 40 °C to room temperature. Bar is 0.2 μm

Fig. 4 Variation of mean diameters and entrap % of HCO-10 vesicles with the temperatures

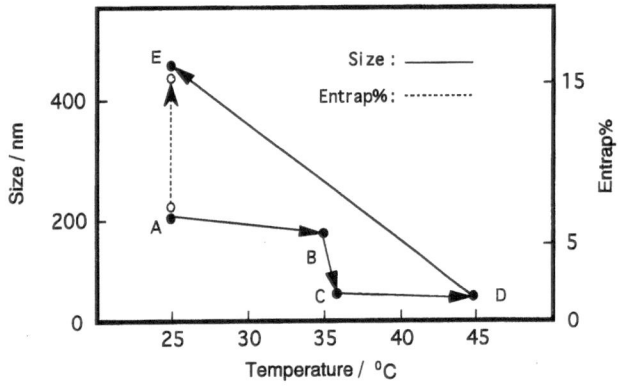

vesicles. However, REVs of HCO-10 showed a low entrapment efficiency despite having a large size (Table 1). This low entrapment efficiency might be due to the residual solvents, which might affect the formation of HCO-10 vesicles, because the solvents could not be removed completely only by the evaporation [15].

In an early investigation, Horiuchi and Tajima observed with QLS that, when HCO-10 vesicle suspension was heated above 37 °C, those vesicles were transformed into small ones. They emphasized the influences of the dehydration of EOs on the vesicular membrane of HCO-10 [12]. We found that, when those small vesicles were subsequently cooled to room temperature, the larger vesicles formed, which can entrap a high amount of the solute molecule (Fig. 4). Hofland et al. prepared the large

Progr Colloid Polym Sci (1997) 106: 281–286
© Steinkopff Verlag 1997

multilamellar NSVs using a sonication method followed by incubating at 60 °C for 5 min for increasing the entrapment efficiency [17, 18]. The heating–cooling cycle, which is useful for an alternative of the preparations for HCO-10 vesicles, was established by the combination of Horiuchi and Tajima's observation and Hofland's way of the preparation.

The heating–cooling cycle is based on the thermal transformation of HCO-10 vesicles. For a detailed understanding, it is necessary to demonstrate and discuss the process of the vesicle change. According to the freeze-fracture electron microscopic appearances, it was suggested that during the heating the vesicles were transformed into the small ones in Figs. 3A and B and during the cooling the small vesicles aggregated and fused into the larger vesicles in Figs. 3C and D. While HCO-10 suspension was being heated, the calcein release from HCO-10 vesicles was accelerated and reached a maximum at 37 °C (Fig. 5). The formation of the small vesicles, or the high bilayer curvature, might be related to the changes in the hydrophile–lipophile balance (HLB) of HCO-10 molecules and/or molecular packings in the membrane matrix caused by the dehydration of EOs [14], which gave rise to the membrane destabilization and acceleration of the calcein release. On the other hand, during the cooling those small HCO-10 vesicles became larger and effectively sandwiched the solute molecules. The size increase of the vesicles composed of HCO-10 might be ascribed to the fusion, which is different from that of the liposomes modified by PEG [19, 20]. It is interesting that after cooling any small vesicles such as those observed above 37 °C disappeared. There are two factors promoting the fusion of those small HCO-10 vesicles: (1) the thin hydration shell on the small vesicles due to the dehydration of EOs, by which the bilayers might be mediated to sufficiently close proximity

for a complex [21, 22]; and (2) the high curvature of the bilayer, which is responsible for the fusion of the vesicles to form less curved, and thermodynamically more stable vesicles [19]. Therefore, the aggregation and fusion of those small vesicles could occur easily during the cooling, producing the larger vesicles of HCO-10. Furthermore, a complete fusion of the small vesicles was dependent on the method of cooling. The cooling directly from 40 °C to room temperature proceeded at room temperature. There is the reason that the larger separate vesicles could be formed if the complete fusion of the small vesicles occurred spontaneously. However, we did not see larger separate vesicles but aggregates of the vesicles of about 150 nm (Fig. 3D). On the other hand, if the vesicle suspension was incubated for a few minutes at some temperatures during cooling, larger separate vesicles were formed, as shown in Fig. 3C. It suggested that the complete fusion of forming the larger separate vesicles is conditioned by the incubation and might be related to the narrow size distribution (Fig. 2).

We also noted that the heating–cooling cycle for HCO-10 vesicles is supposedly similar to the dehydration–rehydration cycle for liposome formation [23]. They were both used to improve the preparation for vesicles. But the dehydration–rehydration cycle involves lyophilization (freeze–drying) of Egg PC bilayers to induce liposome fusion and solute encapsulation [24], which is apparently different from the fusion of HCO-10 vesicles with a thin hydration shell and a high-curvature bilayer. By the heating–cooling cycle, not only the entrapment efficiency and sizes were increased but also the size distribution of the vesicles was improved (Fig. 2). It is not clear why a relatively narrow size distribution of HCO-10 vesicles formed after the heating–cooling cycle, but it may be related to the rates of cooling. When HCO-10 vesicles were cooled directly from 40 °C to room temperature, the autocorrelation functions in QLS measurement showed a curve, indicating a wide size distribution (Figs. 2C and F); whereas, when HCO-10 vesicles were cooled step by step, the autocorrelation functions in QLS measurement showed a straight line or nearly a straight line, giving rise to a narrow size distribution or improving the size distribution well (Figs. 2B and E). The formation of the relative narrow size distribution was dependent on the temperature and time of the incubation during cooling.

In summary, the results reported here show that the heating–cooling cycle could be a useful preparative method for HCO-10 vesicles. We took advantage of the dehydration and rehydration of HCO-10 bilayers near the phase transition temperature to improve the preparation for HCO-10 vesicles. From a methodological point of view, it is significant that the physicochemical features of NSVs are used to develop their preparations.

Fig. 5 Temperature dependence of the calcein release from HCO-10 vesicles

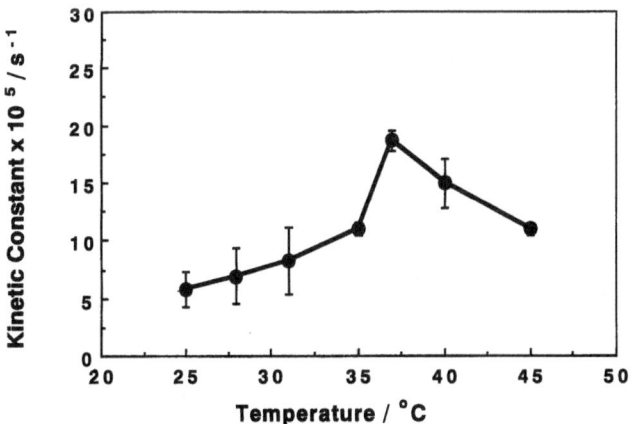

References

1. Hofland HEJ, Bouwstra JA, Talsma H, Junginger HE (1988) Proc Int Symp Control Rel Bioact Mater 15:406–407; Okahata Y, Tanamuchi S, Nagai M, Kunitake T (1981) J/Colloid Interface Sci 82:401

2. Baillie AJ, Florence AT, Hume LR, Muirhead GT, Rogerson A (1985) J Pharm Pharmacol 37:863

3. Kiwada H, Niimura H, Kato Y (1985) Chem Pharm Bull 33:753–759

4. Yoshioka T, Skalko N, Gursel M, Gregoriadis G, Florence AT (1995) J Drug Targeting 2:533–539

5. Duzgunes N, Wilschut J, Hong K, Fraleg R, Perry C, Friend DS, James TL, Papahadjopoulos D (1983) Biochem Biophys Acta 732:289

6. Bosworth ME, Hunt CA, Paratt D (1982) J Pharm Sci 71:806

7. Abra RM, Hunt CA (1981) Biochim Biophys Acta 666:493

8. Juliano RL, Stamp D (1975) Biochem Biophys Res Commun 63:651

9. Seiden A, Lichtenberg D (1979) J Pharm Pharmacol 31:414

10. Ueno M (1986) Medicine J (Japan) 10:2045–2051

11. Sun CQ, Ueno M (1997) Colloid Polym Sci 275:in print

12. Horiuchi T, Tajima K (1992) J Jpn Oil Chem Sci 41:1197–1202

13. Horiuchi T, Tajima K (1992) J Jpn Oil Chem Sci 41:1191–1196

14. Tanaka M, Fukuda H, Horiuchi T (1990) J Am Oil Chem 67:55–60

15. Szok F, Papahadjopoulos D (1978) Proc Natl Acad Sci USA 75 9:4194–4198

16. Sternberg B (1992) In: Gregoriadis G (ed) Liposome Technology. Vol 1, CRC Press Inc, Boca Raton, pp 363–283

17. Hofland HEJ, Bouwstra JA, Ponec M, Boddé HE, Spies F, Verhoef JC, Junginer HE (1991) J Control Rel 16:155–168

18. Hofland HEJ, Geest RV, Boddle HE, Junginer HE, Bouwstra JA (1994) Pharm Res 11:659–664

19. Lentz BR, McIntyre GF, Parks DJ, Yats JC, Massenburg D (1992) Biochemistry 31:2643–2653

20. Anold K, Lvov YM, Szogyi M, Gyorgri S (1986) Stud Biophys 113–7

21. Wilschut J, Nir S, Scholma J, Hoekstra D (1985) Biochemistry 24:4630–6

22. Rupert Leo AM, Engberts Jan BFN, Hoekstra D (1986) J Am Chem Soc 108:3920–5

23. Shew RL, Deamer DW (1985) Biochim Biophys Acta 816:1–8

24. Kirby CJ, Gregoriadis G (1984) In: Gregoriadis G (ed) Liposome Technology. Vol 1, CRC Press Inc, Boca Raton, pp 19–29

Progr Colloid Polym Sci (1997) 106:287–290
© Steinkopff Verlag 1997

OTHER RELATED TOPICS

Y. Enomoto
R. Kato

Annihilation dynamics of two-dimensional magnetic vortex systems

Received: 13 October 1996
Accepted: 14 April 1997

Dr. Y. Enomoto (✉)
Department of Physics
Nagoya Institute of Technology
Gokiso
Nagoya 466, Japan

R. Kato
Department of Electronics
Kagoshima University
Kagoshima 890, Japan

Abstract We numerically study the pair-annihilation process of magnetic vortices in two-dimensional type-II superconductors. The dynamics for interacting vortices is described in terms of the Langevin-type stochastic equation. Carrying out the Langevin dynamics simulation, we find that the power-law behavior of the mean distance among vortices exhibits a crossover phenomenon as a function of time t from the $t^{1/2}$-law at short times to $t^{1/3}$ at long times, and the crossover time is an increasing function of temperature. The present results are also found to be explained by a simple dimensional analysis.

Keywords Magnetic vortex – superconductor – annihilation – power law – crossover

Introduction

Quantized magnetic vortex in type-II superconductors is a representative example of topological defects, which play an important role in the pattern formation dynamics [1]. Recent study on the magnetic vortex dynamics has revealed that point magnetic vortices in two-dimensional systems are analogous to two-dimensional colloid systems, while magnetic vortex lines in three-dimensional systems are analogous to polymer systems [2]. Indeed, from such analogy (although there are many differences between them), new concepts on the magnetic vortex systems have been introduced in the field of superconductor science, such as melting, entanglement, vortex solid, vortex liquid, vortex glass, and so on.

We have studied both static and dynamic properties of the vortex systems, such as Abrikosov vortex lattice formation, vortex lattice melting, the KT transition, and the vortex glass transition, from the point of view of an analogy between our system and colloid, polymer systems [3]. Our study is based on computer simulations of two efficient model equations, recently developed by us.

One is the time-dependent Ginzburg–Landau equation which is described by a complex order parameter and vector potential. The other is the Langevin-type stochastic equation of motion for magnetic vortices in two and three dimensions, which is described in terms of vortex position variables.

The aim of the present work is to discuss the annihilation process of two-dimensional magnetic vortices in type-II superconductor film with the thickness h and area A. Such annihilation might be realized for the system following a quench below the critical temperature from a normal state at zero external field.

Model

Our starting equation to study the two-dimensional annihilation dynamics is as follows with conventional notations [4]:

$$\gamma^{-1} \frac{d\mathbf{r}_i(t)}{dt} = -\frac{\delta}{\delta \mathbf{r}_i} \sum_{j(\neq i)} \delta_i \delta_j V(r_{ij}) + \mathbf{f}_i(t) , \qquad (1)$$

where $\mathbf{r}_i(t)$ denotes the two-dimensional position vector of the ith vortex at time t with $r_{ij} = |\mathbf{r}_i - \mathbf{r}_j|$, $\gamma^{-1} \equiv \sigma H_{c2}(T)\phi_0/c^2$ is the friction coefficient with the normal state conductivity σ and the upper critical field $H_{c2}(T) = \phi_0/(2\pi\xi(T)^2)$, and $\delta_i = 1$ and -1 correspond to a vortex and an antivortex, respectively. In the thin-film superconductors considered here, the vortex pair potential, V, is given by [5]

$$V(r_{ij}) = \frac{\phi_0^2}{8\pi\Lambda}\left(H_0(r_{ij}/\Lambda) - Y_0(r_{ij}/\Lambda)\right), \quad (2)$$

where $\Lambda(T) = 2\lambda(T)^2/d$ denotes the effective penetration depth including the demagnetization effect, and $H_0(x)$ the Struve function of order zero and $Y_0(x)$ the Neumann function of order zero, The last term in Eq. (1) describes the thermodynamic random force, characterized by the fluctuation–dissipation theorem, i.e., $\langle\langle f_{i\alpha}(t)\rangle\rangle = 0$ and $\langle\langle f_{i\alpha}(t)f_{j\beta}(t')\rangle\rangle = 2k_B T\gamma\delta_{ij}\delta_{\alpha\beta}\delta(t - t')/h$ where $f_{i\alpha}$ is the α-component of $\mathbf{f}_i$ and $\langle\langle \cdots \rangle\rangle$ denotes the average over the thermal fluctuation. Note that in the present simulations no creation of additional vortices is allowed. Moreover, we assume the temperature dependence of the coherence length $\xi(T)$ and the magnetic penetration depth $\lambda(T)$ as $\xi(T) = \xi(0)(1 - T/T_c)^{-1/2}$ and $\lambda(T) = \lambda(0)(1 - T/T_c)^{-1/2}$, respectively with the superconducting transition temperature T_c at zero field. In this case, the Ginzburg–Landau parameter, κ, becomes independent of temperature and is defined by $\kappa = \lambda(0)/\xi(0)$. We also use the relationship obtained from the microscopic theory as $4\pi\lambda(T)^2\sigma/c^2 = t_0(1 - T/T_c)^{-1}$ with $t_0 = h/(96k_B T_c)$ [4].

Results and discussion

In the following simulations we set $A = (10^3\xi(0))^2$, $h = \lambda(0)/10$, the GL parameter $\kappa = 50$, and the energy scale ratio $k_B T_c/(\varepsilon(0)h) = 0.01$ with $\varepsilon(0) = \phi_0^2/(4\pi\lambda(0))^2$, which are of almost the same order to those for typical high$-T_c$ materials. The total initial vortex number is taken to be $N(0) = 1000$ with $N(0)/2$-vortices and $N(0)/2$-antivortices at $t = 0$. The dimensionless time increment scaled by t_0 is chosen to be 0.01. Moreover, vortex and antivortex with their distance less than $\xi(0)/2$ are interpreted to annihilate and thus are removed from the system. Since the procedures to solve Eq. (1) are the same as our previous work [3], we will not discuss the numerical techniques any more.

First, to visualize typical dynamical behavior of the annihilation process, trajectories of vortices and antivortices at $T/T_c = 0.2$ are shown in Fig. 1. Corresponding snapshots of the system are plotted in Fig. 2. In these figures we can see the elementary process of the pair

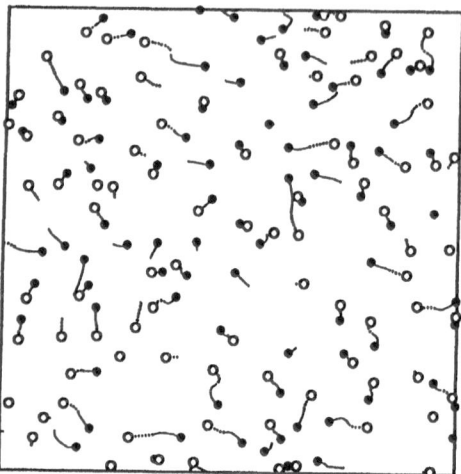

Fig. 1 Trajectories of vortices (solid lines) and antivortices (dashed lines) during the annihilation process at $T/T_c = 0.2$ from $t/t_0 = 75$ to 125 with $t_0 = h/(96k_B T_c)$. Black and white circles denote the initial positions of vortices and antivortices, respectively

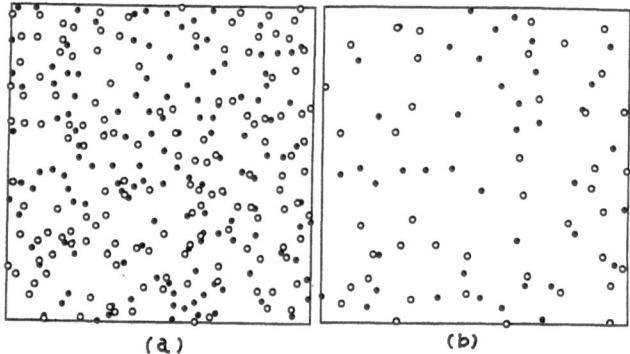

Fig. 2 Time evolution of the annihilation process for $T/T_c = 0.2$ at (a) $t/t_0 = 12.5$ and (b) 50. Notations are the same as Fig. 1

annihilation and the coarsening of the vortex system due to the annihilation. Then, to make this phenomenon clear, we discuss the time evolution of the mean distance between vortices, $w(t)$, defined by $w(t) = \sqrt{N(0)/N(t)}$ with the total vortex number $N(t)$ at time t. The mean distance $w(t)$ is regarded as the time-dependent characteristic length scale of the system. In Fig. 3 we show the mean distance $w(t)$ as a function of time for $T/T_c = 0.2$ and 0.6. Our results, with detailed simulation data performed by changing T in the range from 0 to 0.8, show that (1) at short times, $w(t)$ grows in time as $t^{1/2}$, (2) at long times, $w(t) \propto t^{1/3}$, and (3) the onset time of such crossover is an increasing function of temperature. Similar scaling behavior, although the crossover from $t^{1/2}$ to $\ln t$, has been obtained for three dimensional systems [6].

Progr Colloid Polym Sci (1997) 106:287–290
© Steinkopff Verlag 1997

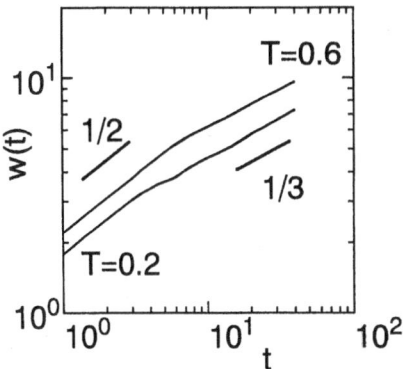

Fig. 3 Mean distance, $w(t)$, between vortices as a function of time t/t_0 for $T/T_c = 0.2$ and 0.6. Two straight lines are also shown with their slopes indicated

These results are understood by a simple dimensional analysis as follows. During the above coarsening process, the pair annihilation due to the attractive force between vortex and antivortex is dominant, but both the repulsive forces between vortices with the same sign and the thermal noise cause only a trigger of the subsequent annihilation. From this point of view, the system dynamics is dominated as $dw/dt \sim dV(w/\Lambda)/dw$. Moreover, $V(x) \sim \ln x$ for $x \ll 1$ and $\sim 1/x$ for $x \gg 1$ (no screening at long-length scales due to the demagnetizing effects), and the crossover for the functional form of $V(x)$ occurs near $x \simeq 1$. Combining these arguments and using the fact that $\Lambda \sim (1 - T/T_c)^{-1}$, we can reproduce the above scaling behavior for $w(t)$. In the next section, we discuss an attempt to explain the annihilation dynamics beyond the dimensional (or mean-field) analysis.

Hydrodynamical approach to the annihilation

The above dimensional analysis is at rather phenomenological and/or macroscopic levels, although it has succeeded in explaining the simulation results. Indeed, in the above analysis, we have neglected mesoscopic effects such as the system inhomogeneity, density fluctuations, and collective vortex motion. Thus, to understand the annihilation dynamics in a mesoscopic level, the statistical-mechanical treatment is needed. Here, as one of such attempts, we review the hydrodynamical approach for the annihilation dynamics due to Ginzberg et al. [7].

We consider a d-dimensional system consisting of vortices and antivortices interacting via the pairwise force whose magnitude is proportional to R^{-n} with the intervortex distance R. We restrict ourselves to systems with weak long-range forces, i.e. the power exponent of the long-range force, n, satisfies the condition $d - 1 \leq n < \infty$

(especially, the case $n = d - 1$ is called the Coulombic system). We label the vortex (antivortex) concentration as $n_1(\mathbf{r}, t)$ $(n_2(\mathbf{r}, t))$, and impose the condition $\langle n_1(\mathbf{r}, t) \rangle = \langle n_2(\mathbf{r}, t) \rangle$ where $\langle \cdots \rangle$ denotes the ensemble and the spatial average. From the generalized law of mass conservation, taking into account only the two-body annihilation mechanism, we can write the evolution equation for the concentration field $n_i(\mathbf{r}, t)$ $(i = 1, 2)$:

$$\frac{\partial}{\partial t} n_i(\mathbf{r}, t) + \nabla \cdot \mathbf{J}_i = -2K n_1(\mathbf{r}, t) n_2(\mathbf{r}, t) , \tag{3}$$

where

$$\mathbf{J}_i = -D \nabla n_i(\mathbf{r}, t) - Q \delta_i n_i(\mathbf{r}, t) \nabla V(\mathbf{r}, t) \tag{4}$$

and the intervortex interaction $V(\mathbf{r}, t)$ is given by

$$V(\mathbf{r}, t) = \int d^d \mathbf{R} \frac{n_1(\mathbf{R}, t) - n_2(\mathbf{R}, t)}{|\mathbf{R} - \mathbf{r}|^{n-1}} \tag{5}$$

with D, Q and K being diffusion, electrostatic and annihilation constants.

It is convenient to describe the system in terms of the vortex density $\rho(\mathbf{r}, t) = n_1(\mathbf{r}, t) + n_2(\mathbf{r}, t)$ and the charge density $f(\mathbf{r}, t) = n_1(\mathbf{r}, t) - n_2(\mathbf{r}, t)$. Rewriting the above equations using these variables, we obtain:

$$\frac{\partial \rho}{\partial t} - D \nabla^2 \rho = -K(\rho^2 - f^2) - Q \nabla \left(f \nabla \int d^d \mathbf{R} \frac{f(\mathbf{R}, t)}{|\mathbf{R} - \mathbf{r}|^{n-1}} \right), \tag{6}$$

$$\frac{\partial f}{\partial t} - D \nabla^2 f = -Q \nabla \left(\rho \nabla \int d^d \mathbf{R} \frac{f(\mathbf{R}, t)}{|\mathbf{R} - \mathbf{r}|^{n-1}} \right). \tag{7}$$

Next, we make an important assumption in order to analytically treat the above Eqs. (6) and (7), i.e., we choose to ignore fluctuations of the vortex density and thus concentrate only on fluctuations of the charge density [8]. A reasonable justification of the proposed assumption lies in a simple observation that while the average vortex density at any time is nonzero so that $\langle (\rho - \langle \rho \rangle)^2 \rangle / \langle \rho \rangle^2$ is finite and likely to be small, the average charge density is always zero so that $\langle (f - \langle f \rangle)^2 \rangle / \langle f \rangle^2$ is infinite. Thus, we can expect that the vortex-density fluctuations are less important than the charge fluctuations, and that we can approximate the vortex density $\rho(\mathbf{r}, t)$ by its spatially averaged value, denoted by $\rho(t)$. Taking into account the above approximation, we rewrite Eq. (7) in Fourier representation:

$$\left(\frac{\partial}{\partial t} + D k^2 + Q \rho(t) k^{2-\sigma} \right) f(\mathbf{k}, t) = 0 \tag{8}$$

with $\sigma = d + 1 - n$. Similarly, Eq. (6) is rewritten as

$$\frac{d\rho(t)}{dt} + K\rho(t)^2 = K \int \frac{d^d\mathbf{k}}{(2\pi)^d} f(\mathbf{k},t) f(-\mathbf{k},t) \left[1 - \frac{Q}{K}k^{2-\sigma} \right]. \tag{9}$$

Note that we have used dimensionless variables in Eqs. (8) and (9), scaled by the initial vortex density ρ_0.

These equations can be solved taking the ensemble average with respect to the initial state to yield

$$f(\mathbf{k},t) = f(\mathbf{k},0) \exp\left[-Dk^2t - Qk^{2-\sigma}\int_0^t \rho(\tau)\,d\tau \right], \tag{10}$$

$$\frac{d\rho(t)}{dt} + K\rho(t)^2 = K \int \frac{d^d\mathbf{k}}{(2\pi)^d} \left[1 - \frac{Q}{K}k^{2-\sigma} \right]$$
$$\times \exp\left[-2Dk^2t - 2Qk^{2-\sigma}\int_0^t \rho(\tau)\,d\tau \right], \tag{11}$$

where we have used the fact that the initial vortex positions are randomly distributed and thus $\langle f(\mathbf{k}_1,0)f(\mathbf{k}_2,0)\rangle = \delta((k_1+k_2)/(2\pi)^d)$. Although several possible scenarios for the system evolution might exist depending on the value of σ, only the asymptotic behavior is obtained analytically.

To demonstrate the usefulness of the present approach, we discuss the asymptotic behavior for the simple case $Q = 0$ (e.g., no long-range force systems), using the above Eqs. (10) and (11). In this case, the evolution Eq. (8) for f is the diffusion equation and the solution is given by Eq. (10) with $Q = 0$, that is, $f(\mathbf{k},t) = f(\mathbf{k},0)\exp[-Dk^2t]$. From Eq. (11) with $Q = 0$, we obtain

$$\frac{d\rho(t)}{dt} + K\rho(t)^2 = \frac{K}{(1 + 2Dt)^{d/2}}. \tag{12}$$

It can be easily shown that Eq. (12) has the well-known asymptotic solution: $\rho(t) \simeq (Kt)^{-1}$ for $d > 4$ and $\rho(t) \simeq (Dt)^{-d/4}$ for $d < 4$ [8].

Finally, rather than describing details of the asymptotic analysis, we simply summarize all results of the asymptotic large t behavior for the vortex density $\rho(t)$ in the general case. Three different regions are found:

(a) Fluctuation-dominated region or strong diffusion case for $n \geq 1 + d/2$ and $d < 4$. Long-range forces are irrelevant and the only power-law asymptotic is Toussaint–Wilczek solution [8] for non-interacting systems described by $\rho \simeq (Dt)^{-d/4}$.

(b) Intermediate region or strong deterministic force case for $n < 1 + d/2$ and $n \geq d - 1$. The asymptotic behavior for $\rho(t)$ is given by $\rho \simeq (Qt)^{-v}$ with $v = d/(2 - d + 2n)$.

(c) Mean-field region for $d \geq 4$. Fluctuations do not slow down annihilation, and classical-field solution is the only asymptotic power-law $\rho \simeq (Kt)^{-1}$.

Recalling that $w(t) \propto \rho(t)^{-1/d}$ and $d = 2$, we can reproduce the $\frac{1}{2}$ power-law behavior for the characteristic length scale $w(t)$ as the case (b) with $n = 1$. However, we cannot explain the $\frac{1}{3}$ power law for the $n = 2$ case. An evident physical reason for this fault in the hydrodynamical description is not yet clear. One can expect that the ignored effects, such as many-body effects, dimensionality effects and the coupling between fluctuations of the number density and the charge density, influence the present predictions. Further, careful study is needed both theoretically and computationally.

Conclusions

In summary, performing the Langevin dynamics simulations, we have discussed the annihilation dynamics of two-dimensional magnetic vortex systems in the absence of the external field. We have found that following a temperature quench from normal to superconducting phase the mean distance among vortices behaves in time t as $t^{1/2}$ at short times, eventually crossing over to $t^{1/3}$ at long times. The late stage $\frac{1}{3}$ power-law behavior is in contrast to the logarithmic behavior at long times observed for three-dimensional systems [6].

More detailed simulations, including realistic effects such as random impurities and external field, will be discussed elsewhere.

The authors thank Dr. A. Dorsey for leading their interest to the related work [5] and for valuable comments.

References

1. Cross MC, Hohenberg PC (1993) Rev Mod Phys 65:851–1112
2. Blatter G et al. (1994) Rev Mod Phys 66:1125–1388
3. Enomoto Y et al. (1993) In: Narlikar AV (ed) Studies of High Temperature Super-conductors. Vol. 11 Nova Science, New York, pp. 309–330
4. Tinkham M (1975) Introduction to Super-conductivity. McGraw-Hill, New York
5. Fetter AL, Hohenberg PC (1967) Phys Rev 159:330–343
6. Enomoto Y (1991) Mod Phys Lett B5:1639–1644
7. Ginzberg VV, Radzihovsky L, Clark NA, (1996), preprint
8. Toussaint D, Wilczek F (1983) J Chem Phys 78:2642–2647

Progr Colloid Polym Sci (1997) 106:291
© Steinkopff Verlag 1997

Progr Colloid Polym Sci (1997) 106:292–293
© Steinkopff Verlag 1997

SUBJECT INDEX

Progr Colloid Polym Sci (1997) 106:292–293
© Steinkopff Verlag 1997